Encyclopedia of Photonic Crystals

Encyclopedia of Photonic Crystals

Edited by **Jason Penn**

CLANRYE
INTERNATIONAL

New Jersey

Published by Clanrye International,
55 Van Reypen Street,
Jersey City, NJ 07306, USA
www.clanryeinternational.com

Encyclopedia of Photonic Crystals
Edited by Jason Penn

International Standard Book Number: 978-1-63240-197-7 (Hardback)

Printed in the United States of America.

Contents

Preface

This book comprehensively introduces photonic crystals and their applications including design and modeling characteristics. Photonic crystals are described as the periodic optical nanostructures that are designed to affect the motion of photons in a similar way that periodicity of a semiconductor crystal affects the motion of electrons. Photonic crystals are described as attractive optical materials for the purpose of manipulating and controlling the flow of light. Specifically, photonic crystals are of significant interest for basic as well as applied research, and two dimensional photonic crystals are being fast employed to commercial applications like micro electro-mechanical systems (MEMS), optical logic devices, and sensors. The first commercial products involving two-dimensionally periodic photonic crystals already exist in the form of photonic-crystal fibers, which employ a microscale structure to limit light with radically distinct features in comparison to traditional optical fiber for applications in guiding wavelengths and nonlinear devices.

The information shared in this book is based on empirical researches made by veterans in this field of study. The elaborative information provided in this book will help the readers further their scope of knowledge leading to advancements in this field.

Finally, I would like to thank my fellow researchers who gave constructive feedback and my family members who supported me at every step of my research.

Editor

Part 1

Introduction to Photonic Crystals

Basics of the Photonic Crystal Gratings

Andriy E. Serebryannikov
Hamburg University of Technology, E-3
Germany

1. Introduction

More than twenty years have passed since that time when the analogy between solid-state physics and optics led to the concept of photonic crystals (PCs) (Yablonovitch, 1987). Fast progress in theory and applications of PCs has been stimulated to a large extent by their unique properties that allow increasing the potential of light controlling. Slabs of PC have mainly been studied in case of linear virtual interfaces and a noncurvilinear lattice. A rich variety of the fascinating physical phenomena has been demonstrated for these structures, which include superprism, subwavelength imaging, focusing, collimation, and negative refraction with and without left-handed behaviour (Inoue&Ohtaka, 2004; Luo, 2002). They all appear while using only conventional linear isotropic metals and dielectrics due to the specific dispersion of Floquet-Bloch waves in PCs, which is not obtainable for solid pieces of the same materials. Breaking of periodicity in PCs results in the appearance of defect modes, which manifest themselves in the transmission and waveguide regimes (Joannopoulos, 1995). On the other hand, PCs with curvilinear virtual interfaces but still linear lattice have been investigated in the context of such applications as planoconcave lenses (Gralak, 2000; Vodo, 2005), mirrors (Saado, 2005), and splitters (Wu, 2005). PCs having both curvilinear interfaces and lattice, e.g., coaxial PCs (Schleuer&Yariv, 2004) and atoll resonators (Nojima, 2007) are also known.

New operation regimes can be obtained due to *merging* effects of dispersion and diffraction, e.g., in two-dimensional PCs. In the mid 2000's, PCs with the corrugated interfaces have been proposed to redirect the reflected waves to the side directions (Collardey, 2005), obtain unusual order of the cutoff wavelengths for higher diffraction orders (Serebryannikov, 2006), and realize a new mechanism of negative refraction due to the umklapp refracted beams (Lu, 2007). Later, PCs with the corrugated interfaces have been called *photonic crystal gratings* (PCGs) (Serebryannikov, 2009). Strong asymmetry in transmission has been demonstrated in dielectric two-dimensional PCGs theoretically (Serebryannikov, 2009) and in the microwave experiment (Cakmak, 2010). Recently, a similar effect has been studied in the two-dimensional sonic crystals (Li, 2011). The structures with a corrugated interface and a defect-mode waveguide, which is perpendicular to the interface, have been used for obtaining of the beaming, that is connected with the excitation of surface waves due to corrugations (Caglayan, 2008; Smigaj, 2007). A structure that is excited by a defect-mode waveguide located along the virtual interfaces of the corresponding noncorrugated PC has been suggested (Le Thomas, 2007), where the corrugations provide coupling of an otherwise uncoupled defect mode to an outgoing wave in air.

In this chapter, we focus on the transmission and reflection regimes with strong directional selectivity that appear in PCGs owing to the additional periodic corrugations arranged at the virtual interface(s) of a defect-free slab of PC with linear virtual interfaces and noncurvilinear lattice. First of all, additional corrugations enable downshifting of the frequency range where higher diffraction orders may propagate in air, so that it corresponds to the range of existence of lower-order Floquet-Bloch waves, which are well studied in the context of the above-mentioned phenomena. Theoretical background and numerical results will be presented with the focus on new operation regimes, which can be used in optical devices that require strong directional selectivity. Consideration is restricted here to the two-dimensional square-lattice PCs composed of dielectric rods, while the virtual interfaces of the corresponding noncorrugated PC are assumed to be along Γ-X direction in $\mathbf{k}$ space, and the incident plane wave is s-polarized. Figure 1 illustrates a possible evolution from the slab of PC with the noncorrugated interfaces (a) to the PCG with the one-side (b) and, then, to the PCG with the two-side asymmetric (c) corrugations.

The *first class* of the considered regimes (Sec. 2) includes those related to the unidirectional, i.e., extremely asymmetric transmission. High transmittance from one half-space to the other can be obtained if a PCG is illuminated from the corrugated side, but it is vanishing if illumination is in the opposite direction, within a wide range of the frequency variation (Serebryannikov, 2009). This is probably the most interesting regime obtainable in the dielectric PCGs. Breaking of the spatial inversion symmetry, i.e., introducing nonsymmetry with respect to the midplane of the corresponding noncorrugated PC is required for obtaining of such a forward-backward unidirectional transmission. The necessary condition is that zero diffraction order is not coupled to any Floquet-Bloch wave, but at least one higher diffraction order may propagate in air due to the one-side corrugations. Transmission from the noncorrugated side towards the corrugated side is forbidden, while that from the corrugated side is possible owing to higher diffraction order(s). Single-beam unidirectional deflection and two-beam unidirectional splitting belong to the most typical unidirectional diode-like transmission regimes. The main attention will be paid to the PCs with the noncircular (non-isotropic type) isofrequency dispersion contours (IFCs), which are located in $\mathbf{k}$ space near either M or X point, and the circular (isotropic type) IFCs, which are located near Γ point and correspond to the effective index of refraction $0 < |N_{\text{eff}}| < 1$.

The *second class* of the operation regimes (Sec. 3) is connected with the Fabry-Perot type resonances that can appear in the nonsymmetric PCGs so that zero and higher diffraction orders simultaneously contribute to the transmission. Classical resonances, i.e., those with a single (zero) order in transmission, are well known for the noncorrugated slabs of PC (Sakoda, 2001; Serebryannikov, 2010). In the PCGs, strong asymmetry of the Fabry-Perot resonance transmission occurs at normal illumination. It can be obtained even if zero order is only coupled to a Floquet-Bloch wave, despite that the higher orders may also propagate in air due to the one-side corrugations. In this case, the higher orders, which appear at the corrugated exit side and propagate in the exit half-space, can mainly contribute to the transmission, if the PCG is illuminated from the noncorrugated side, but they remain evanescent in the exit half-space, if the PCG is illuminated from the corrugated side. At the same time, a nondominant zero-order transmission is symmetric, i.e., it does not depend on the illumination side.

The *third class* of the operation regimes (Sec. 4) is associated with total reflections that involve at least one higher order. They can be obtained inside a band gap of the PC, if the corrugated side is illuminated. In turn, only zero order contributes to the reflection, if the noncorrugated side is illuminated, that leads to strong asymmetry. It will be demonstrated that the corrugations enable transformation of a desired part of the incident wave energy into that of higher reflected orders. The presented numerical results are obtained by using the fast coupled integral equation technique (Magath&Serebryannikov, 2005).

2. Unidirectional transmission

Diode is one of the main elements required in various optical and microwave circuits. Obtaining of unidirectional diode-like transmission is usually associated with nonreciprocity and, hence, with the use of anisotropic, e.g., gyromagnetic (Figotin&Vitebskiy, 2001; Yu, 2007; Wang, 2008), or nonlinear (Scalora, 2004; Shadrinov, 2011) materials, that allows breaking time reversal symmetry. Furthermore, spatial inversion symmetry should be broken, i.e., the resulting structure must be nonsymmetric with respect to the midplane similarly to Figs. 1(b) and 1(c). Nonreciprocal transmission can be obtained when the symmetry of the parity-time operator is broken (Rüter, 2010) that can be obtained, for example, in a two-channel structure owing to a proper choice of the real and imaginary parts of the index of refraction.

Various manifestations of directional selectivity in the structures that are reciprocal, because of being made of isotropic linear materials only, but allow asymmetric transmission due to transformation of all or significant part of the incident wave energy into either another polarization or higher diffraction orders, have been a subject of the extensive study for a few years. For example, chiral structures are considered to be perspective for achieving isolation for certain polarization states (Plum, 2009; Singh, 2009). It has recently been demonstrated that the complete optical isolation can be achieved dynamically in a linear photonic system with temporal modulation of the refractive index (Yu&Fan, 2009). Nonmagnetic optical isolators can be obtained in the structures that contain two modulators, in which a desired phase shift appears for the co- and counter-propagating waves due to the temporal modulation of bias voltages (Ibrahim, 2004). In this context, PCGs present another big but yet weakly studied class of the reciprocal structures for asymmetric transmission. Contrary to the chiral structures, neither polarization transformation nor rotation of polarization plane occurs in PCGs at asymmetric transmission, provided that they are made of linear isotropic materials.

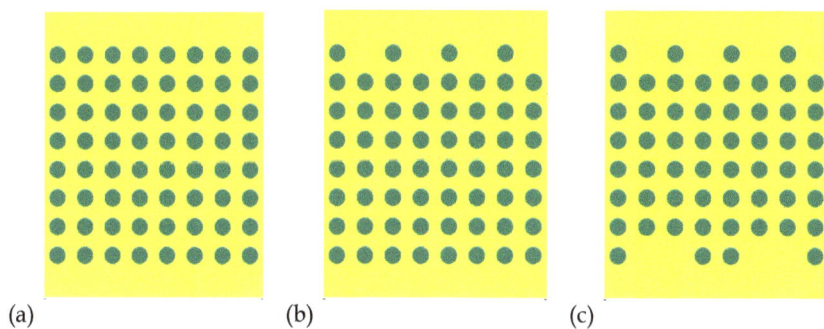

(a)   (b)   (c)

Fig. 1. Noncorrugated slab of square-lattice PC (a) and the corresponding PCGs which have one-side (b) and two-side (c) corrugations.

2.1 Physical background

First, let us briefly consider the effect of breaking spatial inversion symmetry in the conventional dielectric gratings. In line with the classical theory of diffraction gratings (Petit, 1980), transmission and reflection are characterized in terms of the nth order transmission and reflection efficiencies, which are given by $t_n = |\tau_n|^2$ and $r_n = |\rho_n|^2$ and take into account a part of the incident wave energy in the corresponding propagating orders. Here, τ_n and ρ_n are the nth order transmission and reflection coefficients. In turn, transmittance T and reflectance R are given by a sum over all the orders propagating in air. According to the grating formula (Petit, 1980), the nth order diffraction angle, ϕ_n, is determined from

$$\sin\phi_n = \sin\theta + 2\pi n / kL , \qquad (1)$$

where θ is angle of incidence and L is grating period, so that ϕ_n takes a real value for an order that propagates in air. All the diffraction angles for the transmitted beams are measured in the counter-clockwise direction with respect to the normal to the exit side. The angle of incidence is measured in the counter-clockwise direction with respect to the normal to the input side.

Figure 2 presents the transmission spectra for the two nonsymmetric dielectric gratings. One can see that $t_0^\rightarrow = t_0^\leftarrow = t_0$ but $t_n^\rightarrow \neq t_n^\leftarrow$ at $|n|=1$, while $\theta = \theta^\rightarrow = \theta^\leftarrow = 0$, where $\rightarrow$ and $\leftarrow$ stand for the forward (here – from the top) and backward (from the bottom) illumination, respectively. A partially asymmetric transmission with higher orders being responsible for the asymmetry, while the zero-order transmission is symmetric, can be observed in Fig. 2. This is one of the fundamental properties of the nonsymmetric gratings. The ratios $\chi^\rightarrow = \sum_{(m\neq 0)} t_m^\rightarrow / t_0$ and $\chi^\leftarrow = \sum_{(m\neq 0)} t_m^\leftarrow / t_0$ may strongly depend on frequency and geometrical and material parameters of the grating. Replacing a dielectric with a PC enables a wide range where $t_0 = 0$ and $t_m^\leftarrow = 0$ and, hence, $\chi^\rightarrow = \infty$ and $\chi^\rightarrow \gg \chi^\leftarrow$. In the other words, the "strength" of asymmetry can be enhanced, so that asymmetric transmission becomes unidirectional, i.e., $T^\rightarrow \neq 0$ and $T^\leftarrow = 0$. Owing to the band gaps, unidirectional transmission can appear inside wide frequency and incidence angle ranges. This is distinguished from a solid dielectric grating, where these conditions might hypothetically be realized only for a pair of frequency and angle values, but not inside wide ranges. In fact, unidirectional transmission like that in PCGs should not appear in nonsymmetric dielectric gratings, where zero order is always coupled to a wave propagating in the dielectric.

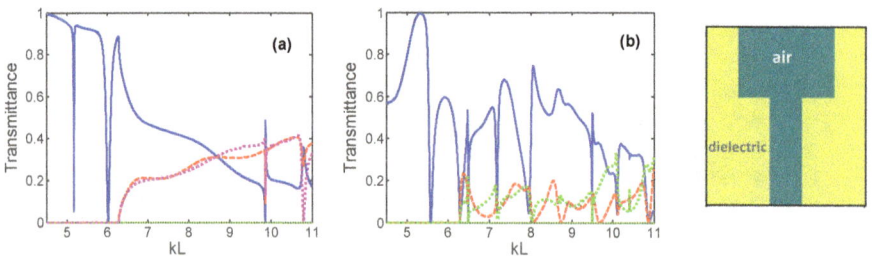

Fig. 2. Transmittance for two nonsymmetric gratings made of dielectric with permittivity (a) $\varepsilon_d = 2.1$ and (b) $\varepsilon_d = 5.8$; solid line - $t_0^\rightarrow = t_0^\leftarrow = t_0$, dotted line - $t_{\pm 1}^\rightarrow$, and dashed line - $t_{\pm 1}^\leftarrow$; right panel - geometry of a grating period.

The use of gyromagnetic and nonlinear materials enables unidirectional devices with a single transmission channel like in conventional electronic diodes. Transmission mechanism in the PCGs needs at least two transmission channels, every being associated with a certain in-air propagating order. In the contrast with the nonreciprocal structures that contain anisotropic or nonlinear materials and reciprocal chiral structures, all the diffracted beams in PCGs show the same linear polarization. In line with the previous studies of PCGs (Serebryannikov, 2009), zero order being uncoupled to any Floquet-Bloch wave is required for obtaining of the unidirectional transmission, that in turn dictates the allowed values of θ and IFC shapes. Accordingly, at the corrugated-side illumination, the *umklapp* refractions are only possible.

Two typical coupling scenarios are demonstrated in Fig. 3. Conservation of the wave vector component that is parallel to the interface, k_x, requires that the IFC crosses a construction line, in order to couple the corresponding order to a Floquet-Bloch wave of the PC (Foteinopoulou&Soukoulis, 2005; Lu, 2007). Locations of the construction lines are determined by the following condition:

$$k_x^{(n)} = (\omega / c)\sin\theta + 2\pi n / L .$$ (2)

Figures 3(a)-3(c) illustrate the coupling mechanism in case of the simplest, i.e., narrow circular IFCs around Γ point, for which the diffraction relevant unidirectional transmission may appear. They correspond to an isotropic material with the index of refraction $0<N<1$, i.e., are narrower than the IFC in air at the same frequency. Hence, similar asymmetry in transmission can be observed, for example, in the nonsymmetric gratings made of a material with $0<N<1$, e.g., a Drude metal above the plasma frequency, or a wire medium above the effective plasma frequency (Serebryannikov&Ozbay, 2009). Figures 3(d)-3(f) illustrate the coupling mechanism in case of near-square IFCs located around M point, which can be obtained in dielectric PCs. Construction lines are plotted for a value of θ, at which at least one higher diffraction order is coupled to a Floquet-Bloch wave due to the corrugations.

In Figs. 3(a) and 3(d), only zero order may propagate in air regardless of whether the incidence is forward or backward. This case is assumed to correspond to a noncorrugated slab of PC, as in Fig. 1(a), or to a PCG in the frequency range where all higher orders are evanescent. In Figs. 3(b) and 3(e), the first order(s) may propagate in air and is allowed to couple to the Floquet-Bloch wave, for the corrugated-side illumination, but should remain evanescent in the input half-space and uncoupled to the Floquet-Bloch wave at the noncorrugated-side illumination. Thus, transmission is not vanishing in the former case only. Assuming that we initially have a noncorrugated slab of PC like that in Fig. 1(a), and then removing some rods from one of the interface layers, we obtain a PCG like that in Fig. 1(b), which is nonsymmetric with respect to the midplane. The simplest corrugations can be obtained by removing every second rod from an interface layer, so that the lateral period of the PCG is $L=2a$, where a is PC lattice constant. In Figs. 3(b) and 3(e), it is assumed that $P=2$ is the minimal integer value of P in $L=Pa$, which provides such a location of the construction lines with respect to the IFC at a given frequency that unidirectional transmission can be obtained. In fact, P depends on the concrete performance of PCG and, thus, may be a rather arbitrary integer. A PCG can still be nonsymmetric and, thus, might support asymmetric transmission, while having corrugations at the both sides, e.g., see Fig. 1(c). However, a larger difference between the periods of the two interfaces should provide a stronger asymmetry in terms of the number of the orders contributing to the transmission. In order to obtain unidirectional transmission with rather strong forward and zero backward transmission, the value of a must be chosen so that higher diffraction orders may not propagate due to the effect of the noncorrugated interface. In Figs. 3(c) and 3(f), the same IFCs are presented as in Figs. 3(b) and 3(e), respectively, but now $L = 4a$. Hence, the distance between the neighbouring construction lines is reduced by factor of 2. As a result, now more diffraction orders may propagate in air due to the corrugated interface, and more orders among them may be unidirectionally coupled to a Floquet-Bloch wave.

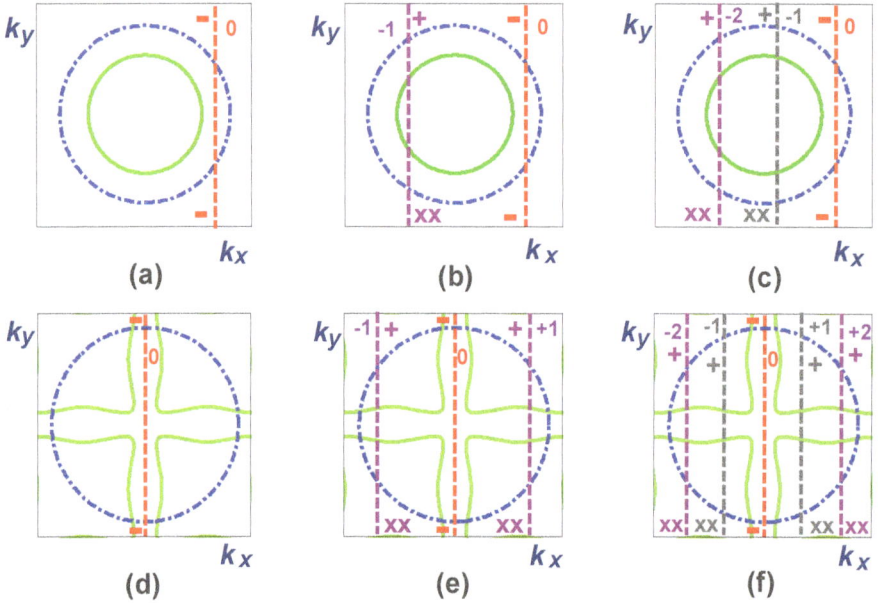

Fig. 3. Coupling scenarios leading to unidirectional transmission: green solid line - IFC of PC; blue dash-dotted circle - IFC in air; dashed lines - construction lines; "0" , "-1", "+1", "-2" and "+2" denote the order index (n); "+" and "-" at plot top indicate that the corresponding order is either coupled or not coupled to a Floquet-Bloch wave, if the corrugated side is illuminated; "-" and "xx" at plot bottom indicate that the corresponding order is either uncoupled at the noncorrugated interface or does not propagate in the input half-space, at the noncorrugated-side illumination; IFCs of PC are assumed to be located around (a-c) Γ point and (d-f) M point; plots (a,d): noncorrugated slab of PC, $L=a$; plots (b,e): PCG with the simplest one-side corrugations, $L=2a$; plots (c,f): PCG with the one-side corrugations, $L=4a$.

From this consideration, it is clearly seen why at least the first negative order for positive nonzero θ and the first positive and first negative orders for zero θ must propagate in air and be coupled to a Floquet-Bloch wave. The scenario shown in Fig. 3(b) corresponds to the regime of *single-beam unidirectional deflection*. The angle between the directions of the incident and single transmitted beams is given by $\Omega=\theta-\phi_{-1}$. At a larger number of the orders propagating in air, as in Fig. 3(c), there may be multiple open transmission channels, every being connected with a certain order, so that splitting occurs in the unidirectional deflection regime. In the general case, we have $t_n \neq t_m$ and $\phi_n \neq \phi_m$, $n \neq m$, for all the propagating orders. In the situation shown in Fig. 3(e) for $\theta=0$, two beams are allowed to propagate in air, $t_{+1} = t_{-1} \neq 0$ and $\phi_{+1} = -\phi_{-1}$. The number of the transmitted beams at normal incidence is always even, since they propagate symmetrically with respect to the normal, $t_{+n} = t_{-n}$ and $\phi_{+n} = -\phi_{-n}$. Thus, *two-beam unidirectional splitting* appears in this case, while deflection of the beams with $n=m$ and $n=-m$ is symmetric regarding the normal. It is noteworthy that the IFC shapes, which can be obtained in two-dimensional PCs but are distinguished from those in Fig. 3, can also be consistent with the requirements to the diffraction relevant mechanism of unidirectional transmission (Serebryannikov, 2009). Furthermore, this remains true for IFCs that are typical for one-dimensional PCs (Kang, 2010) and anticutoff (indefinite) media (Schurig&Smith, 2003). Hence, this mechanism is quite flexible regarding the choice of materials/structures that might be utilized to create a nonsymmetric grating.

Since the structures we consider are assumed to be composed of isotropic linear materials only, transmission remains reciprocal in sense of the Lorentz Lemma (Kong, 2005). This results in the equal transmittances while replacing source and observation point with each other, i.e., (i) when the PCG is illuminated from the corrugated side at $\theta=\theta^\rightarrow$ and (ii) when the PCG is illuminated from the noncorrugated side but at $\theta = \theta^\leftarrow = \phi_n^\rightarrow$, where $\phi_n^\rightarrow$ is the diffraction angle for the nth order transmitted beam at the corrugated-side illumination. For example, if the beam of the order $n=-1$ is the only higher-order propagating beam, as can appear at nonzero θ, and $T^\rightarrow = t_{-1}^\rightarrow \approx \tilde{T}$ at $\theta=\theta^\rightarrow$, then $T^\leftarrow = t_{-1}^\leftarrow \approx \tilde{T}$ at $\theta=\theta^\leftarrow = \phi_{-1}^\rightarrow$. This does not contradict with the fact that the transmission is unidirectional for the two *opposite* directions of incidence.

2.2 Asymmetry in threshold location

In the conventional dielectric gratings, each higher order ($|n|>0$) has a cutoff wavelength and, hence, a threshold frequency, i.e., it propagates if

$$k > | \alpha_0 + 2\pi n / L |, \tag{3}$$

where $\alpha_0 = k \sin \theta$, $k=\omega/c$ (Petit, 1980). In the gratings made of Drude metals or composites, the actual thresholds have different locations as compared to the classical case that is associated with dielectric gratings (Serebryannikov & Ozbay, 2009). In the PCGs with either dielectric or metallic rods, the actual thresholds can also be affected by location of the stop bands of the PC (Serebryannikov, 2006; Serebryannikov, 2009).

Let denote the k thresholds which correspond to the boundary between the propagation and evanescent regimes for the nth diffraction order in a dielectric grating by

$$k_{\pm n} = 2\pi n / [L(1 \mp \sin \theta)], \tag{4}$$

where $n \geq 0$ and $\theta \geq 0$. In the vicinity of $k=k_{\pm n}$, rapid variations in t_n and r_n often appear, which are assigned to the Rayleigh-Wood anomalies (Hessel&Oliner,1965). In turn, the actual thresholds for a PCG with one-side corrugations at the exit interface are denoted by $\tilde{k}_n^{(S)}$. The actual thresholds for a PCG with one-side corrugations at the input interface are denoted by $\hat{k}_n^{(S)}$. Here, $S=T$ for transmission and $S=R$ for reflection. Finally, $k^{(u)}$ and $k^{(l)}$ denote k values that correspond to the upper and lower boundaries of the first stop band. Figure 4 schematically shows the stop and pass bands, the idealized transmission spectrum of PC, the threshold values of k, and the ranges of propagation of the lowest higher order(s) in transmission, for the both PCG and dielectric grating with one-side corrugations. According to Fig. 4(a),

$$k^{(l)} < k_m < k^{(u)} = \tilde{k}_m^{(T)},\tag{5}$$

where $m=-1$ if $\theta \neq 0$ and $m = \pm 1$ if $\theta = 0$. In this case, location of the actual cutoff is determined by the upper boundary of the stop band. The mth order(s) propagate in the exit half-space due to the corrugated interface, starting from this boundary. In Fig. 4(b),

$$k^{(l)} < k_m < k^{(u)} < \hat{k}_m^{(T)}.\tag{6}$$

Hence,

$$\tilde{k}_m^{(T)} \neq \hat{k}_m^{(T)},\tag{7}$$

that is distinguished from the classical grating theory, which gives

$$\tilde{k}_m^{(T)} = \hat{k}_m^{(T)}.\tag{8}$$

The situation in Fig. 4 is realized if zero order is only coupled to the second lowest Floquet-Bloch wave of the PC, i.e., at $k^{(u)} < k < \hat{k}_m^{(T)}$, leading to that $t_0^{\rightarrow} = t_0^{\leftarrow}$ and $T^{\rightarrow} = t_0^{\rightarrow}$ while $T^{\leftarrow} > t_0^{\leftarrow}$. In fact, $\hat{k}_m^{(T)}$ is determined in Fig. 4(b) by the lower boundary of the third lowest passband, for which the mth order(s) are assumed to be coupled to the Floquet-Bloch wave.

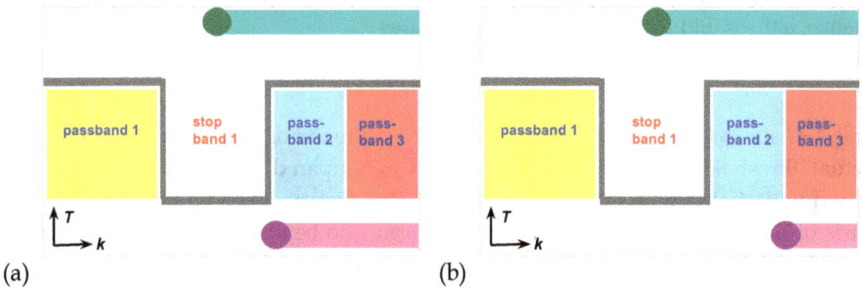

(a)   (b)

Fig. 4. Example of composition of pass and stop bands of PC that affects k-domain threshold of higher order(s) in case when $k_m < k^{(u)}$; circles – the actual thresholds in transmission for the dielectric grating (top) and PCG (bottom); the adjacent rectangles show the k ranges where the mth order(s) may propagate; gray line – the idealized transmission spectrum; plot (a) – noncorrugated-side illumination, plot (b) – corrugated-side illumination.

Figure 5 schematically shows the stop and pass bands of the PC and reflection spectrum together with the thresholds and ranges of contribution of the lowest higher order(s) in reflection for the both PCG and dielectric grating with one-side corrugations. In Fig. 5(a), we have $k_m < k^{(u)} < \tilde{k}_m^{(R)}$ where m is the same as in Fig. 4. Thus, location of the actual threshold is determined here by the lower edge of the third lowest passband, i.e., $\tilde{k}_m^{(R)} = \hat{k}_m^{(T)}$. In turn, $\tilde{k}_m^{(T)} \neq \tilde{k}_m^{(R)}$. In Fig. 5(b), $\hat{k}_m^{(R)} = k_m$, as in the dielectric grating case. Then, $\hat{k}_m^{(R)} \neq \tilde{k}_m^{(R)}$, $\hat{k}_m^{(T)} \neq \hat{k}_m^{(R)}$, and $\tilde{k}_m^{(T)} \neq \hat{k}_m^{(R)}$. Hence, in the contrast with nonsymmetric dielectric gratings, asymmetry in threshold location may appear owing to peculiar types of PC dispersion. In fact, the difference in location of the ranges of $t_m^{\rightarrow} = 0$ and $t_m^{\leftarrow} = 0$, on the one hand, and $r_m^{\rightarrow} = 0$ and $r_m^{\leftarrow} = 0$, on the other hand, is a key feature that is connected with the expected asymmetry in transmission and reflection at least for lower-order stop and pass bands. For higher-order bands, it can be explained in terms of the generalized cutoffs/thresholds.

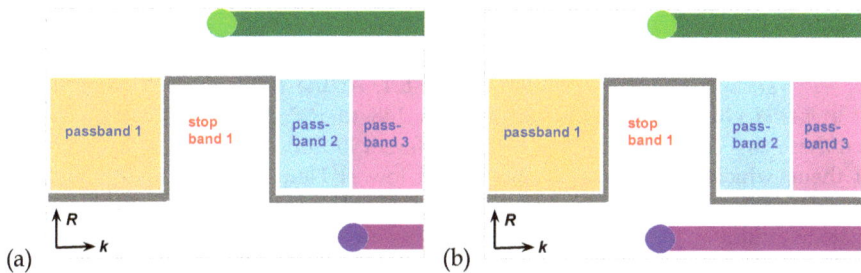

(a)   (b)

Fig. 5. Same as Fig. 4 but for reflection.

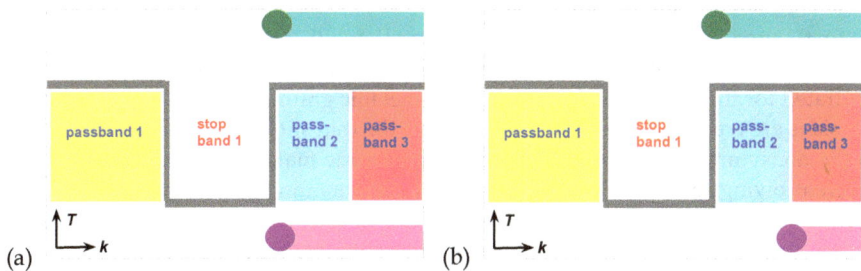

(a)   (b)

Fig. 6. Same as Fig. 4 but in case when $k_m = k^{(u)}$.

Figures 6 and 7 are analogous to Figs. 4 and 5, respectively, but now $k_m = k^{(u)}$. In the contrast with Figs. 4 and 5, we have simultaneously $\hat{k}_m^{(T)} = \tilde{k}_m^{(R)} > k_m$ and $\tilde{k}_m^{(T)} = \hat{k}_m^{(R)} = k_m$. In turn, $\tilde{k}_m^{(T)} \neq \hat{k}_m^{(T)}$ and $\tilde{k}_m^{(R)} \neq \hat{k}_m^{(R)}$. Hence, the different combinations of locations of $\tilde{k}_m^{(S)}$ and $\hat{k}_m^{(S)}$ with respect to each other and to k_m can be obtained by adjusting the corrugation and PC lattice parameters.

2.3 Forward vs backward transmission

Let us consider the effect of variation in L on the appearance of higher orders in the transmission, in both cases of the corrugated-side and the noncorrugated-side illumination, at normal incidence. An example is shown in Fig. 8 for typical values of the rod-diameter-to-lattice-constant ratio, d/a, relative permittivity of the rod material, ε_r, and an intermediate number of the rod layers, Q. Figures 8(a) and 8(b) partially correspond to the case of the k thresholds location, as in Figs. 5(a) and 5(b).

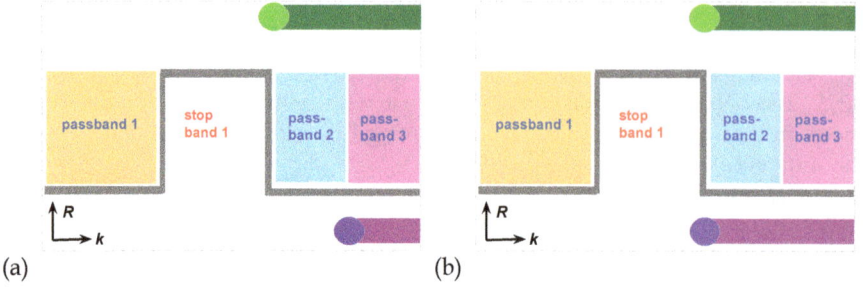

Fig. 7. Same as Fig. 5 but in case when $k_m = k^{(u)}$.

Indeed, $T^{\rightarrow} = t_0$ but $T^{\leftarrow} = t_0 + t^{\leftarrow}_{-1} + t^{\leftarrow}_{+1}$ at $6<kL<8.4$. At the same time, $T^{\rightarrow} = t^{\rightarrow}_{-1} + t^{\rightarrow}_{+1}$ and $T^{\leftarrow} = 0$ at $8.4<kL<10.5$, where $\phi_{\pm 1}$ is varied from ± 48 to ± 37 degrees. In fact, the difference between these two cases originates from the different locations and shapes of IFCs. For the first of them, which corresponds to the second lowest Floquet-Bloch mode, the IFCs are located around Γ point, but narrower than in air. Thus, the orders with $n = \pm 1$ may propagate in air due to the corrugated interface, but are not coupled. For the second of them, which corresponds to the third lowest Floquet-Bloch mode, the IFCs are located around M point, so that the orders with $n = \pm 1$ may propagate in air due to the corrugations and are coupled at the corrugated-side illumination. In turn, zero order is not coupled.

In the first range, we observe *one-way transmission* with the both symmetric (t_0) and asymmetric (t_n, $|n|>0$) components being nonzero. This case is similar to that demonstrated for the nonsymmetric microwave metallic gratings, where different periods at the two sides are created by a proper branching of the thin slit waveguides (Lockyear, 2006), and in the nonsymmetric gratings that contain the Drude material layers, while the periods are different owing to the one-side corrugations (Serebryannikov&Ozbay, 2009). In all the figures, the most representative ranges with $T^{\rightarrow} = t_0$ and $T^{\leftarrow} > t_0$ are denoted by "One-Way". In the second range, *unidirectional transmission* takes place in the form of unidirectional splitting, i.e., $t^{\rightarrow}_{-1} = t^{\rightarrow}_{+1}$, $T^{\rightarrow} = t^{\rightarrow}_{-1} + t^{\rightarrow}_{+1}$, while $T^{\leftarrow} = 0$. In Figs. 8(a) and 8(b), one more range of unidirectional splitting is seen at $11.2<kL<12.4$, where $\varphi_{\pm 1}$ is varied from ± 34 to ± 30 degrees. It is connected with the fifth lowest Floquet-Bloch mode. Here, we again have $T^{\rightarrow} \neq 0$ and $T^{\leftarrow} = 0$, while even a higher transmittance is achieved than at $8.4<kL<10.5$. Furthermore, $T^{\rightarrow} = 1$ at the lower edge of the passband, so that exactly the diode regime is realized. In the figures, the most representative ranges of unidirectional transmission ($T^{\rightarrow} \neq 0$ and $T^{\leftarrow} = 0$) are denoted by "UD".

Increasing the number of the rod columns per grating period might lead to that the actual thresholds are shifted towards smaller ka and, hence, lower passbands. An example is shown in Figs. 8(c) and 8(d) at $P=3$. Contrary to the case of $P=2$, $t_{+n} \neq t_{-n}$ at $|n|>0$. Unidirectional transmission with $T^{\rightarrow} = t^{\rightarrow}_{-1} + t^{\rightarrow}_{+1}$ and $T^{\leftarrow} = 0$ appears already for the lowest Floquet-Bloch mode, i.e., at the edge of the lowest passband for $7<kL<7.9$, where ϕ_{-1} is varied from -63.8 to -52.7 degrees. The IFCs are located now around M point, so that zero order is uncoupled. In turn, at smaller kL, the IFCs are located around Γ point that enables coupling of zero order. For the third lowest Floquet-Bloch mode, we obtain unidirectional transmission with $T^{\rightarrow} = t^{\rightarrow}_{-1} + t^{\rightarrow}_{+1} + t^{\rightarrow}_{-2} + t^{\rightarrow}_{+2}$ and $T^{\leftarrow} = 0$ at $13.2<kL<15.8$.

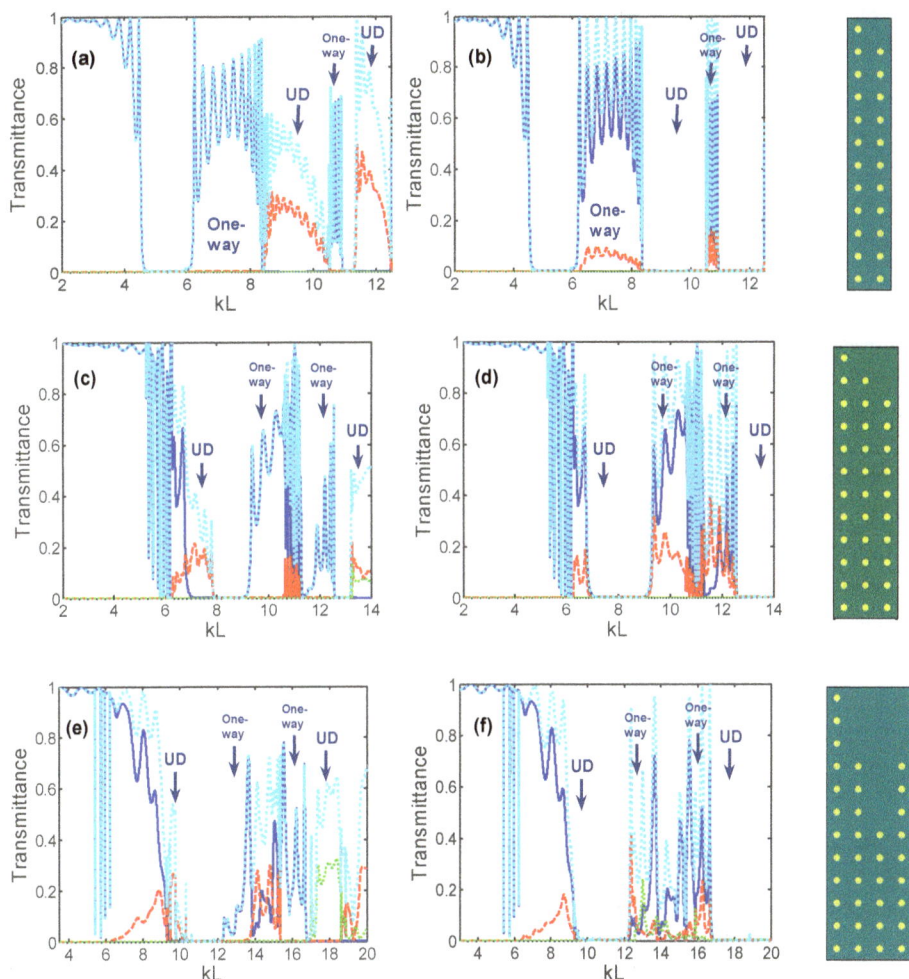

Fig. 8. Transmittance for PCGs with (a,b) $L=2a$, (c,d) $L=3a$, (e,f) $L=4a$, for the corrugated-side (a,c,e) and noncorrugated-side (b,d,f) illumination; d/a=0.31, $\varepsilon_r = 5.8$, Q=12, θ=0; blue solid line - t_0 ; red dashed line - $t_{\pm1}^{\rightarrow}$ (a, e), $t_{\pm1}^{\leftarrow}$ (b, f), $t_{-1}^{\rightarrow}$ (c), and $t_{-1}^{\leftarrow}$ (d); green dotted line - $t_{\pm2}^{\rightarrow}$ (e), $t_{\pm2}^{\leftarrow}$ (f), $t_{-2}^{\rightarrow}$ (c), and $t_{-2}^{\leftarrow}$ (d); cyan dotted line - $T^{\rightarrow}$ (a,c,e) and $T^{\leftarrow}$ (b,d,f); right panels – geometry of PCG within a period.

At P=4, unidirectional splitting takes place for the first (now at 9.6<kL<10.4) and third lowest Floquet-Bloch waves. - See Figs. 8(e) and 8(f). The main difference as compared to the case of P=3 is probably that the regime with $T^{\rightarrow} = t_{-2}^{\rightarrow} + t_{+2}^{\rightarrow}$ can be realized at the band edge for the third lowest Floquet-Bloch wave (at kL>16.9). It is noteworthy that one-way transmission with $T^{\rightarrow} = t_0$ and $T^{\leftarrow} > t_0$ can appear also at P>2, e.g., at 9.2<kL<10.5 in Figs. 8(c) and 8(d) and at 15.6<kL<16.9 in Figs. 8(e) and 8(f). A proper choice of the PC lattice parameters is important from the point of view of obtaining of the *switching* between different regimes. For example, in Fig. 8(a), the ranges of $T^{\rightarrow} = t_0$ and $T^{\rightarrow} = t_{-1}^{\rightarrow} + t_{+1}^{\rightarrow}$ are adjacent but do not superimpose near kL=8.4.

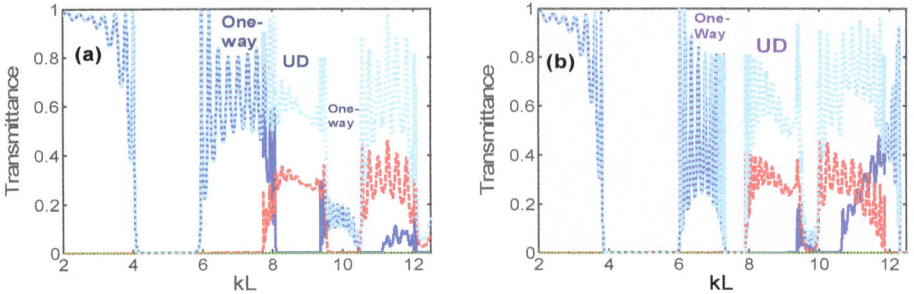

Fig. 9. Transmittance for PCGs with (a) d/a=0.4, $\varepsilon_r = 5.8$, Q=12, and (b) d/a=0.31, $\varepsilon_r = 9.61$, Q=12, at L=2a and θ=0, for corrugated-side illumination; blue solid line - t_0, red dashed line - $t_{\pm 1}^{\rightarrow}$, and blue dotted line - $T^{\rightarrow}$.

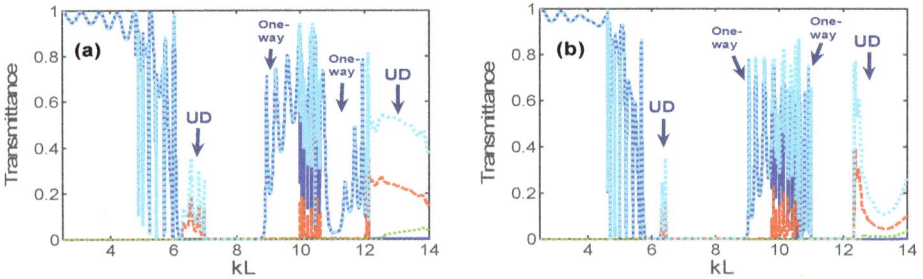

Fig. 10. Same as Fig. 9 but at L=3a, green dotted line – $t_{\pm 2}^{\rightarrow}$, cyan dotted line - $T^{\rightarrow}$.

Two more cases are shown in Figs. 9(a) and 9(b). In Fig. 9(a), the one-way and unidirectional transmission ranges superimpose at 7.7<kL<8.1. In Fig. 9(b), there is a gap between them at 7.3<kL<7.9. Hence, the PCG in Fig. 8(a) is optimal for the switching realizable by varying frequency. In Fig. 8, it has been shown that unidirectional transmission can be observed at rather small kL. Figure 10 demonstrates, in addition, that the ranges of $T^{\rightarrow} = t_0$ and $T^{\rightarrow} = t_{-1}^{\rightarrow} + t_{+1}^{\rightarrow}$ can do not superimpose at the edge of the lowest passband. Here, the corrugations are obtained by removing one rod from every two of three columns in the interface layer in Fig. 9(a) and from every third column in the interface layer in Fig. 9(b). Now, $t_{+n} = t_{-n}$ due to the used corrugation shape. Comparison of Figs. 8-10 shows that the maximal transmittance achievable in the unidirectional transmission regime for a certain higher-order Floquet-Bloch mode is strongly dependent on the PC lattice and corrugation parameters, as well as the contribution of individual higher orders.

For a PCG with fixed parameters, variation in θ gives an efficient tool for tuning. Strong modification of the transmission spectrum can be achieved even at a rather weak variation. An example is presented in Fig. 11 for the PCG illuminated from the corrugated side, for which transmittance at θ=0 is shown in Fig. 8(a). At θ=10 degrees (Fig. 11(a)), transmittance is two-way, i.e., $t_{-1}^{\rightarrow} \neq 0$, $t_{-1}^{\leftarrow} \neq 0$, $T^{\rightarrow} > t_0$, and $T^{\leftarrow} > t_0$, for the most part of the one-way transmission range of Fig. 8(a) that belongs to the second lowest passband (here at 6.2<kL<8.2). At the same time, transmission remains unidirectional for the third lowest passband, but now $t_{-1}^{\rightarrow} \neq t_{+1}^{\rightarrow}$ and a stop band appears between the second and third passbands. Appearance of a narrow unidirectional transmission band in the vicinity of kL=5.4, where $T^{\rightarrow} = t_{-1}^{\rightarrow}$ and $T^{\leftarrow} = 0$, is probably the most interesting feature seen in Fig. 11(a). The lower boundary of this band is determined by $k=k_{-1}$ (kL=5.35), according to (4), so that the rapid increase of $T^{\rightarrow} = t_{-1}^{\rightarrow}$ is connected with the Rayleigh-Wood anomaly. The upper boundary is due to the narrowing and further disappearance (near kL=5.47) of the IFCs at increasing kL. The transmitted beams in this case are strongly deflected. For example, $\phi_{-1} \approx -82$ degrees at kL=5.4.

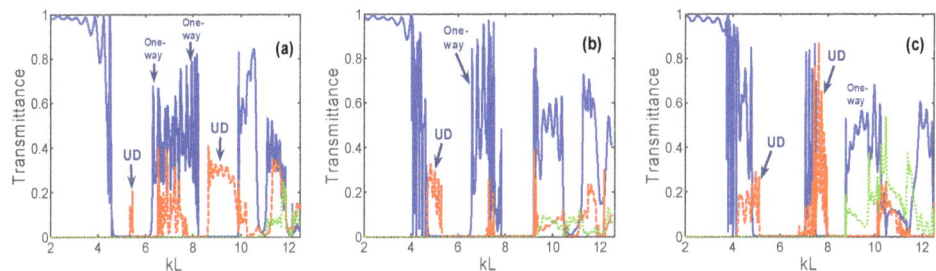

Fig. 11. Same as Fig. 8(a), but at θ=10 degrees (a), θ=20 degrees (b), and θ=30 degrees (c); blue solid line - t_0, red dashed line - $t_{-1}^{\rightarrow}$, and green dotted line - $t_{-2}^{\rightarrow}$.

Increase of θ up to 20 degrees leads to that this band becomes wider, and the switching between the regimes of $T^{\rightarrow} = T^{\leftarrow} = t_0$ (two-way, symmetric) and $T^{\rightarrow} = t_{-1}^{\rightarrow}$ (unidirectional) occurs at $kL \approx 4.7$ – See Fig. 11(b). To compare, (4) gives $k_{-1}L = 4.68$, i.e., the rapid increase of $t_{-1}^{\rightarrow}$ is again due to the Rayleigh-Wood anomaly. At the same time, the unidirectional transmission range observed in Fig. 11(a) in the vicinity of kL=9 disappears. The obtained results show that a rather wide range of $T^{\rightarrow} = t_{-1}^{\rightarrow}$ can appear at the edge of the lowest passband in the both unidirectional splitting (θ=0) and unidirectional deflection ($\theta \neq 0$) regimes. Further increase of θ can result in the appearance of the high-T unidirectional deflection range, as occurs for the second lowest Floquet-Bloch mode at kL=7.6 and θ=30 degrees, where $\max T^{\rightarrow} = \max t_{-1}^{\rightarrow} > 0.6$. – See Fig. 11(c). However, in this case, switching of such a kind as in Fig. 11(b) at the edge of the lowest passband cannot be obtained. Instead, there are three consequent ranges at 4<kL<5, which are similar to those in Fig. 8(c): $T^{\rightarrow} = t_0$ (two-way, symmetric), $T^{\rightarrow} = t_0 + t_{-1}^{\rightarrow}$ (one-way, asymmetric), and $T^{\rightarrow} = t_{-1}^{\rightarrow}$ (unidirectional). Besides, two new ranges of one-way transmission appear at 8.7<kL<10.1 and 11.3<kL<12.4, where $T^{\rightarrow} = t_0 + t_{-2}^{\rightarrow}$ and $T^{\leftarrow} = t_0 + t_{-1}^{\leftarrow} + t_{-2}^{\leftarrow}$.

Figure 12 presents the transmission spectra at the three values of θ, for the same PCG as in Fig. 9(a). The defect-mode-like unidirectional peak in Fig. 12(b) and switching between the regimes of $T^{\rightarrow} = t_0$ and $T^{\rightarrow} = t_{-1}^{\rightarrow}$ in Fig. 12(c), both being connected with the first lowest Floquet-Bloch mode, are obtained now at larger θ than in Fig. 11. Here, the rapid increase of $t_{-1}^{\rightarrow}$ has the same nature as in Fig. 11. Other features observed are similar, too. Hence, various operation regimes can co-exist in the adjacent frequency ranges, at a proper choice of the PC lattice and corrugation parameters and a value of θ.

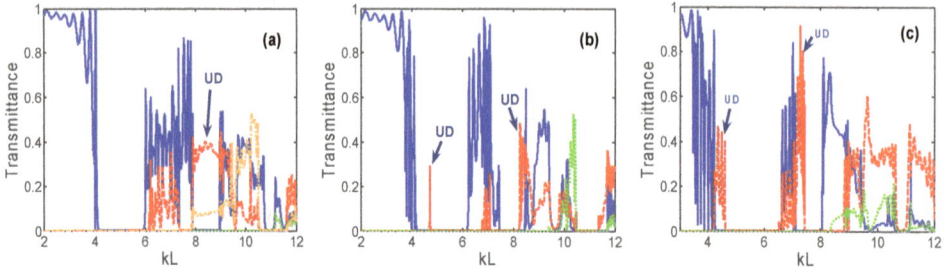

Fig. 12. Same as Fig. 9(a), but at θ=10 degrees (a), θ=20 degrees (b), and θ=30 degrees (c); blue solid line - t_0 , red dashed line - $t_{-1}^{\rightarrow}$, orange dash-dotted line - $t_{+1}^{\rightarrow}$, and green dotted line - $t_{-2}^{\rightarrow}$.

There are several problems to be solved in order to design such PCGs that are consistent with the requirements and limitations regarding the realistic nanofabrication process and illumination characteristics. For example, the requirement to the frequency range of unidirectional transmission to be wide, which is connected with possible fabrication inaccuracies, should be fulfilled simultaneously with the requirement to this range to show high transmittance within a wide range of θ variation, which is important for the incident beams with a wide plane-wave angular spectrum. Figure 13 presents the transmission spectra for a PCG with the selected parameters, which is expected to better fulfil the above-mentioned requirements. The wide unidirectional transmission range with $T^{\rightarrow} = t_{-1}^{\rightarrow} > 0.8$ and $T^{\leftarrow} = 0$ is located near kL=5.6. Obtaining of θ-independent unidirectional ranges with $T^{\rightarrow} \approx 1$ should be the next step towards practical diode-type devices.

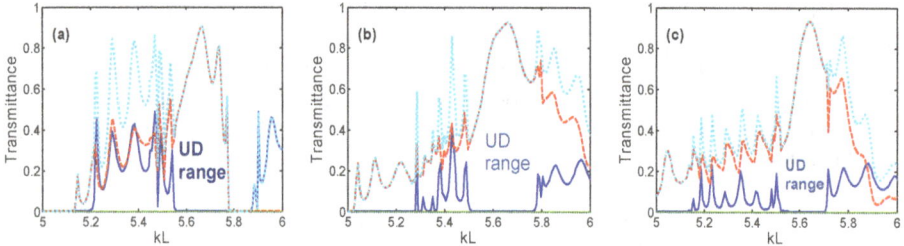

Fig. 13. Transmittance at the corrugated-side illumination for PCGs with L=2a, d/a=0.5, ε_r = 9.61 , Q=12, θ=40 degrees (a), θ=50 degrees (b), and θ=60 degrees (c); blue solid line - t_0 , red dashed line - $t_{-1}^{\rightarrow}$, cyan dotted line - $T^{\rightarrow}$.

3. Fabry-Perot type transmission

The alternating total-transmission maxima and zero-transmission minima, which can be interpreted in terms of the Fabry-Perot resonances, belong to the main features of the transmission spectra of the lossless dielectric slabs. Transmittance is given in this case by the well-known formula (Born&Wolf, 1970)

$$T = \frac{(1-\tilde{R})^2}{(1-\tilde{R})^2 + 4\tilde{R}\sin^2(N'kD\cos\theta')} \tag{9}$$

where θ' is angle of refraction, D is thickness of the slab, N' is index of refraction of the slab material, and $\tilde{R}$ is reflectance of a dielectric-air interface. The peaks of $T=1$ appear at $N'kD\cos\theta' = \pi m$, $m = 1,2,3,...$. Fabry-Perot resonances can also appear in the noncorrugated slabs of PCs (Sakoda, 2001; Serebryannikov, 2010). In the contrast with the dielectric slabs, in PCs we have $v_g \neq v_{ph}$, where v_g and v_{ph} are group and phase velocity, respectively. Location of the minima and maxima of T depends, in fact, on v_g. On the other hand, the *equivalent group index* can be estimated at $\theta=0$ from the locations of the peaks of T (Sakoda, 2001):

$$N'_g = \pi c / (\Delta\omega D),\qquad(10)$$

where $\Delta\omega$ is spectral distance between the neighbouring peaks. Clearly, characterization of the finite-thickness slabs of PCs in terms of N'_g is ambiguous, at least because of the unavoidable uncertainty in location of the virtual interfaces. Besides, it is assumed that $N'_g > 0$, that is not always the case. Nevertheless, this approach usually gives the estimates of N'_g that are qualitatively correct within sign, for thick slabs. Obtaining of accurate (intrinsic) values of the group index needs post-processing of the dispersion results. The corresponding formulas can be found in the literature (Foteinopoulou & Soukoulis, 2005).

In the PCGs, Fabry-Perot resonances can appear while higher orders contribute to the transmission. Since this contribution is asymmetric, i.e., dependent on the illumination side, there may be asymmetry in the appearance of the resonances, which manifests itself in a high contrast between the backward and forward transmittances. From the point of view of demonstration of such asymmetry, the regimes with nonzero transmittance in the both directions are most interesting. In particular, this is related to the one-way transmission regime with $T^{\rightarrow} = t_0$ and $T^{\leftarrow} > t_0$.

Figure 14 presents an example of strong asymmetry, which is observed in the one-way transmission regime for the fourth lowest Floquet-Bloch mode in the PCG from Fig. 9(b). Asymmetry appears here owing to that the contribution of $t_{\pm1}^{\leftarrow}$ to $T^{\leftarrow}$ is more significant than that of t_0, while $t_{\pm1}^{\rightarrow} = 0$. A high contrast can be achieved, e.g., $T^{\leftarrow}/T^{\rightarrow} = 11$ at $kL \approx 9.78$. Furthermore, the peaks of $T^{\leftarrow} \approx 1$ are observed, like in the case of a noncorrugated PC, or a dielectric slab. The values of N'_g obtained from (10) are given in Table 1. k_lL and k_sL mean the larger and smaller values of kL for each pair of the neighbouring peaks. The smaller the distance between the peaks, the larger the value of N'_g is.

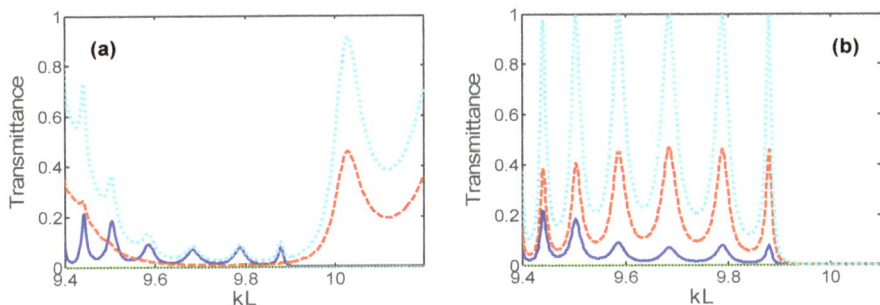

Fig. 14. Fragment of the transmission spectrum for the PCG with $L=2a$, $d/a=0.31$, $\varepsilon_r = 9.61$, $Q=12$, at $\theta=0$, for corrugated-side (a) and nocorrugated-side (b) illumination; blue solid line - t_0, red dashed line - $t_{\pm1}^{\rightarrow}$ (a) and $t_{\pm1}^{\leftarrow}$ (b), cyan dotted line - $T^{\rightarrow}$ (a) and $T^{\leftarrow}$ (b).

	1	2	3	4	5
$k_s L$	9.44	9.504	9.587	9.684	9.79
$k_l L$	9.504	9.587	9.684	9.79	9.88
N_g'	8.18	6.30	5.40	4.94	5.82

Table 1. Equivalent group index for the transmission peaks in Fig. 14(b).

Figure 15 illustrates the case when the same diffraction orders contribute to $T^{\leftarrow}$ and $T^{\rightarrow}$, but the contributions of individual orders strongly depend on the illumination direction. Here, two-way transmission occurs at $14<kL<15.8$, while unidirectionality with $T^{\rightarrow} \approx t_{-1}^{\rightarrow} + t_{+1}^{\rightarrow}$ takes place in the adjacent range, i.e., at $12<kL<14$. For example, the order with $n=-1$ is the main contributor at $14.1<kL<14.4$ at the noncorrugated-side illumination, but its effect tends to vanish in the vicinity of $kL=14.4$ at the corrugated-side illumination.

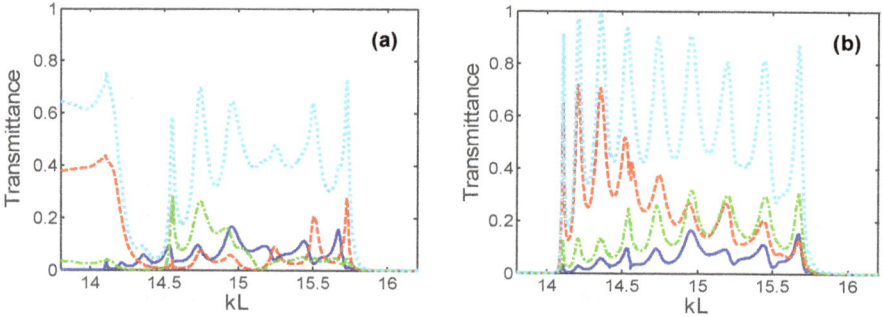

Fig. 15. Fragment of the transmission spectrum for the PCG with $L=3a$, $d/a=0.4$, $\varepsilon_r = 5.8$, $Q=12$, at $\theta=0$, for corrugated-side (a) and nocorrugated-side (b) illumination; blue solid line - t_0, red dashed line - $t_{-1}^{\rightarrow}$ (a) and $t_{-1}^{\leftarrow}$ (b), green dash-dotted line - $t_{-2}^{\rightarrow}$ (a) and $t_{-2}^{\leftarrow}$ (b), cyan dotted line - $T^{\rightarrow}$ (a) and $T^{\leftarrow}$ (b); corrugations are the same as in Figs. 8(c) and 8(d); note that $t_{-n}^{\rightarrow} \neq t_{+n}^{\rightarrow}$ and $t_{-n}^{\leftarrow} \neq t_{+n}^{\leftarrow}$, $|n|>0$.

4. Reflection regime

Band gaps and relevant total reflections belong to the main effects known in PCs. Corrugations may lead to that the higher diffraction orders contribute to reflection starting from the frequency and kL values, which correspond to the lowest stop band of the PC. Furthermore, if corrugations are placed at one side only, reflections can be asymmetric, so that the different diffraction orders play the different roles, depending on the illumination direction, although $R^{\rightarrow} = R^{\leftarrow} = 1$. Figure 16 presents the reflection spectra for the same PCG as in Figs. 8(a) and 8(b), at $\theta=0$. Now, $r_{\pm1}^{\rightarrow}>0$ at $kL>2\pi$ and $r_{\pm1}^{\leftarrow}>0$ at $kL>4\pi$. It is noteworthy that, in the contrast with transmission, the reciprocity principle requires that $r_0^{\rightarrow} = r_0^{\leftarrow}$ only at $kL \leq 2\pi$. Hence, zero-order reflection is itself asymmetric, provided that higher order(s) are allowed to propagate in air. In fact, the possibility of contribution of higher orders to $R^{\rightarrow}$ and $R^{\leftarrow}$ mainly depends on the period of the illuminated interface, i.e., $L_1 = 2a$ and $L_2 = a$, for the corrugated and noncorrugated interfaces, respectively.

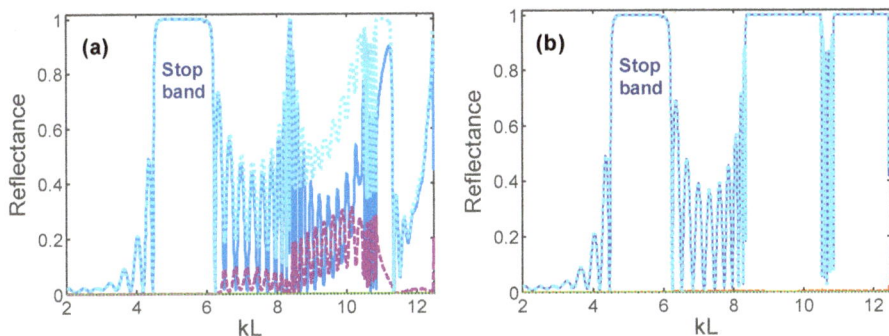

Fig. 16. Reflectance for the PCG with $L=2a$, $d/a=0.31$, $\varepsilon_r = 5.8$, $Q=12$, at $\theta=0$, for (a) corrugated-side and (b) nocorrugated-side illumination; blue solid line - r_0, violet dashed line - $r_{-1}^{\rightarrow} = r_{+1}^{\rightarrow}$ (a) and $r_{-1}^{\leftarrow} = r_{+1}^{\leftarrow}$ (b), cyan dotted line - $R^{\rightarrow}$ (a) and $R^{\leftarrow}$ (b).

At $P>2$, some new features can be observed in the reflection spectrum as compared to Fig. 16. Figure 17 presents an example for a PCG with $P=4$, while the PC lattice parameters are the same as in Figs. 9(b), 10(b), and 14. Corrugations are obtained here by removing two rods from every second column, and four rods from every fourth column, so that they are similar to but not so deep as those in Figs. 8(e) and 8(f). If the corrugated side is illuminated, the orders with $n=\pm1$ contribute to $R^{\rightarrow}$ also in the total-reflection regime at $8.5<kL<11.7$. In particular, splitting with $r_0^{\rightarrow} = r_{-1}^{\rightarrow} = r_{+1}^{\rightarrow}$ and $R^{\rightarrow} = 1$ takes place at $kL=9.29, 9.68, 10.54, 11.33$, and 11.55. If the noncorrugated side is illuminated, zero order is the main contributor to $R^{\leftarrow}$ within the entire kL-range considered. Comparing to Fig. 16(b), the orders with $n=\pm1$ now do not vanish but slightly contribute to $R^{\leftarrow}$ in the vicinity of $kL=7$ and $kL=14$. In these ranges, $t_{\pm1}^{\leftarrow} \neq 0$ due to the effect of the exit (here – corrugated) interface, since higher orders may appear in $R^{\leftarrow}$ due to the input (here – noncorrugated) interface starting from $kL=8\pi$ only.

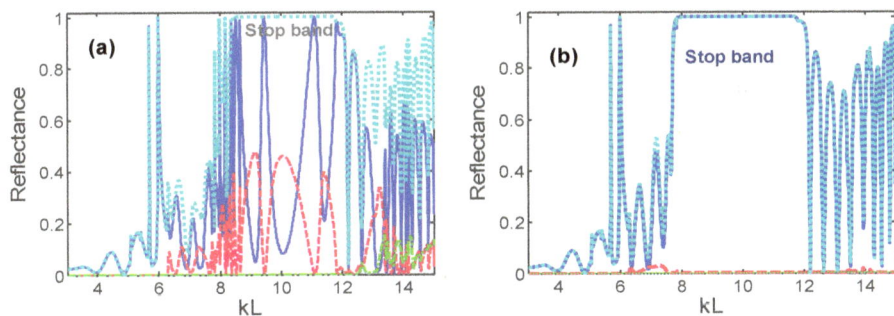

Fig. 17. Same as Fig. 16 but for $L=4a$ and $\varepsilon_r = 9.61$; blue solid line - r_0, red dashed line - $r_{-1}^{\rightarrow} = r_{+1}^{\rightarrow}$ (a) and $r_{-1}^{\leftarrow} = r_{+1}^{\leftarrow}$ (b), green line - $r_{-2}^{\rightarrow} = r_{+2}^{\rightarrow}$ (a) and $r_{-2}^{\leftarrow} = r_{+2}^{\leftarrow}$ (b), and cyan dotted line - $R^{\rightarrow}$ (a) and $R^{\leftarrow}$ (b).

Tilting leads to that the higher orders can strongly contribute to the ranges of $R^{\rightarrow} = 1$ at smaller P than at $\theta=0$. For example, splitting with $r_0^{\rightarrow} = r_{-1}^{\rightarrow}$ and the relatively small values of $|dr_n / d(kL)|$ can be obtained in the first stop band at $P=2$. Besides, the order with $n=-1$ can dominate in $R^{\rightarrow}$ at a stop band edge, where $\phi_{-1} \approx -\theta$, i.e., reflection is nearly backward.

Two examples are shown in Fig. 18. Here, $\vec{r_0} = \vec{r_{-1}}$ at kL=5.57 in Fig. 18(a), and at kL=5.66 and kL=8.29 in Fig. 18(b). Tilting can be an efficient tool of tuning in the reflection regime. Varying θ, one can change k_{-1} and, hence, obtain $\vec{r_{-1}} > 0$ for the entire, or a desired part of the lowest stop band.

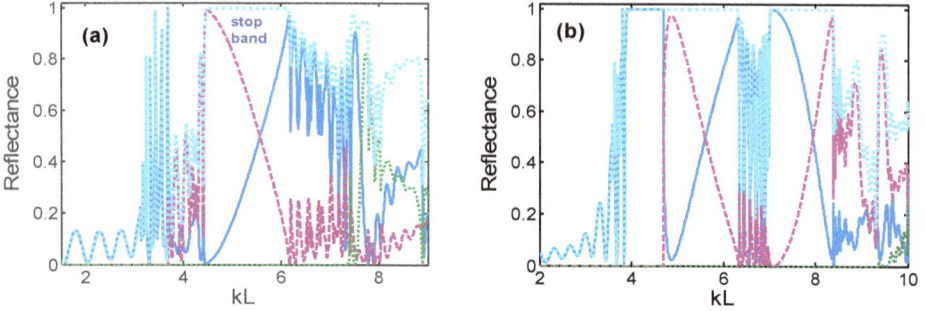

Fig. 18. Reflectance (a) for the same PCG as in Fig. 9(a) at θ=45 degrees, and (b) for the same PCG as in Fig. 9(b) at θ=20 degrees; corrugated-side illumination; blue solid line - r_0, violet dashed line - $\vec{r_{-1}}$, dark green line - $\vec{r_{-2}}$, and cyan dotted line - $R^{\rightarrow}$.

5. Conclusion

Multiple examples can be given to demonstrate that a combination of different physical phenomena can create a possibility of substantial extension of the variety of the obtainable regimes and new ideas for low-cost and/or compact designs. Thus, hybridization is a rather general approach in modern optics and physics. In this chapter, it has been shown how the effects, which are well known for the gratings, on the one hand, and those for the PCs, on the other hand, can be combined in the nonsymmetric PCGs composed of dielectric rods in such a way that new operation regimes can be obtained, which are not realizable in dielectric gratings or noncorrugated PCs. The most interesting transmission and reflection regimes of PCGs originate from the nonsymmetry, i.e., from the broken spatial inversion symmetry. The studied mechanism is characterized by absence of polarization conversion, while the extreme redistribution of the incident wave energy into that of of higher orders plays a key role. In particular, PCGs promise new solutions for unidirectional diode-like devices, splitters, deflectors, mirrors, and nano- and microwave antennas. From the point of view of the theory of PCs, introduction of corrugations while dispersion is known enables new coupling scenarios owing to diffractions. From the point of view of the grating theory, using a PC with the alternating pass and stop bands and substantially different properties of various Floquet-Bloch modes, instead of a homogeneous linear material, enables new diffraction scenarios as compared to those typical for dielectric gratings. Finally, from the point of view of the asymmetric and unidirectional transmission, PCGs demonstrate a high potential in obtaining of strong directional selectivity without breaking time reversal symmetry and, hence, without using anisotropic or nonlinear materials. A new direction in the studies of PCGs concerns asymmetric transmission for defect modes that might appear in chains of the cavity defects or/and line defects, which are parallel to the interfaces.

6. Acknowledgment

The author thanks the Deutsche Forschunsgemeinschaft for partial support of this work under Project Nos. SE1409-2/1 and SE1409-2/2, and Prof. E. Ozbay and members of his research group for fruitful discussions.

7. References

Born, M. & Wolf, E. (1970). *Principles of Optics*, Pergamon Press, Oxford

Caglayan, H.; Bulu, I. & Ozbay, E. (2008). Off-Axis Beaming From Subwavelength Apertures, *J. Appl. Phys.*, Vol. 104, No. 7, 073108

Cakmak, A.O.; Colak, E., Serebryannikov, A.E. & Ozbay, E. (2010). Unidirectional Transmission in Photonic-Crystal Gratings at Beam-Type Illumination, *Opt. Express*, Vol. 18, No. 21, pp. 22283-22298

Collardey, S.; Tarot, A.-C., Pouliguen, P. & Mahdjoubi, K. (2005). Use of Electromagnetic Band-Gap Materials for RCS Reduction, *Microwave Opt. Technol. Lett.*, Vol. 44, No. 6, pp. 546-550

Figotin, A. & Vitebsky, I. (2001). Nonreciprocal Magnetic Photonic Crystals, *Phys. Rev. E*, Vol. 63, No. 6, 066609

Foteinopoulou, S. & Soukoulis, C.M. (2005). Electromagnetic Wave Propagation in Two-Dimensional Photonic Crystals: A Study of Anomalous Refraction Effects, *Phys. Rev. B*, Vol. 72, No. 16, 165112

Gralak, B.; Enoch, S. & Tayeb, G. (2000). Anomalous Refractive Properties of Photonic Crystals, *J. Opt. Soc. Am A*, Vol. 17, No. 6, pp. 1012-1020

Hessel, A. & Oliner, A.A. (1965). A New Theory of Wood's Anomalies on Optical Gratings, *Appl. Opt.*, Vol. 4, No. 10, pp. 1275-1297

Ibrahim, S.K.; Bhandare, S., Sandel, D., et al. (2004). Non-Magnetic 30dB Integrated Optical Isolator in III/IV Material, *Electron. Lett.*, Vol. 40, No. 20, pp. 1293-1294

Inoue, K. & Ohtaka, K., Eds. (2004). *Photonic Crystals: Physics, Fabrication, and Applications*, Springer, Berlin Heidelberg New York

Joannopoulos J.D. (1995). *Photonic Crystals: Molding the Flow of Light*, Princeton Univ. Press, Princeton

Kang, X.-B.; Tan, W., Wang, Z.-S., et al. (2010). High Efficiency One-Way Transmission by One-Dimensional Photonic Crystals with Gratings on One Side, *Chin. Phys. Lett.*, Vol. 27, No. 7, 074204

Kong, J.A. (2005). *Electromagnetic Wave Theory*, EMW Publishing, Cambridge, MA

Le Thomas, N.; Houdre, R., Frandsen, L.H., et al. (2007). Grating-Assisted Superresolution of Slow Waves in Fourier Space, *Phys. Rev. B*, Vol. 76, No. 3, 035103

Li, X.-F.; Ni, X., Feng, L., et. al. (2011). Sonic-Crystal-Based Acoustic Diode, *Phys. Rev. Lett.*, Vol. 106, No. 8, 084301

Lockyear, M.J. ; Hibbins, A.P., White, K.R., and Sambles, J.R. (2006). One-Way Diffraction Gratings, *Phys. Rev. E*, Vol. 74, No. 5, 056611

Lu, W.T.; Huang, J.Y., Vodo, P. et al. (2007). A New Mechanism for Negative Refraction and Focusing Using Selective Diffraction from Surface Corrugation, *Opt. Express*, Vol. 15, No. 15, pp. 9166-9175

Luo, C.; Johnson, S.G., Joannopoulos, J.D., & Pendry, J.B. (2002). All-Angle Negative Refraction Without Negative Effective Index, *Phys. Rev. B*, Vol. 65, No. 20, 201104(R)

Magath, T. & Serebryannikov, A.E. (2005). Fast Iterative, Coupled-Integral-Equation Technique for Inhomogeneous Profiled and Periodic Slabs, *J. Opt. Soc. Am. A*, Vol. 22, No. 11, pp. 2405-2418

Nojima, S. (2007). Long-Sojourning Light in a Photonic Atoll , *J. Opt. A, Pure Appl. Opt.*, Vol. 9, No. 9, S425

Petit, R., Ed. (1980). *Electromagnetic Theory of Gratings*, Springer, Berlin Heidelberg New York

Plum, E.; Fedotov, V.A. & Zheludev, N.I. (2009). Planar Metamaterial with Transmission and Reflection that Depend on the Direction of Incidence, *Appl. Phys. Lett.*, Vol. 94, No. 13, 131901

Rüter, C.E.; Makris, K.G., El-Ganainy, R., et. al. (2010). Observation of Parity-Time Symmetry in Optics, *Nature Phys.*, Vol. 6, No. 3, pp. 192-195

Saado, Y.; Golosovsky, M., Davidov, A. & Frenkel, A. (2005). Near-Field Focusing by a Photonic Crystal Concave Mirror, *J. Appl. Phys.*, Vol. 98, No. 6, 063105

Sakoda, K. (2001). *Optical Properties of Photonic Crystals*, Springer, Berlin Heidelberg New York

Scalora, M.; Dowling, J.P., Bowden, C.M. & Bloemer, M.J. (1994). The Photonic Band Edge Optical Diode, *J. Appl. Phys.*, Vol. 76, No. 4, pp. 2023-2026

Schleuer, J. & Yariv, A. (2004). Circular Photonic Crystal Resonators, *Phys. Rev. E*, Vol. 70, No. 3, 036603

Schurig, D. & Smith, D.R. (2003). Spatial Filtering Using Media with Indefinite Permittivity and Permeability Tensors, *Appl. Phys. Lett.*, Vol. 82, No. 14, pp. 2215-2217

Serebryannikov, A.E.; Magath, T. & Schuenemann, K. (2006). Bragg Transmittance of s-Polarized Waves Through Finite-Thickness Photonic Crystals With Periodically Corrugated Interface, *Phys. Rev. E*, Vol. 74, No. 6, 066607

Serebryannikov, A.E. & Ozbay, E. (2009). Unidirectirectional Transmission in Nonsymmetric Gratings Containing Metallic Layers, *Opt. Express*, Vol. 17, No. 16, pp. 13335-13345

Serebryannikov, A.E. (2009). One-Way Diffraction Effects in Photonic Crystal Gratings Made of Isotropic Materials, *Phys. Rev. B*, Vol. 80, No. 15, 155117

Serebryannikov, A.E.; Ozbay, E. & Usik, P.V. (2010). Defect-Mode-Like Transmission and Localization of Light in Photonic Crystals without Defects, *Phys. Rev. B*, Vol. 82, No. 16, 165131

Shadrinov, I.V.; Fedotov, V.A., Powell, D.A., et al. (2011). Electromagnetic Wave Analogue of an Electronic Diode, *New J. Phys.*, Vol. 13, No. 3, 033025

Singh, R.; Plum, E., Menzel, C. et. al. (2009). Terahertz Metamaterial with Asymmetric Transmission, *Phys. Rev. B*, Vol. 80, No. 15, 153104

Smigaj, W. (2007). Model of Light Collimation by Photonic Crystal Surface Modes, *Phys. Rev. B*, Vol. 75, No. 20, 205430

Vodo, P.; Parimi, P.V., Wu, W.T. & Sridhar, S. (2005). Focusing by Planoconcave Lens Using Negative Refraction, *Appl. Phys. Lett.*, Vol. 86, No. 20, 201108

Wang, Z.; Chong, Y.D. & Joannopoulos, J.D. & Soljacic, M. (2008). Reflection-Free One-Way Edge Modes in a Gyromagnetic Photonic Crystal, *Phys. Rev. Lett.*, Vol. 100, No. 1, 013905

Wu, L.; Mazilu, M., Gallet, J.-F. & Krauss, T.F. (2005). Dual Lattice Photonic-Crystal Beam Splitters, *Appl. Phys. Lett.*, Vol. 86, No. 21, 211106

Yablonovitch, E. (1987). Inhibited Spontaneous Emission in Solid-State Physics and Electronics, *Phys. Rev. Lett.*, Vol. 58, No. 20, pp. 2059-2062

Yu, Z.; Wang, Z. & Fan, S. (2007). One-Way Total Reflection with One-Dimensional Magneto-Optical Photonic Crystals, *Appl. Phys. Lett.*, Vol. 90, No. 12, 121133

Yu, Z. & Fan, S. (2009). Complete Optical Isolation Created by Indirect Interband Photonic Transitions, *Nature Phot.*, Vol. 3, No. 2, pp. 91-94

2

Resonant Guided Wave Networks

Eyal Feigenbaum, Stanley P. Burgos and Harry A. Atwater
Thomas J. Watson Laboratory of Applied Physics
California Institute of Technology Pasadena, CA
USA

1. Introduction

In the last two decades, the development of new photonic material design paradigms has opened up new avenues for designing photonic properties based on different underlying physics. For example, photonic crystals, as described elaborately throughout this book, are based on dispersive Bloch wave modes that arise in periodic index structures. Different in operation than photonic crystals, metamaterials (Smith 2004, Shalaev 2007) are based on subwavelength resonant elements (or "meta-atoms") that interact with incident radiation to give rise to complex refractive indices. In this chapter, we introduce a new approach to optical dispersion control based on resonant guided wave networks (RGWNs) in which power-splitting elements are arranged in two- and three-dimensional waveguide networks.

A possible framework for comparing and classifying photonic design paradigms is according to their basic resonating elements with which light interacts to give the desired artificial dispersion. Under this classification scheme, we can think of materials that operate based on the local interaction of waves with sub-wavelength resonating elements (i.e. metamaterials), structures based on the nonlocal interference of Bragg periodic waves (i.e. photonic crystals), and arrays of coupled resonator optical waveguides (CROWs) where adjacent resonators are evanescently coupled (Yariv 1999). Different from these existing concepts, the dispersion that arises in RGWNs is a result of the multiple closed-path loops that localized guided waves form as they propagate through a network of waveguides connected by wave splitting elements. The resulting multiple resonances within the network give rise to wave dispersion that is tunable according to the network layout. These distinctive properties, that will be described here, allow us to formulate a new method for designing photonic components and artificial photonic materials.

A RGWN is comprised of power splitting elements connected by isolated waveguides. The function of the splitting element is to distribute a wave entering any of its terminals between all of its terminals, as illustrated in Fig. 1a. The waves are then propagated in isolated waveguides between the splitting elements, where the local waves from different waveguides are coupled together. For example, four splitting elements arranged in a rectangular network layout form a 2x2 RGWN (see Fig. 1b). When one of the terminals is excited, the multiple splitting occurrences of the incident wave within the network form closed path resonances that reshape the dispersion of the emerging waves according to the network layout and is different from the dispersion of the individual waveguides. Properly designing this network layout reshapes the interference pattern and the optical function of the RGWN, as will be exemplified later in this chapter. The 2x2 RGWN consists of one closed loop resonance, however larger two- or three-dimensional networks can support multiple resonances, which give rise to more design possibilities.

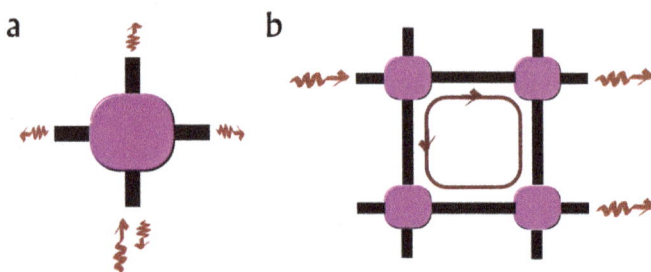

Fig. 1. Schematic illustration of (a) a 4-terminal equal power-splitting element and (b) a local resonance in a 2x2 RGWN.

Although the concept of RGWNs is quite general, we will first illustrate the underlying physics of this paradigm using plasmonics since it allows for a simple topological implementation. After introducing this implementation, in the following sections we will demonstrate how the local wave interference can be designed to engineer small (2x2) energy storage RGWN resonators, and also how we can program the optical transmission function of inhomogeneous RGWNs using transfer matrix formalism. We will also address how the same design principles can be utilized to control the optical dispersion properties of infinitely large RGWNs that behave like artificial optical materials. After addressing other possible implementation and practical issues we will conclude with possible future directions and a more detailed comparison to other optical design paradigms.

2. Plasmonic RGWN components

The operation of RGWNs is based on two basic components: power splitting elements and isolated waveguides. While the waveguides could easily be implemented using dielectric waveguides, the power splitting elements at the intersection of two such waveguides could not be achieved using dielectrics alone. Nevertheless, this splitting operation, which is the key enabler of this technology, is native to the intersection of two plasmonic waveguides. Consequently, a possible implementation of a RGWN is by using plasmonics via a mesh of intersecting sub-wavelength air gaps in a metal matrix.

Surface plasmon polaritons (referred here to as plasmons for brevity) are slow surface waves that propagate at metal-dielectric interfaces. Adding another metal-dielectric interface to this system, results in a metal-insulator-metal (MIM) waveguide, which supports a highly confined plasmon wave (the lowest order transverse magnetic mode - TM0) that does not get structural cut-off as the dielectric gap between the metal layers becomes vanishingly small. The existence of this lowest-order plasmonic mode in MIM waveguides allows for such plasmonic components as power splitters (Feigenbaum 2007-1) and high transmission sharp waveguide bends (for a review of MIM waveguides and their possible applications see Feigenbaum 2007-2). However, the existence of metal in the MIM waveguide configuration does add a source of a modal attenuation to the system as a result the usual loss mechanisms present in any real metal-containing system. This results in a trade off between the compression of the modal cross-section and the modal attenuation as the air gap size is decreased. Since the loss in metals is strongly frequency and material dependent, the focus here will be on RGWNs composed of Au-air-Au MIM waveguides operating at telecommunication frequencies where the modal propagation lengths are on the order of tens of microns, which are substantially larger than the propagation lengths at visible frequencies. The optical properties of the materials throughout this chapter are based on tabulated data (Palik 1998).

In this implementation, the intersection of two sub-wavelength MIM waveguides forms an X-junction that functions as the power splitting elements in the network (Feigenbaum 2007-1) and the MIM segments between the intersections serve as the isolated waveguides connecting the X-junctions. Through this implementation, X-junctions can be tuned to split power equally at infrared wavelengths both for continuous waves and for short pulse waves consisting of only a few optical cycles while conserving the shape of the input signal. The observed equal-power split is a result of the subwavelength modal cross-section of the input plasmonic waveguide that excites the junction with a broad spectrum of plane waves. As such, equal four-way optical power splitting is enabled for transmission lines (e.g., MIM and coaxial configurations) but cannot be easily achieved using purely dielectric waveguides due to their half-wavelength modal cross-section limit. Thus, through a plasmonic implementation, the strong coupling to all four neighboring X-junctions gives the plasmonic RGWN structure an optical response different from a cross-coupled network of purely dielectric waveguides, where most of the power would be transmitted in the forward direction, with only weak coupling to perpendicular waveguides.

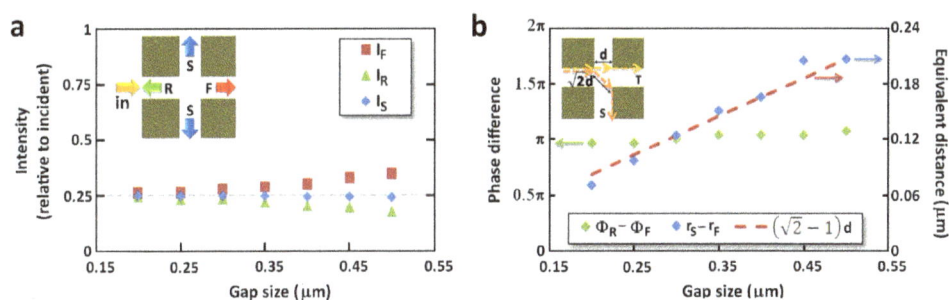

Fig. 2. Power splitting properties of the emerging pulses in an X-junction: (a) intensity relative to the exciting pulse, and (b) phase difference at $\lambda_0=1.5\mu m$ (Feigenbaum 2010).

As the MIM waveguide air gap thickness is varied, the power-split between the X-junction terminals can be tuned both in terms of amplitude and phase (Feigenbaum 2010). This, in addition to determining the phase accumulation in the waveguide segments, sets independent controls in designing the interference pattern that governs the operation of a RGWN. The power splitting in the Au-air X-junction was investigated using the 2D finite-difference time-domain (FDTD) method with short pulse excitation and two equal thickness intersecting MIM waveguides. Through this study, it was found that for small (0.25μm) MIM gaps, these plasmonic X-junctions exhibit equal power splitting with the reflected pulse being out-of-phase (i.e., approximately π-phase shifted) with respect to the sideways and forward transmitted pulses. As illustrated in Fig. 2a, as the MIM gap size is increased, the optical power flow deviates from equal power splitting between the terminals towards dominant power transmission directly across the X-junction, which resembles the wavelength-scale photonic mode limit. Furthermore, in these calculations, the phase shift between the sideways (S) and the forward (F) transmitted pulses is consistent with the geometrical difference in their pulse propagation trajectories (see Fig. 2b).

3. Resonators

After characterizing the properties of the RGWN building blocks, we illustrate the working principles of RGWNs by investigating the dynamics of a compact 2x2 RGWN resonator. In order to form a resonance, the network is designed such that when an X-junction is excited from the internal ports, the exciting waves are out-of-phase, resulting in constructive interference inside the network, as illustrated in Fig. 3. For such out-of-phase excitation the fields in the external terminals interfere destructively, and the power is coupled back into the resonator, enhancing the energy storage quality factor (Q-factor).

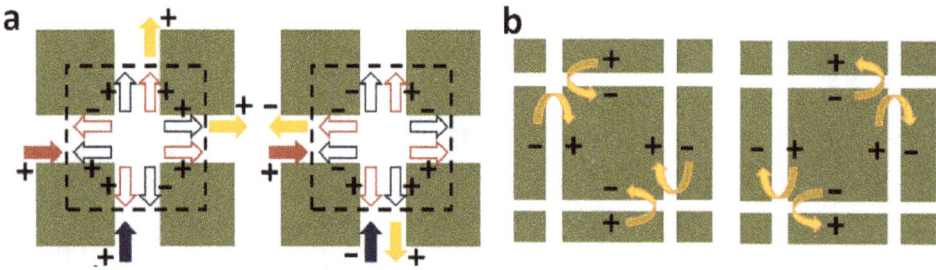

Fig. 3. Resonance build-up in a 2×2 RGWN. (a) Two in/out-of phase input pulses result in destructive/constructive interference inside the network. (b) Steady-state of waves resonating in a 2×2 network where each pair of pulses excites the X-junctions out of phase (Feigenbaum 2010).

When the 2×2 RGWN is excited from the lower-left arm (see Fig. 4), after a transient that includes the first five splitting events, the resonant state is reached as pairs of pulses resonate between junctions 1 and 3 (exemplified by snapshot t_6) and junctions 2 and 4 (exemplified by snapshot t_7). However, before this steady state is reached, it is instructive to follow the dynamics that lead up to this resonance. Starting with the third power split, this event occurs as junctions 1 and 3 are both simultaneously excited by two waveguides. The incoming pulses arrive at both junctions in-phase, which would result in destructive interference inside the network if the R and S split components of each pulse were exactly π-phase shifted. However, the interference is not completely destructive due to the finite size of the waveguides, which causes the phase difference to deviate from a perfect π-phase shift (in accordance with Fig. 2). This power splitting event determines how much power couples into the network. For all future power splitting events after the third one, the two pulses arriving simultaneously at each junction are out-of-phase and therefore interfere constructively inside the resonator. The trade-off between coupling power into the resonator and maintaining it inside suggests that MIM gap sizes that are subwavelength, but not arbitrarily small, will maximize the network resonance. To interpret the FDTD observations and arrive to the conclusion described above, a simplified analytical description of pulse propagation in the network is derived in which only a few parameters are tracked: phase, amplitude, position and direction. The pulses are assumed to travel in the waveguides and split into four new pulses upon arrival at an X-junction. This model also illustrates the compactness of the possible mathematical representation of RGWNs, and the importance of this advantage becomes more substantial when considering the dynamics of larger 2D and 3D network topologies.

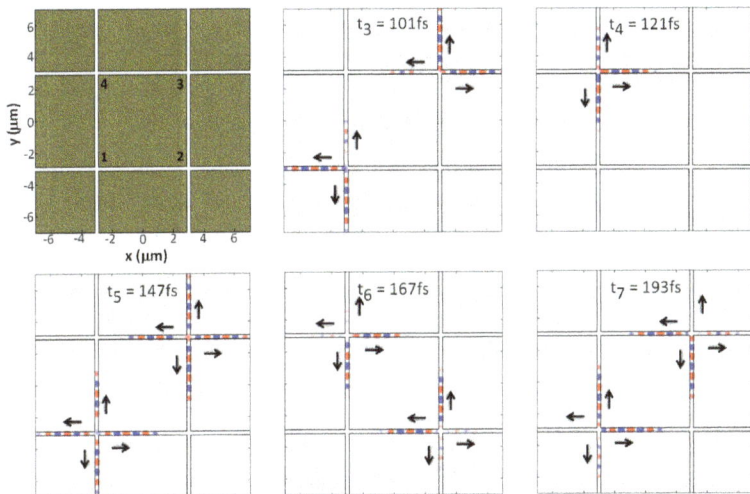

Fig. 4. Time snapshots of H_z (normalized to the instantaneous maximum value) in a 2x2 plasmonic RGWN recorded at the third to the seventh power splitting events for a 2D-FDTD simulation. The MIM waveguides are 0.25μm thick and 6μm long (Feigenbaum 2010).

Calculating the Q-factor of such 2x2 RGWN resonators (Fig. 5) illustrates the role of interference in generating a strong network resonance, which causes the network Q-factor to be an order of magnitude larger than what would be expected if optical power splitting in the X-junctions operated incoherently, i.e. we lost half the power in each splitting event. Increasing the MIM gap size causes the phase of the interfering waves to deviate from being π-phase shifted, resulting in a degradation of the constructive interference inside the resonator and a decrease in the overall network Q-factor. On the other hand, as the gap size is decreased, the plasmonic mode attenuation increases due to metallic losses in the waveguides. Between these two competing effects, the maximal Q-factor value is obtained for a gap size of 250nm. These RGWN Q-factor values are considerable for plasmonic resonators and even comparable to typical values of wavelength-size dielectric resonators that are dominated by radiation loss (e.g., a cylindrical dielectric cavity of radius 1.3λ with a purely real refractive index of n=2.5 surrounded with air has a Q~100). If we were to artificially decrease the Au loss at 1.5μm (or alternatively go to longer wavelengths), the Q-factor of the resonator would increase appreciably (e.g., Q ~ 750 for a 200 nm gap width), indicating that the resonator Q-factor is primarily limited by the material loss.

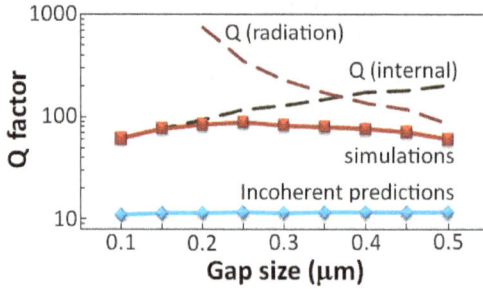

Fig. 5. Q-factor of 2×2 RGWN resonator from simulation results compared with those resulting from incoherent power splitting (Feigenbaum 2010).

4. Tailoring the optical properties of artificial materials

After studying the resonance effects in a small RGWN, we now investigate the dispersion characteristics of infinitely large 2D periodic RGWNs by modelling the structure unit cell in FDTD with Bloch boundary conditions. Through this analysis, we find that RGWNs exhibit wave dispersion and photonic bandgaps due to interference effects, and that their band structure can be controlled by modifying the network structural parameters. Two different length-scales control the network dispersion: the subwavelength width of the MIM gaps determines the phase shift at each X-junction, and the wavelength-order distance between the nodes along with network topology determine the interference scheme.

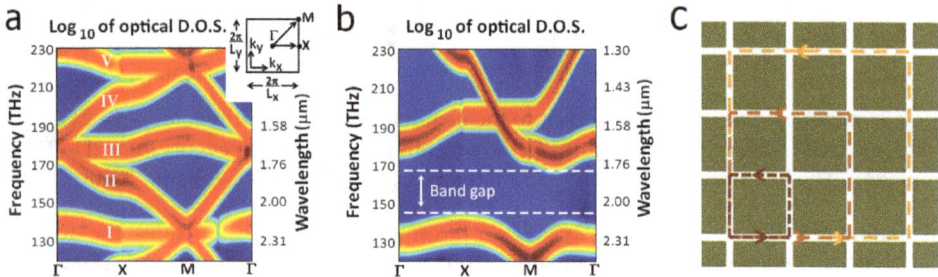

Fig. 6. Photonic band structure of infinitely large periodic RGWNs (Feigenbaum 2010).

The same interference dynamics that govern the energy storage in finite-size 2x2 RGWN resonators are the same that control the optical properties of artificially designed RGWN materials of infinite-size. If the network parameters are chosen such that a planewave excitation at a given incidence angle results in a resonance effect similar to the one demonstrated for the 2x2 network, then this would correspond to a forbidden state of propagation in the photonic band diagram. Examining the optical density of states (DOS) for different wave vectors over frequencies in the near infrared range, where the material (Au in this case) dispersion is small, we observe a photonic band structure which is only due to dispersion resulting from the network topology, as shown in Fig. 6a. The functionality of the infinitely large RGWN is not hindered by loss since its dispersion depends on the waveguide decay length being much larger than the size of the largest resonant feedback oop that has dominant contribution to the RGWN dispersion. Further possibilities for achieving band dispersion control are illustrated in Fig. 6b, showing a flat bands over a wide range of wavevectors at 130 and 170 THz, as well as the formation of a photonic bandgap between 140-160 THz, for appropriately chosen network parameters.

The infinitely large RGWN is illustrated in Fig. 6c along with a few schematic resonance orders that represent the resonances that could arise within the network. The operating mechanism of the RGWN is very different from that of photonic crystals composed of metal-dielectric alternating materials. Although the schematic layout might look similar, the difference between the two classes of artificially designed optical materials becomes clear when considering the difference in the length scales of their composite elements. Whereas photonic crystals operate based on non-local interaction of Bloch waves with the entire array, RGWNs rely on the interference of local waves. Therefore RGWNs are not sensitive to the actual topology of waveguides between junctions but only to its trajectory length, whereas the properties of photonic crystals would greatly depend on the shape of the periodic metallic islands. Additionally, RGWNs do not necessarily have to be periodic to operate as resonant guided wave networks, and for the same reason, planting a defect in a RGWN would not have the same effect as it would in a photonic crystal.

5. Programming the optical properties of a network

Because the underlying physics of RGWNs is based on the interference of local waves, it allows for layouts that are inhomogeneous and non-periodic across the network. Unlike photonic crystals, which are restricted to Bragg wave effects in periodic structures, the flexibility of RGWNs open up design possibilities where the wave properties are varied across the structure. With respect to metamaterials, which could inherently be non-homogeneous due to the local nature of the interaction between light and the meta-atoms, RGWNs have the advantage of having interference effects within the network, which allows for frequency spectrum reshaping designs through these effects.

An additional unique feature of RGWNs relates to the constraints on wave propagation within the structure. Unlike other photonic designs, RGWNs have a limited number of modes that are allowed to propagate within the structure (e.g. only the TM0 mode for the case of the plasmonic implementation described previously). Furthermore, the waves can propagate only inside the waveguides connecting the splitting elements. The different waveguides are coupled only by X-junctions, which each have only a limited number of terminals. This level of control is beneficial for several reasons. First, the interference pattern in the network can be controlled more directly. Second, it allows for a comprehensive mathematical representation of the RGWN by scattering matrix (S-matrix) formalism that greatly reduces the computational complexity of programming the network. Third, since the waveguides are isolated from each other, their only contribution to the network is to serve as phase retardation elements between the splitting elements. As a result, the waveguide length is the only effective parameter in its contour, as long as the bending is not too severe. This waveguide feature allows for the network to maintain its engineered function even when distorted. Additionally, the ability to utilize curved or bent waveguides to accommodate long contours is useful when designing the interference pattern of RGWNs.

These distinctive RGWN characteristics open up new opportunities for designing photonic devices by programming the entire network rather than by assembling interconnected discrete components with traceable functions. The usual way of designing photonic devices is to target the desired subsystem functions, map them logically into sub-functions, and then assembling components that carry out these sub-functions in the desired system. For example, a wavelength router could be designed using add/drop ports where the input and output waveguides are coupled by wavelength sensitive ring resonators (Little 1997) or by defects in a photonic crystal (Fan 1998). Similarly, in free space optics, this function could be achieved through the use of collinear beam splitters, each designed to deflect a desired wavelength band. In these schemes, the couplers and waveguides are discrete components that are associated with a specific function, and are combined in a logical way to carry out the overall system function. An alternative approach is to use a network of components that carries out the desired function but, unlike traditional designs, there is no specific logical sub-function associated with any individual component. While the inner connectivity of the device will be less intuitive, it has the potential to result in more efficient designs of complex and compact devices.

One possible way of representing a system function in a RGWN is through the use of a scattering operator that maps the set of local waves entering the device terminals to the set of the waves exiting from the same terminals (Feigenbaum 2010-2). Since a RGWN is composed of a discrete set of components (waveguides and X-junctions) and terminals, the system function is represented by a scattering matrix (S-matrix) connecting the vectors of the waves entering and emerging from the RGWN via the external ports (see Fig. 7). Designing the system function of the RGWN is then mathematically equivalent to designing the S-matrix to yield a desired output, given a set of inputs.

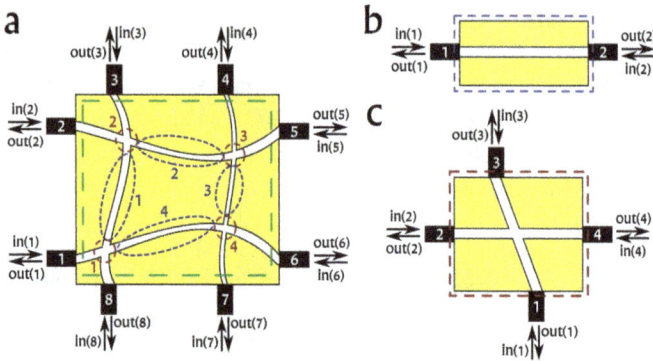

Fig. 7. Mathematical representation scheme of (a) a 2x2 RGWN system and its components, (b) a waveguide component, and (c) an X-junction component (Feigenbaum 2010-2).

Programming an optical function onto a network, according to the design principle described above, will first be demonstrated for a plasmonic 2x2 RGWN, in which the constituent MIM waveguides are allowed to differ in width, length, and contour. The device has eight terminals, numbered from '1' to '8', as illustrated in Fig. 7a. The input vector lists the complex amplitudes of the magnetic fields (H-fields) entering the network in the eight terminals, and similarly the output vector describes the complex H-field amplitudes of the waves exiting the network through these same terminals.

The network S-matrix is assembled from the mathematical representation of its components according to the network layout. As a first step, a 'function library' of mathematical representations is generated for all the possible network components (i.e., waveguides and X-junctions) using finite difference time domain (FDTD) full wave electromagnetic simulations. Once this library is established, the RGWN S-matrix can be assembled according to the network layout. It is worth pointing out that the S-matrix calculation scheme is almost always found to be much faster than resolving the RGWN behavior from full wave electromagnetic simulations, yet reproduces the same information about the network. This becomes significant for optimization tasks and especially as the network size increases.

To carry out this formalism, the two basic RGWN components (waveguides and X-junctions) need to first be represented mathematically. The waveguides are mathematically represented by their complex phase retardation, determined by the complex propagation constant of the wave and the waveguide length. The propagation constants are extracted from FDTD simulations for waveguides with various widths at different frequencies. The X-junctions, which are comprised of two intersecting waveguides with four terminals, are mathematically represented by a (4x4) S-matrix. For a given set of waveguide widths, the complex transmission coefficients of the X-junction ports are extracted from FDTD simulations by measuring the amplitude and phase of the wave transmitted to the different ports when excited from one of the terminals at a given wavelength.

The S-matrix of the 2x2 RGWN is then assembled from the mathematical representation of its constituent components according to the network layout. The phasor representation of the local wave H-fields in the network is represented by three column vectors (transposed for brevity):

$$\underline{A}_{out}^{tr} = \left\{ a_{1(1)}^o, a_{1(2)}^o, a_{2(3)}^o, a_{2(2)}^o, a_{4(1)}^o, a_{4(4)}^o, a_{3(3)}^o, a_{3(4)}^o \right\};$$

$$\underline{A}_{in}^{tr} = \left\{ a_{1(1)}^i, a_{1(2)}^i, a_{2(3)}^i, a_{2(2)}^i, a_{4(1)}^i, a_{4(4)}^i, a_{3(3)}^i, a_{3(4)}^i \right\}; \tag{1}$$

$$\underline{A}_{net}^{tr} = \left\{ a_{1(3)}^i, a_{1(4)}^i, a_{2(1)}^i, a_{2(4)}^i, a_{4(3)}^i, a_{4(2)}^i, a_{3(1)}^i, a_{3(2)}^i \right\};$$

where A_{out} and A_{in} hold the values of the local input and output waves of the RGWN at its ports, and A_{net} represents the input wave on the X-junctions from the internal terminals of the RGWN. The i/o superscripts denote input/output waves with respect to the X-junction, the number subscripts corresponds to the junction number as defined in Fig. 7a, and the bracketed number subscripts label the ports as defined in Fig. 7c. The coupling of the H-field vectors by the network connectivity can then be represented by the system:

$$\begin{cases} \underline{A}_{out} = \underline{\underline{M}}_{FS} \underline{A}_{net} + \underline{\underline{M}}_{RS} \underline{A}_{in} \\ \underline{0} = \left(\underline{\underline{M}}_{RS} - \underline{K} \right) \underline{A}_{net} + \underline{\underline{M}}_{FS} \underline{A}_{in} \end{cases} \tag{2}$$

where M_{FS} and M_{RS} are diagonal 8-by-8 matrices that originate from the splitting relations in the X-junctions and K is a sparse 8-by-8 matrix that stands for the wave propagation in the waveguides. These matrices are defined as:

$$\underline{\underline{M}}_{FS} = Diag\left\{\underline{\underline{S}}^1_{FS}, \underline{\underline{S}}^3_{FS}, \underline{\underline{S}}^2_{FS}, \underline{\underline{S}}^4_{FS}\right\}, \quad \underline{\underline{M}}_{RS} = Diag\left\{\underline{\underline{S}}^1_{RS}, \underline{\underline{S}}^3_{RS}, \underline{\underline{S}}^2_{RS}, \underline{\underline{S}}^4_{RS}\right\}$$

$$\underline{\underline{S}}^i_{FS} = \begin{pmatrix} \left(t^V_F\right)^i & \left(t^H_S\right)^i \\ \left(t^V_S\right)^i & \left(t^H_F\right)^i \end{pmatrix} \quad \underline{\underline{S}}^i_{RS} = \begin{pmatrix} \left(t^V_R\right)^i & \left(t^H_S\right)^i \\ \left(t^V_S\right)^i & \left(t^H_R\right)^i \end{pmatrix} \tag{3}$$

$$\underline{\underline{K}} = \begin{cases} K(1,3) = K(3,1) = \kappa^3_1 \\ K(2,6) = K(6,2) = \kappa^2_1 \\ K(4,8) = K(8,4) = \kappa^4_3 \\ K(5,7) = K(7,5) = \kappa^4_2 \\ \textit{other matrix elements} = 0 \end{cases}, \kappa^m_i = \exp\left\{j\left(\beta L\right)_{i\Leftrightarrow m}\right\}$$

where the V/H superscript index denotes if the transmission coefficient is for excitation of the vertical or the horizontal waveguide of that X-junction.

Algebra of equation set 3 gives the matrix representation of the 2x2 RGWN S-matrix:

$$\underline{\underline{S}}^{2x2RGWN} = \underline{\underline{M}}_{RS} - \underline{\underline{M}}_{FS}\left(\underline{\underline{M}}_{RS} - K\right)^{-1} \underline{\underline{M}}_{FS} \tag{4}$$

When validating the field amplitude predictions of the S-matrix representation with FDTD simulations, less than 5% difference is found for various test cases. The two major contributions to this small deviation result from the interpolation between the parameter space points, where the library components were calculated, and from the error added when the waveguides are bent. For cases where no interpolation or waveguide bending occurs, the FDTD results differ by only 1% from the S-matrix predictions. The ability to accurately predict the RGWN interference using S-matrix representation reduces the complicated task of programing a desired optical function into a RGWN into an efficient optimization of its S-matrix.

For example, the RGWN can be programmed by minimizing the difference between the actual network output and the desired one (for a given input), as the network parameter space is swept across the various waveguide widths and lengths. The optimization process then results in a set of network parameters that can be translated to a network layout and then validated with FDTD simulations.

6. Multi-chroic filters using RGWNs

The S-matrix programming method can be exemplified by designing a 2x2 RGWN to function as a dichroic router (Fig. 8a). Although simple in concept, the exercise of setting a passive device to have different functions at different wavelengths is quite instructive. Explicitly, the required function is to route two different wavelengths (λ_1 and λ_2) to a different set of ports ('1' and '6' for λ_1 and '2' and '5' for λ_2) when the two bottom ports ('7' and '8') are simultaneously excited with equal power. Mathematically, we can represent the device as an 8x8 S-matrix $S(\lambda_1, \lambda_2)$ connecting the input and the output vectors. For both wavelengths, the input vector is nonzero for the bottom ports (i.e. In=(0,0,0,0,0,0,1,1)) and the desired output vectors would be Out(λ_1)=(1,0,0,0,0,1,0,0) for λ_1 and Out(λ_2)=(0,1,0,0,1,0,0,0) for λ_2. Because we do not have enough degrees of freedom in this small 2x2 network to exactly attain the desired outputs, we instead optimize the ratio of power going to the two sets of ports at the different wavelengths.

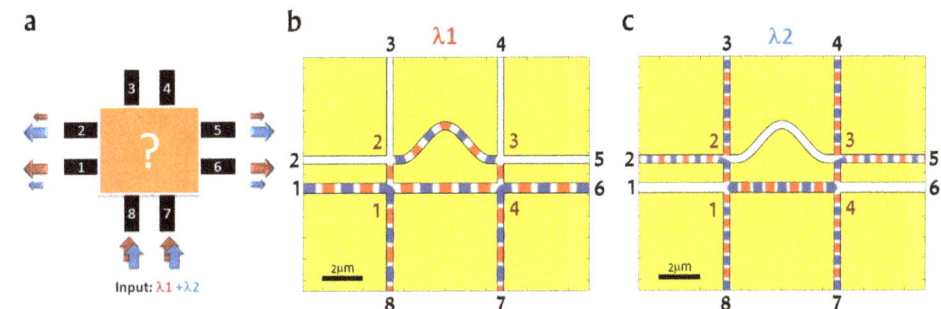

Fig. 8. 2x2 RGWN programmed to function as a dichroic router: (a) schematic drawing, and (b, c) time snapshots of the H-field at the two operation frequencies (Feigenbaum 2010-2).

The optimization procedure is implemented in Matlab using the pre-calculated mathematical representation data set of the RGWN components obtained from full-field electromagnetic FDTD simulations excited with continuous wave sources (see illustration in Fig. 9).

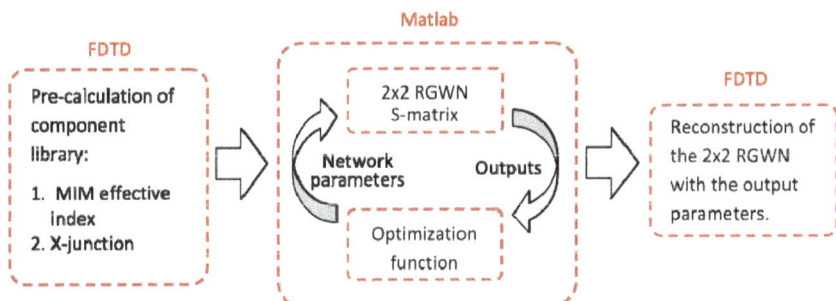

Fig. 9. Flow chart of the RGWN S-matrix optimization procedure (Feigenbaum 2010-2).

The dichroic router network is defined by eight parameters: the length and width of the upper, lower, and side waveguides and the two wavelengths. The waveguide widths determine the effective index in the waveguides as well as the transmission coefficients of the X-junctions. The optimization procedure is conducted in Matlab with the optimization function for the dichroic router defined as follows:

$$O_1 = |Out_6(\lambda_1)| / |Out_5(\lambda_1)|$$
$$O_2 = |Out_5(\lambda_2)| / |Out_6(\lambda_2)|$$
$$\frac{1}{f} = \left[\sqrt{|Out_6(\lambda_1)| \cdot |Out_5(\lambda_2)|} \right] \cdot [O_1 \cdot O_2] \cdot \left[1 - \left| \frac{O_1 - O_2}{O_1 + O_2} \right| \right]$$

(5)

where O_1 and O_2 represent the two terminal output ratios that need to be maximized at the two different wavelengths, λ_1 and λ_2. The function f is used to merge the two ratios together into one weighted optimization parameter, where the first term in squared brackets maximizes the total power routed into the selected terminals, the second term maximizes the two ratios, and the third term is a weighting factor that ensures that the two ratios are maximized equally. The target function is defined as the inverse of these three terms in multiplication for the minimization Matlab function. At each point in the parameter space, the network output vector is calculated as the multiplication of the 2x2 RGWN S-matrix evaluated at the parameter values times the input vector representing excitation only from the two bottom terminals ('7' and '8').

After defining the optimization function, we constrain the parameter space based on practical considerations. The parameter space includes the width and length of the upper, lower, and side waveguides as well as the two wavelengths of operation (λ_1 and λ_2). We decrease the number of parameters to optimize by restricting the device to have left-right symmetry based on the desired operation. We restrain the design to operate in the infrared frequency range ($\lambda_0 = 1.2$-$2\mu m$) where the material dispersion and loss are less pronounced than in the visible. Furthermore, the waveguide thickness is constrained to be small enough to only support the lowest order plasmonic mode (air gap widths 100-500nm).

The optimization procedure yields the network parameters given in Table 1, which reveal that the required RGWN for color routing is distributed inhomogeneously.

Waveguides	Width (μm)	Length (μm)
Lower	0.47	5.4
Side	0.31	1.34
Upper	0.38	6.6

Table 1. Set of optimized parameters for 2x2 RGWN dichroic router operating at $\lambda_1=2\mu m$ and $\lambda_2=1.26\mu m$.

When translating the optimized network parameters into the network layout, we learn that the upper waveguide is longer than the lower one, and therefore needs to be bent. Importing the resulting layout into FDTD, we obtain the steady state H-field distribution shown in Fig. 8b and 8c which show time snap shots at the two operation wavelengths. The FDTD simulation results validate the S-matrix design, with λ_1 and λ_2 clearly routed to a different set of sideways ports as illustrated in Fig. 8b and 8c, respectively. From these FDTD results, it is also possible to observe the build-up of local resonance inside the network, which results in the filtering out of the desired output ports. We note that the transmission ('3' and '4') and reflection ('7' and '8') ports from the device are not identically zero since the device does not have enough degrees of freedom and were therefore not included in the optimization function.

The matrix representation can also be used to understand the interference conditions through which the RGWN accomplishes its desired function. From the known input vector and the network S-matrix, the wave complex amplitudes can be identified at any point in the network. For each wavelength, we resolve the excitation conditions of the X-junctions that have the ports that are to be filtered out. For example, for λ_1 to be filtered out from terminals '2' and '5', we examine the excitation conditions in X-junction '3,' which has four terminals. Two of the terminals are external device ports ('4' and '5') and the other two are internal network terminals. There is no input signal incident on the two external ports, so it s the excitation conditions of the remaining two junction terminals that null the output in erminal '5.' Indeed, the excitation amplitudes of junction '3' obtained from the S-matrix epresentation are 0.23exp(-j0.21π) and 0.34exp(j0.64π), which are close in amplitude and ~ π ohase-shifted. This is consistent with the results from section 3, which show that when an X-junction' is simultaneously excited π phase-shifted from two adjacent terminals, the two other terminals will be filtered out (Fig. 3a). The fact that the excitations are not exactly the ame in amplitude and π phase-shifted is attributed to the additional constraints the design las on the other wavelength as well as the limitations imposed on the parameter space.

similarly, the excitation conditions necessary for filtering out terminals '1' and '6' at λ_2 (Fig. b) are examined by focusing on the S-matrix amplitudes of X-junction '4.' In this case there are three terminals being excited: the lower terminal of the X-junction (port '7') is given by he network excitation, so the excitation of the other two internal ports will determine the iltering out of port '6.' Intuitively, the condition to filter out terminal '6' will be simply a Π ohase-shifted excitation of the upper and lower terminals of junction '4', with zero excitation rom the side port. From the case of λ_1 we also know that additional constraints might ause a residual wave emerging from terminal '6', which could be compensated by a small amplitude excitation at the other side terminal of the junction '4.' Indeed, the excitation amplitudes of junction '4' in the S-matrix representation are 1 in lower terminal,).9exp(j0.82π) in upper terminal, and 0.3exp(-j0.32π) in the side terminal.

To further exemplify the programmability of RGWNs via S-matrix formalism, we consider a 3x3 RGWN programmed to function as a trichroic router. In order to implement the more complex task of routing three wavelengths we allow for more degrees of freedom in the network by increasing the number of components, effectively increasing the amount of data contained. The function is defined as an extension of the dichroic router, but here when the hree bottom terminals are simultaneously excited at three different frequencies, the requencies are filtered out to three different sets of side terminals as illustrated in Fig. 10. The analysis results in the optimal RGWN parameters shown in Table 2.

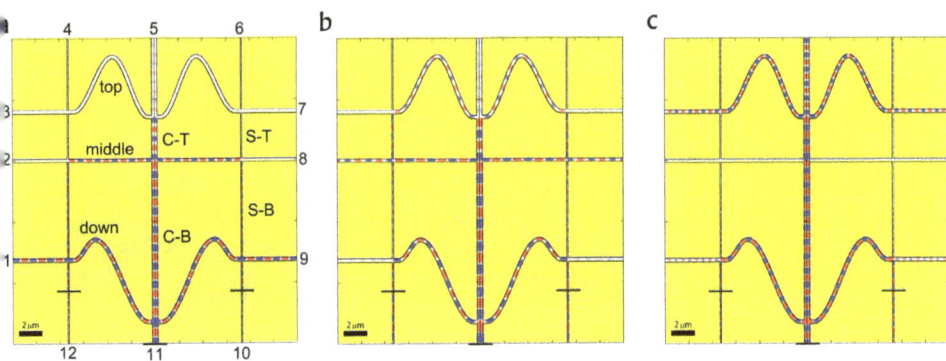

Fig. 10. 3x3 RGWN programmed to function as a trichroic router. Time snapshots of the steady state H-field at the three operation frequencies (Feigenbaum 2010-2): a) λ_1, b) λ_2, c) λ_3.

It is interesting to note that the wavelengths are not mapped monotonically to the output terminals (i.e. from bottom/top ports as the wavelength increases/decreases), which would be the usual case for devices relying on material dispersion, such as a glass prism.

Waveguides		Width (μm)	Length (μm)
Vertical	center-bottom (C-B)	0.45	13.25
	side-bottom (S-B)	0.1	8.15
	center-top (C-T)	0.45	3.55
	side-top (S-T)	0.1	4
Horizontal	Top	0.29	12.8
	Middle	0.26	7.3
	Down	0.3	11.95

Table 2. Set of optimized parameters for a 3x3 RGWN trichroic router operating at λ_1=1.59μm, λ_2=1.97μm, and λ_3=1.23μm.

7. Possible Implementations

The underlying physics and the working principles of the RGWNs were demonstrated in the previous sections with an idealized 2D implementation using MIM waveguides. However, for the same 2D network topology as shown in Fig. 4, but implemented with 3D high aspect ratio Au-air channel plasmon waveguides (Bozhevolnyi 2006), the observed wave dynamics are found to closely resemble that of the 2D MIM waveguide network, as studied with 3D full-field simulations. If the aspect ratio of the channel plasmon waveguide is high enough, the propagating mode within the channels strongly resembles the MIM gap plasmonic mode. This can for instance be seen in the measured quality factors of RGWNs comprised of channel plasmon waveguides (3D simulations) and MIM slot waveguides (2D simulations) which have Q-factor values of 82 and 83, respectively, at a wavelength of 1.5μm. Furthermore, the two power splitting events that define the RGWN resonant state are similar for both the channel and MIM waveguides (Fig. 4).

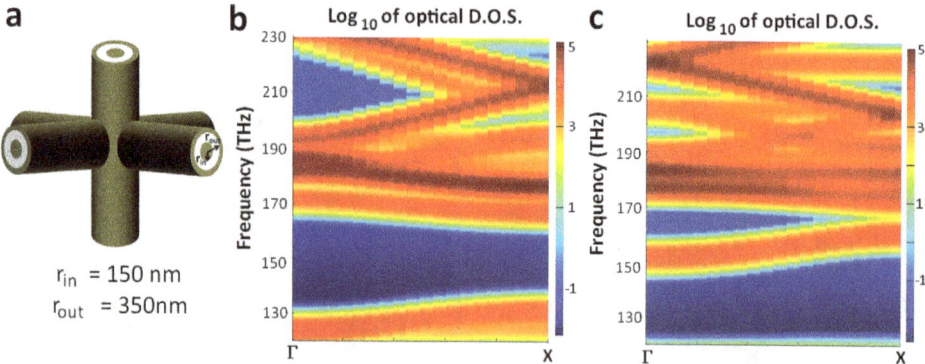

Fig. 11. 3D RGWN: (a) rendering of a 3D RGWN building block (6-arm junction). (b, c) Optical DOS of an infinite 3D network spaced periodically with cubic periodic unit cell with different spacing (Feigenbaum 2010-1).

The dispersion design in a volume can be addressed by 3D-RGWN topologies, for example, constructing an array of orthogonally intersecting 3D networks of coaxial Au-air waveguides aligned in a Cartesian grid (Fig. 11a). In this case, the four-arm X-junction element of the 2D network is replaced by a six-arm 3D junction element. Using 3D FDTD, we have verified that six-way equal power splitting occurs for pulsed excitation in a coaxial Au-air waveguide junction. Like for the 2D-RGWN, the dispersion of the infinitely large periodic 3D-RGWN is predominantly determined by the network parameters rather than the waveguide dispersion. This is demonstrated by the noticeable difference in the band diagrams (Fig. 5b and 5c) obtained for two networks comprised of the same waveguides but with different inter-node spacing.

3. Conclusions and future directions

RGWNs offer a different approach for designing dispersive photonic materials. Whereas photonic crystals rely on the formation of Bloch wave states by interference of waves diffracted from an array of periodic elements, a truly non-local phenomenon; RGWNs rely on the coherent superposition of power flowing along isolated waveguides and splitting at X-junctions. Furthermore, in photonic crystals, the interference pattern of the diffracted waves depends on the nonlocal periodic spatial arrangement of the diffracting elements; and in RGWNs it is the local network topology that determines the dispersion and resonance features. For example, in a RGWN, the coherent wave propagation through the network is determined only by the total path length along the waveguide and the phase shift added at a power splitting event, having no restriction on whether the waveguides are straight or curved. Metamaterials also feature a design approach based on the attributes of localized resonances, but their dispersive properties do not depend on any length scale between resonant elements – thus differing substantially from RGWNs. Arrays of coupled resonator optical waveguides (CROWs) feature discrete identifiable resonators that act as the energy storage elements, and dispersion occurs as modes of adjacent resonators are evanescently coupled. By contrast, in RGWNs, energy is not stored resonantly in discrete resonators, but rather in the network of waveguides that are designed to exhibit a collective resonant behaviour.

The operation of RGWNs was demonstrated in this chapter using plasmonics, which allowed for a simple layout and broadband range of operation; however, this implementation also brought about substantial attenuation due to the fundamental loss of plasmonic modes. As indicated above, the plasmonic MIM modes used here have typical propagation lengths of about 50 microns due to metal loss. Since the RGWN scope is broader than the field of plasmonics, it calls for an all-dielectric implementation to mitigate the losses brought on by plasmonics. Implementing RGWNs using photonic circuitry would also address the coupling loss associated with the difference in the modal overlaps between the plasmonic modes in the RGWN and the interfacing dielectric optics.

This new design paradigm is based on different underlying physics and thus opens up new directions for the design of artificial optical materials and devices. Since the RGWN design relies on the interference of local waves, we can use these accessible design parameters to program optical functions directly onto the network. Furthermore, the constraints on the propagation and coupling of the local waves in RGWNs allow for the device operation to reduce to a simple mathematical representation using S-matrix formalism. This allows for the network programming to take the form of an optimization procedure over a relatively small parameter space. The RGWN S-matrix representation was demonstrated here where the inputs were given and the S-matrix of device was designed to give a desired output (e.g., routing, mode converting). However, this formalism could be extended to different type of functions, such as sensing, in which the inputs are given and the output changes are monitored. In this chapter, dichroic and trichroic RGWN color routing was demonstrated as a proof of concept; however, incorporating more components into the RGWN and therefore increasing the possible degrees of freedom, could allow for more complex devices or alternatively for devices with enhanced performance. Furthermore, we exemplified the RGWN design paradigm using plasmonics, nesting a split element simply by intersecting waveguides, still the concept is broad and implementing the concept using photonic component could open new opportunities in the design of photonic circuitry devices.

9. Acknowledgements

This work was supported by the DOE 'Light-Material Interactions in Energy Conversion' Energy Frontier Research Center under grant DE-SC0001293 and by the National Science Foundation under the Graduate Research Fellowship Program.

10. References

Bozhevolnyi, S.I. (2006), Volkov, V.S., Devaux, E., Laluet J.Y., Ebbesen, T.W. Channel plasmon subwavelength waveguide components including interferometers and ring resonators. Vol.440, (2006), pp. 508-511

Fan, S. (1998), Villeneuve, P., Villeneuve, J., Haus, H.A. Channel drop filters in photonic crystals. *Opt. Express*, Vol.3, (1998), pp. 4-11

Feigenbaum, E (2007-1), Orenstein, M. Perfect 4-way splitting in nano plasmonic X-junctions. *Opt. Express*, Vol.15, (2007), pp. 17948-17953

Feigenbaum, E (2007-2), Orenstein, M, Modeling of Complementary (Void) Plasmon Waveguiding, *J. Lightwave Tech.*, Vol.25, (2007), pp. 2547- 2562

Feigenbaum, E. (2010). Atwater, H.A., Resonant Guided Wave Networks, *Phys. Rev. Lett.* Vol.104, (2010), 147402.

Feigenbaum, E. (2010-2), Burgos, S.p., Atwater, H.A. Programming of Inhomogeneous Resonant Guided Wave Networks. *Opt. Express* Vol.18, (2010), 25584-25595.

Little, B.E (1997), Chu, S.T., Haus, H.A., Foresi, J., Laine, J.P. Microring resonator channel dropping filters. *J. Lightwave Tech.*, Vol.15, (1997), pp. 998-1005

Palik, E.D. (1998), *Handbook of optical constants of solids*, 2'nd Ed. San-Diego: Academic 1998.

Shalaev, V.M. (2007). Optical negative-index metamaterials. *Nature Photon.*, Vol.1, (2007), pp. 41-48

Smith, D.R. (2004), Pendry, J.B., Wiltshire, M.C.K. Metamaterials and Negative Refractive Index. *Science*, Vol.305, (2004), pp. 788-792

Yariv, A (1999), Yong L., R.K., Scherer, A. Coupled-resonator optical waveguide: a proposal and analysis. *Opt. Lett.*, Vol.24, (1999), pp. 711-713

3

How Nature Produces Blue Color

Priscilla Simonis and Serge Berthier

Institut des Nanosciences de Paris (INSP), University Pierre et Marie Curie, Paris, France

1. Introduction

Today, blue is a very fashionable color in European countries. This has not always been the case (Pastoureau, 2000), as cultural perceptions have slowly evolved since prehistoric times. In cave paintings, white, red and black have been the only available tones and these colors remained basic for Greek and Latin cultures, where blue was neglected or even strongly devalued. The word *caeruleus*, which is often used for brightly blue species, in naming plants and insects, is etymologically related to the word *cera*, which designates wax (not to the world *caelum* – sky – as often believed): it meant first white, brown or yellow (André, 1949), before being applied to green and black, and much lately, to a range of blues. Latin and Greek philosophers were so diverted from blue that they even did not notice its presence in the rainbow: for *Anaximenes* (585-528 BC) and later for *Lucretius* (98-55 BC), the rainbow only displayed red, yellow and violet; *Aristotle* (384-322 BC) and *Epicurus* (341-270 BC) described it as red, yellow, green and violet. *Seneca* (ca. 4 BC - 65 AD) only mentioned red, orange, green, violet but, strangely, also added purple, a metameric color not found in the decomposition of white light. Later in the Middle-Ages, *Robert Grosseteste* (ca. 1175-1253) revisited the rainbow phenomenon in its book *"De Iride"* and still did not find there any blue color (Boyer, 1954). Blue emerged slowly in minds and art, only after the advent of technological breakthroughs in stained glass fabrication (as introduced in the 12th century rebuild of St Denis Basilica) and after the progressive use of blue dyes, which followed the extension of woad cultivation, all after the 13th century.

Another slow emergence of blue has been observed in the development of an efficient blue light-emitting diode. Red, yellow and green solid-state diodes appeared early after the development of the first device by Nick Holonyak Jr and S. Bevacqua in 1962, but the blue diode did not become practical until the work of Shuji Nakamura, in 1993. Since then, the blue and ultraviolet diodes have gained maturity and give rise to the emergence of powerful white sources that appear to be the future of all lightening devices.

If the blue color has been slow to emerge in human culture and technology, it was not so in nature. Blue flowers, birds, fishes, reptiles, insects, spiders, shrimps... have been observed very frequently. Blue colored structures have even been found on fossil beetles. The objective of this chapter is to discover how blue colorations are achieved in living organisms.

A classification of *natural* photonic structures is not straightforward: these structures are complex, with multiscale effects and disorder. A useful classification requires some mathematical idealization of the structures. Our scheme is based on the number of dimensions in which we can assume a total translational invariance. One-dimensional structures are only inhomogeneous in one dimension, with, perpendicularly, complete invariance for two independent translations. These one-dimensional structures are then described as "layered". Thin films, thin film stacks and Bragg mirrors (with the repetition of identical layers) are examples of one-dimensional structures. Two-dimensional structures are totally invariant under a single direction. A straight optic fiber is a two-dimensional structure. A periodic array of parallel fibers, such as the bunch of cilia in ctenophores or the aligned melanin rods in peacock or other bird's feathers, is also a two-dimensional, as well as gratings engraved on flat surfaces. In three-dimensional structures, no direction shows total invariance under translations. This is the most general geometry for a photonic structure.

When inhomogeneous, the refractive index can be periodic, in which case the propagation acquires special features that will be examined later. A one-dimensional periodic structure is the basis for a Bragg mirror that produces well-defined reflection bands around specific frequencies. Obtaining blue colors with such a system is relatively tricky and, as will be discussed in this chapter, requires particularly thin layers in order to avoid producing a metameric purple color. Blue two-dimensional photonic crystal also requires special scatterer's spacing and specific conditions: the blue coloration of the wing feathers in the magpies is a very instructive example. Somewhat more complex, when neglecting cross-ribs, microribs and lamellae slant, the *Morpho rhetenor* ribs structure is another example of a two-dimensional photonic structure that produces a vivid blue under most directions. Gratings, as found in butterflies can also produce blue iridescence for specific grating periods. Finally, blue three-dimensional photonic crystals are observed in weevils and longhorns.

2. Tyndall diffusion

Tyndall scattering by relatively distant particles in the range 40 nm – 900 nm generally produces a bluish diffuse coloration.

2.1 Theoretical background

The elastic scattering of light by isolated particles is an important chapter of electrodynamics. The physical mechanism of light diffusion is simple: the electric field which accompanies an incident light beam penetrates and disturbs the polarizable material in the scatterer, which responds by charge oscillations. The sustained acceleration of these oscillating charges produces a reemission of light, at the same frequency, but its directional distribution is much wider than in the incident light. The distribution of the scattered light essentially depends on the polarizability of the material at the incident frequency, but also on the shape of the particle – in just the same way as the relative location of antennas influences the emission direction of radio waves.

An important parameter, for classifying the scattering mechanisms is the ratio between the scatterer's size r to the wavelength λ, $x = 2\pi r/\lambda$. For $x \ll 1$, we encounter a mechanism of diffusion called "Rayleigh scattering". This type of light redirection leads to the following distribution of intensities:

$$I = I_0 \frac{1 + \cos^2 \theta}{2R^2} \left(\frac{2\pi}{\lambda}\right)^4 \left(\frac{n^2 - 1}{n^2 + 2}\right)^2 \left(\frac{d}{2}\right)^6 \tag{1}$$

where I is the intensity scattered at an angle θ from the incident direction, R is the distance from the particle's center, n is the refractive index of the particle and d its diameter. Typically, for visible light, the diameter of the scatterer should be smaller than about 50 nm to warrant a good quantitative accuracy of the scattering. This expression was probably first derived by John William Strutt (third Baron Rayleigh), based on dimensional ("similitude") arguments (Hoeppe, 1969). Assuming that the scattering is proportional to the number of atoms in the particle – which is the atom concentration times the volume V , and inversely proportional to the distance R between the particle center and the detector used for measurement, the ratio between the scattered amplitude A and the incident amplitude A_0 can be expressed as

$$\frac{A}{A_0} \propto \frac{V}{R} \lambda^x c^y ,$$
(2)

which should be a dimensionless quantity. The light speed c and the wavelength λ should also enter the formula because this is an optical phenomenon, but we do not yet know the exponents x and y . The right-hand side of the equation, in terms of time $[T]$ and length $[L]$, has dimensions $[L]^{2+x+y}[T]^{-y}$. For this to become dimensionless, we must have $y = 0$ and $x = -2$. This means (as the volume V is proportional to the cube of the particle diameter d)

$$\frac{I}{I_0} \propto \left(\frac{A}{A_0}\right)^2 \propto \frac{d^6}{R^2\lambda^4} .$$
(3)

Much of the physics of the Rayleigh scattering is already present in this result, based on this simple reasoning. In particular, the essential point is the so-called "inverse fourth power law", stating that the scattered intensity is inversely proportional to the fourth power of the wavelength. This means that the short wavelengths in the visible white light (violet-blue) are scattered much more efficiently than the long wavelengths (orange-red). The sunlight scattered by small particles appears essentially blue because the solar spectrum contains less violet than blue and because we are less sensitive to violet than to blue.

The Irish physicist John Tyndall contributed to this question as early as in 1860. He noticed the appearance of blue scattering by a vapor of hydrochloric acid, as the particles condensed into larger size droplets and its desaturation, reaching white color, when the particles became too large. Indeed the blue Rayleigh scattering is reinforced as the volume of the scattering center is increased, and continues to do so until the particle becomes larger than the illuminating wavelength. Then, standing waves and resonances start affecting the wavelength dependence of the scattering, giving rise to a much more complex scatter color. Typically, the range of particle sizes that produce a strong blue scattering is between 50 nm to 900 nm and, for this range, where the characteristics of Rayleigh scattering are still qualitatively useful, the scattering is usually called "Tyndall scattering". For spherical particles of small, medium or large sizes, a general treatment exists: Mie scattering (Mie, 1908).

This is not quite the end, as Rayleigh, Tyndall and Mie scattering only describe the scattered intensity by a single isolated particle. When considering aggregated particles, things also become more complicated. When the distance between the particles is much larger than the coherence length of the illuminating light, the collective scattering is incoherent, which means that the diffused intensity is merely the sum of the intensities diffused by each scattering center. The incoherent scattering by "Rayleigh particles (in the range of diameters smaller than 50 nm)" and by "Tyndall particles" (in the range 50-900 nm) can still be considered as mechanisms of "Rayleigh" or "Tyndall" scattering, respectively. If the particles are closer, we encounter a case of coherent scattering and we must add vector amplitudes with respective phases rather than intensities. The multiple scattering on nearby particles provides further opportunities for standing waves and resonances and we again lose the inverse fourth power law. The intensity and scatter directions then depend on the spatial distribution of the particles and in particular, the average distance between them.

2.2 Tyndall diffusion in nature

Tyndall scattering has long been recognized to be responsible for blue coloration of the sky (Tyndall, 1869) and the color of blue eyes (Mason, 1924). It appears when small particles or voids with dimensions of the order of the wavelength of blue light (about 500 nm) are present in the propagation medium. In that case, the small wavelengths of the incident white light will be scattered and the longer wavelengths will pass undisturbed through the medium. Thus, the red and yellow wavelengths are transmitted and the blue and violet colors are scattered by the composite medium, giving out a non-iridescent light blue diffusion spectrum.

In this phenomenon, the particle's sizes and refractive indexes control the coloration. As shown here above, the intensity of the reflected light by such a system is inversely proportional to the 4th power of the wavelength. The amplitude of the reflected light and its angular distribution will depend on the particle's sizes.

For incoherent Tyndall or Rayleigh scattering to occur, it is necessary that the diffusers are separated by more than the coherence length of sunlight (about 600 nm). Under this distance, coherent interaction occurs, even if the diffusive particles are randomly arranged and one can no more talk about Tyndall or incoherent scattering. This misleading fact is at the origin of several wrong interpretations of blue coloration in animals.

In living organisms, Tyndall blue is almost always present in association with underneath pigments. The underlying pigment granules absorb the incident light that penetrate through all the structural tissue and prevent desaturation by wavelengths scattered by inner tissues. They allow structural colors to be generated with a limited number of scatterers. In most cases, the pigment granules are made of melanin (Fox, 1976) but carotenoids, antocyanins and pterins (Lee, 1991; Stavenga et al., 200; Walls, 1995) can also be present, giving rise to various coloring effects.

Since the early 20th century, much work has been devoted to discovering the origin of dull blue colorations seen in animals. At first, it was common to distinguish two cases: the *iridescent* blue, synonym of coherent scattering and the *non-iridescent* blue, assumed to be incoherent Rayleigh or Tyndall scattering (Fox, 1976; Mason, 1926; Mason, 1927). At that time, the difference was based on the visual observation and not from the microscopic distances between scatterers.

It is interesting to mention that, in the twenties, Mason attributed all the non-iridescent blue colorations seen in bird feathers to Tyndall scattering but was aware that, in some insects, such coloration could arise from other phenomena (Mason 1923, Mason 1927). As new experimental and imaging techniques developed, new insights showed that blue in bird feathers could also be produced by constructive interference of light waves. Interfaces between keratin and air in the spongy medullar layer of the barbs act as coherent scatterers in that case (Prum et al., 1998; Prum et al., 1999). Blue Tyndall skins also appear in birds. For example, the extinct dodo head skin was found to be showing a diffuse blue color. This skin reveals randomly arranged, fine particles, about the size of the blue light wavelength (Parker, 2005).

Fig. 1. The male dragonfly *Orthetrum caledonicum* (Libellulidae). The blue coloration of the body comes from Tyndall scattering in a waxy layer over the black cuticle (Parker, 2000). (reproduced from GNU free documentation)

Scattered blues have early been assigned to insects. The scattering occurs in the epidermal cells beneath a transparent cuticle. In the odonate order such as aeschnids, agrionids and libelluloids (*Libellula Pulchella, Mesothemis Simplicicollis, Enallagma Cyathigerum, Aeshnea cyanea, Anax walsinghami*) the bright blue diffuse coloration on their body or wings (Mason, 1926; Parker, 2000; Parkcr, 2005; Veron, 1973) originates from scattering centers under the cuticle. Dragonflies (Mason, 1926) and some other adult insects can also develop a waxy bloom on the surface of their cuticle. The Tyndall effect is then produced by this waxy material and coloration can be destroyed by washing it with a wax solvent (Parker 2000): see Fig. 1.

Some butterflies have also been thought to be colored by this mechanism, such as *Papillio zalmoxis* or lycaenids (Huxley, 1976; Berthier, 2006). However, recent research shows that *coherent* interferences could also explain the various observed colors in these butterflies (Wilts et al., 2008; Prum et al., 2006). Tyndall blue has also been recorded in the cuticle of the larvae of some tent caterpillars (Byers, 1975), due to the presence of inhomogeneous transparent cuticular filaments.

Fig. 2. The male grasshopper *Kosciuscola tristis* at 30°C (left) and 5°C (right). The blue coloration accuring at high temperature comes from Tyndall scattering (from K. D. L. Umbers, with permission, Umbers, 2011)

The male grasshopper *Kosciuscola tristis*, also called "chameleon grasshopper" has the ability to change from black to bright sky blue. It has been shown that the mechanism of this color change is completely reversible and regulated by temperature changes (Filshie et al., 1975; Umbers, 2011). It is currently admitted that the blue color arises from Tyndall scattering of light on a suspension of small granules, intensified by the underlying dark background. Intracellular granule migration can explain the color change. However, recent discussion may lead to the conclusion that coherent scattering may play a role much more important than expected (Umbers, 2011).

Tyndall scattering has also been observed in molluscs (Fox, 1976; Herring, 1994) and in nudibranch mollusks (Kawaguti & Kamishima, 1964). These are obtained by small diffusive granules displayed over a pigmentary melanophore layer. Octopus and squids are sometimes able to control the blue hue of their body patterns. This adaptive blue is achieved by varying the melanophore's grains distances in order to change the underneath absorbing screen density (Fox, 1976).

Several blue mammals skins, especially in the primates family were thought to be Rayleigh or Tyndall scattered (Fox, 1976; Price et al., 1976). However, recent research on several structurally colored mammal skins pointed out that these colorations should come from coherent scattering from quasi-ordered arrays of collagen fibers (Prum & Torres, 2004).

As far as we know, it is hard to find in nature true incoherent Tyndall scattering. Diffusive layers are often made of randomly arranged particles too close from each other to assume incoherent scattering. This condition for incoherent diffusion is often misunderstood in papers that attribute to Tyndall scattering an array of disordered particles of the size of the wavelength, whatever the distance between them.

Fig. 3. Blue skins in mammals: Male mandrill facial blue skin (left) and male vervet monkey with blue scrotum (right) (reproduced from GNU free documentation)

3. Pigmentary coloration

3.1 Theoretical background

Pigmentary coloration is based on a spectrally selective absorption of the incident white-light. For a pigment to be useful, the light which has not been absorbed must be diffused in all directions, providing the same color in all directions. This means that a sheet of material colored by absorption and diffusion will appear with roughly the same color in reflection and transmission. This contrasts structural colors obtained by interference, without absorption, where back- and forward scattering colors tend to be complementary.

The physical description of a selectively absorbing material illuminated by a single frequency needs to extend the concept of the refractive index to include refraction and absorption. A simple way to do this is, at a fixed frequency, to accept to replace its real value by a complex number $\tilde{n} = n + ik$. A frequency-dependent complex refractive index (or, equivalently) a frequency-dependent complex dielectric constant can explain the optical response of dyes in a homogeneous material. But pigments require to produce diffuse scattering and this will only take place in a random inhomogeneous material or when the absorbing material appears in the form of concentrated granules. This helps providing a distinction between dyes and pigments.

3.2 Pigmentary coloration in plants

In plants, blue coloration is quite rare. However, it can be seen in some leaves, flowers or fruits. The blue is produced by modified anthocyanin pigments. A wide variety of mechanisms for modifying anthocyanin pigments has been observed in order to get blue or violet colorations. In flowers, they form complexes with flavonoids pigments and are in solution in cellular vacuoles. In leaves, they take place in chloroplasts. The structuration of the leave surface cells can help absorption by increasing high angle incident light to be transmitted through the leave or by focusing light in the pigmentary region. For example, in the velvet-leaved anthurium (*Anthurium warocqueanum*), the surface cells are convexly curved to focus light at some internal distance, just onto chloroplasts area (Lee, 2007).

3.3 Pigmentary coloration in animals

Pigments are very common in insects where they are responsible of almost all yellow, orange, red, brown and blacks. A blue pigmented hue is however very rare and can mainly be obtained by bile pigments such as pterobilin, phorcabilin and sarpedobilin. The two last names are coming from the species from which they were first extracted: Papilio phorcas and Graphium sarpedon (Barbier, 1990; Vuillaume & Barbier, 1969; Choussy et al., 1975).

Light induce cyclisation in bile pigments and transforms pterobilin into phorcabilin which in turn converts to sarpedobilin. In butterflies, pterobilin is widely distributed while the phorcabilin and sarpedobilin remain rare (Barbier, 1981).

Pigmented blue are mainly seen in two genera: *Papilio* (*Papilio weiskei*, *Papilio phorcas*) and *Graphium* where almost all the species contain blue pigments. In the *Graphium* species (*G. agamemnon*, *G. doson*, *G. antiphates*, *G. sarpedon*), pterobilin is responsible of the blue coloration. In the *Graphium sarpedon*, pterobilin is located in the wing membrane. Moreover, the transparent scales of the ventral side of the wing improve this blue coloration by further diffusing and polarizing light (Stavenga et al., 2010). This situation is rare among butterflies. Generally, the coloration of the wing originates from the scales covering the wing. Pigments can be embedded in the scales, absorbing part of the visible light spectrum. Alternatively or in addition, interferences can provoke structural coloration or modify pigmentary colors, as explained in the next section.

Fig. 4. Swordtail *Graphium sarpedon*. The blue coloration comes from the bile pigment sarpedobilin.

Pigmented blue is also seen in the dull blue stripes of the *Nessaea* genus (Nimphalidae). This coloration has been attributed to pterobilin (Vane-Wright, 1979).

To our knowledge, pigmented blue has not been found in mammals and in other insects. However, the presence of blue pigment remains very difficult to prove, partly because of their weak solubility. Extracting and characterizing very weakly soluble pigments is a complex task that restraints the possibilities of analysis. Moreover, determining the concentrations and localization of pigments within the tissues is still a real challenge.

Pigments in bird feathers are assumed to be present since the ages of dinosaurs. Studies on a *Sinosauropterix* (125 million years old), showed that their feathers would be filled with melanosomes and thus should appear dark (Vinther et al. 2008, Zhang et al., 2010). This work was, however, taken with cautious and discussions (Lingham-Soliar, 2011).

4. One-dimensional photonic structures

One-dimensional, planar or curved, photonic structures are frequent in nature. Insects, in particular, have frequently evolved this kind of structure for the purpose of coloration, as part of signaling or camouflage strategies. The reason may be that the process of fabrication of the outer part of a cuticle by epidermal cells, layer by layer, is compatible with the formation of such structures, even if we cannot claim at the moment that these mechanisms have been understood in all details.

Many different cases of one-dimensional photonic crystal have been seen in animalia, maybe because this is the most direct way to produce a metallic and/or iridescent color and, in this way, improve specific intra or interspecific functions. We will essentially examine two cases of physical designs: the single layer film and the Bragg mirrors.

The multilayer is the most common type of iridescent structure found in beetles and is also very common in butterflies (Kinoshita et al., 2008; Noyes et al., 2007; Parker et al., 1998). In many cases, these multilayers are composed of alternating layers of chitin and air partially filled with a chitinous compound. This produces a high/low index bilayer. Constructive interferences between light reflected by different layers produce one or several colors. The dominant reflected wavelength can be determined by the thicknesses of the layers and the average refractive index (see formula 11). The wavelengths that are not reflected are transmitted and the transmission spectrum is the exact complement of reflection if the system is considered non-absorbing. The color arising from a multilayer also varies with the angle of observation. As the reflection angle increases (starting from normal), the color shifts to lower wavelengths (blue shift).

The reflectors can be epicuticular (as in cicindelinae or in some chrysomelidae) (Kurachi et al., 2002), while others are endocuticular (Hinton, 1973).

Iridescence in bird feathers comes sometimes from 1D structure. They are located in the barbules like in satin bowerbirds *Ptilonorhynchus violaceus minor* that shows a violet to black iridescence coming from a single layer of keratin on the top of a layer of melanin (Doucet et al., 2006). In European starlings *Sturnus vulgaris*, multiple layers of keratin and melanin give a green-blue iridescence (Cuthill et al., 1999, Doucet et al., 2006).

4.1 Thin films

The single self-supported thin film and the optical overlayer covering a substrate have been known since a long time, in the planar and some other geometry. Constructive and destructive interference of light waves in thin films (soap bubbles or oil films on water) show colorful patterns. The interference occurs between light waves reflecting off the top surface of a film with the waves multiply reflected from the bottom surface. In order to obtain a nice colored pattern, the thickness of the film has to be on the order of the wavelength of the incident light.

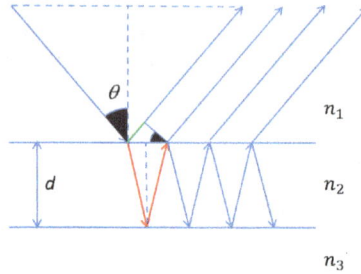

Fig. 5. Interference in a planar homogeneous slab. The phase shift between one reflected wave and its successor determine the intensity, for a given incidence angle and wavelength.

A single thin film illuminated from air reflects light at a wavelength λ as a Fabry-Pérot etalon. For a given thickness d of the film and incidence angle θ, the phase delay between the two first successively emerging rays from the multiply reflected beams is (we assume, for instance, a dense slab $n_2 > n_1$ and $n_2 > n_3$ and we name the angle of refraction inside the film: $n_2 \sin t = n_1 \sin \theta$)

$$\Delta \phi = \left\{ n_2 \frac{2\pi}{\lambda} \left(\frac{2d}{\cos t} \right) \right\} - \left\{ [\pi] + n_1 \frac{2\pi}{\lambda} (2d \tan t \sin \theta) \right\} \qquad (4)$$

The first term is the phase change of the transmitted electric field wave travelling one round trip in the film before its next exit, while the last term is the progress of the reflected beam in air before joining back the other path wave. The "optional" half-wavelength phase delay (which can be written $[\pi]$ or $[-\pi]$, without consequences) occurs only when an electric wave reflects on a medium with higher refractive index. All other phase delays between successive emerging rays are the same. Maximal reflections occur when all the exiting beams are in phase, which means $\Delta \phi = m(2\pi)$, where m is an integer. This condition allows determining, under a specific incidence angle, the reinforced wavelengths, which turn out to be

$$\lambda = \frac{2n_2 d}{m + \left[\frac{1}{2} \right]} \cos t \qquad (5)$$

Or, equivalently,

$$\lambda = \frac{2d \sqrt{n_2^2 - n_1^2 \sin^2 \theta}}{m + \left[\frac{1}{2} \right]} \qquad (6)$$

Note that, if the refractive index n_3 is larger than n_2 (itself larger than n_1), the addition of 1/2 at the denominator must be skipped. The dependence of this dominant reflected wavelength on the incidence angle θ, which means a change of color with the angle under which the surface is viewed, is the phenomenon of iridescence, which often signals a structural color.

A typical blue color is perceived for a dominant reflected wavelength of 475 nm. Take, specifically, a thin film of thickness 200 nm and refractive index 1.5. The branch $m = 0$ reflects infrared radiation, from 1470 nm under normal incidence to 1150 nm under grazing angles. The $m = 1$ order extends in the visible, from 490 to 385 nm: providing short-wavelength blue coloration. The $m = 2$ and higher orders are deeper in the ultraviolet.

A natural example comes from the study of the iridescent wing of a giant tropical wasp, *Megascolia procer javanensis* (Sarrazin et al., 2008). In this particular case, the wing is shown to be made of rigid structure of melanized chitin, except for an overlayer, on each side of the wing. The overlayer can be shown to act as a transparent interference thin film with a thickness of 286 nm. The refractive index of the material in this layer is not precisely known, so that its analysis requires examining the reflectance spectrum in detail, for various angles of incidence. The substrate supporting this layer is better known, as a solid mix of chitin and melanin. This mix was studied by de Albuquerque (de Albuquerque et al., 2006), including the dispersion related to melanin absorption. This absorption is strong here, as can be seen from the opacity of the wing. An adjustment of the refractive index of the overlayer allows the reflectance spectra to be fitted quite nicely at all incidence angles, and provides a value $n \approx 1.76$, which is very reasonable. The iridescence is weak: from bluish green near-normal incidence to greenish blue under a grazing incidence.

4.2 Bragg mirrors

The stacking of multiple planar layers (or "multilayers structure") is another type of structure that produces selective reflection. Among these, one important class is the periodic stack, where a group of two layers is repeated a finite number of times. This structure is also known as a Bragg mirror. When the number of periods is large (but, in practice, it does not need to be in excess of, say, 3 or 4 in the kind of structures examined here), the optical response can be approached by assuming an infinite number of periods, which can be called a one-dimensional photonic crystal. In this case, it is not very difficult to predict the dominant color that will be reflected.

In this limit, the incident frequency is conserved in all scattered waves, but not the wave vector in the stacking direction. A wave with wave number k_z will change this wave number so that its output value is a choice of any of the quantities $k_z + m(2\pi/a)$, where m is an integer and a is the period thickness:

$$k_z' = k_z + m\frac{2\pi}{a} \tag{7}$$

This means that, in a periodic multilayer stack, waves propagate normally, unless their wave vector k_z' obeys the above relation. For waves at the conserved incident frequency ω, this implies that

$$k_z = \pm\frac{\pi}{a}, \pm 2\frac{\pi}{a}, \pm 3\frac{\pi}{a}... \tag{8}$$

This corresponds to wave number values that match the so-called Brillouin-zone boundaries. For low contrasts of refractive indexes, defining an average refractive index $\bar{n}$ for the whole structure is a good starting point, taking the perturbation point of view. In this average material the following dispersion relation holds,

$$\omega = \sqrt{k_y^2 + k_z^2}\,\frac{c}{\bar{n}} \tag{9}$$

where c is the light velocity in vacuum and k_y the wave vector component parallel to the layers. This quantity is conserved across the interfaces, so that anywhere in the structure,

$$k_y = \frac{\omega}{c}\sin\theta_i, \tag{10}$$

where θ_i refers to the incidence angle in the incidence medium (with refractive index assumed to be 1). At the zone boundaries, the k_z and $-k_z$ modes are degenerate, but with the appearance of the refractive index contrasts, the degenerescence is lifted because of the formation of standing waves that adopt different configurations relative to high and low refractive-index regions. Then, a gap (forbidden frequencies) appears when the unperturbed dispersion curve crosses the zone boundaries, and this occurs for incident wavelengths in narrow bands centered on (m integers) (Vigneron et al., 2006)

$$\lambda = \frac{2a\sqrt{\bar{n}^2 - \sin^2\theta_i}}{m} \tag{11}$$

An incident light wave with a frequency in these ranges, impinging on the surface of this semi-infinite photonic crystal, will be totally reflected. We note that, on a photonic crystal surface, total reflection can occur under normal incidence, and also when air is the incidence medium, contrasting our usual knowledge of total reflection conditions.

In order to produce blue (say 480 nm) under normal incidence and violet under grazing incidence (say 350 nm - many living organisms have UV vision), this equation fixes the refractive index average (1.46) and the period a (162 nm). This, however, still leaves ample flexibility in choosing the actual bilayer that defines a period. These values, calculated for the first gap ($m = 1$), are easily produced with biopolymers such as chitin or keratin, with refractive indexes close to 1.56.

It is also possible to produce long-wavelength blue (say 480 nm) under normal incidence with the second gap $m = 2$ because the fundamental reflection ($m = 1$) would not appear in the visible, but in the infrared, near 960 nm. However, under larger incidences, iridescence may bring in red contribution to the spectrum, turning violet into some extraspectral metameric color in the purple range.

A good example of this structure is provided by the "blue beetle" *Hoplia coerulea*. This structure could have been classified as a two-dimensional photonic crystal, to be described later, but the optical response is in fact close to that expected for a Bragg mirror, for reasons that will become clear in a moment.

Hoplia coerulea, has evolved a cuticle bearing scales Fig. 6 (left). The inner region of these scales is structured to filter out a spectacular blue-violet iridescence on reflection (Vigneron et al., 2005). The cuticle, as seen in scanning electron microscopy is shown in Fig. 6 (right). The scales are attached by a single peripheral point to the underlying cuticle. These scales are easily removed by breaking this binding.

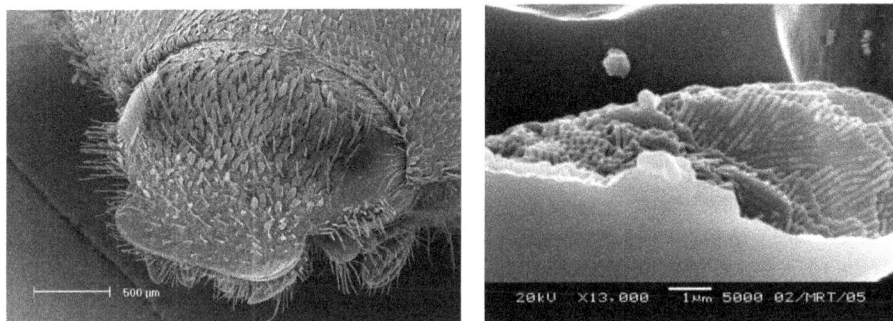

Fig. 6. *Hoplia coerulea* (right). The beetle's cuticle is covered by scales. The scales take the shape of a disk, with a diameter of about 50 μm and a thickness of 3.5 μm. These scales render a blue or violet color. The scanning electron microscope images (right) shows the coloring structure inside the scales (with permission).

The structure in each scale can be interpreted as a stack of some 20 sheets, roughly parallel to the cuticle. Each sheet is actually composed of a very thin plate of bulk chitin, bearing, on one side, a network of parallel rods with a rectangular section. The lateral corrugation associated with the rods has a period of 170 nm, just too small to produce the diffraction of light in the visible range. This acts as a zero-order grating: for visible wavelengths, the rods array appears to be a homogeneous layer, and the concept of an average refractive index is adequate. The average refractive index of the whole structure was evaluated to $\bar{n} = 1.4$ for unpolarized light near normal incidence. As the vertical period turns out to be 120 nm + 40 nm =160 nm, it fulfills perfectly the conditions described above for the production of weakly iridescent blue. The *Hoplia coerulea* structure gives some iridescence, ranging from blue to violet, and effectively behaves as a flat multilayer structure, in spite of the lateral structuring of the rods layers. This structure was recently shown to have an optical response modifiable in presence of humidity, because water can infiltrate the voids. Strangely, the structure's materials turn out to be hydrophilic (Rassart et al., 2009).

Under a crude approximation, the structure carried by the ridges of *Morpho* butterflies (for instance *M. menelaus*) can be viewed as a stack of alternating chitin and air layers, with a period of the order of 180 nm and an average refractive index well under 1.4 (Berthier et al., 2003, Berthier et al., 2006). This can explain the normal-incidence bright blue coloration and the shift of the reflected wavelength to the violet as the angle of incidence increases. This simple model has limits: it is unable to explain the off-specular variation of the scattering and the polarization effects observed in the directional reflectance pattern (Berthier, 2010).

Plants can also produce coloring multilayers for displaying a blue coloration. Examples can be found in the genus *Selaginella*, for example *S. willdenowii* and *S. uncinata* (Lee, 1997). These plants live in the understory of Central- and South-American rainforests and, strangely, display blue on freshly grown shadowed leaves. The blue coloration arises from a one-dimensional multilayer in the moistened cellulose of outer cell walls. The refractive index of moistened cellulose is, in the average, 1.45 and the multilayer found has a period of 160 nm (two layers of different refractive indexes and equal thicknesses, 80 nm). We again find the exact conditions to provide a blue coloration with a one-dimensional photonic crystal.

5. Two-dimensional photonic structures

In two-dimensional photonic structures, one only observes a total translational invariance in one dimension, the other two dimensions being structured by refractive index inhomogeneities. We will include gratings in this category, besides fibrous photonic crystals, both of them being encountered in nature and able to select blue reflectance.

Actually, fibrous photonic crystals can be viewed as a combination of a grating and a multilayer. From the symmetry point of view, we can view a 2D photonic crystal as totally invariant in the "fibers" directions and periodic in two directions perpendicular to the fibers. Defining the surface parallel to the fibers, these two directions are adequately defined as the surface plane and the direction of the normal. The periodicity parallel to the surface produces diffraction similar to that produced by a grating, while the periodicity along the normal, deep under the surface, produces a color selection with a Bragg mirror. Being a combination of both, a 2D photonic crystal tends to be more flexible than either a grating or a Bragg mirror to produce a color such as blue. As explained below, a short-period grating will cease diffracting as its associated lateral inhomogeneity is smaller than the shortest visible wavelength. For this wavelength and larger, the grating will act as a homogeneous average material that can only generates a specular reflection. However, with a 2D photonic crystal, the "normal" periodicity can still be there to produce color selection. If, on the other extreme, the normal periodicity is constrained by weak refractive index contrasts or a tight film thickness, the Bragg mirror will not be effective, but the lateral grating can take over and still manages to produce blue. Additionally, intermediate – less easily described – mechanisms involving simultaneously cooperating diffraction and interference color selection add new channels for producing blue.

5.1 Gratings

A grating is usually a superficial structure, periodic in a single direction (say x), with a period b. The characteristic rule which describes light scattering by a grating arises from the observation that an incident wave with wave vector k_x which travels through a medium with periodicity b comes out with a series of possible wave vectors $k'_x = k_x + m(2\pi/b)$, where m is a negative, null or positive integer.

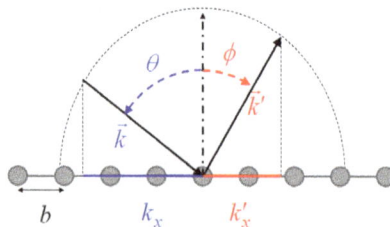

Fig. 7. Geometry of a grating, for an incident beam falling at right angle with the parallel lines.

This implies the following relationship between the incidence and emergence angles

$$\frac{\omega}{c}\sin\phi = \frac{\omega}{c}\sin\theta + m(2\pi/b) \tag{12}$$

And, more explicitly,

$$\sin\phi = \sin\theta + m\frac{\lambda}{b} \tag{13}$$

The integer m is the diffraction order. The order zero is a reflection, with emergence and incidence angles identical, with no dependence on the incidence wavelength λ. By contrast, for $m \neq 0$, the emergence angle changes with the wavelength, which means that an incidence white beam is decomposed in a colored spectrum after being scattered by the grating. Several orders may be simultaneously present, but the actual number depends on the grating period. In order to produce an acceptable emergence angle, the condition $-1 - \sin\theta \leq m(\lambda/b) \leq 1 - \sin\theta$ must be fulfilled. A given wavelength λ starts appearing in the order $m = -1$ (it will then emerge for a grazing illumination $\theta = 90°$) when

$$b = \frac{\lambda}{2} \tag{14}$$

For a larger period, the same wavelength will be observed for a range of incidence angles which contain $\theta = 90°$. This means that we actually can build gratings that produce only "blue" colors, i.e. wavelengths smaller than 490 nm (then including blue, purplish blue and violet), if the period is the rather precisely defined: $b = 245$ nm . They must be illuminated under incidence angles larger than 33°.

An example of such a grating is provided by the array of flutes found on the ridges of the scales on the butterfly *Lamprolenis nitida* (Ingram, 2008). This butterfly is special because it is equipped with two types of gratings on the same scale. One, with a large period, produces a full decomposition of the visible white light, when illuminated from the front. All colors from red to green are shown, but in this configuration, blue light is scattered with a very low intensity. The grating responsible for this coloration is shown in Fig. 8, at the tip of the arrow C. The lamellae, repeated at 700 nm spacing, are slanted in such a way as to maximize the emission in the $m = -1$ order, from red to green and to reduce the scattering in the $m = 0$ order. This can be understood as a blazed grating and the lack of blue in this coloration is the result of the precise slant angle. However, slanted in the reverse direction, the so-called "flutes" are separated by about 235 nm, not far from the period $b = 245$ *nm* mentioned above. The result is, as observed, a grating that produces only a purplish blue color, under large illumination angles.

5.2 Two-dimensional photonic crystals

Two-dimensional photonic crystals are fibers with two-dimensional periodic variations of the refractive index in the cross-section. In much the same way as with one-dimensional multilayers, the colored reflections originate from the formation of directional band gaps in the photonic band structure of these crystals. Producing blue alone from an ideal structure which fulfills these rules is difficult, because each stack of reticular plane in the two-

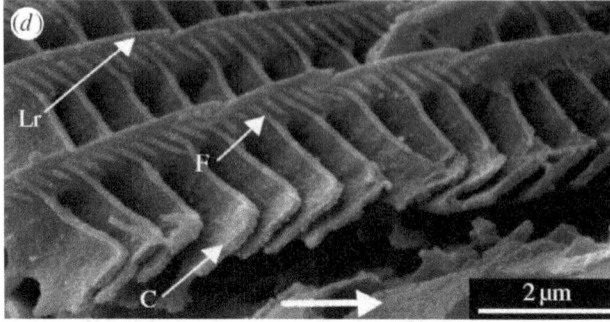

Fig. 8. The dorsal wings of male *Lamprolenis nitida* appear matt brown under incident light, normal to the wing surface, but shows various colorations under large incidences. These visual effects are due to the presence of two interspersed gratings on the scales of the rear wings. When illuminated in a postero-anterior direction and observed in backscatter, blue to violet is observed with increasing angle from the wing surface.

dimensional lattice can be considered as a Bragg mirror and the variety of values of stacking periods easily leads to the production of a wide range of colors. The coloration of fibrous organs in some marine animals, such as the Aphrodite sea mouse (Parker et al., 2001) or the Ctenophore *Beroë cucumis* (Welch et al., 2005; Welch et al., 2006), is accompanied with a broad iridescence, covering a spectral range from red to far in the ultraviolet. The high refractive index of water, compared to air, partly explains the iridescence richness, but in both cases, the structure can be viewed as a bunch of parallel fibers and a simple two-dimensional photonic crystal. Nature, however has found unexpected ways to produce blue coloration from these fibrous structures and we will give here an account of the way a bird such as the magpie (*Pica pica*) (Vigneron et al., 2006a; Lee, 2010) produces the blue reflection on some of their wing feathers.

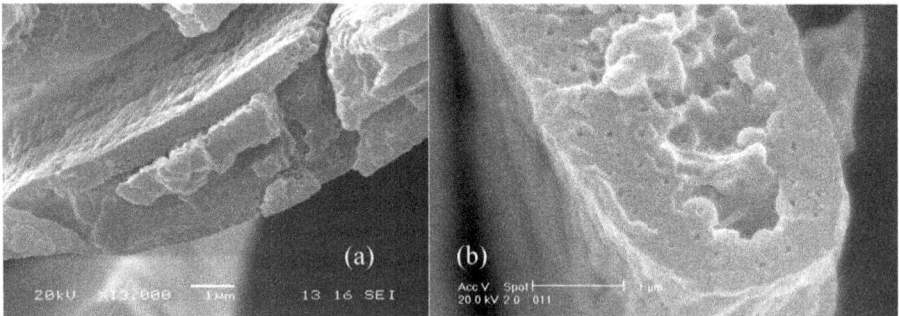

Fig. 9. The coloring structure in the tail (a) and wing (b) feather of the common magpie (*Pica pica*). The structure is formed by cylindrical melanine bars distributed to form a hexagonal two-dimensional lattice in the cortex of the barbules. The scattering centers are thin cylindrical cavities in the melanine granules and the color is related to the distance between these centers. The green color on the tail depends on the distance between the granules centers, 180 nm. Strangely, the blue photonic crystal has a larger lattice parameter (270 nm) : the coloration is controlled by a "second gap", at high frequency.

Many (but not all) birds show structural coloration at the level of the feather's barbules. A feather is rigidified by a rachis, an array of barbs attached to the rachis, and an array of barbules attached to the barb. The barbules have the topology of a sack, with an envelope (a hard cortex) containing a medullar medium. As in the Peacock's feathers (Zi et al., 2003), the coloring structure on the blue feathers of the magpie lies on the barbule's cortex. It is constituted of elongated melanine cylinders, disposed parallel to each other with the symmetry of a two-dimensional triangular lattice. These cylinders are the scatterers that produce the coherent reflection in the blue. Strangely, the distance between these scatterers is 270 nm, much too large to explain the blue coloration. In fact, simulation shows that such a fibrous crystal should produce a fundamental gap in the near infrared, and a blue scattering as a "harmonic" of this gap. Indeed, a second band of forbidden propagation exists at higher frequency. At this frequency, the diffraction is less dispersive and the blue coloration produced is relatively saturated. The coloration hue could also be spoiled by the addition of red, arising from the line width of the fundamental reflection in the infrared, producing an extraspectral purple. This is not the case: the structured cortex is thin enough to avoid producing long-wavelengths resonances and the result is a dark blue color, easily visible under a bright sunshine.

6. Three-dimensional photonic crystals

Three-dimensional photonic crystals are also encountered in nature, especially in butterflies, weevils and longhorns. In view the very high diversity of living organisms and the frequency of structural colors, it can be speculated that other families of insects will soon reveal a similar evolution. Three-dimensional photonic crystals are periodic in all three dimensions of space. Illuminated by white light in a well-defined direction, an ideal structure produces several colored beams, each of them corresponding to a stack of reticular planes with its appropriate spacing. A good example of the visual effect produced by an ideal photonic crystal is provided by the so-called Brazilian "diamond" weevil, which displays a green color when viewed from a distance, but under an optical microscope, shows individual scales with a variety of very saturated (pure) colors. Most other weevils and longhorns show less iridescence, and this is usually explained in term of orientation disorder: natural three-dimensional photonic crystals most often appear under the form of photonic polycrystals, with well-defined domains bringing short-range order and long-range orientation disorder.

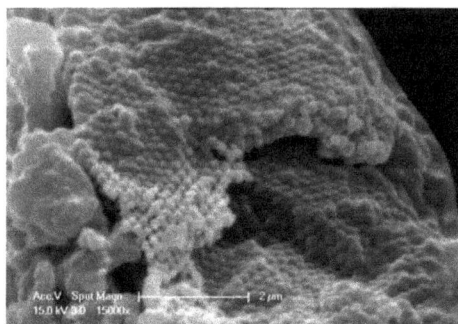

Fig. 10. The internal structure of a scale on a blue area from the cuticle of the weevil *Eupholus schoenherri*. The structure, with a face-centered cubic symmetry, can be described as an "opal" structure.

Blue weevils are frequent, in particular in the *Eupholus* genus, as *E. loriai* (completely blue) or *E. bennetti*, *E. magnificus* or *E. schoenherri* (partially blue). At the moment, the structures that produce this blue colors can be described as a photonic polycrystal with grains locally organized in a face-centered cubic symmetry. A typical "blue" structure from a weevil is shown in Fig. 10, which shows a scale from a blue area on an elytron of *Eupholus schoenherri*. The same kind of structure has been encountered in a previous work (Parker et al., 2003) for a different weevil displaying green spots. This structure is generally referred to as an "opal" structure, making a parallel with the assembly of monodispersed spheres constituting the iridescent stone. The present photonic structure is also an arrangement of non-absorbing spheres but the constituting material is a chitinous compound, with a refractive index of the order of 1.6. In order to produce short-wavelength photonic gaps, the size of the spheres is kept small and the compactness is maximized. In such a structure, the light scatterers are effectively the tiny air-filled interstices left between the spheres, not easily seen in electron microscopy images. In weevils however, the "inverse opal" structure like the one shown in Fig. 11 is the most common case. This structure corresponds to an arrangement of spherical hollows in a chitinous matrix.

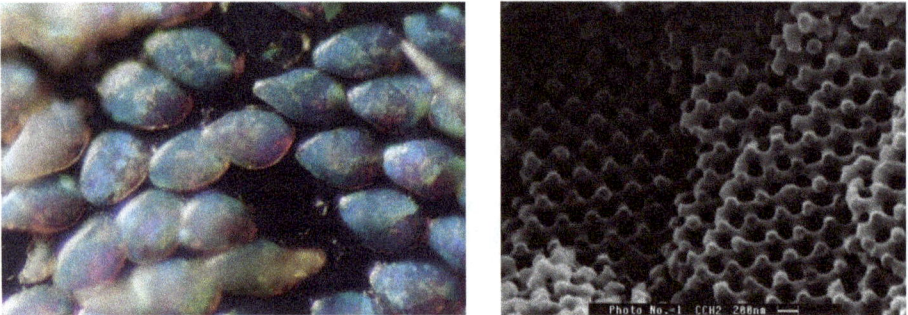

Fig. 11. Optical microscope view of the blue scales of the weevil *Cyphus hancoki*. The different colors correspond to different crystal grains (left). On the right, electron microscope image of one grain. The 3D array of spherical hollows is described as an "inverse opal" structure (Berthier, 2006).

A blue longhorn, with a three-dimensional photonic-crystal structure has also been described (Simonis et al., 2011). *Pseudomyagrus waterhousei* shows a slightly desaturated purplish blue color. These colorations arise from a dense layer of droplet-shaped scales covering the dorsal parts of the cuticle. These colors are caused by structural interferences and produced by an aggregate of internally ordered photonic-crystal grains. As in the weevils' case, the structure is built with spherical diffusion centers arranged according to a face-centered cubic symmetry. Domains are also present, with long-range orientation disorder, a complex structure which partly explains the lack of iridescence in the visual effect, in spite of a structural coloration. Theoretical considerations suggests that the contents of the observed reflectance dominantly arise from photonic crystallites with (111) reticular plane parallel to the cuticle surface. Another source of disorder lies in the observation that, in this structure, internally ordered photonic crystal grains can be separated by regions of amorphous arrangement, with spheres diameters varying over a rather wide range, from 170 to 300 nm in diameter. Here also, the short-wavelength blue coloration is reached by increase of the structure compactness, with the production of spheres with an average diameter of 212 nm, arranged in a compact cubic structure.

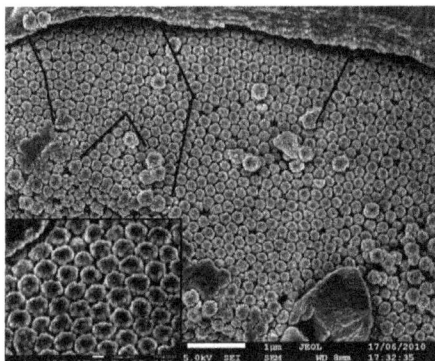

Fig. 12. The array of spherical centers found in scales of the Malaysian longhorn *Pseudomyagrus waterhousei*. The compactness of the structure leads to the production of short-wavelength scattering, providing a slightly desaturated blue-violet color.

7. Conclusion

This brief survey of the production of blue color on living organisms shows that all broad categories of structural mechanisms can be put to use to produce short-wavelengths scattering. We have seen that all structures known to be at the root of a structural coloration in nature (Vigneron & Simonis, 2010) can actually provide a blue coloration. Each device has its own rules for providing scattering solely on the short-wavelength end of the visible spectrum.

This particular objective is not always easy and often requires a multiscale solution. For living organisms that has undergone evolution over many million years, this is not a problem: the "modification-selection" algorithm, which is the engine of the past and present biodiversity, has no reluctance for complexity. Even if the range of refractive indexes in biologically prepared materials is rather narrow (typically 1.3 to 1.8), complexity in geometrical structure can provide a very wide range of functions that turn out to be an advantage for species population increase.

We do not always know what can be the biological advantage of producing blue. We understand that a male metallic blue *Morpho* can be seen from far away, which is an advantage for accelerating productive mates encounters. But the answer is less obvious for the formation of iridescent blue plants, as blue is one of the spectral components of the light captured by chlorophyll molecules to achieve photosynthesis. While an answer to the physical "how" question - referring to a description of the production mechanisms - is relatively easy, an answer to the biological "why" question is far less obvious.

8. Acknowledgment

Priscilla Simonis, on leave from the University of Namur, Belgium, acknowledges the hospitality of the "Institut des Nanosciences de Paris", Université Pierre et Marie Curie – Paris 6, where this work was carried out.

9. References

Andre, J. (1949) *Etude sur les termes de couleurs dans la langue latine*, Librairie C. Klincksieck, Paris

Barbier, M. (1981) The status of blue-green bile pigments of butterfiles, and their phototransformations, *Experientia* 37, pp1060-1062

Barbier, M. (1990) A new sarpedobilin-containing butterfly - *Papilio-Graphium-stresemani-stresemani* and its bioecological situation within the species, *J. Chem. Ecol.* 16, pp 743-748

Berthier, S., Charron, E. & Da Silva, A. (2003) Determination of the cuticle index of the scales of the iridescent butterfly Morpho Menelaus, *Opt. Comm.* 228, pp 349-356

Berthier, S. (2006) *Iridescences: the physical colors of insects*. In Springer New York, ISBN: 978-0387341194

Berthier, S., Charron, E. & Boulenguez, J. (2006) Morphological structure and optical properties of the wings of Morphidae, *Ins. Sci.* 13, pp 145-157

Berthier, S. (2010) *Photonique des Morphos (French Edition)*, Springer; 1st Edition, ISBN-13: 978-2287094071

Boyer, C.B. (1954) Robert Grosseteste on the rainbow, *Osiris* 11, 247-258

Byers, J. R. (1975) Tyndall blue and surface white of tent carterpillars, Malacosoma spp. *J. Insect. Phys.* 21 pp 401-415

Choussy, M., Barbier, M. & Vuillaume, M. (1975) Biosynthesis of phorcabilin, a blue bile pigment from *Actias-Selene* (Lepidoptera,Aattacidae), *Biochimie* 57, pp 369-373

Cuthill, I. C., Bennett, A. T. D., Partride, J. C. & Maier, E. J. (1999) Plumage reflectance & the objective assessment of avian sexual dichromatism *Am. Nat.* 53, pp 183-200

De Albuquerque, J., Giacomantonio, C., White, A. & Meredith, P. (2006) Study of optical properties of electropolymerized melanin films by photopyroelectric spectroscopy *Eur. Biophys. J.* 35, 190

Doucet, S. M., Shawkey, M. D., Hill, G. E. & Montgomerie, R. (2006) Iridescent plumage in satin bowerbirds: structure, mechanisms and nanostructural predictors of individual variation in colour *J. Exp. Biol* 209, pp380-390

Filshie B. K., Day M. F. & Mercer E. H. (1975) Colour and colour change in the grasshopper, Kosciuscola tristis *J. Insect Phys.* 21 pp 1763-1770

Fox, D. L. (1976) *Animal Biochromes and Structural colors*, University of California Press, ISBN 978-0520023475, Berkeley, CA

Herring, P. J (1994) Reflective systems in aquatic animals Comput. *Biochem.Physiol.* 109A pp 513-546

Hinton, H. E. (1973) Some recent work on the colours of insects and their likely significance *Proc. Br. Ent. Nat. Hist. Soc.* 6, pp 43-54

Hoeppe, G. (1969) *Why the sky is blue: discovering the color of life*, Princeton University Press, ISBN 0-691-12453-1

Huxley, J. (1976) The coloration of Papilio zalmoxis and P. antimachus and the discovery of Tyndall blue in butterflies, *Proc. R. Soc. Lond. B* 193 pp441-453

Ingram, A. L, Lousse, V., Parker A. R. & Vigneron, J.P. (2008) Dual gratings interspersed on a single butterfly scale, *J. R. Soc. Interface*, 5, 1387-1390

Kawaguti, S. & Kamishima, Y. (1964) Electron microscopic study on the iridophores of opisthobranchiate mollusks *Biol. J. Okayama Univ.* 10, pp 83-91

Kinoshita, S., Yoshioka, S. & Miyazaki, J. (2008) The physics of structural colors, *Rep. Prog. Phys.* 71, 30 pp, 076401

Kurachi, M., Takaku, Y., Komiya, Y. & Hariyama, Y. (2002) The origin of extensive colour polymorphism in Plateumaris sericea (Chrysomelidae, Coleoptera) *Naturwissenschaften* 89, pp 295–298

Lee, D. W. (1991) Ultrastructural basis and function of iridescent blue colour of fruits in Elaeocarpus, *Nature* 349, pp 260-262

Lee, D.W. (1997) Iridescent blue plants, *Am. Sci.* 85, 56-63

Lee, D.W. (2007) *Nature's Palette: the Science of Plant Color*, The University of Chicago Press, USA, ISBN: 9780226470528

Lee, E., Lee, H., Kimura, J. and Sugita, S. (2010) Feather Microstructure of the Black-Billed Magpie (Pica pica sericea) and Jungle Crow (Corvus macrorhynchos), J. Vet. Med. Sci. 72(8): 1047–1050

Lingham-Soliar, T. (2011) The evolution of the feather: Sinosauropteryx, a colourful tail, *J. Ornithol.* 152, pp 567-577

Mason, C. W. (1923) Structural colours of feathers I, *J. Phys. Chem* 27, pp 201-251

Mason, C. W. (1924) Blue eyes, *J. Phys. Chem.* 28, pp498-501

Mason, C. W. (1926) Structural colours of insects I, *J. Phys. Chem* 30, pp 383-395

Mason, C. W. (1927) Structural colours of insects II, *J. Phys. Chem* 31, pp 321-354

Mie, G. (1908) Beiträge zur Optik trüber Medien, speziell kolloidaler Metallösungen *Ann. Phys., Lpz.* 25, pp. 377–445

Myiamoto, A & Kosaku, A. (2002) Cuticular Microstructures and Their Relationship to Structural Color in the Shieldbug Poecilocoris lewisi Distant, *Forma* 17 pp 155-167

Noyes, J. A., Vukusic, P. & Hooper, I. R. (2007) Experimental method for reliably establishing the refractive index of buprestid beetle exocuticle, *Opt. Express* 15, pp 4351-4358

Parker, A. R., Mc Kenzie, D. R. & Large, C. J. (1998) Multilayer reflectors in animals using green and gold beetles as contrasting examples *J. Exp. Biol.* 201, pp 1307-1313

Parker, A. R. (2000) 515 million years of structural colour *J. Opt. A* 2 pp R15-R28

Parker, A. R. (2001) et al., Photonic engineering - Aphrodite's iridescence *Nature* 409, 36

Parker, A. R., Welch, V.L., Driver D. & Martini N. (2003) Structural colour: Opal analogue discovered in a weevil, *Nature* 426, 786-787

Parker, A. R. (2005) A geological history of reflecting optics, *J. Roy. Soc. Interface* 2, pp 1-17

Pastoureau M. (2000) *Bleu, histoire d'une couleur*, Editions du Seuil, ISBN 978-2-02-086991-1

Price, J. S., Burton, J. L., Shuster, S. & Wolff, K. (1976) Control of scrotal color in vervet monkey, *J. Med. Primat.* 5, pp 296-304

Prum, R. O., Torres, R. H., Williamson, S. & Dyck, J. (1998) Constructive interference of light by blue feather barbs *Nature* 396, pp 28-29.

Prum, R. O., Torres, R., Williamson, S. & Dyck J. (1999) Two-dimensional Fourier analysis of the spongy medullary keratin of structurally coloured feather barbs *Proc. R. Soc. Lond. B* 266, pp 13-22

Prum, R. O. & Torres, R. H. (2004) Structural colouration of mammalian skin: convergent evolution of coherently scattering dermal collagen arrays, *J. Exp. Biol.* 207, pp 2157-2172

Prum, R. O., Quinn, T. & Torres, R. H. (2006) Anatomically diverse butterfly scales all produce structural colours by coherent scattering *J. Exp. Biol* 209, pp 748-765

Rassart, M., Simonis, P., Bay, A., Deparis, O. & Vigneron, J.P. (2009) Scale coloration change following water absorption in the beetle Hoplia coerulea (Coleoptera), *Phys. Rev. E* 80, 031910

Sarrazin, M., Vigneron, J.P., Welch, V. & Rassart, M. (2008) Nanomorphology of the blue iridescent wings of a giant tropical wasp Megascolia procer javanensis (Hymenoptera), *Phys. Rev. E* 78, 051902.

Simonis, P. & Vigneron, J.P. (2011) Structural color produced by a three-dimensional photonic polycrystal in the scales of a longhorn beetle: *Pseudomyagrus waterhousei* (Coleoptera: Cerambicidae), *Phys. Rev. E* 83, 011908

Stavenga, D. G., Stowe, S., Siebke, K., Zeil, J. & Arikawa, K. (2004) Butterfly wing colours: scale beads make white pierid wings brighter, *Proc. R. Soc. Lond. B* 271 pp1577-1584

Stavenga, D. G., Giraldo, M. A. & Leertouwer, H. L. (2010) Butterfly wing colors: glass scales of *Graphium sarpedon* cause polarized iridescence and enhance blue/green pigment coloration of the wing membrane, *J. Exp. Biol.* 213, pp 1731-1739

Tyndall, (1869) On the blue colour of the sky, the polarization of skylight, and on the polarization of light by cloudy matter generally, *Proc. Roy. Soc. London.* 17, pp 223-233

Umbers, K. D. L. (2011) Cues for colour change in the chameleon grasshopper (Kosciuscola tristis) *J Insect Physiol.* 57, pp 1198-1204

Verron, J. E. N. (1973) Physiological control of the chromatophores of Austrolestes annulosus (Odonata), *J. Insect. Phys.* 19, pp 1689-1693

Vigneron, J.P., Colomer, J.F., Vigneron, N. & Lousse V. (2005) Natural layer-by-layer photonic structure in the squamae of Hoplia coerulea (Coleoptera), *Phys. Rev. E* 72, 061904

Vigneron, J.P. & Virginie Lousse, V. (2006) Variation of a photonic crystal color with the Miller indices of the exposed surface, *Proc. SPIE* 6128, 61281G

Vigneron, J.P., Colomer, J.F. Rassart, M., Ingram, A.L. & Lousse V. (2006a) Structural origin of the colored reflections from the black-billed magpie feather, *Phys. Rev. E* 73, 021914

Vigneron J.P. & Simonis P. (2010) Structural Colours, in Jérôme Casas and Stephen J. Simpson, editors: *Advances in Insect Physiology*, 38, pp. 181-218, Academic Press (Burlington) , ISBN: 978-0-12-381389-3

Vane-Wright, R. I. (1979) The coloration, identification and phylogeny of Nessaea butterflies (Lepidoptera: Nymphalidae) *Bull. Brit. Mus. Nat. Hist. Entomol.* 38, pp 27-56

Vinther, J., Briggs, D. E. G., Prum, R. O. & Saranathan, V. (2008) The colour of fossils feathers *Biol. Lett.* 4, pp 522-525

Vuillaume, M. & Barbier, M., (1969) Tetrapyrrole pigments of lepidoptera, *C. R. Acad. Sci. Paris* 268, pp 2286-&

Walls, J. (1995) *Fantastic Frogs* T.F.H. Publications, Neptune City, NJ ISSN 978-0793801312

Welch, V. L., Vigneron J. P. & Parker A. R. (2005) The cause of colouration in the ctenophore Beroë cucumis, *Current Biology*, 15 pp R985-R986 Supplement

Welch, V. L., Vigneron J. P., Parker A. R & Lousse V. (2006) Optical properties of the iridescent organ of the comb-jellyfish Beroe cucumis (Ctenophora), *Phys. Rev. E* 73, 041916

Wilts, B. D, Leertouwer, H. L & Stavenga, D. G. (2008) *J. Roy. Soc. Interface* 6, pp s185-s192

Zhang, F., Kearns, S. L., Orr, P. J., Benton M. J., Zhou, Z., Johnson D., Xu, X. & Wang, X. (2010) Fossilized melanosomes and the colour of Cretaceous dinosaurs and birds, *Nature* 463, pp 1075-1078

Zi, J., Yu, X., Li, Y., Hu, X., Xu, C., Wang, X., Liu, X. & Fu, R. (2003) Coloration strategies in peacock feathers *Proc. Nat. Acad. Sci. USA* 100, pp 12576-12578

Part 2

Photonic Crystals and Applications

4

Photonic Crystal Waveguides and Bio-Sensors

Alessandro Massaro

Italian Institute of Technology IIT
Center for Bio-Molecular Nanotecnology, Arnesano, Lecce
Italy

1. Introduction

Photonic crystals (PCs) are actually implemented as biosensors [Ganesh et al., 2007], optical resonators [Karnutsch et al., 2007] and wavelength filters [D'Orazio et al., 2008; Pierantoni et al., 2006]. Other kinds of photonic crystals can be implemented by considering a periodic structure with defect line and/or central cavities. Several architectures of micro-cavities (see examples in Fig. 1 (a), (b)and (c)) have been studied in the past by using triangular and square lattices layouts [Joannopoulos, 1995] oriented on optoelectronic technology. Optoelectronic technologies are often affected by cost and space problems that prevent them from being used even more widely. The development and implementation of photonic integrated circuits (PICs) could provide a solution to these two major obstacles. Couplers such as tapered waveguides and photonic crystal (PhC) devices can be integrated in the same chip in order to reduce the space, especially concerning complex optical switch systems, and, to provide high transmitted power and high efficiency of the PICs. For example, the use of tapered waveguides is necessary in order couple the light into a W1 PhC waveguide (illustrated in Fig. 1 (a)). This kind of W1 PhC waveguide is object of much interest because of its potential for controlling and manipulating the propagation of light. In particular, sharp bends, junctions, couplers, cavities, add-drop filters, and multiplexers have been experimentally demonstrated or theoretically predicted, thus making these devices very attractive for highly integrated photonic circuits [Mekis et al., 2008; Pottier et al., 2003; Johnson et al., 2002; Sanchis et al., 2002; Chau et al., 2004; Chietera et al., 2004; Xing et al., 2005; Camargo et al., 2004; Talneau et al., 2004; Marki et al., 2005; Camargo et al., 2004; Sanchis et al.,2004; Khoo et al.,2006]. The in-plane coupling of W1 PhC is also an important issue for bio-sensors implemented by micro- and nanofabrication technologies. In fact, the development of micro- and nanofabrication technologies, biomolecular patterning and micro-electromechanical systems (MEMS), has greatly contributed to the realization of miniaturized laboratories applied to genomic and proteomic analysis. The application fields of these biochips are extremely broad, and they have been referred as several different terms (gene-chip, gene-array, DNA microarray, protein chip, and lab-on-chip). Essentially, these chips, developed both in simple stand-alone configurations and integrated devices/architectures, consist of planar structures, realized on several substrates such as glass or plastic materials, where (bio)molecules (such as DNA, proteins or cells, which selectively conjugate with target molecules) can be immobilized on them through chemical surface modification or in situ synthesis [Fan et al., 2006] as happens DNA sensors. These chips require the use of suitable micro-reactors and/or capillary systems, and the detection of complemental reaction between biomolecular is performed in a solution. Biochip technology has revolutionized the field of molecular biology, finding broad application regarding the study of gene and protein expressions in several fields such as experimental and clinical diagnostics, biomarker detection, and pharmacogenomics. Actually, several chip setups have been used, such as enzyme assays [Hadd et al., 1997], immunochemistry assays [Wang, et al. 2001], polymorphism detection in genetic variations [Dunn et al., 2000], nucleic acids sequencing [Scherer et al., 1999], chips for the realization of ligase reaction [Cheng et al., 1996], and DNA amplification on micro-volumetric scale [Kopp et al., 1998; Daniel et al., 1998]. In particular, due to the high specificity of the hybridization reaction among the oligonucleotides sequences (complementary base-pairing between adenine and thymine, and guanine and cytosine), chips based on biomolecular interactions among DNA filaments have been developed more rapidly than chips based on proteins. In the latter case, despite of keen interest among the scientific community, it slowed down due to the complex bio-recognition mechanism of proteinaceous molecular species [Bodovitz, 2005].

Fig. 1. (a) W1 PhC waveguide. (b) Triangular lattice layout. (c) Circular photonic crystals.

Concerning the discussed topics, we propose to provide examples and design criteria useful to address the reader on the implementation of PhC oriented on bio-applications. The main goal of the chapter is to present an overview about the basic principles of light coupling, light emission and detection approaches of photonic crystals behaving as bio-sensors. In particular, we list below the sections proposed in this chapter.

- The first section analyzes the in plane coupling of tapered waveguides with a PhC waveguide around the working wavelength of 1.31 μm. We first analyze and characterize the coupling between two tapered waveguides, and, then, we model the coupling between tapered waveguides and PhC with micro-cavity. The analysis and the experimental results show the peak frequency shift obtained by varying the taper length. A maximum efficiency of the coupling is reached by a compromise between electromagnetic field confinement and low reflectivity at the input of the coupled photonic crystal. A good agreement with experimental results validates the 2D and 3D numerical results. The proposed in-plane coupling system can be used for W1 bio PhC.
- The second section provides technical advantages of a photonic crystal optical read-out in bio-molecular detection systems (deoxyribonucleic acid (DNA) chips, protein chips, micro-array, and lab-on-chip systems) for genomics/proteomics applications. The proposed method is based on arrays of PhC resonators which contribute to improve a detection efficiency of bio-samples marked with luminescent substances. The detection efficiency is characterized in terms of sensitivity of the analysis, the signal/noise ratio, and speed of the optical read-out process.
- The third section introduces an accurate modeling regarding PhC diffraction efficiency in bio-detection systems. The approach optimizes the detection enhancement of a

luminescent emitting substance located on the photonic crystal which is characterized by a specific emission band. By starting with the analysis of the periodic passive structures it is possible to define design maps in which are reported the diffraction efficiency versus the incidence angles. The PhC is designed to provide a high diffraction efficiency in the emission band of the luminescent substance. In this way the emission of the luminescent substance is enhanced through the high intensity of the zeroth-order backward diffracted wave. These maps could be used to define an admissible error margin due to the uncertainty of fabrication process. The proposed technique can be utilized in different spectral ranges starting from ultra violet to infrared wavelengths, and can be applied to different PhC layouts.

- In the last section we introduce the design criteria for a bio-compatible based polymer PhC suitable as a bio-sensor.

2. In-plane coupling of photonic crystal waveguides

W1 PhC can be obtained by introducing a line defect within the periodic lattice, usually realized through a triangular lattice layout of air holes etched into the substrate. This configuration is compatible with standard planar-semiconductor processing technology. A way to couple efficiently in-plane the light is the use of tapered waveguides. The best geometrical configurations of the tapered profiles are found by performing a good electromagnetic field confinement and low losses. In order to define the frequency response and the electromagnetic coupling of the tapered waveguides, we consider two numerical approach: the finite difference time domain (FDTD) method, and finite element method (FEM). The first one defines accurately the scattered light and the field coupled inside the device, and the second one analyzes the peak frequency resonance and provides the frequency shift versus different taper lengths. This section is organized as follows:

i. we design and model, according with the technological aspects, the optimum tapered waveguide layout (see Fig. 2 (a) and (b)) which couples the electromagnetic field around a working wavelength of $\lambda_0=1.31$ μm;

ii. by considering the optimum geometrical configuration of the tapered couplers, we simulate the W1 PhC illustrated in Fig. 2 (c) and (d).

We design integrated tapered waveguides coupled in-plane with an external source ($\lambda_0=1.31$ μm) and able to focus the energy in a small waveguide region (ridge of the waveguide).

According with the technological limits (technology resolution) we fix the optical and the geometrical parameters indicated in Fig. 2 (a) and (b) as: D=2μm, d=1.2μm, s=0.3μm, n(GaAs)=3.408, n(AlGaAs)=3.042, w=0.5μm, L_s=16.77 μm. We analyze the bandpass behaviour around the working λ_0=1.31 μm, and evaluate the energy density at the output of the two coupled tapered waveguides. This procedure allows to calculate the best optimum length L of the tapered profile. The inset of Fig. 3 reports a schematic representation of the employed transmission experimental set-up. In this set-up a light probe beam (tungsten broad band lamp) is launched from a tapered fibre and directly injected into the ridge waveguide. The light exiting the waveguide is collected and collimated by a microscope objective with high numerical aperture. The real image of the output facet of the waveguide is then formed on the common focal plane of a telescopic system, where a horizontal slit is placed. This allows us to separate the light coming from the ridge waveguide from the radiation freely propagating in the air and through the substrate. The transmitted light is collected by a multimode fibre with its free end lying on the focal plane of a lens (end-fire coupling) and brought to an N-cooled InGaAs OMA (optical multichannel analyzer).

As reported in Fig. 4 (2D FDTD simulation) and in Fig. 5 (3D FEM simulation) we have predicted the measured shift of the central wavelength. In our analysis, we have considered four lengths L of the tapered profile (in particular L=25 μm, L=50 μm, L=100 μm, L=200 μm): the central wavelength decreases for the cases L=25 μm to L=50 μm and increases from L=50 μm to L=200 μm. As reported in the 3D FEM results of Fig. 4 and in the 2D FDTD results of Fig. 5, the case of L=50 μm is characterized by a central wavelength far from the working wavelength. Moreover, we observe in the same figures other peaks due to backscattering interference phenomena of the slanted profile. In Fig. 6 we show the comparison between measured, 2D FDTD and 3D FEM spectra by considering L=100 μm: a low error about the central wavelength is observed (the different band amplitudes are due to different kind of light sources). By evaluating the coupled energy at the output of both tapered waveguides, we observe that the cases of L=100 μm and L=200 μm represent the best coupling condition (see Fig. 7 where the energy is defined by the subtended area of the electric field density). in these cases, the losses due to the radiation in the external space are low and, consecutively, the light coupling inside the guiding region is strong (compromise between high transmittivity and low losses at the working frequency). The integral used to evaluate the electric field density of Fig. 7 is:

$$S(t) = \int_V \varepsilon(\mathbf{r})|\mathbf{E}(t)|^2 \, dV \tag{1}$$

Where E is the electric field, ε is the spatial dependent permittivity index, and V is the volume of calculus.

We use for the calculus of (1) as source a carrier modulated by an exponential signal expressed by

$$\Psi_{source} = \exp(-(t \cdot dt / T_0)^2) \cdot \cos(\omega_0 \cdot t \cdot dt) \tag{2}$$

where ω_0 is the angular frequency at λ_0=1.31 μm, T_0 is a constant, and dt is the time step.

We note from numerical results that the 3D FEM results provide better the accuracy of the frequency shift according with the experimental spectra.

We conclude that good choices of tapered waveguides working at λ_0 =1.31 μm are the profiles with L=100 μm and L=200 μm. The field losses can be estimated by introducing a PhC waveguide between the two tapered waveguides as the W1 PhC illustrated in Fig. 8 (a), where the input is coupled through the tapered waveguide with the defect region, and, the signal is coupled with a cavity (as shown by the simulations of Fig. 8 (b) and Fig. 9 (a) illustrating different perspectives of the E_y field component). We analyze as filter a triangular lattice structure characterized by a lattice constant of 0.95 μm and air hole radius of 0.399 μm. In order to confine better the electromagnetic field inside the cavity, the radius of the holes near the cavity are reduced to 0.304 μm (optimization process). In Fig. 9 (b) are illustrated the radiation losses of the whole device using L=100 μm. Moreover the comparison between Fig. 9 (c) and Fig. 9 (d) shows the losses distribution for tapered waveguides with L=200 µm and helps to discriminate the part of energy irradiated by the photonic crystal from the part irradiated by the tapered profile.

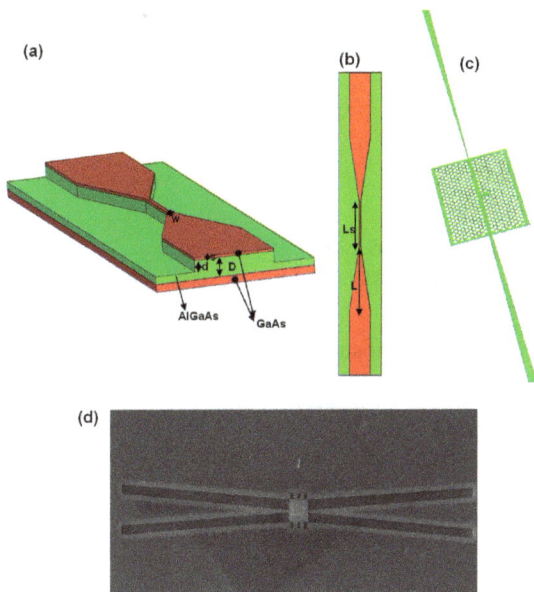

Fig. 2. a) 3D Coupled tapered waveguides; b) top view of coupled tapered waveguide; c) W1 PhC and tapered waveguides. (d) SEM image of the W1 PhC coupled to the input and output by tapered waveguides.

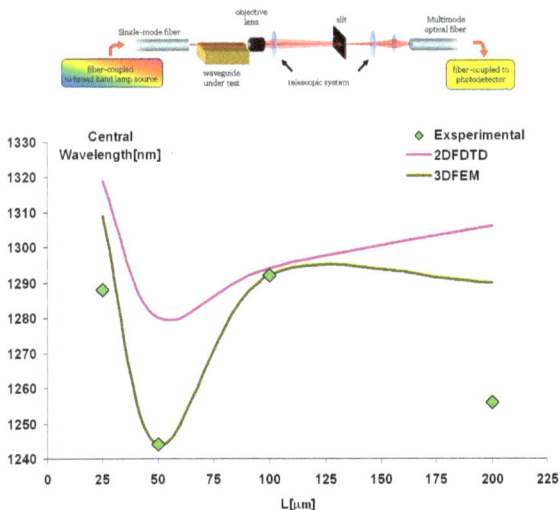

Fig. 3. Central wavelength: comparison between experimental, 2D FDTD and 3D FEM method. Inset: Experimental setup which measures the optical transmittivity of the coupled tapered waveguides of Fig. 1 (a).

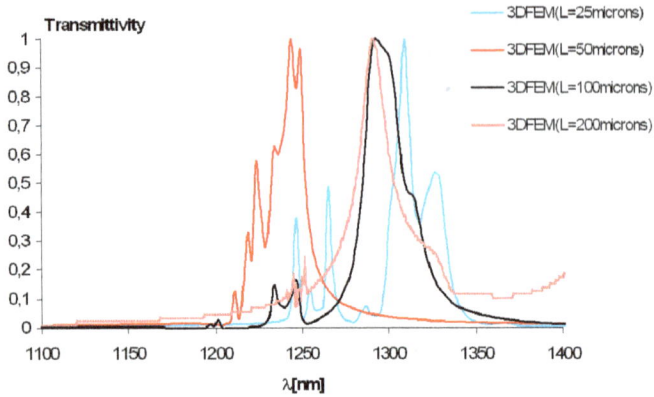

Fig. 4. Wavelength shift: 3D-FEM transmittivity for different lengths L.

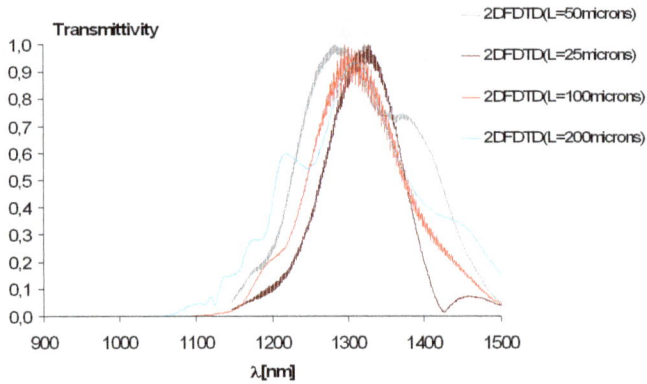

Fig. 5. 2D FDTD spectra comparison.

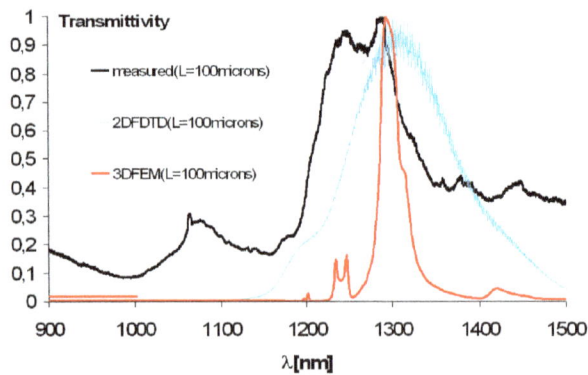

Fig. 6. Comparison between measured, 2D FDTD and 3D FEM spectra of tapered waveguides with L=100 μm.

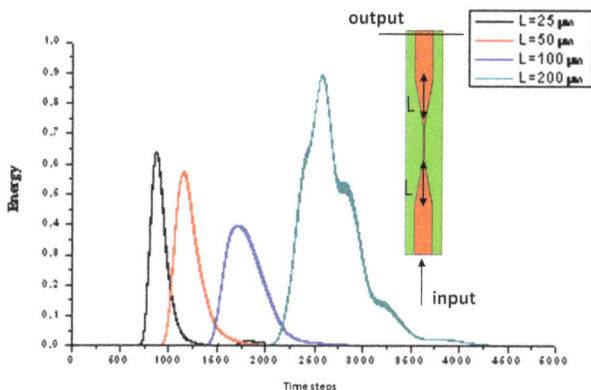

Fig. 7. Density of energy for different taper lengths L at λ_0=1.31µm.

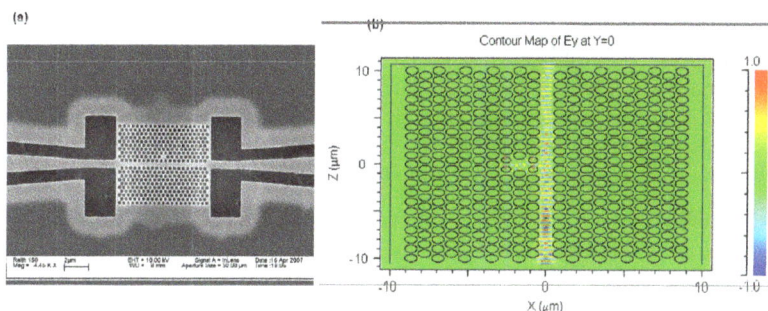

Fig. 8. (a) SEM image of the simulated add drop filter. (b) 2D FDTD simulation of the photonic crystal by using a continuous wave at λ_0=1.31 µm as source.

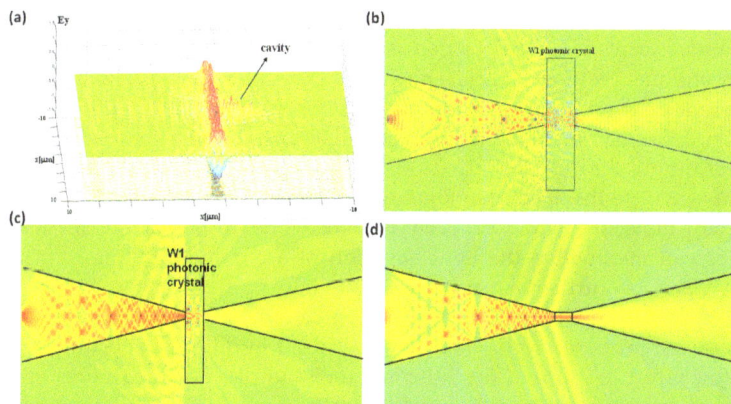

Fig. 9. (a) FDTD simulation: E_y component along the W1 region and into the cavity defect. (b) Top view of E_y distribution for tapered waveguides with L=100 µm. (c) Top view of E_y distribution for tapered waveguides with L=200 µm. (d) Top view of E_y distribution for L=200 µm and L_s= 16.77 µm.

3. Photonic crystal resonators array chip for improved optical sensing of (bio) molecules in genomics and proteomics

One of the main goals of this chapter is to analyze the detection efficiency of an optical signal coming from a bio-sensor chip based on luminescence emission. Our approach is based on the employ of arrays of photonic crystals which resonate with emitted lights from a luminous marker associated to a bio-molecule target at a predetermined wavelength. A proper design of these photonic crystal arrays and a spectral analysis of resonant peak emissions allow us to unambiguously associate each analyte to a peak emission, and pick out a useful signal from source reflection/diffraction noise. The use of properly fabricated PhC patterns in an optical read-out region provides us a freedom to modify local spectral distribution of allowed optical modes [Scully et al., 1997]. This concept has been used to increase both excitation and emission efficiencies of optical markers attached on a PhC pattern in bio-molecule detection devices [Mathias et al., 2007]. In particular, in this section we propose to use PhC resonators for a selective enhancement of a luminescent marker-emission only at a specific resonant wavelength. By fabricating several photonic crystal resonators in one bio-chip, each one characterized by a different resonant wavelength and a specific bio-recognition sample attached on each resonator, it is possible to spectrally detect bio-molecule targets in certain positions on a chip. Through the analysis of the total emission spectra collected from the whole read-out area it is thus possible to detect the presence of a certain analyte in a bio-specimen. In comparison with traditional detection systems, based on the spatial scanning of an optical read-out area, this approach permits us to:

a. decrease analysis time. Since the detection of analytes is based on detection of resonant peak emissions, simultaneous measurement on the whole read-out area is possible. This feature allows us to detect analytes through a spectral scanning alone. As the bandwidth of a resonant peak becomes narrower, more peaks can be arranged in a spectrum of a luminous marker. Then, it becomes possible to analyze a larger number of analytes in one time;

b. drastically decrease the reading error caused by diffused, reflected and/or diffracted light source signal;

c. significantly increase the detection sensitivity through the enhancement of the luminescent marker-signal. This enhancement is generated by the coupling of the luminescent light with a resonant mode of PhC patterns. In the proposed system, the unambiguous assignment of a resonant emission to each analyte does not require spatial scanning and/or the use of different fluorescence markers emitting light at different wavelengths. The unambiguous detection is guaranteed by combining each target (bio) molecule with a properly designed PhC resonator. In this way each PhC resonator can manipulate the light emission of a common luminescent marker. This approach can be further extended by utilizing, in the same chip, more than one fluorescent substances simultaneously, with different resonant emission frequencies. Moreover, the proposed technique can be utilized in different spectral ranges, starting from ultra violet to infrared (by utilizing, for example, organic fluorescent substances and/or colloidal nanocrystals) according to the scalability of target wavelengths and scale order of corresponding photonic crystal design [Yablonovitch, 1987; Joannopoulos et al., 1995]. The optical properties of the structure can be modified with a good accuracy in each desired frequency range, by simply changing the geometrical layout of the PhC.

Our optical signal detecting device featuring high sensitivity and multiplexing detection is composed of an array of PhC resonators with specific probes (e.g. single-stranded DNA (ssDNA) sequences, antibodies, receptors, aptamers, etc…) for analytes (e.g. DNA, proteins, ligands, etc.) attached on them. Analytes are trapped with high spatial precision through chemical, physical, electrostatic techniques, and so on (Fig. 10). The target analytes can be directly (e.g. through syntesis) or indirectly marked by conjugation with one or more fluorophores. The basic schemes for a detection of biomolecules (proteins, ligands, etc.) and nucleic acids are shown in Figs. 11 (a) and (b), respectively. To eliminate background noise caused by scattering of excitation light, excitation light can be selectively provided to read-out regions via PhC waveguides. The proposed system is based on a unique optical detection scheme. The detection is performed through the collection of the emission spectrum coming from the whole bio-recognition area of the chip. As previously discussed, by applying a suitable matrix composed by PhC resonators possessing different resonant wavelengths, each bio-recognition element of the device is unambiguously associated to a different resonance peak. By detecting resonant peak emissions at certain wavelengths, it is therefore possible to detect the presence of specific target (bio) molecules contained in the analyzed sample. Thus, it is possible to collect signals from several resonators in a single analysis, increasing the signal collection speed. Moreover, the PhC strongly inhibits the excitation radiation that is diffused or reflected towards the detection direction. Elimination of radiation of excitation light together with the increase in the intensity of the emission signal increases significantly the overall signal-to-noise ratio. This feature helps to reduce reading errors, allowing operators skip the step of complex post-processing to correct read-out errors. Photonic crystals can control the light propagation by introducing a 1D, 2D or 3D periodicity in materials having high optical transparency in the frequency range of interest. Light can be trapped, for instance, by introducing a defect in the periodicity. Summarizing, photonic crystals technology could, therefore, be applied to biochip technology in order to provide the following advantages:

a. controllability of the resonant wavelength of each resonator in the matrix through the accurate material and geometry design. Specifically, it becomes possible to enhance the emission spectrum of the fluorophore conjugated to an analyte. This allows us to perform an optical detection not only on the basis of a spatial discrimination of the different contributions but also on the basis of a spectral discrimination, since each pixel contains a specific optical resonator working at a different frequency.

b. increase of the fluorophore emission efficiency in specific spectral bands through the Purcell effect (e.g. micro resonators with high quality factor (Q-factor) and small modal volume), thus increasing the signal-to-noise ratio.

c. possibility to selectively excite light-emitting marker via waveguides or a resonance of a certain optical mode of photonic crystal to suppress diffused, reflected or diffracted excitation light from the substrate. This can be achieved by controlling the angle of emission or excitation by properly engineering a photonic crystal. This property can be exploited to spatially separate the excitation radiation and the emission band.

d. Regarding point (a), an external excitation light sent towards the matrix at a proper angle will excite the marker bound to the captured analyte, which will emit its typical broad signal. Then, the broad emission is peculiarly amplified in specific spectral bands by the underlying pixel with a photonic crystal resonator (Fig. 12 (a)). The variation from pixel to pixel (that is, from analyte to analyte), of the frequencies contemporarily emitted from the matrix allows a high degree of parallelism on a high number of different analytes, thus also fastening the recognition of the examined samples.

Regarding points (b) and (c), the proper choice of both the materials and the photonic crystal geometry can lead to the realization of a photon energy bandgap. As already mentioned, it is possible to localize specific optical modes to confine them in a small defect region in the photonic crystal. Only the modes which resonate in this small region will be amplified. Photonic crystal cavities are, therefore, designed to separate the useful signal coming from the biological assay from the noise coming from excessive source light. Then, we can obtain very sharp optical signals which are easily recognized thanks to their amplified intensity together with the spectral scanning of the detection system. Figs. 12 (b) and (c) show examples of an intact emission spectrum from a read-out region and signals collected from several read-out regions assisted by the array of photonic crystal resonators, respectively. The presence of each peak in the ensemble spectrum of Fig. 12 (c) reveals the presence of the corresponding target analyte in the analyzed assay. A further spatial separation between the scattered excitation light and a signal can be achieved by providing the excitation light to a read-out region or extracting the excited light from a read- out area through suitable waveguides. In this sense, the PhC can be designed to selectively guide light between resonators and detectors [Aoki et al., 2009].

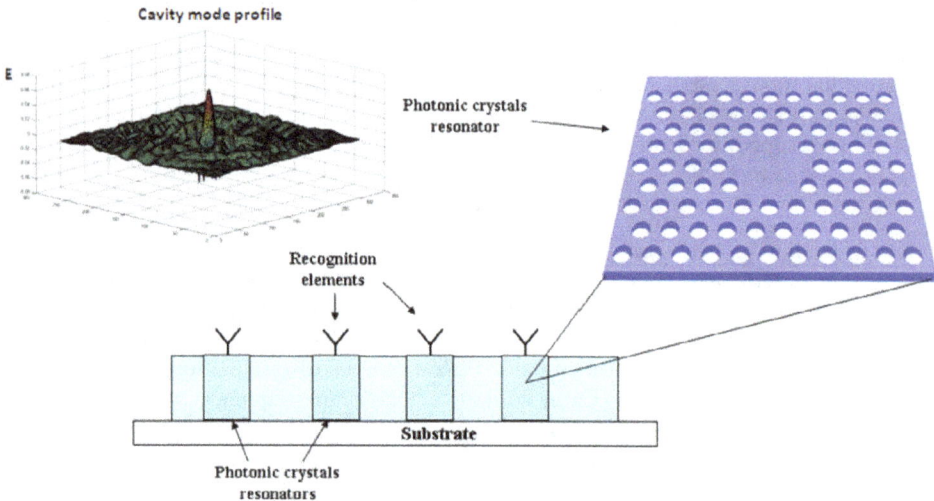

Fig. 10. Schematic of the optical transducer for biomolecular analysis in genomics/proteomics. The proposed device is essentially characterized by a substrate on which are realized arrays of resonant photonic crystals. On the surface of each resonator specific bio-molecules are fixed (for example, through chemical functionalization), which behave as target probes for the analytes to be detected. In the sketch there have been shown, as an example, arrays of photonic crystal and bio-molecules of the same typology. It is obviously possible to consider, in the same chip array, different photonic crystal structures with different wavelength resonances. Analogously, each bio-recognition element bound on the single resonator can be different from each other, and peculiar for each analyte.

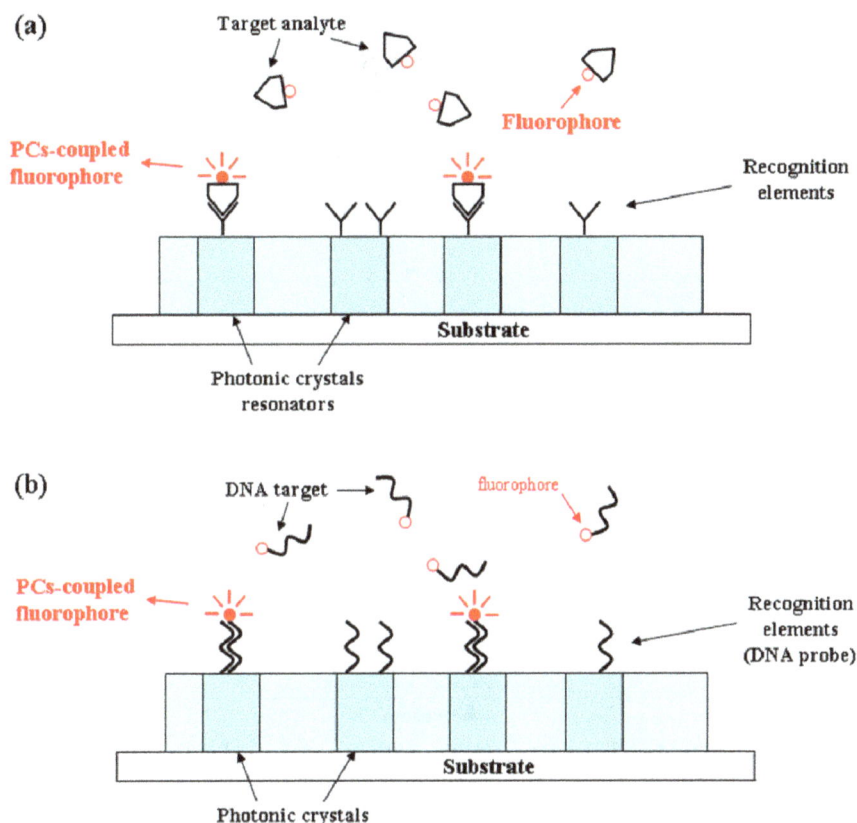

Fig. 11. Schematics of the optical transducer for the analysis of biomolecules in genomics/proteomics. (a) General sketch including a selective bio-recognition element for a specific target analyte (oligonucleotides, proteins, ligands, etc.). (b) Example of a devices suitable for DNA analysis: in this case, the probe bound to the resonator surface is a specific ssDNA sequence. Similarly to the above description, the resonators can differ from each other, as far as the (bio) recognition elements (for instance, the DNA probe) bound to each resonator, in order to perform spectral multiplexing analysis.

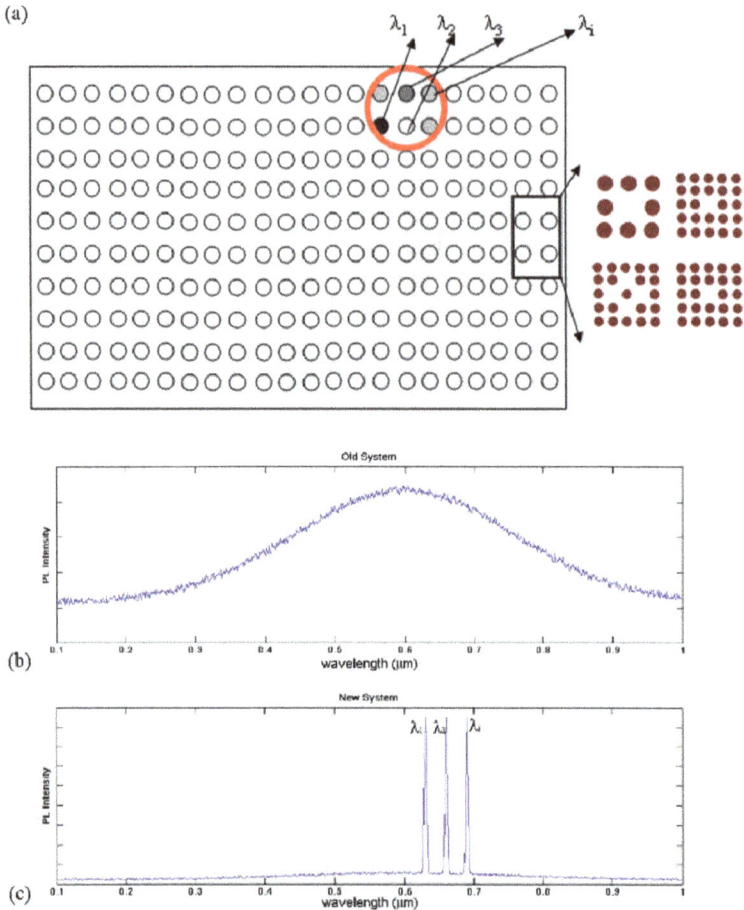

Fig. 12. (a) Schematics of a photonic crystal resonators matrix used in the optical DNA micro array chip. Each resonator of the matrix (bounded to a specific bio-recognition element) is designed in order to show a different resonant wavelength. (b) A typical example of the emission collected from a read-out area not assisted by photonic crystal resonators. The line-shape is typical of the original marker, with the presence of a significant noise due to scattered excitation light. (c) Example of the emission signal detected on the whole read-out area, where λ_1, λ_2 and λ_i are the spectral peculiar modifications due to the coupling of the fluorophores with the 1st, 2nd and ith resonator in the matrix. The presence of each peak in the ensemble spectrum reveals the presence of the corresponding target analyte in the analyzed assay.

4. Diffraction efficiency modeling of 2D photonic crystals for biosensing applications

Optical bio-sensing approach usually consists of light intensity detection systems. Typically, the luminescent signal is emitted by a luminescent marker conjugated to the bio-target. A good enhancement of this emitted signal can be obtained by combining a PhC structure with the luminescent substance. The signal enhancement can be optimized by analyzing the diffraction efficiency of the light emitted from the luminescent substance, characterized by the **K** wave-vector (see Fig. 13 (a) and (b)). In general, by exciting a properly designed PhC structure with an optimum incidence angle, it is possible to detect a high-intensity diffracted signal. The presented diffraction modeling takes into account this optimum incidence condition by defining sets of possible incidence angles (working regions) in the emission luminescent band. In this section the diffraction efficiency of a square lattice PhC Si_3N_4 membrane covered with a luminescent substance is modelled. The structure is excited by a plane wave and the light coming out from the luminescent substance is also modeled as a plane wave with the **K** vector shown in Fig. 13 (a), characterized by θ and ϕ angles. In Fig. 14 (a) and (b) we report the diffraction efficiency map versus the wavelength and the ϕ angle for different launch θ-angles. By fixing the θ and ϕ angles in the working region (see Fig. 14 where $\phi=\theta=30°$), it is possible to analyze the sensitivity of the reflectivity response. This allows to estimate the error margins due to the limits of the fabrication technology and of the experimental setups. As example, in Fig. 15 (a) is reported the sensitivity of the reflectivity response by varying hole radius in steps of 5 nm. Transverse electric (TE) and transverse magnetic (TM) PhC radiation modes define the diffraction concerning different PhC Brillouin directions. Figure 15 (b) shows the TM PhC radiation modes (diffraction modes) in the luminescent emission band $0.56\mu m \leq \lambda \leq 0.58\mu m$. The PhC Si_3N_4 membrane structure is excited by a plane wave and the light coming out from the luminescent substance, is collected in a detection system.

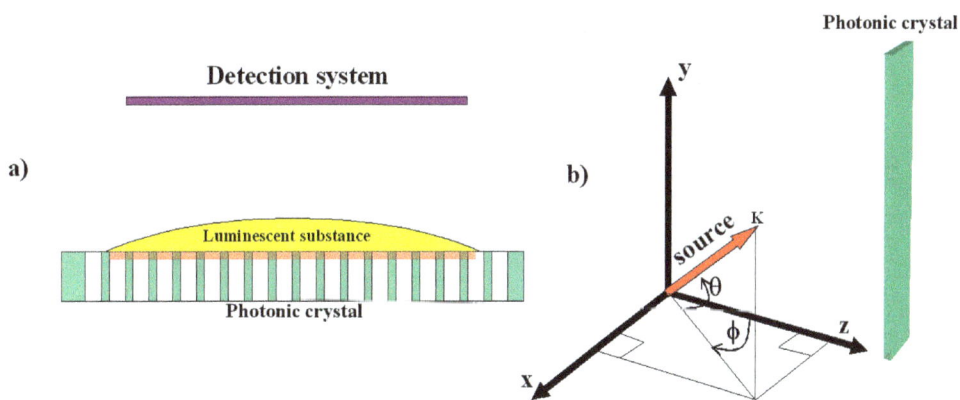

Fig. 13. (a) Photonic crystal for bio-sensing applications. (b) Source launch coordinate system of the model.

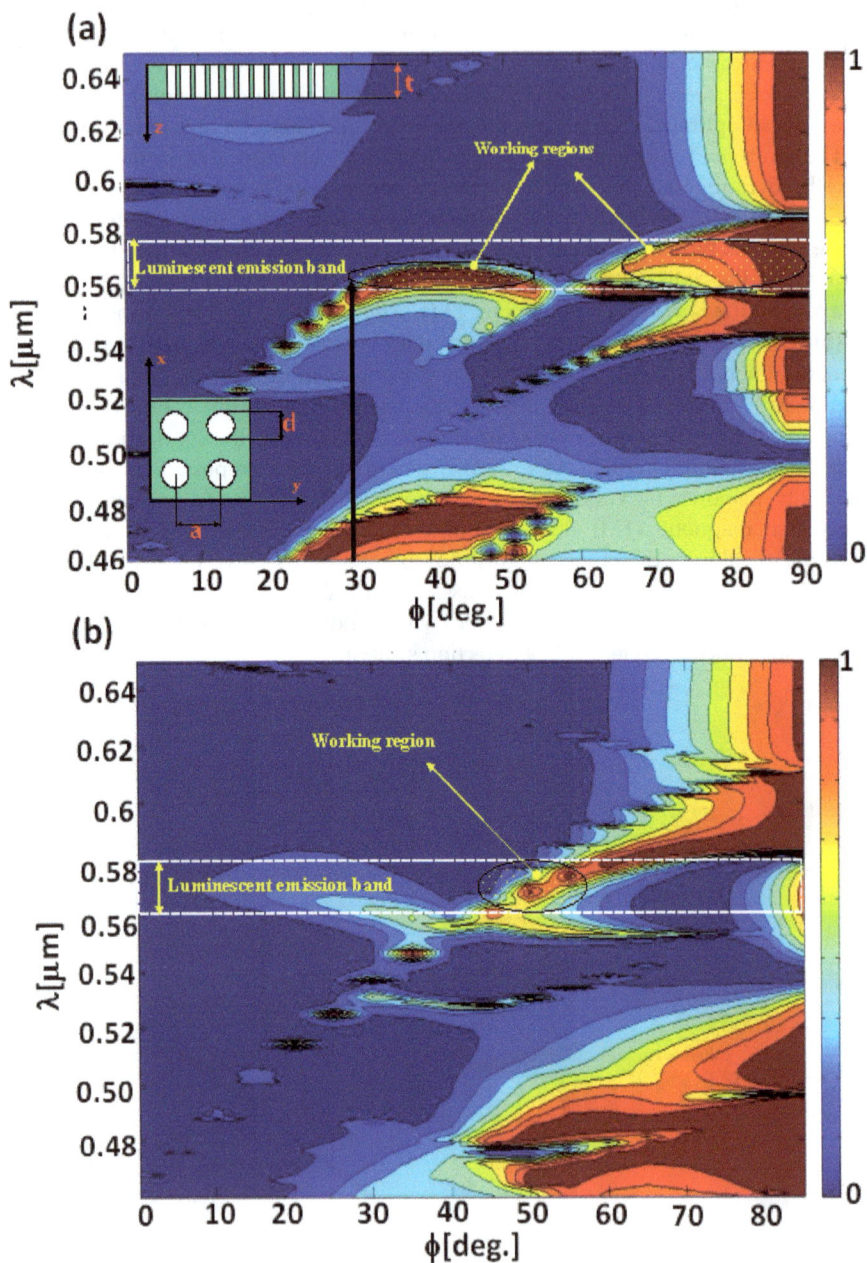

Fig. 14. Diffraction efficiency map versus λ and φ angle: d=180nm, a=300nm, t=300nm, θ=30 deg. Diffraction efficiency map versus λ and φ angle: d=180nm, a=300nm, t=300nm, θ=45 deg.

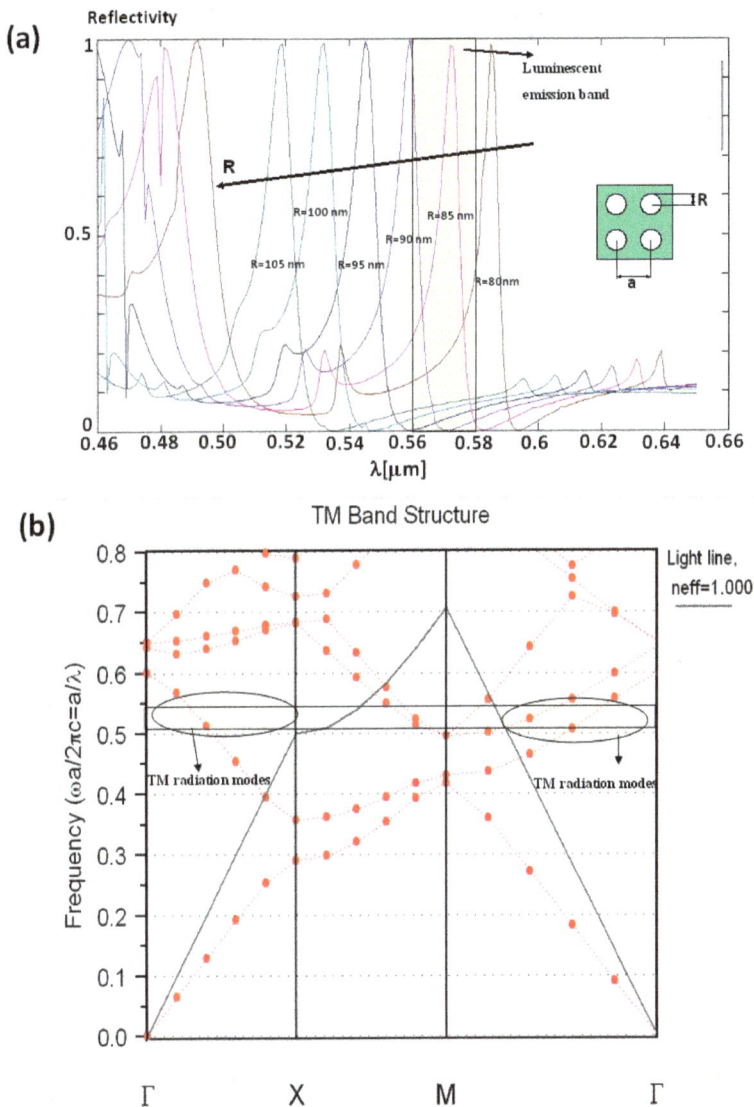

Fig. 15. Reflectivity of a 2D periodic structure versus the air hole radius R (a=300nm, t=300nm). Example of TM radiation modes for a square lattice PhC with a=300nm.

5. Design criteria of a biocompatible polymeric photonic crystal sensor and discussions

Concerning bio compatible PhC, we consider in the example of this section, gold pillars growth on Polydimethylsiloxane (PDMS) polymer. In order to improve resonant emitting peaks, we fix as example a square lattice layout with a central micro-cavity defect (see Fig.

16 (a)) [Massaro, 2011]. The input is a laser beam working at a wavelength λ. The source excites in the PhC waveguides TE and TM modes characterized by the electromagnetic field components reported in Fig. 16 (b). The PDMS material ($n_{background}$=1.4) represents the background where will be growth the gold pillars. The design criteria in order to improve efficient emitting cavities are listed by the following points:

1. we define the band gaps of the PhC without micro-cavity defect;
2. we calculate the band gaps of the same PhC with central defect obtained by omitting the central pillar;
3. we define the mode distribution of the cavity modes.

By focusing on TE modes we calculate the band gaps illustrated in Fig. 17 (a). Then we introduce the central defect as indicated in Fig. 16 (a) and calculate the new band gaps by observing that one of previous band will be divided into two band gaps. In the analyzed case, the band gap found around $a/\lambda = 0.3$ is divided into two ones as shown in Fig. 17 (b). The cavity modes are defined between the two new band gaps. The cavity modes are confined inside the cavity as proved by Fig. 18 which illustrates the modal profiles of the electric field component E_y. It is possible to select the best cavity modes (characterized by the best quality factor) by tuning the source wavelength around the selected mode. In order to improve strong power energy inside the cavity, we excite the PhC slab by a TE polarized source. The light emitting property along the direction orthogonal to the layout plane, could be improved by considering a 3D membrane type configuration or a central defect pillar characterized by a different size or material. The 2D approach allows to mainly fix the geometrical layout and to define the working wavelength. An accurate study of the 3D model will supply information about the best geometrical parameters such as the height of the pillars, the PDMS slab core thickness, and the dimensions of the membrane [Massaro et al., 2008]. Possible measurements of the designed PhC can be performed by Micro-photoluminescence setup [Massaro et al., 2008], by Fourier transform infrared (FTIR) and by UV visible analysis.

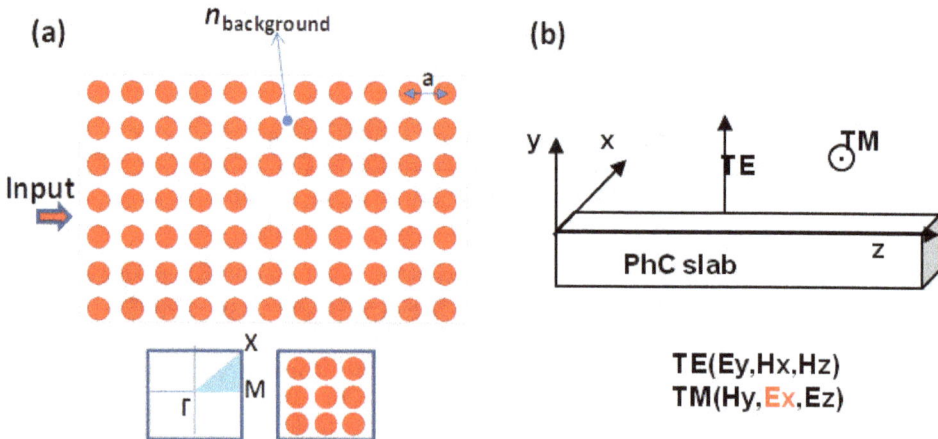

Fig. 16. (a) Example of square lattice 2D layout of a PhC with central micro-cavity and gold pillars. (b) TE and TM mode classification.

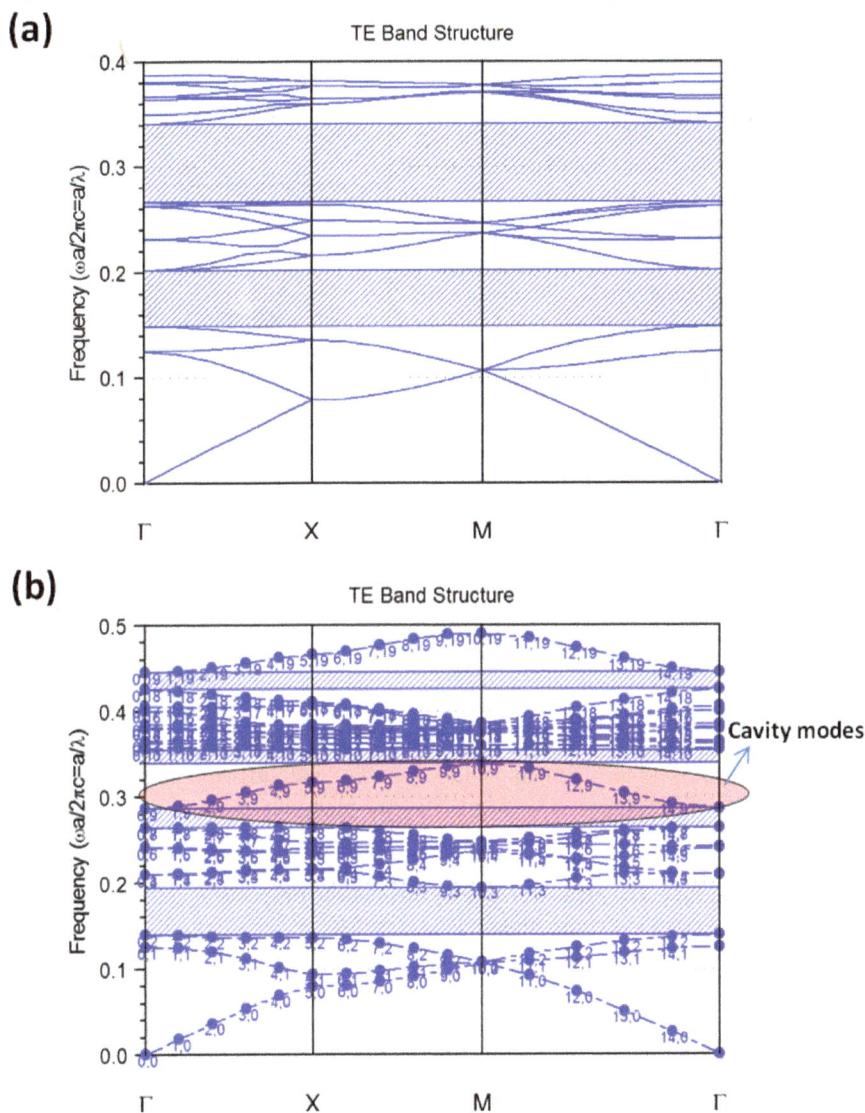

Fig. 17. (a) TE Band diagram of square lattice PhC without (a) and with (b) micro-cavity central defect, respectively.

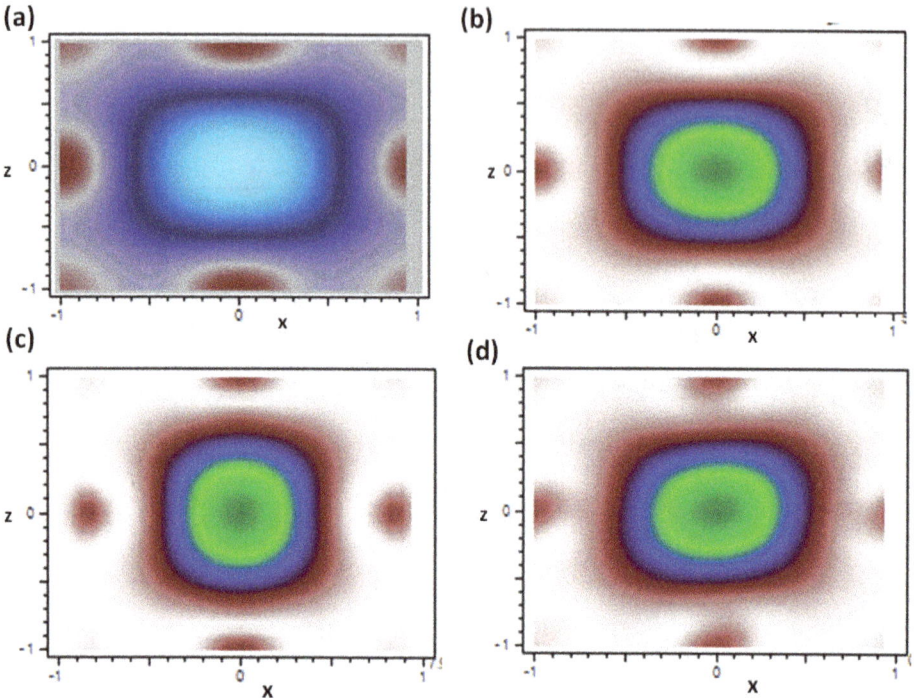

Fig. 18. Cavity mode profiles of modes indicated in Fig. 17: (a) mode 0.9; (b) mode 2.9; (c) mode 11.9; (d) mode 14. 9.

6. Conclusion

Examples of photonic crystal waveguides and bio-sensors are presented. The main goal of the proposed chapter is to provide information about methods and approaches of bio-sensing systems. In the first part of the chapter we focus on the in-plane coupling of a light source by analyzing different tapered waveguide profiles able to couple W1 photonic crystals. Then we analyze different bio-sensors such as DNA sensor chip and PhC Si_3N_4 membrane, by discussing the irradiation and diffraction properties. In the last part we provide design criteria of a bio-compatible polymeric PhC.

7. Acknowledgment

The author would like to appreciate the NNL, Istituto di Nanoscienze - CNR for the support received during my past research activity.

8. References

Aoki K., Vittorio M., Stomeo T., Pisanello F.,Massaro A., Martiradonna L., Sabella S., Rinaldi R., Arakawa Y., Cingolani R., Pompa, P., (2009), EP 09166989.5-1234.
Bodovitz, S., Joos, T., & Bachmann, *J. DDT*, Vol. 10, (2005), No. 4, pp. 283.

Camargo E. A. Chong, H. M. H. & De La Rue, R. Four-port coupled channel-guide device based on 2D photonic crystal structure. *Photon. Nanostruct.*, (2004), pp. 207.

Camargo, E. A.; Chong, H. M. H. & De La Rue, R. M. 2D Photonic crystal thermo-optic switch based on AlGaAs/GaAs epitaxial structure. *OSA Opt. Express*, Vol. 12, (2004),No. 4, pp. 588.

Chau, Y.-F.; Yang, T.-J.; & Gu, B.-Y. Significantly enhanced coupling Efficiency in 2D Photonic crystal waveguides by using cabin-side-like tapered structures at two terminals. *J. Appl. Phys.*, Vol. 43, (2004), pp. 1064.

Cheng, J., Shoffner, M.A., Mitchelson, K.R., Kricka, L.J., & Wilding, P., *J. Chrom. A* , Vol. 732, (1996), No. 1, pp. 151.

Chietera, G.; Bouk, A. H.; Poletti, F; Poli, F.; Selleri, S. & Cucinotta, A. Numerical design for efficiently coupling conventional and photonic-crystal waveguide. *Microwave Opt. Technol. Lett.*, Vol. 42, (2004), No. 3, pp. 196.

D'Orazio, A.; De Sario, M.; Marrocco, V.; Petruzzelli, V. & F. Prudenzano. Photonic crystal drop filter exploiting resonant cavity configuration. *IEEE Trans. Nanotechnol.*, Vol. 7, (2008), pp. 10.

Daniel, J.H et al., Sensors and Actuators A: Physical ,Vol. 71,(1998) ,No. 1-2, pp. 81-88.

Dunn, W.C. et al., *Anal. Biochem.* Vol. 277, (2000), No. 1, pp. 157.

Fan, J.-B., Chee, M.S., & Gunderson K.L., *Nature Review Genetics* Vol. 7, (2006), pp. 632.

Ganesh, N.; Zhang, W.; Mathias, P. C.; Chow, E.; Soares, J. A. N. T.; Malyarchuk, V.; Smith, A. D. & Cunningham B. T. Enhanced fluorescence emission from quantum dots on a photonic crystal surface. *Nature Nanotechnol.*, Vol. 2, (2007), pp. 515.

Hadd, A.G. et al., *Anal. Chem.* Vol. 69, (1997) ,No.17, pp. 3407.

Joannopoulos, J. D., Meade, R. D., and Winn J. N. Photonic Crystal –Modeling the Flow of Light. Princeton University Press, (1995).

Johnson, S. G. ; Bienstman, P.; Skorobogatiy, M. A.; Ibanescu, M.; Lidorikis, E.; & J. D. Joannopoulos. Adiabatic theorem and continuous coupled-mode theory for efficient taper transitions in photonic crystals. *Phys. Rev. E*, Vol. 66, (2002), pp. 1.

Karnutsch, C.; Stroisch, M.; Punke, M.; Lemmer, U.; Wang, J. & Weimann, T. Laser diode-pumped organic semiconductor lasers utilizing two-dimensional photonic crystal resonators," *IEEE Photon. Technol. Lett.*, Vol. 19, (2007), pp. 741.

Khoo, E. H. ; Liu, A. Q. & Wu J. H. Modified step-theory for investigating mode coupling mechanism in photonic crystal waveguide taper. *OSA Opt. Express*, Vol. 14,(2006), No. 13, pp. 6035.

Kopp, M.U., de Mello, A.J., & Manz, A., *Science* , Vol. 280, (1998), No. 5366, pp. 1046.

Marki, I.; Salt, M.; Herzig, H. P.; Stanley, R.; El Melhaoui, L.; Lyan, P. & Fedeli, J. M. Characterization of buried photonic crystal waveguide and microcavities fabricated by deep ultraviolet lithography. *J. Appl. Phys.*, Vol. 98, (2005), pp. 1.

Massaro A., "Theory, Modeling, Technology and applications of Micro/Nano quantum electronic and photonic devices," Transworld Research Network, IBN: 978-81-7895-498-1, 2011.

Massaro, A., Errico, V., Stomeo, T., Salhi, A., Cingolani, R., Passaseo, A. & De Vittorio M. 3D FEM Modeling and Fabrication of Circular Photonic Crystal Microcavity. *IEEE J. Light. Technol.*, Vol.26, (2008), No. 16, pp. 2960.

Mathias, P.C., Ganesh, N., Chen, L.L. & Cunningham, B.T., *Appl. Opt.* Vol. 46, (2007), pp. 2351.

Mekis, A. & Joannopoulos, J. D. Tapered couplers for efficient interfacing between dielectric and photonic crystal waveguide. *IEEE J. Light. Technol.*, Vol. 19, No.,(2001), pp. 861.

Pierantoni, L.; Massaro, A. & Rozzi, T. Efficient Modeling of a 3D-Photonic Crystal for Integrated Optical Devices. *IEEE Photonics Technology Letters*,Vol.18, No.2, (2006), pp.319.

Pottier, P.; Ntakis, I.; De La Rue, R. M. Photonic crystal continuous taper for low-loss direct coupling into 2D photonic crystal channel waveguides and further device functionality. *Opt. Comm.*, Vol. 223,(2003), pp. 339.

Sanchis, P.; Garcia, J.; Marti, J.; Bogaerts, W.; Dumon, P.; Taillaert, D.; Baets, R.; Wiaux, V.; Wouters, J. & Beckx, S.. Experimental demonstration of high coupling efficiency between wide ridge waveguides and single-mode photonic crystal waveguides. *IEEE Photon. Techn. Lett.*, Vol. 16,(2004), No. 10, pp. 2272.

Sanchis, P.; Martì, J.; Blasco, J.; Martinez, A.; & Garcia, A. Mode Matching technique for highly efficient coupling between dielectric waveguides and planar photonic crystal circuit. *OSA Opt. Express*, (2002), Vol. 10, No. 24, pp. 1391.

Scherer, J.R.et al. *Electrophoresis* Vol. 20, (1999), No. 7, pp. 1508.

Scully, M.O. & Zubairy, M.S. Quantum Optics. *Cambridge University press*, Cambridge, 1997.

Talneau, A.; Mulot, M.; Anand, S.; Olivier, S.; Agio, M.; Kafesaki, M. & C.M. Soukoulis. Modal behavior of single-line photonic crystal guiding structures on InP substrate. *Photon. Nanostruct.*, (2004), pp. 1.

Wang, J., Ibanez, A., Chatrathi, M.P., & Escarpa, A., *Anal. Chem.* Vol. 73, (2001), pp. 5323.

Xing, A.; Devanço, M.; Blumenthal, D. J. & Hu, E. L. Transmission measurements of tapered-single line defect photonic crystal waveguide. *IEEE Photon. Techn. Lett.*,Vol. 17, (2005),no. 10, pp. 2092.

Yablonovitch, E., *Phys. Rev. Lett.* Vol. 58, (1987), pp. 2059.

5

EIT-Based Photonic Crystals
and Photonic Logic Gate Design

Teh-Chau Liau[1], Jin-Jei Wu[2], Jian Qi Shen[3] and Tzong-Jer Yang[2]
[1]Ph. D. Program in Engineering Science, College of Engineering
Chung Hua University, Hsinchu, Taiwan
[2]Department of Electrical Engineering, Chung Hua University, Hsinchu, Taiwan
[3]Centre for Optical and Electromagnetic Research, State Key Laboratory of
Modern Optical Instrumentations,
Zijingang Campus, Zhejiang University, Hangzhou
China

1. Introduction

Over the past two decades, the effects of atomic phase coherence have exhibited a number of physically interesting phenomena such as electromagnetically induced transparency (EIT) (Harris, 1997) and the effects that are relevant to EIT, including light amplification without inversion (Cohen & Berman, 1997), spontaneous emission cancellation (Zhu & Scully, 1996), multi-photon population trapping (Champenois et al., 2006), coherent phase control (Zheltikov, 2006; Gandman et al., 2007) as well as photonic resonant left-handed media (Krowne & Shen, 2009). EIT is such a quantum optical phenomenon that if one resonant laser beam propagates in a medium (e.g., an atomic vapor or a semiconductor-quantum-dot material), the beam will get absorbed; but if two resonant laser beams instead propagate inside the same medium, neither would be absorbed. Thus the opaque medium becomes a transparent one. Such an interesting optical behavior would lead to many applications, e.g., designs of new photonic and quantum optical devices. Since it can exhibit many intriguing optical properties and effects, EIT has attracted extensive attentions of a large number of researchers in a variety of areas of optics, atomic physics and condensed state physics (Harris, 1997), and this enables physicists to achieve new novel theoretical and experimental results. For example, some unusual physical effects associated with EIT include the ultraslow light pulse propagation, the superluminal light propagation, and the light storage in atomic vapors (Schmidt & Imamoğlu, 1996; Wang et al., 2000; Arve, 2004; Shen et al., 2004), some of which are expected to be beneficial (and powerful) for developing new technologies in quantum optics and photonics.

In this chapter, we shall consider a new application of EIT, i.e., EIT-based artificial periodic dielectric: specifically, the EIT medium (an atomic vapor or a semiconductor-quantum-dot material) is embedded in a periodic host dielectric (e.g., GaAs). As is well known, the photonic crystals, which are periodic arrangements of dielectrics, have captured wide attention in physics, materials science and other relevant fields (e.g., information science) due to its capacity of controlling light propagations (Yablonovitch, 1987; Joannopoulos et al., 1995; Joannopoulos et al., 1997). Here, we shall propose some new effects relevant to light propagation manipulation via EIT responses in an artificial periodic dielectric. Such effects result from the combination of EIT and photonic crystals. In this new application of EIT for manipulating light wave propagations, the periodic dielectric can exhibit a tunable reflectance and transmittance (induced by an external control field) and can show extraordinary sensitivity to the frequency of the applied probe field. For example, a change of one part in 10^8 in the probe frequency ω_p would lead to a dramatic change in the reflectance and transmittance of the EIT-based periodic layered medium, and therefore, it can be used for designing sensitive optical switches, photonic logic gates as well as tunable photonic transistors. In the literature, although there have been some investigations that are relevant to the tunable photonic crystals based on EIT media (Forsberg & She, 2006; He et al., 2006; Zhuang et al., 2007; Petrosyan, 2007), yet less attention has been paid to the frequency-sensitive optical behavior that would be the most remarkable property of such a kind of periodic layered media.

We should point out that photonic logic gates designed based on new coherent materials, such as near-field optically coupled nanometric materials (Sangu et al., 2004; Kawazoe et al., 2003) and double-control multilevel atomic media (Shen, 2007; Shen & Zhang, 2007; Gharibi et al., 2009), have been suggested during the past few years. It should be emphasized that the mechanism presented in this chapter can be considered an alternative way to realize such a kind of photonic and quantum optical devices. Very recently, Abdumalikov et al. reported an experimental observation of EIT on a single artificial atom, and found that the propagating electromagnetic waves are allowed to be fully transmitted or backscattered (Abdumalikov et al., 2010). We will demonstrate in the present chapter that such a full controllability of optical property of artificial media could also be achieved in the EIT-based layered structure, of which the reflectance can be either zero or large depending sensitively on the intensity of the external control field applied in the EIT system. We believe that this would open a good perspective for its application in some new fields such as photonic microcircuits (or integrated optical circuits).

This chapter is organized as follows. In Sec. 2 we shall discuss the characteristic optical property of an EIT medium (e.g., an atomic vapor), and in Sec. 3 we review a formulation for treating the electromagnetic wave propagation in a periodic layered medium. The frequency-sensitive tunable band structure as well as the behavior of frequency-sensitive reflectance and transmittance of such an EIT-based periodic layered medium are presented in Sec. 4 and Sec. 5, respectively, where the spectrum of the reflectance as well as the transmittance of the EIT-based periodic structure (when the TE wave of the probe beam is normally incident on the layered medium) versus the normalized Rabi frequency Ω_c / Γ_3 of the control field and the normalized probe frequency detuning Δ_p / Γ_3 will be addressed. The frequency-sensitive tunable band structure of TM wave in the EIT-based periodic structure containing a left-handed medium is discussed in Sec. 6, where the reflection coefficient exceeding unity would occur in some frequency ranges. This will lead to a negative transmittance (so-called photonic analog of Klein tunneling in an LHM-EIT-based periodic layered medium). In Sec. 7 and Sec. 8, a potential application, i.e., photonic transistors and logic gates (tunable photonic logic gates) are suggested by taking full advantage of the effect of such an optical switching control. In Sec. 9 we close the chapter with some concluding remarks.

2. Optical properties of an EIT medium

Here we shall address the intriguing optical behavior of an EIT atomic vapor. Consider a Lambda-configuration three-level atomic system with two lower levels $|1\rangle, |2\rangle$ and one upper level $|3\rangle$ (see Fig.1 for its schematic diagram). This atomic system interacts with the electric fields of the two applied light waves (probe and control fields), which drive the $|1\rangle$ - $|3\rangle$ and $|2\rangle$ - $|3\rangle$ transitions, respectively. Note that the parity of level $|3\rangle$ needs to be opposite to levels $|1\rangle$ and $|2\rangle$, since the level pairs $|1\rangle$ - $|3\rangle$ and $|2\rangle$ - $|3\rangle$ can be coupled to the electric fields of the probe and control waves, respectively. Such a three-level system can be found in metallic alkali atoms (e.g., Na, K, and Rb). The off-diagonal density matrix elements ρ_{21} and ρ_{31} can form a closed set of equations under the condition of weak probe field (Scully & Zubairy, 1997), and the atomic system can be characterized by an SU(2) time-dependent model when the control field intensity varies adiabatically. The present atomic system interacting with two light fields (Ω_c and Ω_p) is governed by

$$\frac{\partial}{\partial t}\begin{pmatrix} \rho_{21} \\ \rho_{31} \end{pmatrix} = \begin{pmatrix} -\left[\dfrac{\gamma_2}{2} + i\left(\Delta_p - \Delta_c\right)\right] & \dfrac{i}{2}\Omega_c^* \\ \dfrac{i}{2}\Omega_c & -\left(\dfrac{\Gamma_3}{2} + i\Delta_p\right) \end{pmatrix}\begin{pmatrix} \rho_{21} \\ \rho_{31} \end{pmatrix} + \begin{pmatrix} 0 \\ \dfrac{i}{2}\Omega_p \end{pmatrix}. \tag{1}$$

It can be verified that the atomic microscopic electric polarizability of the $|1\rangle$ - $|3\rangle$ transition is of the form

$$\beta = \frac{i|\wp_{13}|^2}{\varepsilon_0\hbar} \frac{\dfrac{\gamma_2}{2} + i\left(\Delta_p - \Delta_c\right)}{\left(\dfrac{\Gamma_3}{2} + i\Delta_p\right)\left[\dfrac{\gamma_2}{2} + i\left(\Delta_p - \Delta_c\right)\right] + \dfrac{1}{4}\Omega_c^*\Omega_c}. \tag{2}$$

Here, Γ_3 and γ_2 stand for the spontaneous emission decay rate and the collisional dephasing rate, respectively. The Rabi frequency Ω_c of the control field is defined by $\Omega_c = \wp_{32}E_c/\hbar$ with E_c the slowly-varying amplitude (envelope) of the control field. The two frequency detunings are defined as $\Delta_p = \omega_{31} - \omega_p$, $\Delta_c = \omega_{32} - \omega_c$ with ω_p and ω_c the mode frequencies of the probe field and the control field, respectively. By using the Clausius-Mossotti relation (governing the local field effect due to the dipole-dipole interaction between neighboring atoms), the relative electric permittivity of the EIT vapor at probe frequency ($\omega_p = \omega_{31} - \Delta_p$) is given by

$$\varepsilon_r = 1 + \frac{N_a\beta}{1 - \dfrac{N_a\beta}{3}}, \tag{3}$$

where N_a denotes the atomic concentration (atomic number per unit volume) of the EIT atomic vapor.

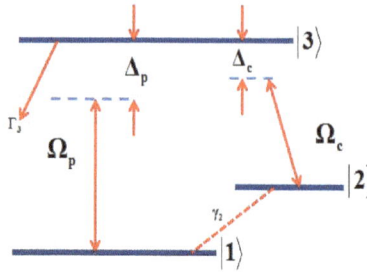

Fig. 1. The schematic diagram of a three-level EIT atomic system. The parity of upper level $|3\rangle$ is opposite to that of lower levels $|1\rangle$ and $|2\rangle$. The control and probe laser beams drive the $|2\rangle$-$|3\rangle$ and $|1\rangle$-$|3\rangle$ transitions, respectively. Once the control laser beam Ω_c is switched off, the vapor will be a resonantly absorptive medium for the probe light. However, the vapor would be transparent to the probe light because of the destructive quantum interference between the $|1\rangle$-$|3\rangle$ and $|2\rangle$-$|3\rangle$ transitions when the control laser beam is present.

The tunable dispersive behavior of the bulk EIT atomic vapor is shown in Figs. 2 and 3. The typical atomic and optical parameters chosen for Figs. 2 and 3 are as follows: the atomic number density $N_a = 5.0 \times 10^{20}$ m^{-3}, the electrical dipole moment $|\wp_{31}| = 1.0 \times 10^{-29}$ C$\cdot$m, the frequency detuning of the control field $\Delta_c = 1.0 \times 10^7$ s^{-1}, the spontaneous emission decay rate $\Gamma_3 = 2.0 \times 10^7$ s^{-1} and the dephasing rate $\gamma_2 = 1.0 \times 10^5$ s^{-1}. Fig. 3 shows the three-dimensional behavior of the real part (a) and the imaginary part (b) of the relative electric permittivity of the EIT atomic vapor (bulk). As the dispersive curve of the refractive index of the EIT bulk is a function of Δ_p and Ω_c, in the section that follows, we shall consider a band structure (versus both Δ_p and Ω_c) of the EIT-based periodic medium (see Fig. 4 for its schematic diagram).

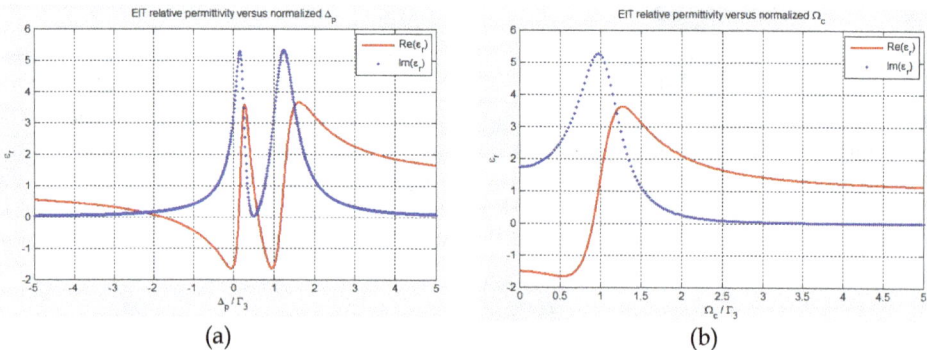

(a)   (b)

Fig. 2. The relative electric permittivity of the three-level EIT atomic vapor as a function of the probe frequency detuning Δ_p and the Rabi frequency Ω_c of the control field. In (a) the Rabi frequency of the control field is $\Omega_c = 2.0 \times 10^7$ s^{-1}. In (b) the probe frequency detuning is $\Delta_p = 3.4 \times 10^6$ s^{-1}.

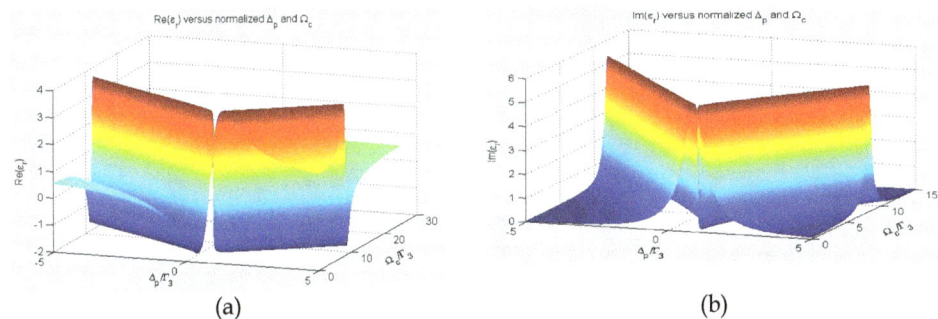

Fig. 3. The dispersion of the relative electric permittivity of the EIT atomic medium versus the frequency detuning Δ_p of the probe field and the Rabi frequency Ω_c of the control field.

The 1D periodic (D|E) cells shown in Fig. 4 are composed of two kinds of media: a dielectric (e.g., GaAs dielectric with the relative refractive index $n_1 = 3.54$) and a typical Lambda-configuration three-level EIT medium whose electric permittivity is determined by Eqs. (2) and (3). Here, the characters "D" and "E" in "(D|E)" denote the dielectric (GaAs) and the EIT, respectively. Assume the two materials are both homogeneous along y-direction (i.e. $\partial / \partial y = 0$) and the probe signal wave travels in the (...D|E|D|E...) structure always along x-direction. The reflection coefficient (Yeh, 2005) on the left side interface ($x = 0$) of such an EIT-based periodic medium, which is in fact a 1D N-layer (D|E) layered structure bounded by the GaAs dielectric material, will be addressed.

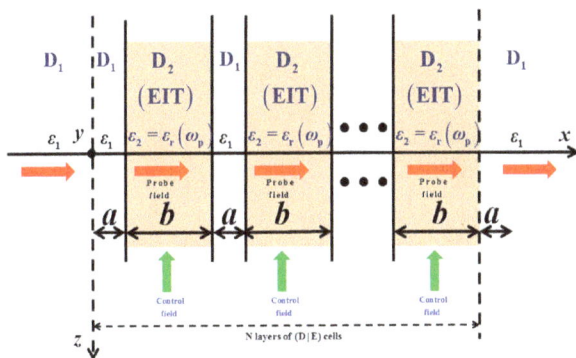

Fig. 4. The 1D N-layer structure of (D|E) cells embedded in GaAs homogeneous dielectric. The dielectrics D_1 and D_2 stand for the GaAs and EIT atomic media, respectively. A (D|E) cell consists of GaAs dielectric (D) and EIT medium (E). The lattice constants of the (D|E) cells are chosen as $a = b = 0.1\mu m$.

3. The electromagnetism of periodic layered medium

In order to make the chapter self-contained, we shall in this section review the formalism for treating the light wave propagation in a periodic layered medium (Readers are referred to e.g. Yeh's reference (Yeh, 2005) for a more complete and detailed formalism). According to the theory of electromagnetism in photonic crystals, the electric field in the mth unit cell can be expressed by (Yeh, 2005)

$$E(x) = \begin{cases} a_m e^{-jk_{1x}(x-m\Lambda)} + b_m e^{jk_{1x}(x-m\Lambda)}, & m\Lambda - a < x < m\Lambda \\ c_m e^{-jk_{2x}(x-m\Lambda+a)} + d_m e^{jk_{2x}(x-m\Lambda+a)}, & (m-1)\Lambda < x < m\Lambda - a \end{cases} \tag{4}$$

Here, the wave vectors $k_{1x} = n_1\omega/c$, $k_{2x} = n_2\omega/c$. By using the matrix formalism for treating the wave propagation in layered media, one can arrive at the equation

$$\begin{pmatrix} a_{m-1} \\ b_{m-1} \end{pmatrix} = \frac{1}{2} \begin{pmatrix} e^{jk_{2x}b}\left(1+\dfrac{k_{2x}}{k_{1x}}\right) & e^{-jk_{2x}b}\left(1-\dfrac{k_{2x}}{k_{1x}}\right) \\ e^{jk_{2x}b}\left(1-\dfrac{k_{2x}}{k_{1x}}\right) & e^{-jk_{2x}b}\left(1+\dfrac{k_{2x}}{k_{1x}}\right) \end{pmatrix} \begin{pmatrix} c_m \\ d_m \end{pmatrix} \tag{5}$$

of electric field amplitudes as well as the eigenvalue equation

$$\begin{pmatrix} A & B \\ C & D \end{pmatrix} \begin{pmatrix} a_m \\ b_m \end{pmatrix} = e^{iK\Lambda} \begin{pmatrix} a_m \\ b_m \end{pmatrix} \tag{6}$$

for the column vector characterizing the electromagnetic field strengths in the periodic layered structure. The matrix elements are given by (Yeh, 2005)

$$A = e^{jk_{1x}a}\left[\cos k_{2x}b + \frac{1}{2}j\left(\frac{k_{2x}}{k_{1x}} + \frac{k_{1x}}{k_{2x}}\right)\sin k_{2x}b\right],$$

$$B = e^{-jk_{1x}a}\left[\frac{1}{2}j\left(\frac{k_{2x}}{k_{1x}} - \frac{k_{1x}}{k_{2x}}\right)\sin k_{2x}b\right], \tag{7a}$$

$$C = e^{jk_{1x}a}\left[-\frac{1}{2}j\left(\frac{k_{2x}}{k_{1x}} - \frac{k_{1x}}{k_{2x}}\right)\sin k_{2x}b\right],$$

$$D = e^{-jk_{1x}a}\left[\cos k_{2x}b - \frac{1}{2}j\left(\frac{k_{2x}}{k_{1x}} + \frac{k_{1x}}{k_{2x}}\right)\sin k_{2x}b\right]. \tag{7b}$$

Note that the eigenvalue equation yields

$$\det\begin{pmatrix} A - e^{jK\Lambda} & B \\ C & D - e^{jK\Lambda} \end{pmatrix} = 0. \tag{8}$$

This can be rewritten as a well-known form

$$\cos K\Lambda = \cos k_{1x}a\cos k_{2x}b - \frac{1}{2}\left(\frac{n_2}{n_1} + \frac{n_1}{n_2}\right)\sin k_{1x}a\sin k_{2x}b. \tag{9}$$

From this relation, one can obtain the Bloch wave number K. Now we are in a position to derive the coefficient of reflection, which is defined as $r_N = b_0/a_0$. It follows that the relation of the column vectors between the left side interface (at $x = 0$) and in the Nth unit cell is given by (Yeh, 2005)

$$\begin{pmatrix} a_0 \\ b_0 \end{pmatrix} = \begin{pmatrix} A & B \\ C & D \end{pmatrix}^N \begin{pmatrix} a_N \\ b_N \end{pmatrix} \tag{10}$$

with

$$\begin{pmatrix} A & B \\ C & D \end{pmatrix}^N = \begin{pmatrix} AU_{N-1} - U_{N-2} & BU_{N-1} \\ CU_{N-1} & DU_{N-1} - U_{N-2} \end{pmatrix}. \tag{11}$$

Here, the explicit expression for U_N is $U_N = \dfrac{\sin\left[(N+1)K\Lambda\right]}{\sin K\Lambda}$. With the help of the relations

$$\begin{aligned} a_0 &= \left(AU_{N-1} - U_{N-2}\right)a_N + BU_{N-1}b_N, \\ b_0 &= CU_{N-1}a_N + \left(DU_{N-1} - U_{N-2}\right)b_N, \end{aligned} \tag{12}$$

the coefficient of reflection of an N-layer periodic medium is given by (Yeh, 2005)

$$r_N = \frac{CU_{N-1}}{AU_{N-1} - U_{N-2}}, \tag{13}$$

where $b_N = 0$ has been substituted (since the present periodic layered medium is composed of N unit cells and is bounded by the medium of the refractive index n_1, the reflected amplitude of the electric field in the last unit cell vanishes). It should be emphasized that the factor of phasor time dependence, $e^{-i\omega t}$, has been adopted for the time harmonic wave in deriving the atomic microscopic electric polarizability (2) of EIT. Such a convention is often used by physicists. In the convention of engineers, however, the time dependence is $e^{+j\omega t}$ (Yeh, 2005; Caloz & Itoh, 2006). As we shall employ the formalism in the reference of Yeh (Yeh, 2005) for treating the wave propagation in the periodic layered medium, we need to convert the convention of physicists to that of engineers. This can be easily accomplished by the imaginary variable substitution, i.e., $i \rightarrow -j$.

In the sections that follow, we shall concentrate our attention on the influence of the external control field on the probe wave propagation inside the EIT-based periodic layered medium. It should be noted that we only consider a passive multilayered structure in this chapter. Although there are control and probe laser beams exciting the two electric-dipole allowed transitions, it is a passive atomic system because of the large spontaneous emission decay from the excited states to the ground state. If, however, there is an extra strong pumping laser beams driving the atomic system (Wu, 2004), we should address its optical response relevant to gain factor. But here such a strong pumping interaction is not taken into account.

4. The frequency-sensitive tunable band structure

In order to show how sensitive (to the probe frequency) the band structure of the EIT photonic crystal is, let us first see the dispersive relation of the 1D infinite periodic (D|E) cells, in which the probe frequency ω_p is far from the resonant frequency of the atomic

$|1\rangle$ - $|3\rangle$ transition. We shall plot the band structure by using Eq. (9), which is an equation of dispersion of a 1D infinite periodic structure. Since the permittivity of EIT also depends upon the Rabi frequency of the control field, Ω_c is a tunable parameter involved in the equation of dispersion. Then we will also present the three-dimensional behavior of the Bloch wave number versus both Ω_c and Δ_p with the help of Eqs. (2), (3) and (9). Here, we choose the typical atomic transition frequency $\omega_{31} = 5.0 \times 10^{15}$ s^{-1}, and the thickness of the two layers $a = 0.1\mu$m (GaAs dielectric) and $b = 0.1\mu$m (EIT medium).

As the probe frequency detuning of TE waves in Fig. 5 is quite large ($\Delta_p \approx \pi c / \Lambda$ with $\Lambda = a + b$), the strong dispersion of EIT cannot be exhibited, and the present (D|E) layered structure behaves like a conventional 1D photonic crystal. However, when the probe frequency detuning Δ_p approaches zero (or negligibly small compared with $\pi c / \Lambda$ having the order of magnitude 10^{15} s^{-1}, e.g., Δ_p is tuned onto resonance, i.e., $\Delta_p \to \Delta_c$ that equals 1.0×10^7 s^{-1}), it would exhibit a band with a fine structure (and hence remarkable frequency-sensitive reflectance and transmittance). The band structure in the probe frequency detuning range $\Delta_p / \Gamma_3 \in [-2.5 \times 10^8, +2.5 \times 10^8]$ is plotted in Fig. 5 (a). The typical atomic and optical parameters such as the atomic number density N_a, the electrical dipole moment $|\wp_{31}|$, the control frequency detuning Δ_c, the spontaneous emission decay rate Γ_3 and the dephasing rate γ_2 are chosen exactly the same as in Figs. 2 and 3 (these typical parameters are also used throughout the chapter). The Rabi frequency of the control field is $\Omega_c = 2.0 \times 10^7$ s^{-1}. Since, seen from Fig. 5 (a), there are some fine structures of the band in the frequency range ($\Delta_p / \Gamma_3 \in [-2.5 \times 10^8, +2.5 \times 10^8]$) that need to be addressed, we present the intricate structures in Fig. 5(b)-(d) and demonstrate them in more details. In Fig. 5(b), for example, as the probe frequency detuning tends to the resonant frequency $\Delta_p \to \Delta_c$ (i.e., Δ_p / Γ_3 approaches almost zero compared with $\pi c / \Lambda$), both the real and imaginary parts of the Bloch wave number K would arise, because the strong dispersion of the EIT medium, of which the relative refractive index is a complex number, plays a key role for creating such a band structure.

In Fig. 5(b) we have shown the fine structure of the band induced by the EIT resonance. However, the detailed fine texture cannot be signified by the coarse curves in Fig. 5(b), since the band structure is plotted within a large range of probe frequency detuning. We shall in what follows treat further the fine structure of the band of EIT-based photonic crystal when the $|1\rangle$ - $|3\rangle$ transition of the EIT atomic levels is on resonance. It follows from Fig. 6(a) that after aligning dielectric GaAs side by side with EIT medium there are three extreme values of imaginary part K'' in the Bloch wave number (or $K'' \to k_0 n''_{r,(D|E)}$). Coincidently, there are also three extreme values of real part K' in the Bloch wave number (or $K' \to k_0 n'_{r,(D|E)}$). However, neither of them reaches the band edge $K = 0$ or $K = \pm 0.5$ (in the units of $2\pi / \Lambda$). Note that K' and K'' simultaneously exist, since the refractive index of the EIT medium has an imaginary part. It should be emphasized that the band structure (i.e., K' and K'' vary as the probe frequency detuning Δ_p changes slightly) is very sensitive to the probe frequency detuning. From Fig. 6 (a) one can see that the real part of the Bloch wave number changes drastically from 0 to 0.5 (in the units of $2\pi / \Lambda$) and the imaginary part changes from -0.5 to 0 (in the units of $2\pi / \Lambda$) within a very narrow probe frequency band (namely, a very small change, e.g., at the level of one part in 10^8 in the probe frequency, gives rise to a large variation in the Bloch wave number). In particular, the slope ($dK / d\Delta_p$) is almost divergent at the position $\Delta_p / \Gamma_3 = 0.5$. The reason for this is because $\Delta_p = 0.5\Gamma_3$ is exactly the two-photon resonant frequency $(\Delta_p = \Delta_c)$. As there is almost divergent dispersion close to $\Delta_p = 0.5\Gamma_3$, the effects of slow light and the negative group velocity in such an EIT-based periodic layered material deserve consideration. This would lead to promising applications in designing devices for slowing down light speed. Besides, the EIT-based band structure is tunable in response to the intensity (characterized by the Rabi frequency Ω_c) of the external control field, since the refractive index of the EIT medium can be controlled by the control field. In Fig. 6 (b) the real part of the Bloch wave number K decreases as Ω_c increases from 0 to $4\Gamma_3$, and then increases when $\Omega_c / \Gamma_3 > 4$; the absolute value of the imaginary part of the Bloch wave number increases first in the range $\Omega_c / \Gamma_3 \in [0,1]$ and then decreases when $\Omega_c / \Gamma_3 > 1$. This, therefore, means that one can use one optical field to controllably manipulate the wave propagation of the other optical field via such an effect of sensitive switching control exhibited in the EIT-based periodic layered structure.

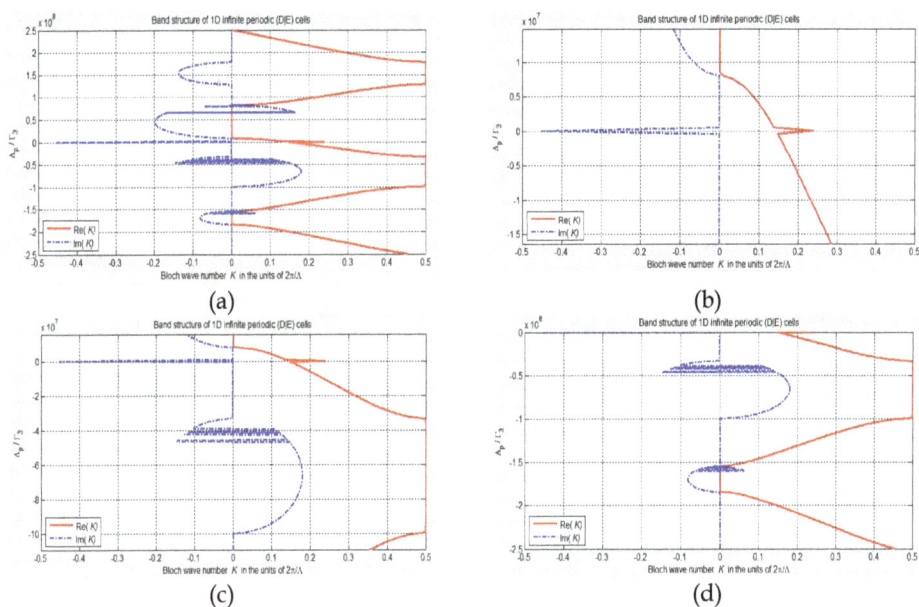

(a)   (b)   (c)   (d)

Fig. 5. The bandgap structure of the 1D infinite periodic (D | E) cells when the probe frequency of TE waves is far from the resonance. In (a) is the band structure in the probe frequency detuning range $\Delta_p / \Gamma_3 \in [-2.5 \times 10^8, +2.5 \times 10^8]$. In (b), (c) and (d) are the fine details exhibited in the EIT-based band structure in the probe frequency detuning ranges (in units of Γ_3), i.e., $\Delta_p / \Gamma_3 \in [-1.7 \times 10^7, +2.0 \times 10^7]$, $[-11.0 \times 10^7, +1.0 \times 10^7]$ and $[-2.5 \times 10^8, 0.0 \times 10^8]$, respectively.

As we have shown the characteristics of both sensitivity and tunability of the EIT-based band structure in Fig. 5(b) and Fig. 6, we shall present its three-dimensional behavior as both the probe frequency detuning and the Rabi frequency of control field vary. The sensitivity and the tunability versus the probe frequency detuning Δ_p and the Rabi frequency Ω_c of control field, respectively, are shown in Fig. 7.

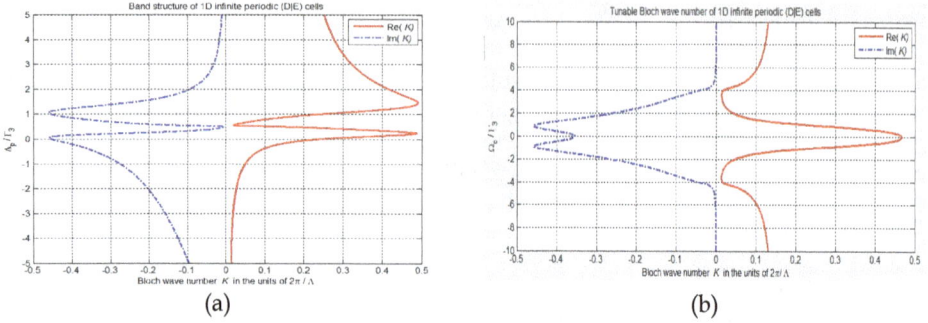

(a)   (b)

Fig. 6. The Bloch wave number K of the 1D infinite periodic (D|E) cells when the EIT atomic transition is on resonance. The curves in (a) indicate the real and imaginary parts of the normalized Bloch wave number K sensitive to the probe frequency detuning Δ_p, where the Rabi frequency of the control field is chosen as $\Omega_c = 2.0 \times 10^7$ s^{-1}. The curves in (b) show the tunable Bloch wave number K at the frequency detuning $\Delta_p = 2.0 \times 10^7$ s^{-1} when the Rabi frequency Ω_c of the control field changes.

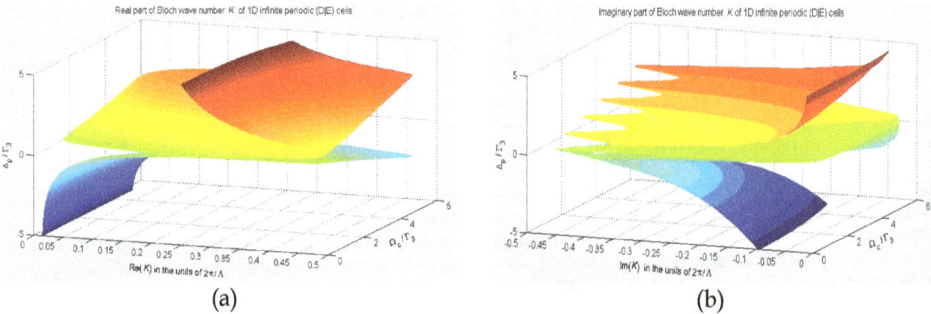

(a)   (b)

Fig. 7. The real part (a) and the imaginary part (b) of the normalized Bloch wave number K (in the units of $2\pi / \Lambda$) of the 1D infinite periodic (D|E) cells versus Δ_p and Ω_c. Both the real and imaginary parts of the Bloch wave number K are sensitive to the small change in the probe frequency (the slope $dK / d\Delta_p$ of the dispersive curve is much more larger than that in a conventional photonic crystal), and both the real and imaginary parts of the Bloch wave number at any fixed probe frequencies can be controllable by the control Rabi frequency Ω_c.

5. Probe-frequency-sensitive and field-intensity-sensitive coherent control effects in an EIT-based periodic layered medium

We shall now show that the reflection coefficient would be sensitive to the probe frequency when it is tuned onto two-photon resonance ($\Delta_p \to \Delta_c$). The typical atomic and optical parameters for the numerical results are chosen exactly the same as those used in the preceding sections. In Fig. 8, the real and imaginary parts of the reflection coefficient r corresponding to N-layer (D|E) cells are presented as an illustrative example, where the layer number $N = 1, 5, 20, 100$. It can be seen that the reflection coefficient changes drastically in the frequency detuning range of concern. We plot in Fig. 8 the dispersive behavior of r in the range of $\Delta_p / \Gamma_3 \in [0.3, 0.7]$, i.e., the probe frequency detuning changes at the level of one part in 10^8 in the probe frequency ω_p (the typical value of the probe frequency $\omega_p \simeq 10^{15} \text{ s}^{-1}$). It follows from Fig. 8 that the real and imaginary parts of r change from about 0.25 to 0.95 and from about -0.25 to 0.40, respectively. As is expected, such a dramatic change in the coefficient of reflection results from the two-photon resonance (because of the destructive quantum interference between the $|1\rangle - |3\rangle$ and $|2\rangle - |3\rangle$ transitions). In general, the more layers there are in the dielectric-EIT cell structure, the more drastic change there would be in the reflection coefficient on the left-side interface of this EIT-based periodic layered medium. Thus, the total number of valleys and peaks in the curve of the reflection coefficient r in a narrow band close to $\Delta_p = 0.5\Gamma_3$ becomes more and more as the total layer number N increases. However, such valleys and peaks in the reflection coefficient are no longer conspicuous for the cases of large N, since the amplitudes of fluctuation become smaller when the layer number N is adequately large. If, for example, the layer number $N = 100$, the small fluctuations tend to efface themselves (see Fig. 8).

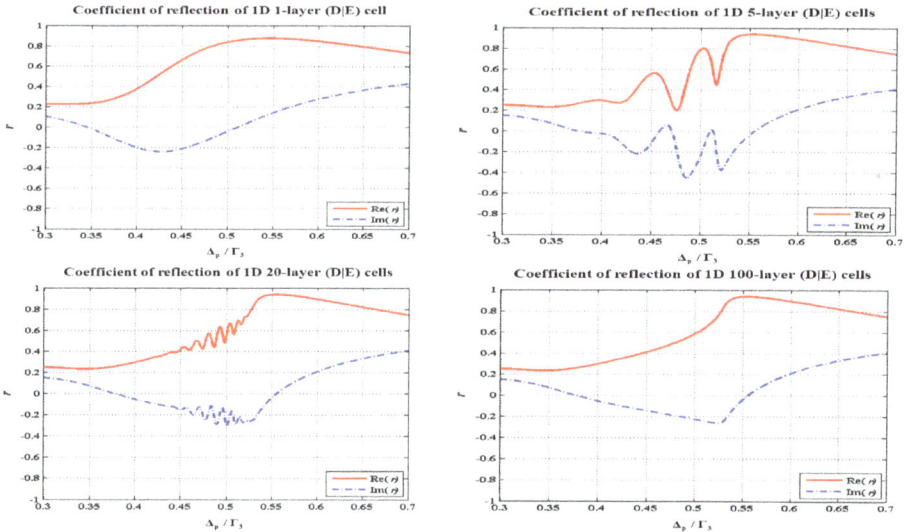

Fig. 8. The real and imaginary parts of the reflection coefficient r versus the normalized probe frequency detuning Δ_p / Γ_3 in the frequency range of two-photon resonance caused by the destructive quantum interference between the $|1\rangle - |3\rangle$ and $|2\rangle - |3\rangle$ transitions (close to $\Delta_p = 0.5\Gamma_3$). The layer number of the EIT-based periodic medium $N = 1, 5, 20, 100$. The Rabi frequency of the control field is chosen as $\Omega_c = 2.0 \times 10^7 \text{ s}^{-1}$.

We have demonstrated the probe *frequency-sensitive* behavior of the EIT-based periodic layered material. It can exhibit another effect (field-controlled *tunable* optical response), where the control field can be used to manipulate the photonic band structure, and therefore the reflection coefficient would vary as we tune the control Rabi frequency Ω_c. It follows from Fig. 9 that the tunable reflection coefficient of the EIT-based periodic layered medium is also sensitive to the Rabi frequency of the control field when the total layer number N increases. This means that the incident probe signal is either reflected or transmitted depending quite sensitively on the intensity of the external control field (characterized by $\Omega_c^* \Omega_c$), and therefore it could be used for designing some sensitive photonic devices (e.g., optical switches, photonic logic gates as well as tunable photonic transistors). In addition, a full controllability of reflection and transmission of the present EIT-based layered structure can also be demonstrated in Fig. 9. It can be readily seen that both the real and imaginary parts of the reflection coefficient r are less than 0.1, and hence the reflectance ($R = r^* r$) approaches zero (or almost zero) when the normalized control Rabi frequency Ω_c / Γ_3 is taken to be certain values, such as $\Omega_c / \Gamma_3 = 6.0$ (for $N = 5$), $\Omega_c / \Gamma_3 = 8.5$ (for $N = 20$) and $\Omega_c / \Gamma_3 = 10, 20$ (for $N = 100$). Thus, a field-intensity-sensitive switchable mirror can be fabricated with the EIT-based layered structure having a large total layer number N (e.g., $N > 100$). The three-dimensional behavior of the reflectance of the EIT-based periodic layered medium as both the Rabi frequency Ω_c and the probe frequency detuning Δ_p change is indicated in Fig. 10. Besides, we also consider the reflectance and transmittance of 1-, 5-, 20-, 100-layer periodic structures at other probe frequency detuning, e.g., $\Delta_p = -10^8 \text{ s}^{-1}$ in Fig. 11 as an illustrative example of tunable field-intensity-sensitive coherent control effect.

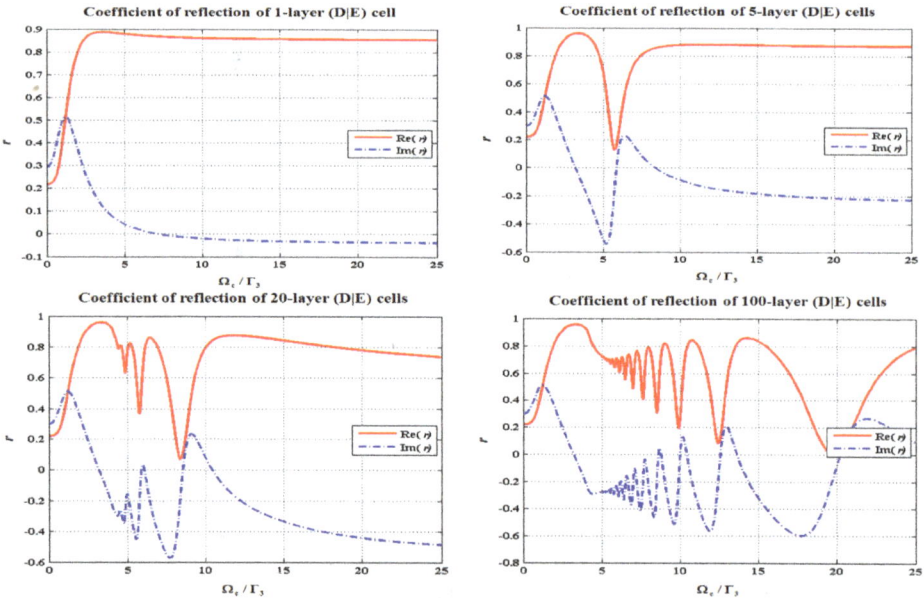

Fig. 9. The real and imaginary parts of the reflection coefficient r versus the normalized Rabi frequency Ω_c / Γ_3 of the control field. The probe frequency detuning is $\Delta_p = 2.0 \times 10^7 \text{ s}^{-1}$. All the atomic and optical parameters such as $\wp_{31}, \Gamma_3, \gamma_2, \Delta_c, N_a$ are chosen exactly the same as those in Fig. 8. In the case of $N = 100$, the reflection coefficient depends quite sensitively on the Rabi frequency of the control field.

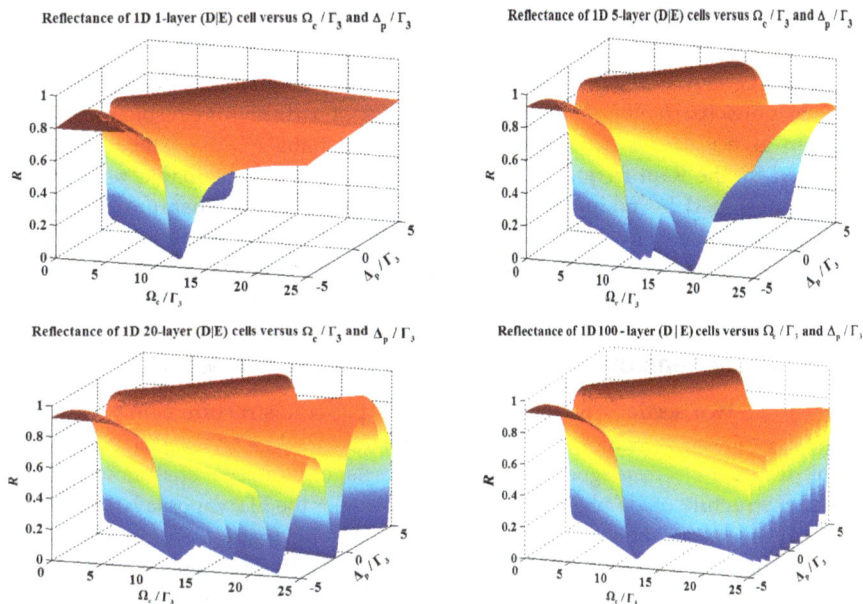

Fig. 10. The three-dimensional behavior of the reflectance of the EIT-based layered medium versus the normalized control Rabi frequency Ω_c / Γ_3 and the normalized probe frequency detuning Δ_p / Γ_3. All the atomic and optical parameters such as $\wp_{31}$, Γ_3, γ_2, Δ_c, and N_a are chosen exactly the same as those in Figs. 8 and 9.

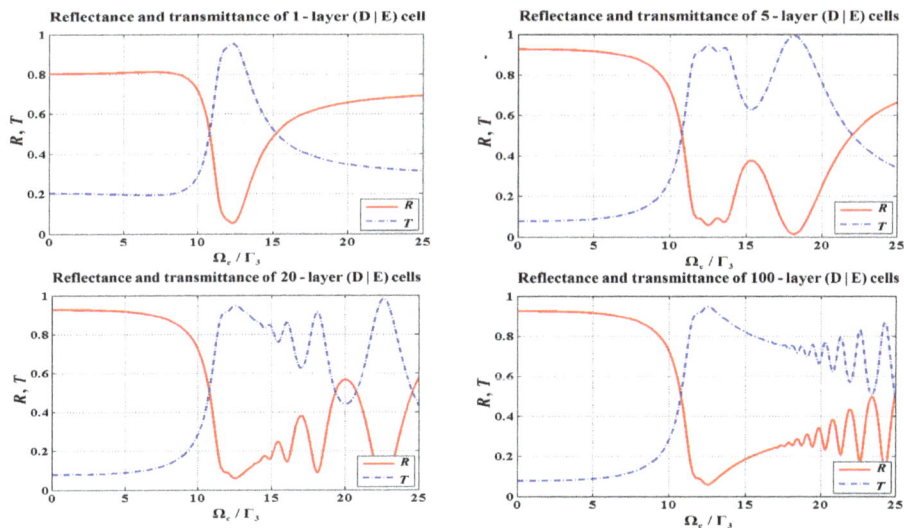

Fig. 11. The reflectance and transmittance versus the normalized Rabi frequency Ω_c / Γ_3 of the control field. The probe frequency detuning is chosen as $\Delta_p = -10^8 \text{ s}^{-1}$. All the atomic and optical parameters such as $\wp_{31}$, Γ_3, γ_2, Δ_c, N_a are chosen exactly the same as those in Fig. 8.

It should be noted that the probe frequency detuning Δ_p does not equal the frequency detuning Δ_c of the control field in Figs. 9-11, which are some typical cases for exhibiting general optical behavior of EIT-based photonic crystals. The quantum interference between atomic transitions (particularly when the condition of two-photon resonance, $\Delta_c = \Delta_p$, is fulfilled) can give rise to a strong dispersion that is tunable by the external control field (characterized by the Rabi frequency Ω_c). The structure of the EIT-based photonic crystal can thus be designed by taking advantage of such an effect of quantum coherence. We expect that the present probe-frequency-sensitive and field-intensity-sensitive coherent control effect with an EIT-based periodic layered structure can be used as a fundamental mechanism for designs and fabrications of new quantum optical and photonic devices.

6. The frequency-sensitive tunable band structure of TM wave

In the preceding two sections we have studied the periodic structure composed of an EIT medium and a normal dielectric (i.e., right-handed material). As a left-handed material (LHM) can exhibit unusual electromagnetic properties (Veselago, 1968), we shall now demonstrate how the layer structure of 1D photonic crystal consisting of the EIT vapor layers and the LHM host dielectric layers can show extraordinary sensitivity to the frequency of the probe field. As the band structure for TM wave seems to be more sensitive to the frequency than that for TE wave (Yeh, 2005), in this section we shall focus our attention on the optical response (e.g., frequency-sensitive band structure induced by two-photon resonance and higher-than-unity reflection coefficients due to the Klein tunneling) of TM wave.

As is well known, the Maxwell curl equations show that the phase velocity of light wave propagating inside a left-handed medium is pointed opposite to the direction of energy flow, that is, the Poynting vector and the wave vector of electromagnetic wave would be anti-parallel (i.e., its wave vector $\mathbf{k}$, electric field $\mathbf{E}$ and magnetic field $\mathbf{H}$ form a left-handed system). There have been some schemes to achieve the left-handed materials in the literature (Veselago, 1968; Shelby et al., 2001; Pendry et al., 1998; Pendry et al., 1996). Note that a right-handed system can be changed into a left-handed one via the operation of mirror reflection. It is thus clearly seen that the permittivity and the permeability of the free vacuum in a mirror world would be negative numbers. We have therefore pointed out that the electromagnetic wave (or a photon field) propagating inside a left-handed medium behaves like a wave of "antiparticle" of photon (Shen, 2003; Shen, 2008). However, as we know, there exist no such "antiphotons" in nature. The theoretical reason for this is that the four-dimensional electromagnetic vector potentials A_μ with $\mu = 0, 1, 2, 3$ are always taking the real numbers. But in a dispersive and absorptive medium, one can utilize an effective medium theory, where the vector potentials A_μ could probably take complex numbers. Such complex vector field theory has been considered previously (Lurié, 1968). The Lagrangian density of a complex electromagnetic field is given by $\ell = -F_{\mu\nu}^{*} F^{\mu\nu} / 2$. The complex four-dimensional vector potentials characterize the propagating behavior of both photons and "antiphotons", and hence both the electromagnetic wave characteristics in left- and right-handed media can be treated in a unified framework.

If the light quanta in a medium of negative refractive index can be considered to be the "antiparticles" of photons, it is of interest to propose an optical (or photonic) analog of the well-known Klein paradox, which appears in regimes of relativistic quantum mechanics and quantum field theory (Calogeracos & Dombey, 1999). In the Klein paradox, the relativistic wave equation can lead to so-called "negative probabilities" induced by certain energy potentials (e.g., the strong repulsive potential barrier with height exceeding the rest energy of particle) (Calogeracos & Dombey, 1999). Such a paradox can be interpreted based on the mechanism of particle-antiparticle pair production, which gives rise to higher-than-unity reflectance and negative transmittance. The Klein tunneling has been expected to be observed in QED regime, where an incoming electron wave function propagates and penetrates through a sufficiently high potential barrier. Though such a counterintuitive effect of relativistic quantum tunneling can be explained by using the notion of creation of electron-positron pairs, which is a physical process at the potential discontinuity, even today it is still referred to as "Klein paradox" in order to indicate its anomalous tunneling characteristics. Since the electron is massive, it is in fact quite difficult to realize the exotic Klein tunneling experimentally. Here, we shall suggest an alternative way to realize this intriguing effect, i.e., the photonic analog of Klein tunneling in an LHM-EIT-based periodic layered medium, where the reflection coefficient exceeding unity will also occurs in some frequency ranges, and this will lead to a negative transmittance.

The 1D periodic LHM-EIT cells are embedded in a left-handed homogeneous dielectric (an LHM-EIT cell consists of a left-handed dielectric and an EIT atomic medium). Fig. 12 indicates the band structure of the 1D infinite periodic LHM-EIT cells (sketched in Fig. 4) when the TM wave of the probe beam whose magnetic field vector is perpendicular to the x-z plane (Yeh, 2005) is incident normally or obliquely on such a periodic layered medium. Here we also choose the typical atomic ($|1\rangle$ - $|3\rangle$) transition frequency $\omega_{31} = 5.0 \times 10^{15}$ s^{-1}, and the thickness of the two layers $a = 0.1\mu$m (left-handed dielectric) and $b = 0.1\mu$m (EIT medium). The thickness of one LHM-EIT cell is $\Lambda = a + b$. We plot in Fig. 12 the dispersive behavior of six typical cases (i.e., the angles of incidence θ_i are $0^o, 15^o, 30^o, 45^o, 60^o$, and 75°, respectively). The tunable Rabi frequency Ω_c of the control field chosen for the present scheme is 2.0×10^7 s^{-1}.

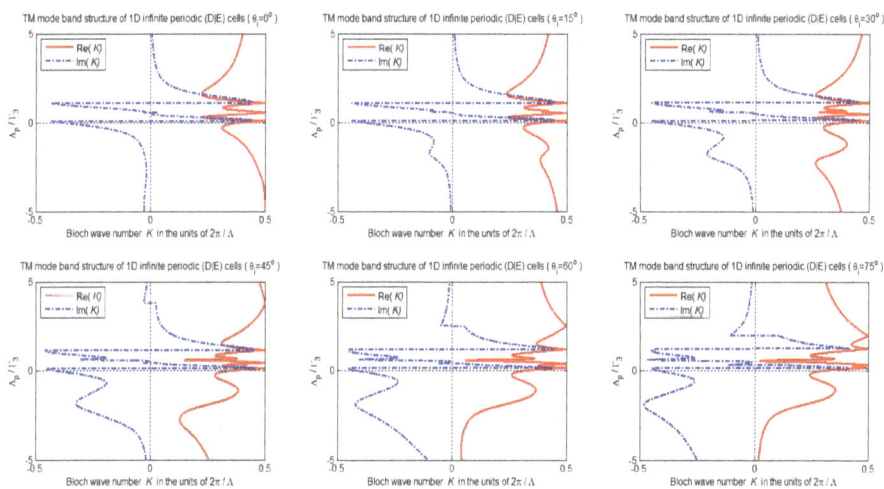

Fig. 12. The band structure of the 1D infinite periodic LHM-EIT cells when the angles of incidence of the TM wave of the probe beam are $\theta_i = 0^o, 15^o, 30^o, 45^o, 60^o, 75^o$, respectively.

As the refractive index of the EIT medium has an imaginary part in the frequency range of concern (see Fig. 12), the real part $\text{Re}(K)$ and the imaginary part $\text{Im}(K)$ of the Bloch wave number simultaneously exist. We emphasize that the band structure (i.e., K vary dramatically as the probe frequency detuning Δ_p changes slightly) is very sensitive to the probe frequency detuning. It follows that in a very narrow frequency band, where Δ_p is close to the resonance position, namely, the EIT two-photon resonance occurs ($\Delta_p \to \Delta_c$, i.e., $\Delta_p / \Gamma_3 \to 0.5$), both the real and imaginary parts of the Bloch wave number change drastically in a wide range, e.g., [0, 0.5] (in the units of $2\pi / \Lambda$) for $\text{Re}(K)$ and [-0.5, $+0.5$] for $\text{Im}(K)$. Since the EIT two-photon resonance arises at $\Delta_p = \Delta_c$ with the control frequency detuning $\Delta_c = 1.0 \times 10^7 \text{ s}^{-1}$, and the probe transition frequency $\omega_{31} = 5.0 \times 10^{15} \text{ s}^{-1}$, a very small change (e.g., at the level of one part in 10^8) in the probe frequency would result in a large variation in the Bloch wave number. For this reason, the slope ($dK / d\Delta_p$) of the Bloch dispersive curves is almost divergent at the position $\Delta_p / \Gamma_3 = 0.5$. Since there is strong dispersion in the curves of Bloch wave number K in the vicinity of $\Delta_p = 0.5\Gamma_3$, the effects of slow light as well as the negative group velocity would arise in the present periodic layered material.

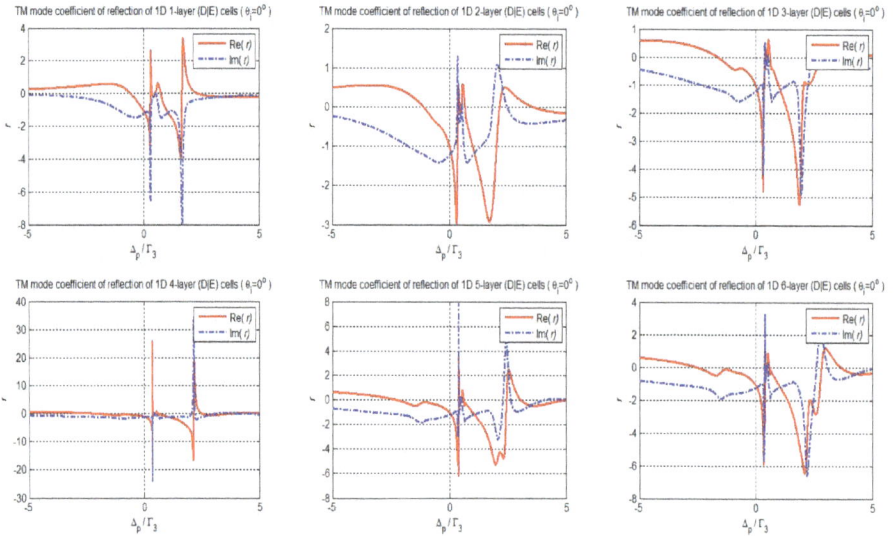

Fig. 13. The real and imaginary parts of the reflection coefficient r corresponding to the N - layer LHM-EIT cells ($N = 1 - 6$), where the relative refractive index of the left-handed medium is $n_1 = -1$. The photonic Klein tunneling occurs, i.e., in some frequency ranges the absolute values of both the real and imaginary parts of the reflection coefficient r are larger than unity.

We are now in a position to address the problem of reflection and transmission of the present photonic crystal. Obviously, the reflectance and transmittance would also be sensitive to the probe frequency when it is tuned onto resonance ($\Delta_p \to \Delta_c$). In Fig. 13 the real and imaginary parts of the reflection coefficient r corresponding to the N -layer LHM-EIT cells are presented as an illustrative example, where the layer number $N = 1 - 6$ and the Rabi frequency of the control field is chosen as $\Omega_c = 2.0 \times 10^7 \text{ s}^{-1}$. The TM wave is incident normally on the periodic layered structure. It can be seen that both the real and imaginary parts of the reflection coefficient r in all the cases (i.e., the layer number $N = 1-6$) change drastically as the probe frequency is close to the resonant frequency (at $\Delta_p = 0.5\Gamma_3$, where the probe frequency is tuned onto the two-photon resonance of the EIT atomic system).

As the layered medium contains the left-handed layers, and the periodic EIT layers act as a potential barrier for the incident electromagnetic wave, the absolute values of the real or imaginary part of the reflection coefficient r in some frequency ranges is larger than unity because of the Klein tunneling. In order to show the exotic and counterintuitive features exhibited in Fig. 13, we will present the behavior of reflection of TM wave by a RHM-EIT-based periodic layered structure, in which the left-handed layers have been replaced with the right-handed medium (or vacuum), for comparison. In the reflection coefficient of the N-layer RHM-EIT cells, in which the relative refractive index of the right-handed medium is $n_1 = +1$, is shown in Fig. 14. In this case, both the host dielectric layers and the EIT layers are right-handed media, so that there is no Klein tunneling, i.e., the absolute values of both the real and imaginary parts of the reflection coefficient r are less than unity. This, therefore, means that the left-handed dielectric is requisite in order to achieve the unusual photonic tunneling in the periodic layer structure containing EIT medium.

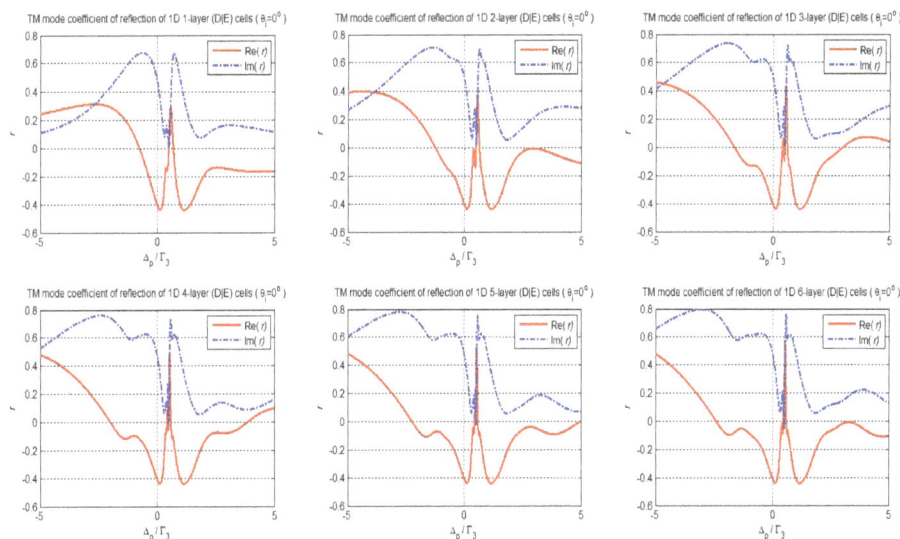

Fig. 14. The real and imaginary parts of the reflection coefficient r corresponding to the N-layer LHM-EIT cells ($N = 1-6$), where the relative refractive index of the right-handed medium is $n_1 = +1$. There is no photonic Klein tunneling, i.e., the absolute values of both the real and imaginary parts of the reflection coefficient r are less than unity.

7. A potential application: Photonic transistors and logic gates

We have shown that the LHM-EIT-based periodic medium can give rise to extraordinary reflection and transmission. Now we shall consider the physical meanings of such a photonic analog of Klein tunneling as well as its photonic application to device design.

The reflectance R and transmittance T on the left interface of the LHM-EIT structures are given in Figs. 15 and 16 as an illustrative example. It follows that the reflectance (and transmittance) is quite sensitive to the probe frequency detuning in some frequency ranges. It seems that the reflected wave intensity is larger than the incident intensity. This is because the additional particles are supplied by the potential barrier (Klein tunneling). Correspondingly, the transmitted wave is opposite in the intensity to the incident wave. In this process, the conservation of both the energy and the photon number is guaranteed. For example, the current density of the complex electromagnetic field can be defined as $J_v = i(A_\mu F_v^{\mu*} - A_\mu^* F_v^\mu)$ whose four-dimensional divergence is given by

$$\partial^v J_v = i\partial^v(A_\mu F_v^{\mu*} - A_\mu^* F_v^\mu) = i(\partial^v A_\mu)F_v^{\mu*} - i(\partial^v A_\mu^*)F_v^\mu. \tag{14}$$

The above equation can be rewritten as $i(F^v{}_\mu + \partial_\mu A^v)F_v^{\mu*} - i(F^{v*}{}_\mu + \partial_\mu A^{v*})F_v^\mu$ $= i(\partial_\mu A^v)F_v^{\mu*} - i(\partial_\mu A^{v*})F_v^\mu$, where the electromagnetic field equations $\partial_\mu F_v^{\mu*} = 0$, $\partial_\mu F_v^\mu = 0$ have been employed. With the help of the electromagnetic field equations, one can arrive at $\partial^v J_v = i\partial_\mu(A^v F_v^{\mu*} - A^{v*}F_v^\mu)$. Thus, the four-dimensional divergence of current density is

$$\partial^v J_v = i\partial_v(A^\mu F_\mu^{v*} - A^{\mu*}F_\mu^v). \tag{15}$$

It follows from Eqs. (14) and (15) that $\partial^v J_v = 0$ (this means that the current density of the complex electromagnetic field obeys the law of conservation). If the Hermitian field operator A_μ can be written as

$$A_\mu \sim \int d^3k\left[a_{(\lambda)}(\mathbf{k},t)e^{ik\cdot x} + b_{(\lambda)}^+(\mathbf{k},t)e^{-ik\cdot x}\right]e_\mu^{(\lambda)},$$
$$A_\mu^+ \sim \int d^3k\left[a_{(\lambda)}^+(\mathbf{k},t)e^{-ik\cdot x} + b_{(\lambda)}(\mathbf{k},t)e^{ik\cdot x}\right]e_\mu^{(\lambda)}, \tag{16}$$

where $e_\mu^{(\lambda)}$ denotes the polarization vector, the "Noether charge" corresponding to the current density J_v of the complex electromagnetic field is given by

$$\int d^3x J_0 \sim \int d^3k \sum_{\lambda=0}^{3}\left[a_{(\lambda)}^+(\mathbf{k},t)a_{(\lambda)}(\mathbf{k},t) - b_{(\lambda)}^+(\mathbf{k},t)b_{(\lambda)}(\mathbf{k},t)\right]. \tag{17}$$

Here, $a_{(\lambda)}(\mathbf{k},t)$, $a_{(\lambda)}^+(\mathbf{k},t)$, $b_{(\lambda)}(\mathbf{k},t)$, $b_{(\lambda)}^+(\mathbf{k},t)$ stand for the annihilation and creation operators of photons and its "antiparticles", respectively. The term $a_{(\lambda)}^+(\mathbf{k},t)a_{(\lambda)}(\mathbf{k},t)$ in Eq. (17) is the total number of photons, while $-b_{(\lambda)}^+(\mathbf{k},t)b_{(\lambda)}(\mathbf{k},t)$ is the total number of the "antiparticle" of photon. It can be seen that the total number of the "antiparticle" is negative.

The photonic analog of the Klein tunneling presented here can be used to design the so-called frequency-sensitive photonic transistors (see Fig. 17(a) for a schematic diagram) that can switch the photonic signals: specifically, the incident probe beam, the reflected probe beam, and the transmitted probe beam can mimic the operation of the three terminals (i.e., base, collector and emitter, respectively) of a bipolar transistor (a semiconductor device for amplifying and switching electronic signals). A small intensity of probe beam at the base terminal can manipulate (or switch) a much larger intensity between the terminals of collector (reflected probe beam) and the emitter (transmitted probe beam), since the incident probe beam can control the reflected wave in proportion to the input signal (incident probe beam).

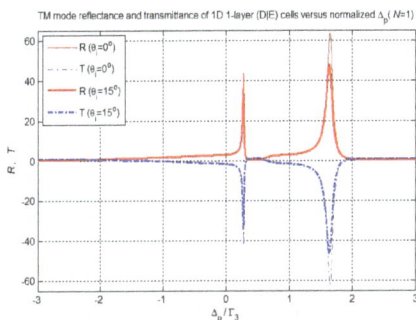

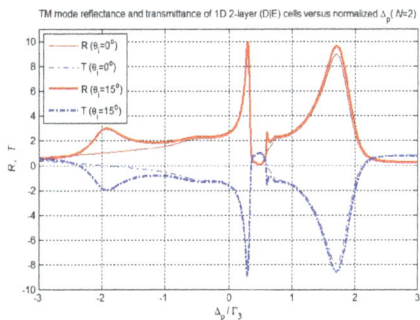

Fig. 15. The reflectance and transmittance of 1-layer and 2-layer LHM-EIT structures (the relative refractive index of the left-handed medium $n_1 = -1$) in the probe frequency range $\Delta_p / \Gamma_3 \in [-3,3]$.

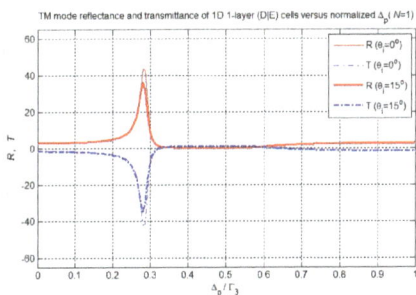

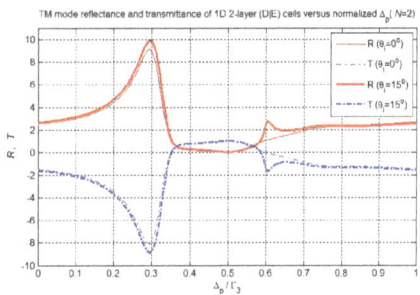

Fig. 16. The reflectance and transmittance of 1-layer and 2-layer LHM-EIT structures (the relative refractive index of the left-handed medium $n_1 = -1$) in the probe frequency range $\Delta_p / \Gamma_3 \in [0,1]$.

As we have pointed out in the preceding sections, the sensitive optical switching control can also be utilized to design photonic logic gates by means of such an EIT-based periodic layered structure. For example, the incident probe beam and the applied control field can act as the two input signals. We suppose that the output signal $Y = 1$ if the probe field can propagate through the periodic layered medium, and the output signal $Y = 0$ if the probe beams cannot be transmitted through the structure (i.e., the reflection and absorption dominates in the probe wave propagation). Then the logic operations of two-input AND gate can be implemented with such a layered structure. Alternatively, we can also apply at least two probe waves at different wavelengths, which correspond to different transmittances. In this new scenario, the two incident probe beams of different frequencies can stand for the two input signals (see Figs. 17 and 18). Therefore, the functional and logic gates, such as NAND, NOR, EXOR and EXNOR gates, can be designed by taking advantage of the effect of sensitive control for optical switching (due to two-photon resonance of EIT) exhibited in such an EIT-based periodic layered medium.

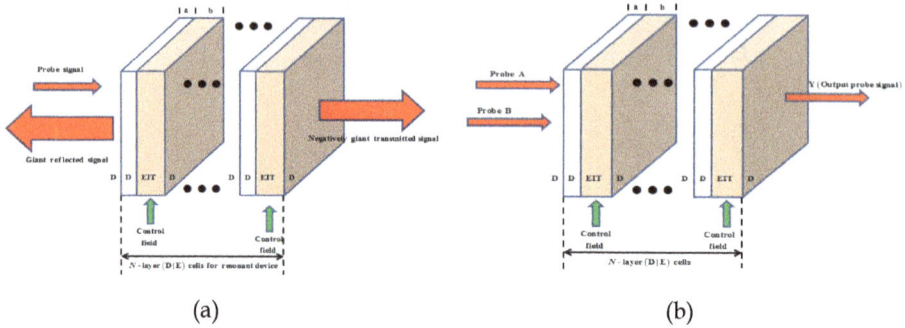

(a)   (b)

Fig. 17. (a) The schematic diagram of a photonic transistor designed based on the LHM-EIT layered structure. The probe field is incident on the structure, and the giant reflected wave with higher-than-unity reflectance and the transmitted wave with negative transmittance will be produced via the intriguing Klein tunneling effect. The incident probe wave, the reflected wave, and the transmitted wave correspond to the terminals of base, collector and emitter, respectively.

(b) The schematic diagram of a two-input photonic logic gate designed based on the EIT-based layered structure. The two incident probe beams at different frequencies represent the two input signals.

8. Design of two-input photonic logic gates

The dramatic reduction and enhancement in the reflectance and transmittance close to $\Delta_p = 0.5\Gamma_3$ is of special interest since the two-photon resonance ($\Delta_p = \Delta_c$) can give rise to the effect of sensitive optical switching control, which would lead to promising applications to new photonic device design. We shall suggest the working mechanism of two photonic logic gates (e.g., OR and NAND gates), which can be fabricated based on such an EIT-based periodic structure. The fine structure of the reflectance and transmittance for showing extraordinary sensitivity to the frequency of the probe field is demonstrated in Fig. 19(a). Here we plot only the reflectance and transmittance of two cases ($N = 4$ and $N = 6$) as an illustrative example. It can be seen that some oscillations in the curves are exhibited in the narrow resonant frequency range $\Delta_p \in [0.2\Gamma_3, 0.7\Gamma_3]$.

One can see from Fig. 18 (a) that there is a minimum (i.e., 0.19) and a maximum (i.e., 0.99) in the transmittance T at $\Delta_p = 0.53\Gamma_3$ and $0.46\Gamma_3$, respectively, for the 6-layer periodic structure. For the 4-layer periodic structure, however, the transmittance T has a maximum (i.e., 0.99) and a minimum (i.e., 0.34) close to $\Delta_p = 0.53\Gamma_3$ and $0.46\Gamma_3$, respectively. Two structures of layer number $N = 4$ and 6 are sketched in Fig. 18(b). Two probe beams with $\Delta_p = 0.46\Gamma_3$ and $0.53\Gamma_3$, which can act as the two input signals, are applied. We suppose that the output signal $Y = 0$ if neither of the two incident probe beams propagates through the periodic layered medium (i.e., the reflection dominates in the wave propagation of the probe field), and the output signal $Y = 1$ if at least one of the probe beams can be transmitted through the structure (i.e., the reflection and absorption can be ignored). Let the probe beams of $\Delta_p = 0.46\Gamma_3$ and $0.53\Gamma_3$ represent the input signals 0 and 1, respectively. Then the logic operations of two-input OR gate and NAND gate can be implemented with the 4 -layer and 6-layer structures, respectively. The truth table of the OR and NAND gates are given as follows:

IN_A	IN_B	$Y = A + B$ (4-layer OR gate)	$Y = \overline{A \cdot B}$ (6-layer NAND gate)
0 $(0.46\Gamma_3)$	0 $(0.46\Gamma_3)$	0	1
0 $(0.46\Gamma_3)$	1 $(0.53\Gamma_3)$	1	1
1 $(0.53\Gamma_3)$	0 $(0.46\Gamma_3)$	1	1
1 $(0.53\Gamma_3)$	1 $(0.53\Gamma_3)$	1	0

Table 1. The truth table of two-input OR gate (fabricated based on the 4-layer periodic structure) and two-input NAND gate (fabricated based on the 6-layer periodic structure).

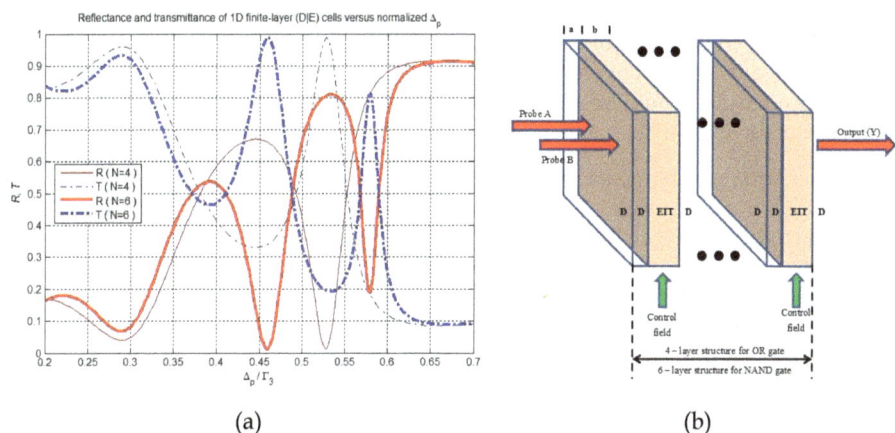

(a)   (b)

Fig. 18. The fine structure of the reflectance and transmittance of the 4-layer and 6-layer periodic (D | E) cells in a narrow probe frequency band (a), and the schematic diagram of photonic logic gates (b). The Rabi frequency of the control field is $\Omega_c = 4.0 \times 10^7 \text{ s}^{-1}$.

9. Conclusions

The quantum optical properties of an EIT medium has been discussed (in Section 2), and the formalism for treating wave propagation in a periodic structure has been reviewed (in Section 3). The band structure and the reflectance of a 1D photonic crystal consisting of both EIT medium layers and host dielectric layers can show extraordinary sensitivity to the frequency of a probe field because of a two-photon resonance relevant to destructive quantum interference between two transition pathways driven by the control and probe fields (in Sections 4 and 5). Such an EIT-based periodic layered material can also exhibit an effect of field-intensity-sensitive switching control (depending quite sensitively on the Rabi frequency of the control field) in the cases of large layer number N. Since the optical responses can be controlled by the tunable quantum interference induced by the external control field via two-photon resonance, the EIT-based layered medium under consideration shows more flexible optical responses than conventional photonic crystals because of the EIT two-photon resonance that gives rise to strong dispersion in the band of transparency window.

As the microscopic electric polarizability as well as the electric permittivity of the EIT medium are caused by the atomic energy level transition processes from the ground state to the excited states, in which the quantum interference relevant to atomic phase coherence is involved, the reflectance and transmittance of an EIT-based periodic layered medium are shown to be quite sensitive to the probe frequency.

The LHM-EIT-based periodic layered medium has also been considered (in Sections 6 and 7). Since there are left-handed layers embedded in the layered medium, and the periodic EIT layers would act as a potential barrier for the incident electromagnetic wave, the absolute values of the real or imaginary part of the reflection coefficient in some frequency ranges would be more than unity due to the Klein tunneling. The present photonic analog of the Klein tunneling might be used for designing frequency-sensitive photonic transistors. We expect that some new photonic devices (e.g., logic and functional gates) and sensitively switchable devices (fundamental building blocks in, e.g., photonic microcircuits on silicon, in which light replaces electrons), which would find new applications in photonic quantum information processing, would be achieved by taking advantage of such an effect of *coherent switching control* (in Section 8).

The present scheme can be generalized to the cases of four-level EIT systems, where two control fields and one probe field drive the atomic level transitions (Shen, 2007; Shen & Zhang, 2007; Gharibi et al., 2009; Shen, 2010). Obviously, the optical response in such a four-level EIT-based photonic crystal would be more sensitive to the probe frequency than in a three-level EIT photonic crystal presented in this paper. Apart from this intriguing property, there are also interesting applications based on the four-level EIT photonic crystal, e.g., some examples of photonic devices (e.g., multi-input logic gates), in which the control fields and the transmitted probe field act as the input and output signals, respectively, can be designed. We expect that all these new optical properties relevant to quantum coherence, including their applications to photonic devices, could be realized experimentally in the near future.

10. Acknowledgments

This work is supported by National Science Council under Grant Nos. NSC 99-2811-M-216-001, NSC 99-2112-M-216-002 and NSC100-2112-M-216-002. The author Shen acknowledges the support of National Natural Science Foundation of China under Grant Nos. 11174250, 60990320, Natural Science Foundation of Zhejiang Province, China, under Grant No.Y6100280, and the Fundamental Research Funds for the Central Universities of China. The author Shen is also grateful to State Key Laboratory of Modern Optical Instrumentations (Zhejiang University, China) for its financial support (2010-2012). Correspondence and requests for materials relevant to the present work can be addressed to Jian Qi Shen (jqshen@coer.zju.edu.cn).

11. References

Abdumalikov, A. A.; Astafiev, Jr., O.; Zagoskin, A. M.; Pashkin, Y. A.; Nakamura, Y. & Tsai, J. S. (2010). Solitons in Weakly Nonlocal Media with Cubic-Quintic Nonlinearity. *Physical Review Letters*, Vol 104, 193601(1-4), ISSN 0031-9007

Arve, P.; Jänes, P. & Thylén, L. (2004). Propagation of Two-Ddimensional Pulses in Electromagnetically Induced Transparency Media. *Physical Review A*, Vol 69, 063809(1-8), ISSN 1050-2947

Calogeracos, A. & Dombey, N. (1999). History and Physics of the Klein Paradox. *Contemporary Physics*, Vol 40, No. 5, pp. 313–321, ISSN 0010-7514

Caloz, C. & Itoh, T. (2006). *Electromagnetic Metamaterials: Transmission Line Theory and Microwave Applications*, Chapt. 2, John Wiley & Sons, Inc., ISBN 978-047-1669-85-2, New Jersey, USA

Champenois, C.; Morigi, G. & Eschner, J. (2006). Quantum Coherence and Population Trapping in Three-Photon Processes. *Physical Review A*, Vol 74, 053404(1-10), ISSN 1050-2947

Cohen, J. L. & Berman, P. R. (1997). Amplification without Inversion: Understanding Probability Amplitudes, Quantum Interference, and Feynman Rules in a Strongly Driven System. *Physical Review A*, Vol 55, pp. 3900-3917, ISSN 1050-2947

Forsberg, E. & She, J. (2006). Tunable Photonic Crystals Based on EIT Media. *Optoelectronic Materials and Devices* (edited by Lee, Y. H., Koyama, F. & Luo, Y.), *Proc. of SPIE*. Vol 6352, 63520S, ISSN 0277-786X

Gandman, A.; Chuntonov, L.; Rybak, L. & Amitay, Z. (2007). Coherent Phase Control of Resonance-Mediated (2+1) Three-Photon Absorption. *Physical Review A*, Vol 75, 031401(1-4), ISSN 1050-2947

Gharibi, A.; Shen, J. Q. & Gu, J. (2009). Tunable Transient Evolutional Behaviors of a Four-Level Atomic Vapor and the Application to Photonic Logic Gates. *Journal of Physics B: Atomic, Molecular and Optical Physics*, Vol 42, 055502(1-8), ISSN 0953-4075

Harris, S. E. (1997). Electromagnetically Induced Transparency. *Physics Today*, Vol 50, No.7, pp. 36-42, ISSN 0031-9228

He, Q.-Y.; Wu, J.-H.; Wang, T.-J. & Gao, J.-Y. (2006). Dynamic Control of the Photonic Stop Bands Formed by a Standing Wave in Inhomogeneous Broadening Solids. *Physical Review A*, Vol 73, 053813(1-6), ISSN 1050-2947

Joannopoulos, J. D.; Mead, R. D. & Winn, J. N. (1995). *Photonic Crystals: Molding the Flow of Light*, Princeton University Press, ISBN 978-069-1124-56-8, Princeton, New Jersey, USA

Joannopoulos, J. D.; Villenenve, P. & Fan, S. (1997). Photonic Crystals: Putting a new Twist on Light. *Nature*, Vol 386, pp. 143-149, ISSN 0028-0836

Kawazoe, T.; Kobayashi, K.; Sangu, S. & Ohtsu, M. (2003). Demonstration of a Nanophotonic Switching Operation by Optical Near-Field Energy Transfer. *Applied Physics Letters*, Vol 82, pp. 2957-2959, ISSN 0003-6951

Krowne, C. M. & Shen, J. Q. (2009). Dressed-State Mixed-Parity Transitions for Realizing Negative Refractive Index. *Physical Review A*, Vol 79, 023818 (1-11), ISSN 1050-2947

Lurié, D. (1968). *Purticles and Fields*, Chapt. 2, Wiley, ISBN 978-047-0556-42-9, New York, USA

Pendry, J. B.; Holden, A. J.; Robbins, D. J. & Stewart, W. J. (1998). Low Frequency Plasmons in Thin Wire Structures. *Journal of Physics: Condensed Matter*, Vol 10, pp. 4785-4809, ISSN 0953-8984

Pendry, J. B.; Holden, A. J.; Stewart, W. J. & Youngs, I. (1996). Extremely Low Frequency Plasmons in Metallic Mesostructures. *Physical Review Letters*, Vol 76, pp. 4773-4776, ISSN 0031-9007

Petrosyan, D. Tunable Photonic Band Gaps with Coherently Driven Atoms in Optical Lattices. *Physical Review A*, Vol 76, (2007), 053823(1-10), ISSN 1050-2947

Sangu, S.; Kobayashi, K.; Shojiguchi, A. & Ohtsu, M. (2004). Logic and Functional Operations Using a Near-Field Optically Coupled Quantum-Dot System. *Physical Review B*, Vol 69, 115334(1-13), ISSN 0163-1829

Schmidt, H. & Imamoğlu, A. (1996). Giant Kerr Nonlinearities Obtained by Electromagnetically Induced Transparency. *Optics Letters*, Vol 21, pp. 1936-1938, ISSN 0146-9592

Scully, M. O. & Zubairy, M. S. (1997). *Quantum Optics*, Chapt. 7, Cambridge Univ. Press, ISBN 0521435951, Cambridge, UK

Shelby, R. A.; Smith, D. R. & Schultz, S. (2001). Experimental Verification of a Negative Index of Refraction. *Science*, Vol 292, pp. 77-79, ISSN 1095-9203

Shen, J. Q. & Zhang, P. (2007). Double-Control Quantum Interferences in a Four-Level Atomic System. *Optics Express*, Vol 15, pp. 6484-6493, ISSN 1094-4087

Shen, J. Q. (2003). Anti-Shielding Effect and Negative Temperature in Instantaneously Reversed Electric Fields and Left-Handed Media. *Physica Script*, Vol 68, pp. 87-97, ISSN 0031-8949

Shen, J. Q. (2007). Transient Evolutional Behaviours of Double-Control Electromagnetically Induced Transparency. *New Journal of Physics*, Vol 15, pp. 374-388, ISSN 1367-2630

Shen, J. Q. (2008). *Classical & Quantum Optical Properties of Artificial Electromagnetic Media*, Chapt. 1, pp. 15-16, Transworld Research Network, ISBN 978-817-8953-56-4, Kerala, India

Shen, J. Q. (2010). Coherence Control for Photonic Logic Gates via Y-Configuration Double-Control Quantum Interferences. *Optics Communications*, Vol 283, pp. 4546–4550, ISSN 0030-4018

Shen, J. Q.; Ruan, Z. C. & He, S. (2004). Influence of the Signal Light on the Transient Optical Properties of a Four-Level EIT Medium. *Physics Letters A*, Vol 330, pp. 487-495, ISSN 0375-9601

Veselago, V. G. (1968). The Electrodynamics of Substances with Simultaneously Negative Values of ε and μ. *Soviet Physics Uspekhi*, Vol 10, pp. 509-514, ISSN 0038-5670

Wang, L. J.; Kuzmich, A. & Dogariu, A. (2000). Gain-Assisted Superluminal Llight Propagation. *Nature*, Vol 406, pp. 277-279, ISSN 0028-0836

Wu, J. H.; Wei, X. G.;, Wang D. F.; Chen, Y. & Gao, J. Y. (2004). Coherent Hole-Burning Phenomenon in a Doppler Bbroadened Three-Level Lambda-Type Atomic System. *Journal of Optics B: Quantum and Semiclassical Optics*, Vol 6, pp. 54-58, ISSN 1464-4266

Yablonovitch, E. (1987). Inhibited Spontaneous Emission in Solid-State Physics and Electronics. *Physical Review Letters*, Vol 58, pp. 2059-2062, ISSN 0031-9007; John, S. (1987). Strong Localization of Photons in Certain Disordered Dielectric Superlattices. *Physical Review Letters*, Vol 58, pp. 2486-2489, ISSN 0031-9007

Yeh, P. (2005). *Optical Waves in Layered Media*, Chapts. 4-6, pp. 83-143, John Wiley & Sons, Inc., ISBN 978-047-1354-04-8, New Jersey, USA

Zheltikov, A. M. (2006). Phase Coherence Control and Subcycle Transient Detection in Nonlinear Raman Scattering with Ultrashort Laser Pulses. *Physical Review A*, Vol 74, 053403 (1-7), ISSN 1050-2947

Zhu, S. Y. & Scully, M. O. (1996). Spectral Line Elimination and Spontaneous Emission Cancellation via Quantum Interference. *Physical Review Letters*, Vol 76, pp. 388-391, ISSN 0031-9007

Zhuang, F.; Shen, J. Q. & Ye, J. (2007). Controlling the Photonic Bandgap Structures via Manipulation of Refractive Index of Electromagnetically Induced Transparency Vapor. *Acta Physica Sinica (China)*, Vol 56, pp. 541-545, ISSN 1000-3290

6

Optical Logic Devices Based on Photonic Crystal

Kabilan Arunachalam[1] and Susan Christina Xavier[2]
[1]Chettinad College of Engineering and Technology, Karur
[2]Mookambigai College of Engineering, Pudukkottai
India

1. Introduction

Optical components that permit the miniaturization of photonic integrated circuits to a scale comparable to the wavelength of light are good candidates for future optical network and optical computing. All-optical communication is one of the solution for the electronic bottleneck viz speed and size, thanks to their ability to process the information at the speed of light. Optical logic gates are the fundamental components in optical digital information processing. In recent years, researchers have demonstrated all-optical logic gates using different schemes based on nonlinear effects in optical fibers (Ahn et al., 1997; Bogoni et al., 2005; Menezes et al., 2007), in semiconductor devices (Kyoung Sun Choi et al., 2010; Kim et al., 2002; Zhihong Li & Guifang Li 2006; Dorren et al., 2004; Stubkjaer, 2000) and in waveguides (Tetsuro Yabu et al., 2002; Yaw-Dong Wu, 2005; Yaw-Dong Wu et al., 2008). But most of the reported works suffer from certain fundamental limitations including big size, low speed and difficult to perform chip-scale integration.

Nowadays, photonic crystals (PhC) draw significant attention as a platform on which to build devices with dimensions in the order of wavelengths of light for future photonic integrated circuits. They are having some unique properties such as compactness, high speed, low power consumption, better confinement which make them promising candidate in photonic integrated circuits (Yablonovitch, 2003; Cuesta-Soto et al., 2004). Logic functions based on photonic crystal can be realized by nonlinear effect (Notomi et al., 2007), ring resonator (Andalib & Granpayeh, 2009), and multimode interference (Hong-Seung Kim et al 2010). They require significant amount of power, nonlinear material, long interaction length and two different wavelengths for probing and input signals. One of the effects of complex spatial dispersion property in PhC namely self collimation provides a mechanism to employ optical switching and logic functions (Zhang et al., 2007; Susan et al., 2010).

2. Theory of photonic crystal

Photonic crystals (PhC) are new class of optical material represented by natural or artificial structure with periodic modulation of the permittivity. Multiple interference of light on a periodic lattice leads to a photonic band gap and anomalous dispersion because light with a wavelength close to the period of modulation cannot propagate in certain directions. This peculiar property leads to an opportunity for a number of applications. Depending on the geometry of the structure, PhC can be classified into one-dimensional (1D), two dimensional (2D) and three-dimensional (3D) structures. Two-dimensional photonic crystals impose periodicity of the permittivity in two directions, while in the third direction the medium is uniform. Because of the ease of fabrication and analyzing, 2D photonic crystals have attractive attention of large number of researchers and engineers.

The properties of photonic crystal can be engineered through the process of doping which is achieved by either adding or removing dielectric material in a certain area. The dielectric materials then act as a defect region that can be used to localize an electromagnetic wave. Upon incident radiation, the periodic scatterers, that is the periodic dielectric materials in the photonic crystal could reflect an incident radiation at the same frequency in all directions. Wherever in space the reflected radiation interferes constructively, sharp peaks would be observed. This portion of the radiation spectrum is then forbidden to propagate through the periodic structure, and this band of frequencies is called photonic bandgap. On the other hand, wherever in space an incident radiation destructively interferes with the periodic scatterers in a certain directions, this part of the radiation spectrum will propagate through the periodic structure with minimal attenuation and this band of frequencies is called pass band. Introducing point defect or line defect, strict periodicity in the PhC is broken and can form optical cavity with high Q factor and low mode volume or lossless optical waveguide.

2.1 Two dimensional square lattice and Brillouin zone

In two dimensional photonic crystals the permittivity is modulated in two directions, say in the x and z plane: $\varepsilon(r) = \varepsilon(x,z)$. Periodicity in two dimensions can be realized in various geometries, the most common being the square and the triangular lattices. In our work, we consider square lattice of silicon rods embedded in air background. This square lattice having a starting period $a' = \sqrt{2}\,a$ along the x axis where 'a' is lattice constant and oriented at 45° with respect to the x axis as shown in Figure 1a. Successive rods are shifted by δx along the x axis and by δz along the z axis, δx and δz values are given as

$$\delta x = \delta y = a'/2 = a/\sqrt{2} \tag{1}$$

In this case, the direct lattice is formed by the primitive vectors a_1 and a_2 in the real space and given by the following equations

$$a_1 = a\left(\frac{\hat{x}}{\sqrt{2}} - \frac{\hat{z}}{\sqrt{2}}\right) \tag{2}$$

$$a_2 = a\left(\frac{\hat{x}}{\sqrt{2}} + \frac{\hat{z}}{\sqrt{2}}\right) \tag{3}$$

Where $\hat{x}$ and $\hat{z}$ are unit vectors along the x axis and z axis respectively.

Reciprocal lattice vectors b_1 and b_2 in the reciprocal space are written as

$$b_1 = \frac{2\sqrt{2}\pi\hat{x}}{a} \tag{4}$$

$$b_2 = \frac{2\sqrt{2}\pi\hat{z}}{a} \tag{5}$$

The first Brillouin zone is defined as the region of the reciprocal space formed by the points which are closer to the origin than to any other vertex of the periodical lattice. Our structure has a diamond shaped Brillouin zone, which is illustrated in Figure 1d. The Irreducible Brillouin Zone (IBZ) is the triangular wedge in the bottom right corner and the rest of the Brillouin zone can be related to this wedge by rotational symmetry. The three special points Γ, M and X correspond to (0, 0), ($\sqrt{2}\pi/a$, 0) and ($\sqrt{2}\pi/a$, $\sqrt{2}\pi/a$) respectively. Due to periodicity structure, the behavior of the entire crystal can be obtained by studying the unit lattice in the IBZ.

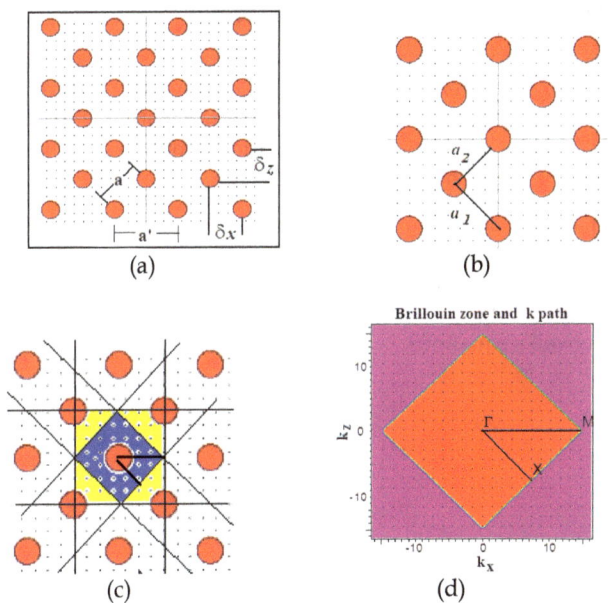

Fig. 1. a) Two-dimensional photonic crystal with square lattice b) primitive vectors a_1 and a_2 in real space c) & d) First Brillouin Zone of square lattice

2.2 Bloch-floquet theorem

The dielectric function in wave vector is a function of spatial co-ordinates r and can be represented by

$$\varepsilon(r) = \int g(k)e^{jkr}dk \tag{6}$$

where g(k) is the dielectric function in wave vector representation and k is the wave vector.

In periodicity condition, the dielectric values of the function are repeated with lattice vector R in each direction and R= ma_1+na_2 where m and n are integers. The dielectric constant in a periodic structure is given as

$$\varepsilon(r) = \varepsilon(r + R) \tag{7}$$

$$\varepsilon(r + R) = \int g(k)e^{jkr}e^{jkR}dk \tag{8}$$

The propagation of a wave in a periodic medium is governed by the Bloch-Floquet theorem which is the product of a plane wave with a periodic function and states that

$$E_k(r) = e^{jkr}u_k(r) \tag{9}$$

where $u_k(r)$ is a periodic envelope function on the lattice and $u_k(r) = u_k(r + R)$.

2.3 Photonic band gap - Plane Wave Expansion Method

Photonic crystals have photonic band gap, which is the gap between the air-line and the dielectric line in the dispersion relation of PhC. Photonic crystals forbid the propagation of the range of frequency in the band gap, and allow the propagation of other frequencies with low loss. Photonic band diagram gives the information about the dispersion characteristics $w(k)$ for the Eigen mode of the PhC.

The Plane Wave Expansion method (PWE), can be used to calculate the band structure using an eigen formulation of the Maxwell's equations, and thus solving for the eigen frequencies for each of the propagation directions of the wave vectors (Igor A. Sukhoivanov & Igor V. Guryev, 2009). The Helmholtz equations for TE and TM polarization can be derived from the fundamental Maxwell's equations,

$$\nabla \times \left[\nabla \times \frac{1}{\varepsilon_r(r)}(\varepsilon_r(r)E) \right] = \frac{\omega^2}{c^2}(\varepsilon_r(r)E) \tag{10}$$

$$\nabla \times \left[\frac{1}{\varepsilon_r(r)}\nabla \times H \right] = \frac{\omega^2}{c^2}H \tag{11}$$

The dielectric function can be expanded to the Fourier series due to the periodicity

$$\frac{1}{\varepsilon(r)} = \sum_G \chi(G).\exp(jG.r) \tag{12}$$

where G is a linear combination of reciprocal vector $G = lb_1 + nb_2$ and $\chi(G)$ is Fourier expansion coefficient which depends on the reciprocal lattice vectors.

Substitution of Eq. 12 in Eq. 10 & 11 gives

$$\nabla \times \left[\nabla \times \sum_G \sum_{G'} \chi(G-G')E(G')\exp(i(k+G).r) \right] = $$
$$\frac{\omega^2}{c^2}\sum_G E(G)\exp(i(k+G).r) \tag{13}$$

$$\nabla \times \left[\nabla \times \sum_{G G'} \chi(G-G')H(G')\exp(i(k+G).r) \right] = \tag{14}$$

$$\frac{\omega^2}{c^2} \sum_{G} H(G)\exp(i(k+G).r)$$

The eigen value equations for the Fourier expansion coefficients of electric field and magnetic field are

$$\sum_{G} \chi(G-G')(k+G) \times [(k+G) \times E(G')] = -\frac{\omega^2}{c^2} E(G) \tag{15}$$

$$\sum_{G} \chi(G-G')(k+G) \times [(k+G') \times H(G')] = -\frac{\omega^2}{c^2} H(G) \tag{16}$$

These are 'Master Equations' for 2D photonic crystals. Here G and G' are in-plane reciprocal lattice vectors, k is in-plane wave vector and ω is the eigen frequency of the TE polarization. The E(G) and H(G) can be projected onto the unit and orthogonal vectors.

Rod type PhC consists of silicon dielectric rods with relative permittivity ε_a periodically embedded in air with a dielectric permittivity ε_b. For simplification, assume only one rod is present in the unit cell and the space dependence of the inverse of the permittivity χ in this elementary cell can be expressed as

$$\frac{1}{\varepsilon_r} = \frac{1}{\varepsilon_b} + \sum_{R} (\frac{1}{\varepsilon_a} - \frac{1}{\varepsilon_b})\theta(r-R) \tag{17}$$

where $\theta(r)$ is the Heaviside function and its value is 1 inside the rod and 0 outside the rod and $\chi_a = 1/\varepsilon_a$, $\chi_b = 1/\varepsilon_b$. The expression for Fourier expansion coefficients of the dielectric function is represented by

$$\chi(r) = \frac{1}{V_0} \int_{V_0} \frac{1}{\varepsilon_r} \exp(-jGr)dr \tag{18}$$

where V_0 is the volume of the unit cell. In our structure, $V_0 = a_1 \hat{x} \times a_2 \hat{z} + a_1 \hat{z} \times a_2 \hat{x}$. Substituting Eq. 17 in Eq. 18, the following equation can be obtained

$$\chi(r) = \frac{1}{\varepsilon_b}\delta_{G,0} + \frac{1}{V_0}\left(\frac{1}{\varepsilon_a} - \frac{1}{\varepsilon_b}\right)\int_{V_0} \theta(r)\exp(-jGr)dr \tag{19}$$

where $\delta_{G,0} = 1$ if G=0 and $\delta_{G,0} = 0$ if G≠0. Using Bessel function Eq. 19 can be written as

$$\chi(r) = \frac{1}{\varepsilon_b}\delta_{G,0} + \left(\frac{1}{\varepsilon_a} - \frac{1}{\varepsilon_b}\right)\frac{2\pi r_a^2 J_1(Gr)}{V_0 Gr} \tag{20}$$

where πr_a^2 is the cross-section area of the rod and $J_1(Gr)$ is the first order Bessel function. The set of the reciprocal lattice vectors should now be selected to provide correct Fourier expansion of the dielectric function and the Bloch functions. Square lattice of silicon rod in air is considered in our structure with $\varepsilon_a = 11.56$ and $\varepsilon_b = 1$.

Thus from Eqs. 15 & 16, for any given value for k leads to an infinite eigen value problem, these truncated by restricting G to a set of M vectors. The k-path within the first Brillouin zone are setting through Γ, M and X correspond to $(0,0)$, $(\sqrt{2}\pi/a,0)$ and $(\sqrt{2}\pi/a,\sqrt{2}\pi/a)$ respectively. The wave vector k describes the edge of the IBZ along the direction ΓX, ΓM and MX for reaching the extrema of $\omega(k)$ and this establishes the dispersion relation.

TM Band Structure

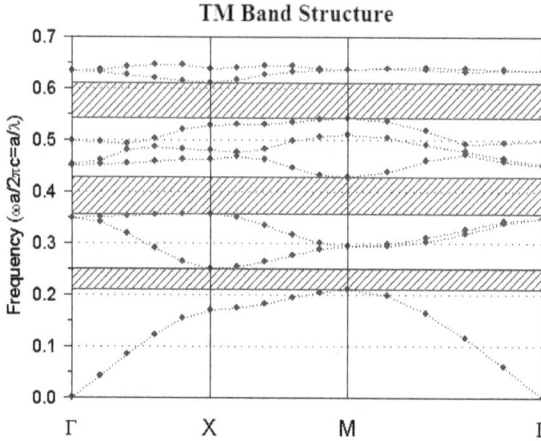

Fig. 2. Photonic band diagram of square lattice of silicon dielectric rod in the air

The Figure 2 shows the photonic band diagram of square lattice of silicon dielectric rod in the air. The radius of the silicon rod (r_a) is 0.3a. In this rod type, only TM polarization exists. The first band gap lies in the normalized frequency region ($\omega a/2\pi c$) 0.21 to 0.25.

2.4 Dispersion properties of photonic crystals

Dispersion of the Bloch modes is one of the most important properties of the photonic crystals and it determines the propagation of modes in the crystal. It depends on many parameters of the PhC such as lattice type, the refractive index contrast between the dielectric material and the host material and distribution of atoms in the primitive cell.

Light pulse in a photonic crystal can be represented as a superposition of the Bloch modes with different Bloch vectors and frequencies

$$u(r,t) = \sum_m \int f(k,m)\psi_{k,m}(r,t)dk \tag{21}$$

where $\psi_{k,m}$ is m-th Bloch mode with the Bloch vector k and $f(k,m)$ is the amplitude of the mode. The motion of the light pulse in photonic crystal is governed by the group velocity $v_g = \nabla_k.\omega(k)$. Since the light pulse is constructed from a superposition of several pulses with different combinations of k and n, let us consider them independently. Each pulse component has the group velocity $v_{g,i} = \nabla_k.\omega(k)|k=k_i$. If the group velocities $V_{g,i}$ are close to each other the distortion of the original pulse will be minimal. If the group velocities are different the original pulse will widen. The group velocity and the group velocity dispersion are obtained from the dispersion diagram. The light propagation inside the PhC is governed by the Equifrequency Surface (EFS) which is the cross section of the band diagram at constant frequency. If the directions of the pulses components group velocities are perpendicular to the EFS, the widening of the original pulse is determined by the shape of the EFS. Each group velocity is locally perpendicular to local EFS. If the curvature of the EFS is large the original pulse will diverge or converge, depending on the sign of the curvature. So, depending on the EFS local curvature as well as on its evolution with the wavelength and the incident wave-vector, there are different types of effect are observed in PhC such as self-collimation, superlensing, negative refraction and superprism. These effects are used to control light propagation inside the PhC.

2.5 Self-collimation effect

Self collimation effect is a linear non-diffraction phenomenon, totally independent of light intensity (Kosaka et al., 1999). PhC are designed to have dispersion properties that allow the beam to propagate without spatial spreading. In the equifrequency contour, flat square contour with zero curvature can be used to latterly confine the light since all the pulse components propagate with the same group velocity. This effect is called self-collimation effect. It provides a mechanism to control the light as in a waveguide.

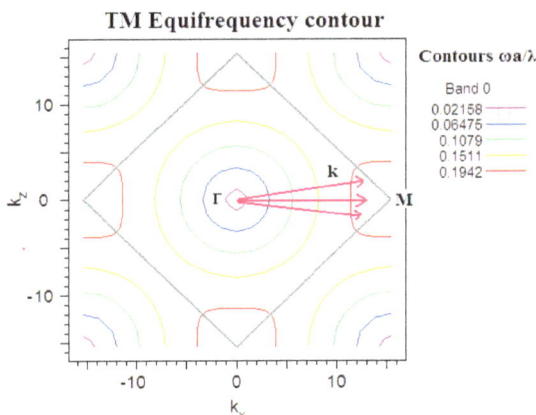

Fig. 3. First band TM Equifrequency contour of photonic crystal dispersion surface in the first Brillouin zone

The Figure 3 shows the frequency contour for the square lattice PhC consisting of silicon rods embedded in air with rod radius $0.3a$. In this contour map, the curves of the frequencies around $0.194(a/\lambda)$ can be identified as squares with round corners centered at the M point, where λ is the wavelength of incident radiation whose value is 1550 nm. So self-collimation phenomenon occurs around the normalized frequency 0.194. RSOFT BandSOLVE tool is used to calculate band diagram and equifrequency contour. When light is incident from a high refractive index (n_h) medium onto a low refractive index (n_l) medium, the incident wave is totally reflected back into the high refractive index medium at the interface, provided the incident angle is larger than the critical angle given by $\theta c = sin^{-1}(n_l/n_h)$ (Chul-Sik Kee et al., 2007) . Self-collimated beams can be totally reflected at the interface of a PhC and air because PhC and air correspond to high refractive and low refractive mediums respectively. When it undergoes total internal reflection, the field amplitude decays very rapidly into air and becomes negligible at a distance within one lattice constant. An air layer created by introducing a line defect by removing a few rods in a row is expected to give rise to total internal reflection. Reflection provides a mechanism for bending and splitting of self-collimated beams.

3. All optical logic gates

All-optical logic gates will be the key elements used in next generation optical computer and optical network. All-optical signal processing can handle large bandwidth signals, large information flows and no need of electrical to optical conversion. All-optical logic gates are capable of performing many logic functions. These are expected to find many applications in optical communication, photonic microprocessors, optical signal processors, optical instrumentation, etc. AND logic gate is used to perform address recognition, packet-header modification, and data-integrity verification. All-optical AND-gates have served as sampling gates in optical sampling oscilloscopes (Westlund, et al., 2005) owing to their ultrafast operation compared to traditional electrical methods. XOR gates can perform a diverse set of processing functions, including comparison of data patterns for address recognition and subsequent packet switching, optical generation of pseudorandom patterns, data encryption/decryption, and parity checking. Threshold detector functionality can be realized by XNOR logic gate. All-optical NOT-gates can be used as inverter and switches. Combination of logic gates may be employed to perform basic or complex computing and arithmetic functions such as binary addition, subtraction, comparison, decoding, encoding and flipflops.

3.1 Structure and optimum values of the proposed logic gates

The proposed logic gate is a square lattice two dimensional PhC that consists of silicon dielectric rods in air background. The structure has a width of $23\sqrt{2}a$ in x-direction and a length of $25\sqrt{2}a$ in z-direction. The dielectric constant and the refractive index of the dielectric rods are 12.0 and 3.46 respectively. In this structure the square lattice is oriented at 45° with the interface parallel to the Γ-M direction with period $a' = \sqrt{2}a$ where 'a' is a lattice constant. Successive cells are shifted by δx along the x axis and δz along the z axis. The amount of shifting is $\delta x = \delta z = a'/2 = \sqrt{2}a/2$. The Figure 4a, 4b and 4c illustrate the 2D PhC lattice used for designing logic gates. The circles represent the silicon rods whose radii are r=0.35a =105 nm, where a=302 nm is the lattice constant.

The schematic circuital layout of the proposed logic gates is shown in Figure 4d. Electrical input signals 1 and 2 activate synchronized laser light sources 1 and 2 respectively to generate optical input signals I_1 and I_2. The reference signal I_{ref} is obtained from reference laser phase locked with lasers 1 and 2. All the input signals including reference have same frequency, polarization, phase, and optical path, with only the reference signal having different amplitude for different gate types. Two input optical signals are coupled using Y coupler and applied to one of the input port of the crystal device and the reference signal

I_{ref} is launched at the second input port. The photonic crystal structure with mirror and splitter performs a specific logic gate function by combining the reflected signal and the partially transmitted reference signal. The optical output is detected and converted into electrical signal by photo detector. This structure can be used for stand alone logic gates. In an integrated circuit the output value will be standardized using a PhC amplifier and given to the input port of the next in sequence logic gate and so on.

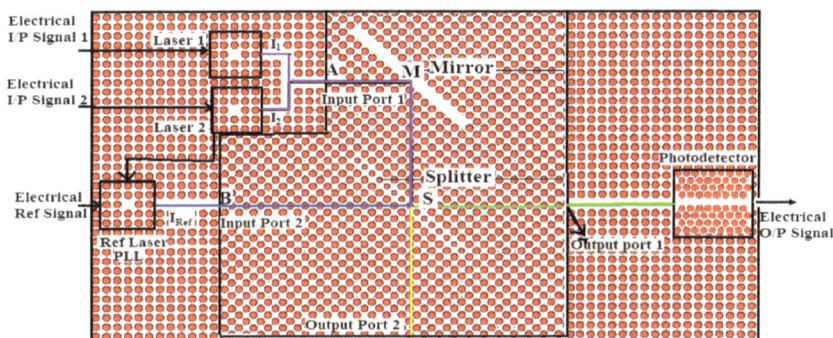

Fig. 4. a Proposed structure of AND, NAND, NOR & XNOR logic gates

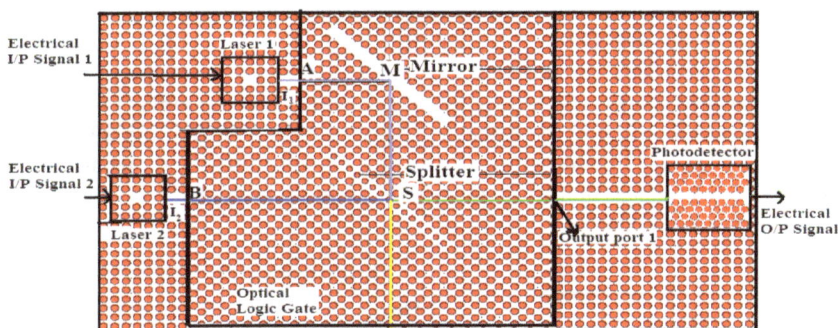

Fig. 4. b Proposed structure of XOR logic gate

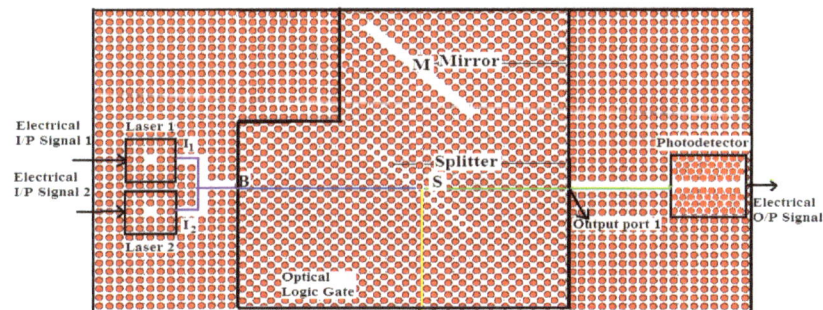

Fig. 4. c Proposed structure of OR logic gate

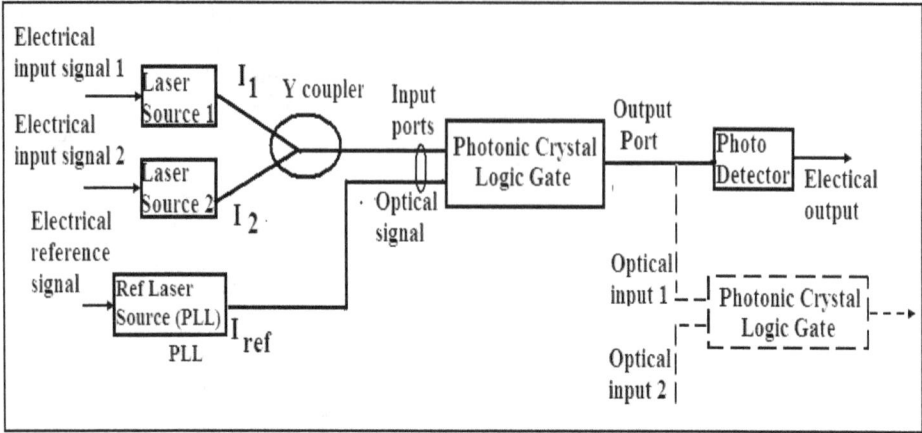

Fig. 4. d Schematic circuital layout of the proposed logic gates

The proposed photonic crystal based logic gate utilizes both bending and splitting mechanisms of self collimating beam. In this structure two line defects are created by reducing the radius of 15 rods in the Γ-X direction. First line defect, in which 15 rods are completely removed act as mirror (M) and second one, in which the radius of the defect rods are reduced act as a splitter (S). When the self collimated beam is incident at rod-air interface, it is partially reflected and partially transmitted and there is a phase change 'ø' occurs in the reflected wave. The power splitting ratio at the line defect and phase difference between the transmitted and reflected signals are dependent on the radius of the rod. From the Figure 5 it is evident that at the defect rod radius r_d=83nm=0.274a the transmitted and the reflected powers are divided equally and at r_d=0 the mirror completely reflects the incident beam. Thus mirror completely reflects the incident beam and the splitter splits the beam with the power ratio 50:50.

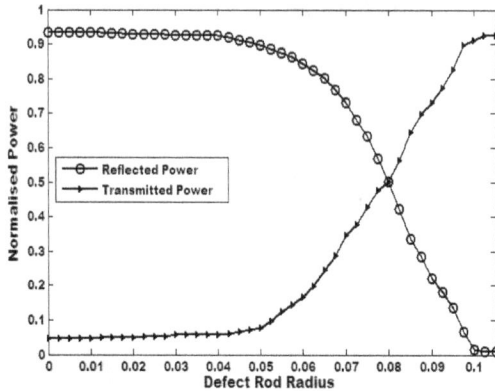

Fig. 5. Normalized transmitted and reflected power with respect to defect rod radius

The phase difference between the transmitted and reflected beam depends on the defect rod radius. If the defect rods radii r_d varied and greater than host rods radii the phase difference between the transmitted and the reflected signal is $-\pi/2$ and if it is less than host rods radii the phase difference is $\pi/2$ (Deyin Zhao et al., 2007).

PhC logic gate structure consists of four ports. Inputs and reference inputs are applied to port 1 and port 2 and the outputs are taken from the output ports 1 and 2. All incident signals including reference signal having the same wavelength of 1.55 μm, phase, and polarization. At the interfering point the path lengths of input signals are equal that is the path length from the AMS is equal to the path length BS which is set as $16\sqrt{2}a$. The reflected input signal from the mirror interferes with another signal at the splitter. This interference is either constructive or destructive depending on their phase difference. The output taken from the output port 1 is destructive and port 2 is constructive.

Based on the relation between the transmitting and the reflecting signal $t^2 + r^2 = 1$, the transmission amplitude of the beam is $e^{i\theta}/\sqrt{2}$ and the reflection amplitude is $e^{i(\theta+\frac{\pi}{2})}/\sqrt{2}$. The propagation of a wave in a periodic medium is governed by the Floquet-Bloch theorem which is given by $E(r) = u(r)Ee^{-i\phi}$, where E is a plane wave and $u(r)$ has the periodicity of the PhC. The transmitted and reflected signals are expressed as

$$R_{12} = \frac{uE}{\sqrt{2}}e^{i\phi_1} \tag{22}$$

$$T_{22} = \frac{uE}{\sqrt{2}}e^{i\phi_2} \tag{23}$$

where R_{12} is reflected input signal 1 at splitter and T_{22} is the transmitted input signal 2 and Φ_1 and Φ_2 are the phase shifts of reflected and transmitted signals respectively. The resultant signal at the output port is a linear combination of reflected and transmitted beam, expressed as

$$O = R_{12} + T_{22} = \frac{uE}{\sqrt{2}}\left(e^{i\phi_1} + e^{i\phi_2}\right) \tag{24}$$

and its corresponding intensity

$$I = |O|^2 = \left|\frac{uE}{\sqrt{2}}\left(e^{i\phi_1} + e^{i\phi_2}\right)\right|^2 \tag{25}$$

Intensity of the reference signal is set at different levels according to the desired logic gates. In our simulations the mesh sizes in the x- and z- directions are set to be $a/16$. The time step for this mesh size is calculated from the Eq. 26

$$\nabla t = \frac{1}{c\sqrt{\frac{1}{\nabla x^2} + \frac{1}{\nabla y^2}}} \tag{26}$$

and it is found to 0.04 femtosecond. The calculated area is surrounded by a Perfectly Matched Layer (PML) boundary.

3.2 Realization and simulation results of logic gates

3.2.1 XOR and OR logic gates

The input signals I_1 and I_2 are applied to the port 1 and 2 and the XOR gate outputs is taken from the output port 1. I_1 and I_2 are equal to P_0. The reflected input signal I_1 from the mirror interfered with either reflected signal or transmitted signal of I_2 at the splitter. Either destructive or constructive interference are obtained at port 1 or 2 respectively depending on their phase difference. When both the input signals are same the output of XOR gate is zero and both are different the output is one.

In the case of OR gate, the input signals are combined and applied to the input port 2 and the output is taken from the output port 1. When both the signals are zero the output is zero and if any one of the input is high, the output is also high. To validate this theoretical prediction, steady state electromagnetic field distribution is simulated using FDTD method. Rsoft FullWAVE software is used to simulate logic gate functions. The field distributions of XOR and OR logic gates are shown in Figure 6 and their corresponding truth table is shown in Table 1.

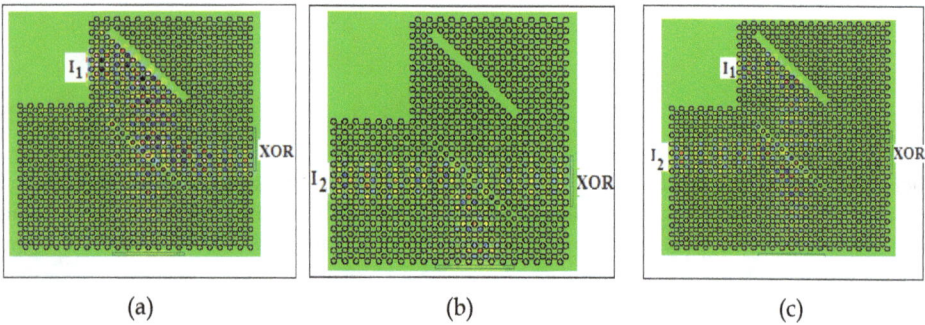

Fig. 6. Electromagnetic field distributions of XOR logic gates. (a) & (b) any one of the input signal is applied (c) both input signals are excited simultaneously

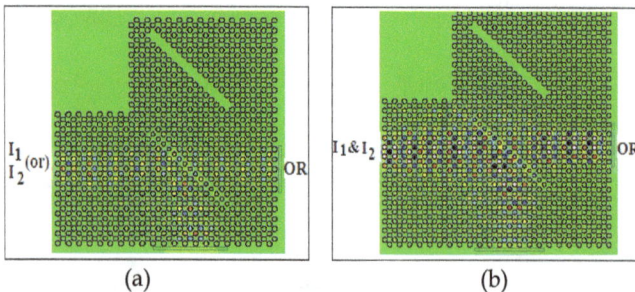

Fig. 7. Field distributions of OR logic gates. a) any one of the input signal is applied b) both input signals are excited simultaneously

I_1	I_2	XOR		OR	
		O/P power	Logic level	O/P power	Logic level
0	0	0	0	0	0
0	1	$0.5P_0$	1	$0.5P_0$	1
1	0	$0.5P_0$	1	$0.5P_0$	1
1	1	$0.001P_0$	0	$1.00 P_0$	1

Table 1. Truth table for XOR and OR gate

3.2.2 AND logic and NOR logic gates

To realize AND logic, inputs I_1 , I_2 and reference beam I_{ref} are taken as P_0 and $0.5P_0$ and combined inputs I_1 and I_2 are applied at input port 1 and I_{ref} is applied at port 2. The destructive interfered signal at the output port 1 is considered as a AND output. In AND logic, when both the input signals are high the output is high otherwise it is zero. In simulation it is found that the output intensity is $0.24P_0$ for separate excitation of I_1 and I_2 and $0.751P_0$ for simultaneous excitation of both I_1 and I_2. If both the signals are not applied the output is $0.25P_0$.In the case of NOR gate, inputs I_1 , I_2 and reference beam I_{ref} are considered as P_0 and $1.5P_0$ respectively. The logic function of this gate is, if both the inputs are low the output is high or else it is zero. The simulation results show that when both the input signals are zero the output is $0.759P_0$ and if any one signal is applied the output is $0.25P_0$. Truth table for AND logic and NOR logic gates is tabulated in Table 2.

I_1	I_2	AND ($I_{ref}= 0.5P_0$)		NOR ($I_{ref} =1.5P_0$)	
		O/P power	Logic level	O/P power	Logic level
0	0	$0.25P_0$	0	$0.759P_0$	1
0	1	$0.24P_0$	0	$0.25P_0$	0
1	0	$0.24P_0$	0	$0.25P_0$	0
1	1	$0.751P_0$	1	$0.25P_0$	0

Table 2. Truth table for AND logic and NOR gate

Steady state electromagnetic field distributions of NOR gate and AND gate for various input combinations are shown in Figure 8 and Figure 9 respectively.

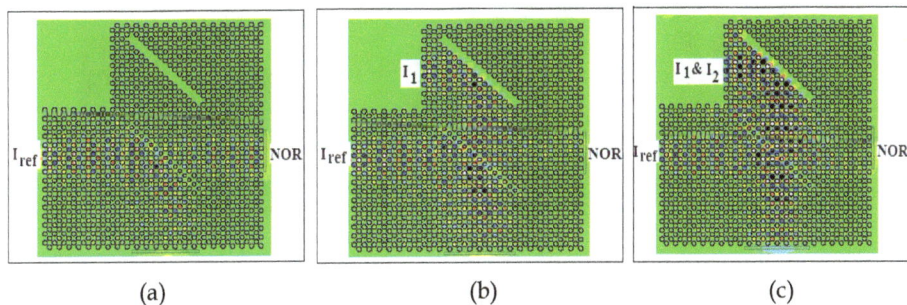

(a)   (b)   (c)

Fig. 8. Electromagnetic field distributions of NOR logic gates. (a) no input signal is applied (b) any one of the input signal is applied c) both input signals are applied simultaneously

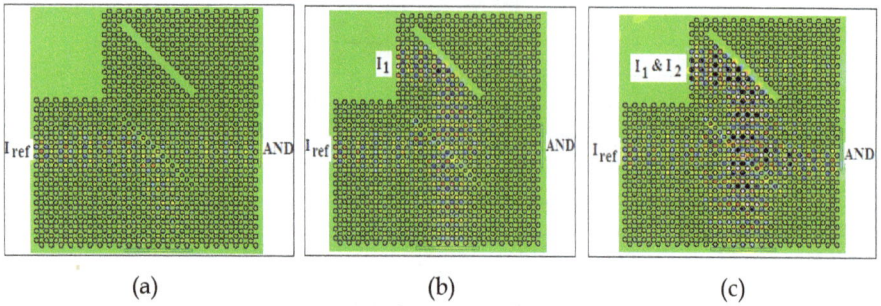

(a)   (b)   (c)

Fig. 9. Electromagnetic field distributions of AND logic gates. a) both the input signals are low b) any one of the input signal is high c) both input signals are high.

3.2.3 NAND and XNOR logic gates

For NAND logic realization, inputs I_1 and I_2 are set as P_0 and reference beam I_{ref} is set as 2.5P_0. These are applied at input port 1 and 2 respectively. In NAND logic, when both the input signals are high the output is zero and any one of the input signal is low the output is   high. It is evident from the simulation that when none of the signal is applied the output is 1.25P_0 and if any one of the signal is applied the output is 0.756P_0.   If both the signals are excited the output is 0.25P_0. The inputs I_1 & I_2 and I_{ref} are considered as P_0 for XNOR gate realization. Logic operation for XNOR gate is known that when both the inputs are same the output is high and if both the inputs are different the output is low. In simulation, it is found that the output is 0.505 P_0 for simultaneous excitation of inputs and also for none of the input signal. When any one of the signal is applied the output is 0.001P_0. The simulated field distributions of XNOR logic gate are illustrated in Figure 10 and NAND logic gate field distribution is shown in Figure 11. Table 3 explicates the truth table for NAND and XNOR logic gates.

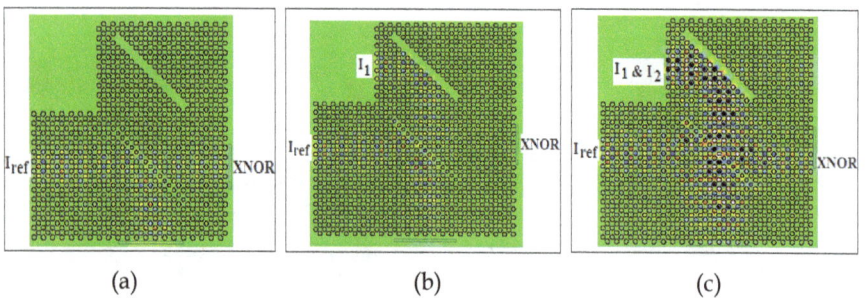

(a)   (b)   (c)

Fig. 10. Electromagnetic field distributions of XNOR logic gates. (a) both the input signals are low b) any one of the input signal is applied (c) both the input signals are applied simultaneously.

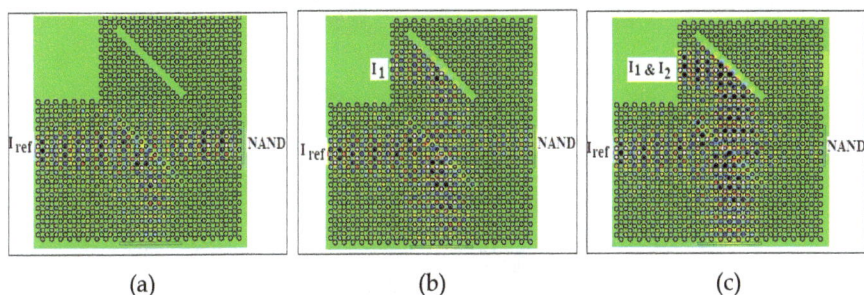

(a)   (b)   (c)

Fig. 11. Field distribution of NAND logic gates a) no input is applied b) any one of the input signal is excited c) both the signals are applied simultaneously

I_1	I_2	NAND(I_{ref}=2.5P_0)		XNOR (I_{ref} =P_0)	
		O/P power	Logic level	O/P power	Logic level
0	0	1.25P_0	1	0.505P_0	1
0	1	0.756P_0	1	0.001P_0	0
1	0	0.756P_0	1	0.001P_0	0
1	1	0.25P_0	0	0.505P_0	1

Table 3. Truth table for NAND logic and XNOR logic gate

4. Photonic crystal optical logic devices for a packaged system

In a complete packaged system, photonic crystal based laser light sources, logic gates and detector are integrated within a single chip. Figure 12 illustrates the integrated photonic crystal based devices. Light source laser is based on a 2D photonic crystal slab patterned with a square lattice. Holes are drilled in GaAs dielectric material. The periodicity of the holes is fixed at 315 nm, and the hole radius is tuned from 105 to 130 nm to change the resonance frequency of the cavities (Hatice Altug et al., 2006). Lasers are driven by the given electrical signals and the corresponding optical output is applied to the all-optical logic gate. Laser 1 & 2 output signals are coupled using coupler and launched to the input port 1 of logic gate and the phase locked reference signal is applied to input port 2. All-optical logic gate performs the logical functions in optical domain.

In the integrated photonic crystal based logic gates, output value of the logic gate will be standardized using a PhC amplifier. The gain of the amplifier is adjusted such that the output level is either "1" or "0". In Figure 12 AND logic gate output is standardized and given to one of the input of XOR gate and other input is getting from the output of another preceding logic gate. Thus the standardized output of one logic gate is given to the input port of the next in sequence logic gate and so on. Finally the output of the last logic gate is applied to the photodetector. The photodetector detects the optical signal. Photo detector is designed using triangular air-hole photonic crystal with lattice constant is 420nm and slab thickness is 204 nm (M Notomi and T Tanabe 2010). The detector converts the optical signal into electrical output.

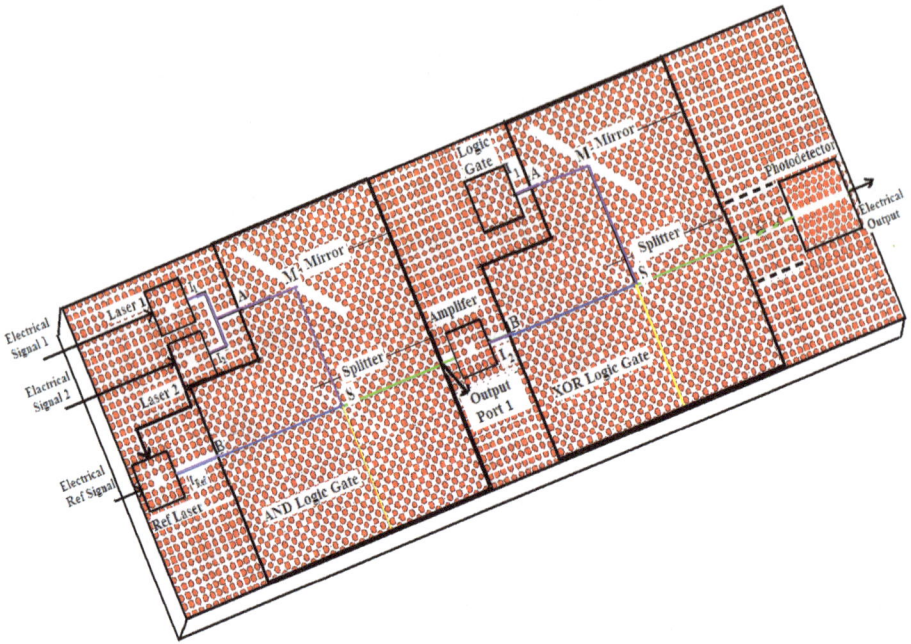

Fig. 12. Integrated photonic crystal based devices

5. Challenges in fabrication

The designing of the beam splitter and the mirror requires high precision of fabrication. Any small deviation from the design of the beam splitters leads to a decay of the optical performance. For instance, a beam splitter would present a wrong phase shift, optical loss, unequal splitting or even mirror like operation if its narrow veins are unintentionally narrower or removed. The Silicon rods must be uniform, smooth and vertical at the side wall in order to maintain the collimation effect throughout the device. The optical path length of the interfering signals should be maintained equal otherwise it will lead to additional phase shift.

6. Conclusion

Thus all the logic gates functions are implemented in non-channel photonic crystal. Self-collimation, bending and splitting effects are used to realize logic gates. The Finite Difference Time Domain method (FDTD) gives fairly accurate results in line with the theoretically predicted concepts. The proposed design exhibits an on-off contrast ratio around 3 dB and a device size of 10x10 μm^2 operating at the optical communication wavelength 1550 nm. The main advantages of all-optical logic gates are small dimensions, simple structure and high speed. These devices may turn out to be good candidate for optical computing and photonic integrated circuits.

7. References

Ahn, K. H.; Cao, X. D. ; Liang, Y.; Barnett, B. C. ; Chaikamnerd, S. & Islam M. N. (1997). Cascadability and functionality of all-optical low-birefringent non-linear optical loop mirror: Experimental demonstration, *Journal Optical Society of America B*, Vol. 14, No. 5, pp. 1228–1236

Andalib, P. & Granpayeh,N. (2009). All-optical ultra-compact photonic crystal AND gate based on nonlinear ring resonators, *Journal Optical Society of America B*, Vol. 26, No.1, pp. 10-16

Bogoni, A. ; Pott`,L.; Proietti, R.; Meloni, G. ; Ponzini, F. & Ghelfi, P.(2005). Regenerative and reconfigurable all-optical logic gates for ultra-fast applications, *Electronics Letters*, Vol. 41, No. 7, pp. 435-436

Chul-Sik Kee; Do-Kyeong Ko; Jongmin Lee; Sun-Goo Lee & Hae Yong Park (2007). Self-Collimated Beams in Two-Dimensional Photonic Crystals: Properties and Applications, *Journal of the Korean Physical Society*, Vol. 51, No. 4, pp. 1479-1483

Cuesta-Soto, F.; Martínez, A.; García, J.; Ramos, F.; Sanchis, P.; Blasco, J. & Martí, J. (2004). All-optical switching structure based on a photonic crystal directional coupler, *Optics Express*, Vol. 12, No. 1 , pp. 161-167

Deyin Zhao; Jie Zhang; Peijun Yao & Xunya Jiang (2007). Photonic crystal Mach-Zehnder interferometer based on self-collimation, *Applied Physics Letter*, Vol. 90, pp. 231114-231116

Dorren, H. J. S. ; Xuelin Yang ; Mishra, A.K.; Zhonggui Li ; Heongkyu Ju ; De Waardt, H.; Khoe, G.D.; Simoyama, T. ; Ishikawa, H. ; Kawashima, H. & Hasama, T. (2004). All-optical logic based on ultrafast gain and index dynamics in a semiconductor optical amplifier, *IEEE Journal of Selected Topics Quantum Electronics*, Vol. 10, No. 5, pp. 1079–1092

Hatice Altug ; Dirk Englund and Jelena Vuckovic(2006). Ultrafast photonic crystal nanocavity laser, Nature Physics, Vol 2, pp. 484-488

Hong-Seung Kim; Tae-Kyung Lee ; Geum-Yoon Oh ; Doo-Gun Kim & Young-Wan Choi (2010). Analysis of all optical logic gate based on photonic crystals multimode interference, *Proceedings of SPIE*, ISBN 9780819480026, San Francisco, 23-28 January 2010

Igor A. Sukhoivanov & Igo r V. Guryev (2009), Photonic Crystals Physics and Practical Modeling, *Springer Series in optical sciences*, ISBN 978-3-642-02645-4, USA

Kim, J.H.; Jhon, Y.M. ; Byun, Y.T. ; Lee, S. ; Woo, D.H. & Kim, S.H. (2002). All-optical XOR gate using semiconductor optical amplifiers without additional input beam, *IEEE Photonics Technology Letters*, Vol.14, No.10, pp. 1436–1438

Kosaka, H. ; Kawashima, T. ; Tomita, A.; Notomi, M. ; Sato, T. & Kawakami ,S. (1999) Self-collimating phenomena in photonic crystals, *Applied Physics Letter*. Vol. 74, pp. 1212-1214

Kyoung Sun Choi; Young Tae Byun; Seok Lee & Young Min Jhon (2010). All-Optical OR/NOR Bi-functional Logic Gate by Using Cross-gain Modulation in semiconductor Optical Amplifiers, *Journal of the Korean Physical Society*, Vol. 56, No. 4, pp. 1093-1096

Menezes, J. W. M.; Fraga, W. B.; Ferreira, A. C.; Saboia, K. D. A.; Filho, A. F. G. F.; Guimarães,G. F.; Sousa, J. R. R.; Rocha, H. H. B.& Sombra, A. S. B.(2007). Logic gates based in two- and three-modes nonlinear optical fiber couplers, *Optical and Quantum Electronics*, Vol. 39, No. 14, pp. 1191-1206

Notomi, M. ; Tanabe, T.; Shinya, A. ; Kuramochi, E. ; Taniyama, H.; Mitsugi, S. & Morita, M.(2007). Nonlinear and adiabatic control of high-Q photonic crystal nanocavities, *Optics Express*, Vol. 15, No.26, pp. 17458-17481

Notomi, M. & Tanabe, T. (2007). Pure-crystal-silicon detector has ultralow dark current, *SPIE Newsroom*, DOI: 10.1117/2.1201004.002936

Stubkjaer, K. E. (2000). Semiconductor optical amplifier-based all-optical gates for high-speed optical processing, *IEEE Journal of Selected Topics in Quantum Electronics*, Vol. 6, No. 6, pp. 1428–1435

Susan Christina,X. ; Kabilan, A.P. & Elizabeth Caroline,P. (2010). Ultra Compact All Optical Logic Gates Using Photonic Band-Gap Materials, *Journal of Nanoelectronics and Optoelectronics*, Vol. 5, No.3, pp. 397-401

Tetsuro Yabu; Masahiro Geshiro; Toshiaki Kitamura; Kazuhiro Nishida & Shinnosuke Sawa (2002). All-Optical Logic Gates Containing a Two-Mode Nonlinear Waveguide, *IEEE Journal of Quantum Electronics*, Vol. 38, No. 1, pp. 37-46

Westlund, M.; Andrekson, P. A.; Sunnerud, H.; Hansryd, J. and Li, J. (2005), High Performance Optical-Fiber-Nonlinearity-Based Optical Waveform Monitoring, IEEE Journal Lightwave Technology, Vol 23, pp. 2012 – 2022

Yablonovitch, E. (2003). Photonic bandgap based designs for nano-photonic integrated circuits, *Proceedings of International Semiconductor Device Research Symposium*, ISBN 0-7803-8139-4, Washington, December 10-12,2003

Yaw-Dong Wu (2005). All-Optical Logic Gates by Using Multibranch Waveguide Structure With Localized Optical Nonlinearity, *IEEE Journal of Selected Topics in Quantum Electronics*, Vol. 11, No. 2, pp. 307-312

Yaw-Dong Wu; Tien-Tsorng Shih & Mao-Hsiung Chen (2008). New all-optical logic gates based on the local nonlinear Mach-Zehnder interferometer, *Optics Express* , Vol. 16, No. 1, pp. 248-257

Zhang, Y. L. ; Zhang, Y. & Li, B. J. (2007). Optical switches and logic gates based on selfcollimated beams in two-dimensional photonic crystals, *Optics Express*, Vol. 15, No.15, pp. 9287-9292

Zhihong Li & Guifang Li (2006). Ultrahigh Speed Reconfigurable Logic Gates Based on Four-Wave Mixing in a Semiconductor Optical Amplifier, *IEEE Photonics Technology Letters*, Vol. 18, No. 12, pp. 1341-1343

7

MEMS Based Deep 1D Photonic Crystal

Maurine Malak and Tarik Bourouina
Université Paris-Est, ESIEE Paris
France

1. Introduction

Since the Bragg layers, also referred as to 1D photonic crystal, lie at the core of many optical devices, this chapter is devoted to the study of the theory underlying the design of multilayered structures [Macleod 2001]. The corresponding analytical model is explained in details in section 2 followed in the next sections, by various design examples for the shake of illustration.

Of special interest are the Silicon-Air Bragg mirrors obtained by DRIE micromachining. They are considered as an important building block leading to a wide variety of applications. First, we elaborate on the use of this building block in resonant cavities and in interferometers (section 3). Then, we apply the multilayered stack theory to a case of study for a special structure: The mode selector, covered in section 3.7). Finally, we conclude this work by highlighting about an advanced architecture of 1D photonic crystals based on curved Bragg mirrors.

2. Theory and modeling of Bragg reflectors

Under specific conditions, a stack of multilayered structure gives rise to nearly perfect optical reflectance, approaching 100 %, as compared to the reflectance from a single interface. This is the main characteristic that makes the interest in such structures, called Bragg reflectors or Bragg mirrors. This phenomenon of enhanced reflectivity might be explained by the fact that the presence of two (or more) interfaces means that a number of light beams will be produced by successive reflections, that may interfere constructively (or destructively, when considering anti-reflective surfaces), and the properties of the multilayered film will be determined by the summation of these beams. This might be the case in thin film assemblies. In thick assemblies however, the later phenomenon does not take place. Before going into the analytical details, we differentiate between thin and thick films. We say that the film is thin when interference effects can be detected in the reflected or transmitted light, that is, when the path difference between the beams is less than the coherence length of the light, and thick when the path difference is greater than the coherence length. Note that no interference can be observed when effects of light absorption dominate within the film, even in the case of thin films. The same film can appear thin or thick depending entirely on the illumination conditions. The thick case can be shown to be identical with the thin case integrated over a sufficiently wide wavelength range or a sufficiently large range of angles of incidence. Normally, we will find that the films on the substrates can be treated as thin while the substrates supporting the films can be considered thick.

In the upcoming treatment, we show analytically a generalized model applicable for an absorbing thin film assembly. The obtained result applies equally well for non-absorbing films.

Let's consider the arrangement shown in Fig. 1 where we denote positive-going waves by the symbol + and negative-going waves by the symbol -. Applying the boundary conditions on the electromagnetic field components at interface B (chosen as the origin of z-axis), we have:

Continuity of the tangential components of the electric filed gives (E_b being the tangential component of the resultant electric field):

$$E_b = E_{1b}^+ + E_{1b}^-$$ (1)

Continuity of the tangential components of the magnetic filed gives (H_b being the tangential component of the resultant magnetic field):

$$H_b = H_{1b}^+ - H_{1b}^- = \eta_1 E_{1b}^+ - \eta_1 E_{1b}^-$$ (2)

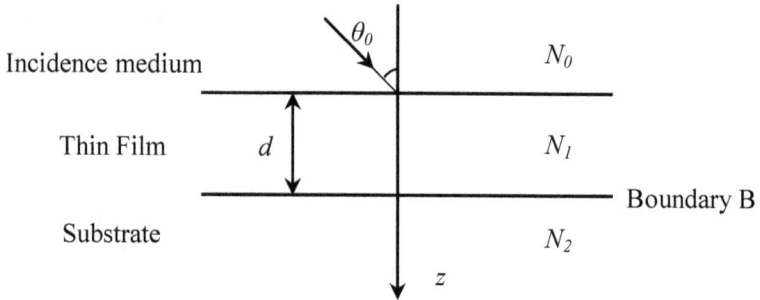

Fig. 1. Plane wave incident on a thin film

where η represents the medium admittance such that: $\eta = H/E$.

The negative sign in (2) comes from the convention used for the field propagation direction such that the right hand rule relating E, H and K (wave vector, along the propagation direction) is always satisfied. In writing equations (1) and (2), we assume that: Common phase factors have been omitted, and the substrate is thick enough such that no field is reflected back from it.

$$E_{1b}^+ = \frac{1}{2}(H_b / \eta_1 + E_b)$$ (3)

$$E_{1b}^- = \frac{1}{2}(-H_b / \eta_1 + E_b)$$ (4)

$$H_{1b}^{+} = \eta_1 E_{1b}^{+} = \frac{1}{2}(H_b + \eta_1 E_b) \tag{5}$$

$$H_{1b}^{-} = -\eta_1 E_{1b}^{-} = \frac{1}{2}(H_b - \eta_1 E_b) \tag{6}$$

The fields at the other interface A at the same instant and at a point with identical x and y coordinates can be determined by altering the phase factors of the waves to allow for a shift in the z coordinate from 0 to $-d$. The phase factor of the positive-going wave will be multiplied by $exp(i\delta)$ while the negative-going phase factor will be multiplied by $exp(-i\delta)$, where,

$$\delta = \frac{2\pi N_1 d}{\lambda}\cos\theta_1 \tag{7}$$

N_1 is the refractive index in medium 1, θ_1 is the angle between the z-axis and the propagation direction in medium 1 and λ is the free space wavelength. The values of E and H at interface A become:

$$E_{1a}^{+} = E_{1b}^{+}e^{i\delta} = \frac{1}{2}(H_b / \eta_1 + E_b)e^{i\delta} \tag{8}$$

$$E_{1a}^{-} = E_{1b}^{-}e^{-i\delta} = \frac{1}{2}(-H_b / \eta_1 + E_b)e^{-i\delta} \tag{9}$$

$$H_{1a}^{+} = H_{1b}^{+}e^{i\delta} = \frac{1}{2}(H_b + \eta_1 E_b)e^{i\delta} \tag{10}$$

$$H_{1a}^{-} = H_{1b}^{-}e^{-i\delta} = \frac{1}{2}(H_b - \eta_1 E_b)e^{-i\delta} \tag{11}$$

So that:

$$E_a = E_{1a}^{+} + E_{1a}^{-} = E_b\left(\frac{e^{i\delta} + e^{-i\delta}}{2}\right) + H_b\left(\frac{e^{i\delta} - e^{-i\delta}}{2\eta_1}\right)$$
$$= E_b \cos\delta + H_b \frac{i\sin\delta}{\eta_1} \tag{12}$$

$$H_a = H_{1a}^{+} + H_{1a}^{-} = E_b\eta_1\left(\frac{e^{i\delta} - e^{-i\delta}}{2}\right) + H_b\left(\frac{e^{i\delta} + e^{-i\delta}}{2}\right)$$
$$= E_b i\eta_1 \sin\delta + H_b \cos\delta \tag{13}$$

In the matrix notation, we finally obtain the following formulation:

$$\begin{bmatrix} E_a \\ H_a \end{bmatrix} = \begin{bmatrix} \cos\delta & (i\sin\delta)/\eta_1 \\ i\eta_1 \sin\delta & \cos\delta \end{bmatrix}\begin{bmatrix} E_b \\ H_b \end{bmatrix} \tag{14}$$

This matrix relates the tangential components of E and H at the incident interface to tangential components of E and H transmitted through the final interface and it is known as the characteristic matrix of the thin film $[M]$.

Now, let us consider an assembly of thin films as shown in Fig. 2

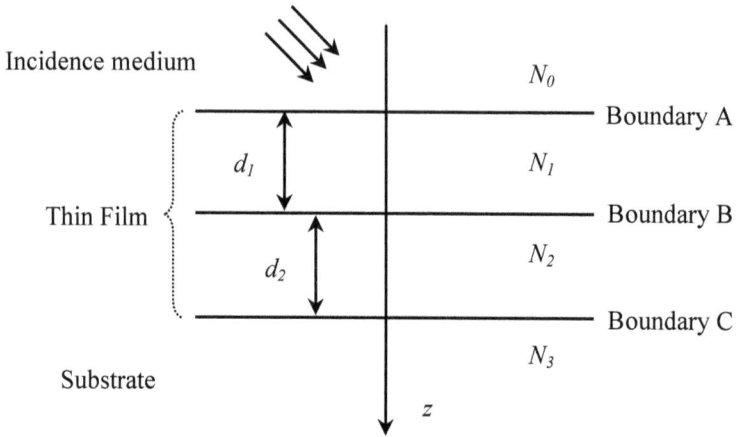

Fig. 2. Plane wave incident on an assembly of thin film

By adding another film to the single film shown in Fig. 1, we can write:

$$\begin{bmatrix} E_b \\ H_b \end{bmatrix} = \begin{bmatrix} \cos\delta_2 & (i\sin\delta_2)/\eta_2 \\ i\eta_2\sin\delta_2 & \cos\delta_2 \end{bmatrix} \begin{bmatrix} E_c \\ H_c \end{bmatrix} \tag{15}$$

Combining with equation (14) to get the field at interface A, then we can write:

$$\begin{bmatrix} E_a \\ H_a \end{bmatrix} = \begin{bmatrix} \cos\delta_1 & (i\sin\delta_1)/\eta_1 \\ i\eta_1\sin\delta_1 & \cos\delta_1 \end{bmatrix} \begin{bmatrix} \cos\delta_2 & (i\sin\delta_2)/\eta_2 \\ i\eta_2\sin\delta_2 & \cos\delta_2 \end{bmatrix} \begin{bmatrix} E_c \\ H_c \end{bmatrix} \tag{16}$$

This result can be immediately extended to the general case of an assembly of q layers, when the characteristic matrix is simply the product of the individual matrices taken in the correct order,

$$\begin{bmatrix} E_a \\ H_a \end{bmatrix} = \left\{ \prod_{r=1}^{q} \begin{bmatrix} \cos\delta_r & (i\sin\delta_r)/\eta_r \\ i\eta_r\sin\delta_r & \cos\delta_r \end{bmatrix} \right\} \begin{bmatrix} E_m \\ H_m \end{bmatrix} \tag{17}$$

where E_m and H_m are the electric and magnetic fields components in the substrate plane.

Dividing both sides of equation (17) by H_m, we get:

$$\begin{bmatrix} B \\ C \end{bmatrix} = \begin{bmatrix} E_a / E_m \\ H_a / E_m \end{bmatrix} = \left\{ \prod_{r=1}^{q} \begin{bmatrix} \cos\delta_r & (i\sin\delta_r)/\eta_r \\ i\eta_r\sin\delta_r & \cos\delta_r \end{bmatrix} \right\} \begin{bmatrix} 1 \\ \eta_m \end{bmatrix} \tag{18}$$

where

$$\delta_r = \frac{2\pi N_r d_r}{\lambda} \cos\theta_r \tag{19}$$

$$\eta_r = \Upsilon N_r \cos\theta_r \text{ for s-polarisation (TE)} \tag{20}$$

$$\eta_r = \Upsilon N_r / \cos\theta_r \text{ for p-polarisation (TM)} \tag{21}$$

$$\Upsilon = \left(\varepsilon_0 / \mu_0\right)^{\frac{1}{2}} = 2.6544 \times 10^{-3} S \text{ (Free space admittance)} \tag{22}$$

Then, the admittance of the thin film assembly is:

$$Y = \frac{H_a}{E_a} = \frac{H_a / E_m}{E_a / E_m} = \frac{C}{B} \tag{23}$$

The amplitude reflection coefficient and the reflectance are then given by:

$$\rho = \frac{\eta_0 - Y}{\eta_0 + Y} \tag{24}$$

$$R = \left(\frac{\eta_0 - Y}{\eta_0 + Y}\right)\left(\frac{\eta_0 - Y}{\eta_0 + Y}\right)^* = \left(\frac{\eta_0 B - C}{\eta_0 B + C}\right)\left(\frac{\eta_0 B - C}{\eta_0 B + C}\right)^* \tag{25}$$

In this part, we look over the case of absorbing layers. Knowing that K is propagation vector:

$$K^2 = K_x^2 + K_z^2 \tag{26}$$

From Snell's law, $K_{x0} = K_{xr} = K_{xm}$ this leads to:

$$N_0 \sin\theta_0 = N_r \sin\theta_r = N_m \sin\theta_m \tag{27}$$

In case of absorbing medium, N_r is complex and in general, it can be expressed as a complex number:

$$N_r = n - ik \tag{28}$$

Then,

$$K_{zr}^2 = K_r^2 - K_{xr}^2 = K_0^2\left(N_r^2 - N_0^2 \sin^2\theta_0\right) = K_0^2\left(n^2 - k^2 - 2ink - N_0^2 \sin^2\theta_0\right) \tag{29}$$

Then,

$$\begin{aligned} \delta_r = K_{zr} d_r &= K_0 d_r \left(n^2 - k^2 - 2ink - N_0^2 \sin^2\theta_0\right)^{\frac{1}{2}} \\ &= \frac{2\pi}{\lambda} d_r \left(n_r^2 - k_r^2 - 2in_r k_r - N_0^2 \sin^2\theta_0\right)^{\frac{1}{2}} \end{aligned} \tag{30}$$

Accordingly, we get:

The admittance of layer r in s-polarization:

$$\eta_{rs} = \Upsilon \left(n_r^2 - k_r^2 - 2in_r k_r - N_0^2 \sin^2 \theta_0 \right)^{\frac{1}{2}} \tag{31}$$

The admittance of layer r in p-polarization:

$$\eta_{rp} = \frac{\Upsilon_r^2}{\eta_{rs}} = \frac{\Upsilon^2 \left(n_r - ik_r \right)^2}{\eta_{rs}} \tag{32}$$

In this last part of the analytical model, we present closed form relations for the transmittance T and the absorbance A using the definition of irradiance. If we consider the net irradiance at the exit of the assembly I_k:

$$I_k = \frac{1}{2} \mathrm{Re}\left(E_k H_k^* \right) = \frac{1}{2} \mathrm{Re}\left(\eta_m^* \right) E_k E_k^* \tag{33}$$

The net irradiance at the entrance of the assembly I_a:

$$I_a = \frac{1}{2} \mathrm{Re}\left(E_a H_a^* \right) = \frac{1}{2} \mathrm{Re}\left(\frac{E_a}{E_k} \cdot \frac{H_a^*}{E_k^*} \right) E_k E_k^* = \frac{1}{2} \mathrm{Re}\left(BC^* \right) E_k E_k^* \tag{34}$$

Let the incident irradiance be denoted by I_i, then the irradiance actually entering the assembly becomes:

$$I_a = (1 - R) I_i = \frac{1}{2} \mathrm{Re}\left(BC^* \right) E_k E_k^* \tag{35}$$

Then,

$$I_i = \frac{\mathrm{Re}\left(BC^* \right) E_k E_k^*}{2(1 - R)} \tag{36}$$

The transmittance is then given by:

$$T = \frac{I_k}{I_i} = \frac{\mathrm{Re}\left(\eta_m \right)(1 - R)}{\mathrm{Re}\left(BC^* \right)} \tag{37}$$

From equation (25), we get:

$$T = \frac{4\eta_0 \, \mathrm{Re}\left(\eta_m \right)}{\left(\eta_0 B + C \right)\left(\eta_0 B + C \right)^*} \tag{38}$$

Knowing that $A + R + T = 1$ from the energy conservation rule, then A becomes:

$$A = 1 - R - T = (1 - R)\left(1 - \frac{\mathrm{Re}\left(\eta_m \right)}{\mathrm{Re}\left(BC^* \right)} \right) \tag{39}$$

From equation (25), we get:

$$A = \frac{4\eta_0 \, \mathrm{Re}\left(BC^* - \eta_m\right)}{\left(\eta_0 B + C\right)\left(\eta_0 B + C\right)^*} \tag{40}$$

3. Microfabricated silicon-Air Bragg as basic building blocks in cavities and interferometers

3.1 Literature survey

Many groups worked on the realisation of silicon-Air Bragg reflectors as basic building blocks in Fabry-Perot (FP) cavities as well as in Michelson interferometers. When considering Fabry-Perot cavities, the use of high reflectance Bragg mirrors is intended to achieve high quality factor Q at the corresponding resonant wavelengths. The use of silicon restricts the wavelength range to the infra-red region. In the same time, light coupling using optical fibers is facilitated by the microfabrication of U-grooves for supporting the fibers with pre-alignment capability. Among the groups working on this topic, we can cite [Lipson & Yeatman 2007] from the imperial college who realized FP cavities obtained by KOH etching on SOI substrate. The best reported value of Q was 2395 in static designs while it was limited to 515 in the MEMS designs due to the rotation associating the translation of the MEMS mirror. Thus, the mirrors became unparallel and the Q-factor degraded. Another group is the group of Ecole Polytechnique de Montreal who realized FP cavity for inertial sensing [Zandi et al 2010] on SOI. Their Q was limited to 662. A third group is that of [Yun and Lee 2003] from Gwangju Institute of Science and Technology, Korea who realized thermally tuned FP cavities, they obtained a Q of 1373. A fourth group is that of [Pruessner et al. 2008] from Naval Research Laboratory, Washington D.C. who realised FP cavities with integrated SOI rib waveguides by cryogenic etching. They recorded a Q factor of 27,000. Lastly, our group at ESIEE Paris was among the pioneers in this domain. In our first achievements, we realized FP cavities with different architectures [Saadany et al. 2006]; the best recorded Q was 1291 for FP structure working as a notch filter. More recently, the performance was improved using cylindrical Bragg mirrors of cylindrical shapes combined with a fiber rod lens, leading to $Q = 8818$ on quite large cavities exceeding $L = 250$ µm [Malak et al. APL 2011], an unreached value for the figure of merit $Q.L$, which is of primary importance for cavity enhancement applications. Table 1 summarizes the specifications of the different designs discussed above.

3.2 Fabrication technology for Si-Air Bragg reflectors (for MEMS and for fixed structures)

In this section, the basic steps of the fabrication process for MEMS structures involving Bragg layers are highlighted. Many techniques can be used to produce vertical structures on silicon substrate as mentioned in [Lipson & Yeatman 2005], [Yun et al. 2006] and [Song et al 2007]. They are based on either dry or wet etching of silicon using KOH. The process described here and shown in Fig. 3 pertains to the (optional) integration of MEMS structures together with the Bragg mirrors using dry etching.

Starting from a raw SOI wafer, we proceed by making thermal oxidation for the whole wafer. In the next step, photoresist (PR) is used to cover the entire wafer where it acts as a

	Lispon and Yeatman Imperial college UK	Zandi et al. Ecole Polytec. de Montréal Canada	Yun and Lee Gwangju Institute of Science and Technology Korea	Pruessner et al. Naval research lab Washington D.C	Saadany et al. ESIEE France	Malak et al. ESIEE France
Cavity length	3.3 µm	27.1 µm	----	12 µm	10 µm	250 µm
Tuning range	Static design : 70 nm MEMS design: 10 nm	30 nm	9 nm	6.7 nm	----	----
Number of silicon layers per mirror	3	3	2	2	3	4
Q-factor	Static design: $Q = 2395$ MEMS design: $Q = 515$ $Q_{MEMS} < Q_{static}$ Because the displacement creates a rotation for the moving mirror.	$Q = 662$	$Q_{average} = 1373$ (estimated from the response of a rejection filter)	$Q = 27000$	$Q_{Trans} = 1291$	$Q = 8818$

Table 1. Summary of the specifications for state-of-the-art FP cavities

mask for photolithography. The PR is then patterned using UV exposure over the DRIE layout mask. Since the PR is a positive type, the areas exposed to UV remain soft while the non-exposed areas become hard and they can not be removed in the development step. Now, the hardened PR acts as a protection mask for the originally oxidized silicon which is patterned using either Reactive Ion etching (RIE) or Buffer HydroFluoric acid (BHF). The role of PR ends here and it is completely removed from the wafer.

The fabrication process continues by the metal deposition over the whole wafer. The metal is patterned by photolithography using the frontside metal layout mask and then etched. In the next step, metal is deposited on the backside of the wafer where it is patterned by the backside layout mask and then etched. We turn again to the front side to make Deep Reactive Ion etching (DRIE) for the silicon structure layer [Marty et al. 2005]. At that level, both the oxide and the aluminum serve as mask materials for silicon etching by DRIE. Processing the backside again, DRIE is done for the backside, in this case, only aluminum serve as mask material for silicon etching by DRIE. The process ends by releasing the MEMS structure in which the insulating oxide is removed by vapor HF.

For the fixed structures involving Bragg mirrors presented in this research work, the process differs from the one detailed above. So, in the next paragraph, we highlight the fabrication process as shown in Fig. 4, used for the realization of the fixed structures.

Starting with an ordinary silicon wafer, a thermal oxidation process is carried for both sides of the wafer to achieve an oxide thickness = 1.7 µm. Next, PR used as a mask for photolithography, is sputtered over the entire wafer. This step is followed by the photolithography for DRIE mask for the front side and the PR is patterned accordingly. The following step is the plasma etching for the oxide. This photolithography ends by PR removal. Then, we start processing for the back side by depositing aluminium. Next, we pattern the aluminium mask by photolithography using the back side layout mask. Then, we proceed by DRIE etching over 300 µm for the back side and the process ends up with DRIE etching for the front side over 100 µm. Note that all steps performed on the backside are optional, depending on the nature of the target device.

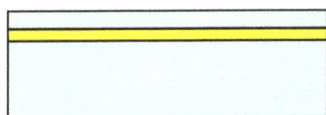

(a) Initial SOI Wafer

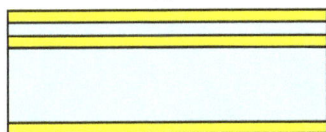

(b) Thermal Oxydation

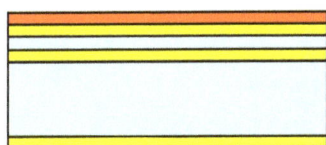

(c) Photoresist spin-coating

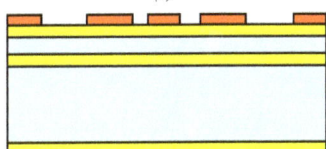

(d) Patterning Photoresist by UV exposure and further development

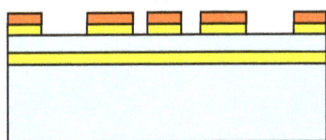

(e) Oxide etching (using Photoresist patterns as a protection mask material) either by BHF or by RIE

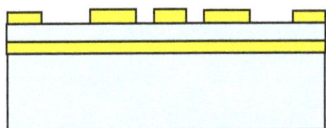

(f) Photoresist removal

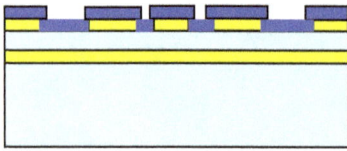

(g) Aluminum deposition (sputtering)

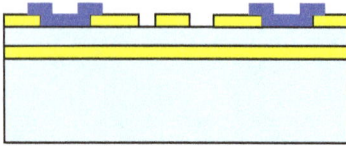

(g)

(h) Lithography patterning and aluminum etching

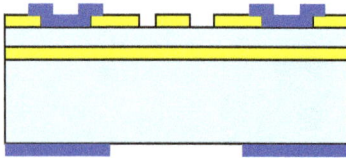

(h)

(i) Metal deposition on the backside, patterning using proper mask and then, etching.

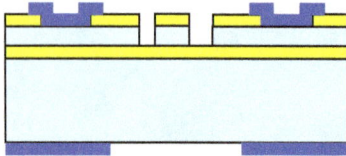

(i)

(j) DRIE Etching for the frontside (Note that here, both oxide and aluminum serve as mask materials for silicon etching by DRIE

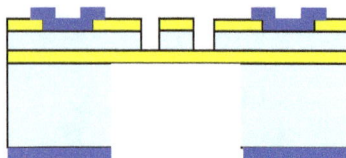

(j)

(k) DRIE Etching for the backside (Note that in this case, only aluminum serve as mask material for silicon etching by DRIE)

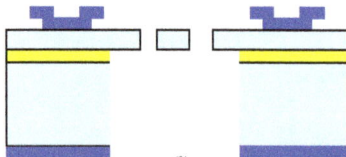

(k)

(l) Release (oxide removal by vapor HF)

(l)

Fig. 3. (a-l) Basic steps of the fabrication process for MEMS structures co-integrated with Si-Air Bragg mirrors.

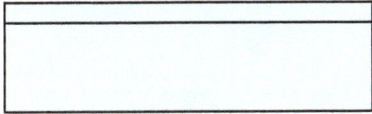

(a) Initial Silicon wafer = 400 μm

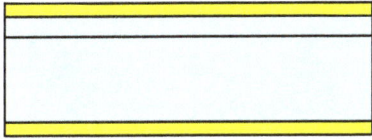

(b) Thermal Oxydation 1.7 μm

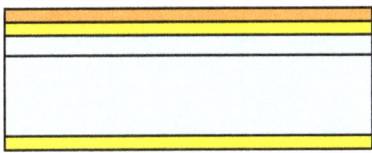

(c) Front side Photolithography for DRIE (PR sputtering)

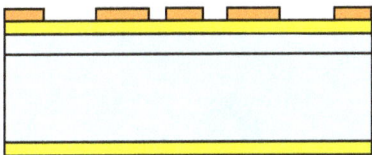

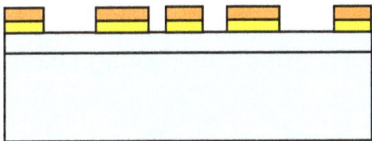

(d, e) Plasma etching for the oxide

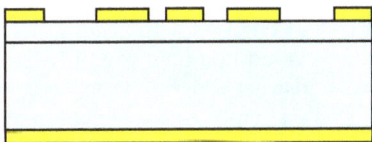

(f) PR removal

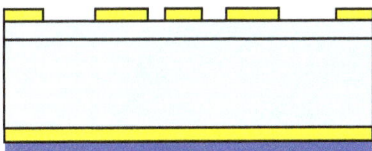

(g) Aluminium deposition – backside

(h) DRIE Photolithography (Oxide + Aluminum)

(h)

(i) DRIE Etching – backside 300 um

(i)

(j) DRIE Etching – frontside 100 um

(j)

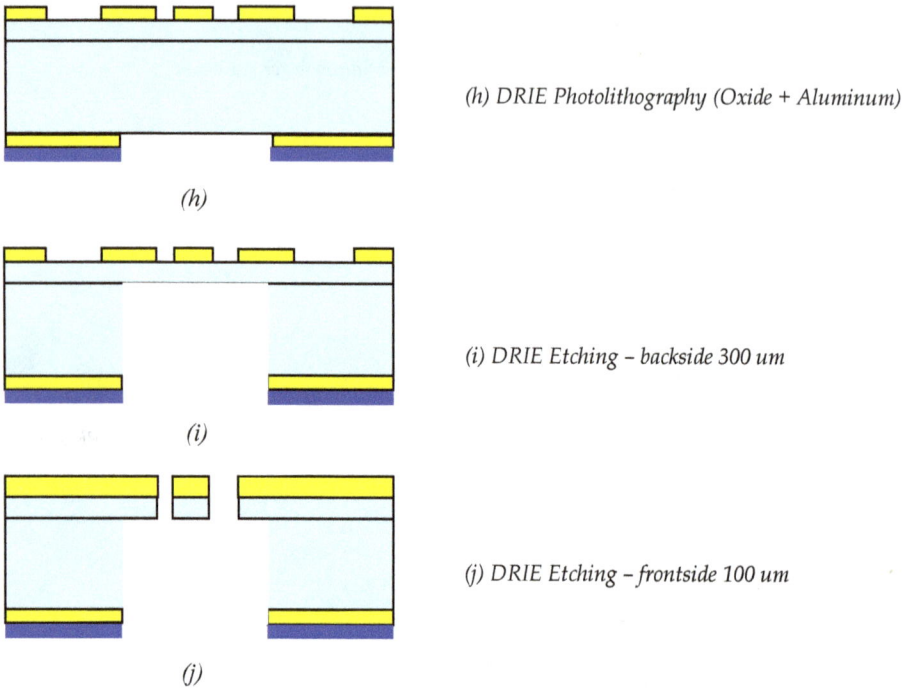

Fig. 4. (a-j) Basic steps of the fabrication process for fixed structures involving Bragg mirrors.

3.3 Modeling and simulation of planar Bragg mirror reflectors

Based on the analytical model presented in section 2, then if we have a single layer whose thickness is an odd number of quarter the wavelength, the characteristic matrix of the layer [M] becomes:

$$[M] = \pm \begin{bmatrix} 0 & i/\eta \\ i\eta & 0 \end{bmatrix} \qquad (41)$$

So, if we stack a combination of several layers alternatively of High refractive index, denote H, and Low refractive index, denoted L, whose thickness is an odd number of $\lambda/4$ (where λ is the wavelength in the corresponding medium), we can construct a high reflectance mirror named Bragg mirror. In the particular case where we stack a combination of five quarter-wave layers which are different, mathematical manipulation of the equivalent characteristic matrix yields an equivalent admittance for the assembly:

$$Y = \frac{\eta_1^2 \eta_3^2 \eta_5^2}{\eta_2^2 \eta_4^2 \eta_m} \qquad (42)$$

where the definitions presented previously are kept unchanged.

For $\eta_m = \eta_0$, and considering similar indices for the high and low layers as well so that, $\eta_1 = \eta_3 = \eta_5 = \eta_H$ and $\eta_2 = \eta_4 = \eta_L$, the reflectance of the assembly becomes:

$$R = \left(\frac{\eta_0 - Y}{\eta_0 + Y}\right)\left(\frac{\eta_0 - Y}{\eta_0 + Y}\right)^* = \left(\frac{1 - \left(\frac{\eta_H}{\eta_L}\right)^6}{1 + \left(\frac{\eta_H}{\eta_L}\right)^6}\right)^2 \tag{43}$$

In general, for a pair (p) of HL layers with similar η_m and η_0, we can write:

$$R = \left(\frac{1 - \left(\frac{\eta_H}{\eta_L}\right)^{2p}}{1 + \left(\frac{\eta_H}{\eta_L}\right)^{2p}}\right)^2 \tag{44}$$

Based on these derivations, we built a MATLAB code to design the Bragg mirrors. In this comprehensive study, we focus mainly on the impact of the number of Bragg layers, the layer thickness and the technological errors on the reflectance and the transmittance of the Bragg mirror. In all the upcoming results, we consider absorption free layers with a silicon refractive index $n_{Si} = 3.478$ and air refractive index $n_{air} = 1$.

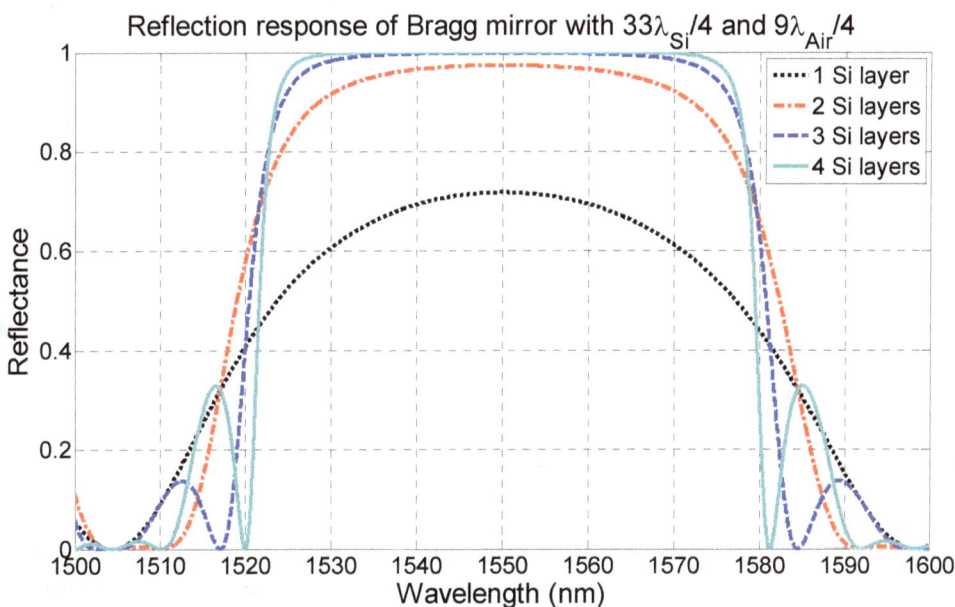

Fig. 5. Reflectance of Bragg mirrors for different numbers of HL pairs

For the simulation results shown in Fig. 5, we choose a silicon thickness = 3.67 μm ($33\lambda_{Si}/4$) and air thickness = 3.49 μm ($9\lambda_{air}/4$) because they are relatively easy to obtain using the affordable fabrication technology (as compared to single quarter wavelength $\lambda_{Si}/4 = 0.111$ μm and $\lambda_{air}/4 = 0.388$ μm at the communication wavelength $\lambda = 1550$ nm). We notice that the reflectance in the mid-band increases as the number of layers increases which goes in accordance with relation (44). In fact, the reflectance increases from 71.8 % (single Si layer) up to 99.98 % (4 Si layers) when the number of *HL* pairs increases from single to four. Also, the mirror response becomes sharper and its bandwidth (BW) decreases as the number of layers increases. In the case of single layer, the BW is about 65 nm and it goes down to 58 nm as the *HL* pairs increases to 4.

In the next simulation, we study the impact of the silicon layers thickness on the mirror bandwidth. For this shake, we consider 4 *HL* pairs with fixed air layers thickness 3.49 µm ($9\lambda_{air}/4$), while the thickness of the silicon layers is increased from $\lambda_{Si}/4$ up to $25\lambda_{Si}/4$ in steps of $2\lambda_{Si}$. Simulation results, depicted in Fig 6, show that the mirror BW decreases as the thickness of the silicon layers increases. For silicon thickness $\lambda_{Si}/4$, the 3dB-BW = 238 nm and it decreases to 73 nm at a thickness of $25\lambda_{Si}/4$.

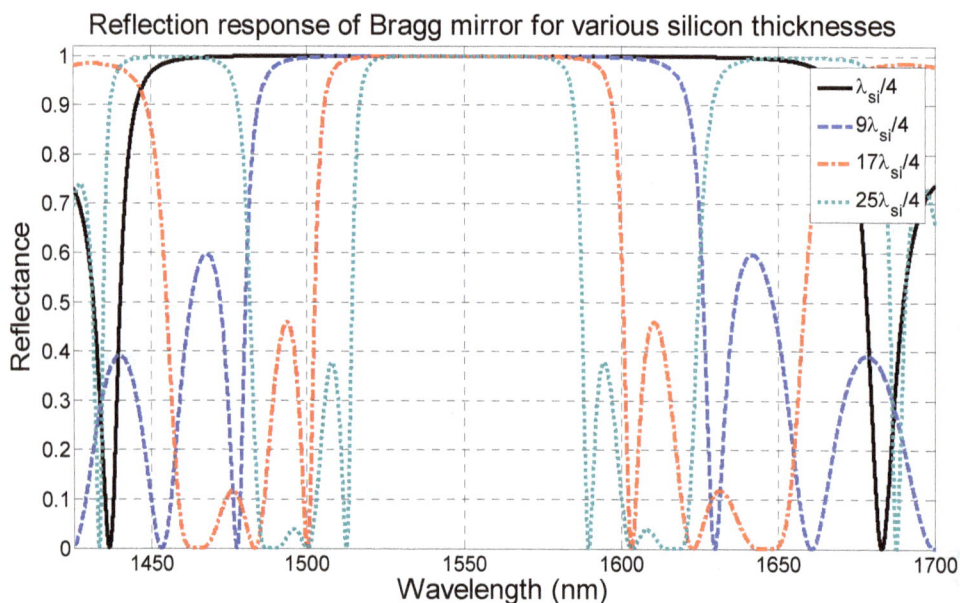

Fig. 6. Reflectance of 4 *HL* pairs for various *H* thicknesses, *L* thickness is fixed to $9\lambda_{air}/4$

If on the other hand, we fix the thickness of the silicon layers to $33\lambda_{Si}/4$, for the same 4 *HL* pairs, and increase the thickness of the air layers from $\lambda_{air}/4$ up to $13\lambda_{air}/4$ in steps of λ_{air}. A similar effect is noticed but on a smaller BW scale since the BW decreases from 65 nm at L thickness = $\lambda_{air}/4$, to 55 nm at L thickness = $13\lambda_{air}/4$. The corresponding results are shown in Fig. 7. Comparing between both results, we can say that the decrease in the H thickness is more pronounced than the decrease in the L thickness in terms of the bandwidth. Good control of the H thickness can give rise to Bragg mirrors with large BW.

Another point of interest for the Bragg mirror is the technological error. The critical dimension, defined as the minimum feature size on the technology mask, can not be maintained as drawn in the original design and thus, it translates into reduced layer thicknesses on the fabricated device. In fact, the thickness of the silicon layer may vary

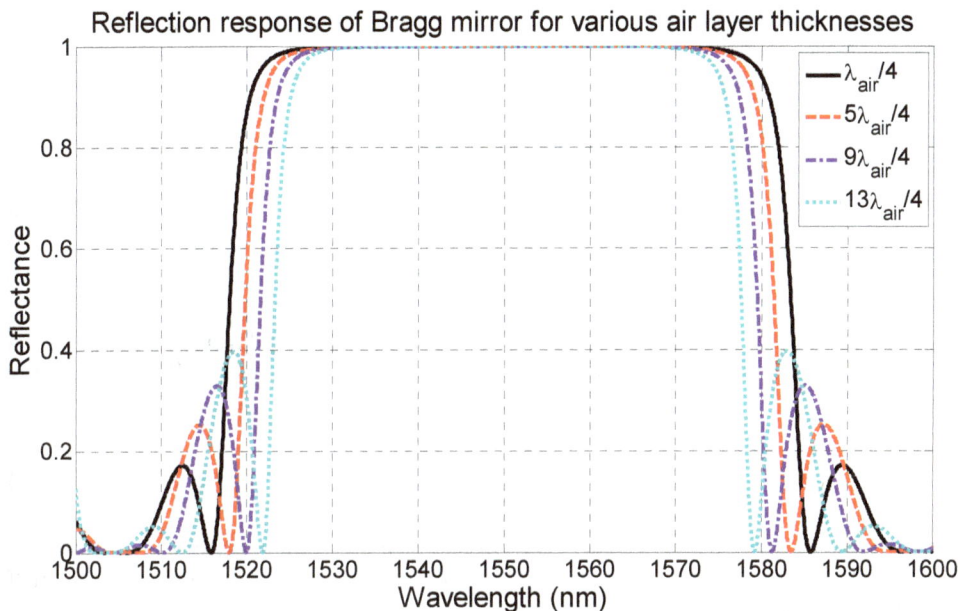

Fig. 7. Reflectance of 4 HL pairs for various L thicknesses, H thickness is fixed to $33\lambda_{Si}/4$

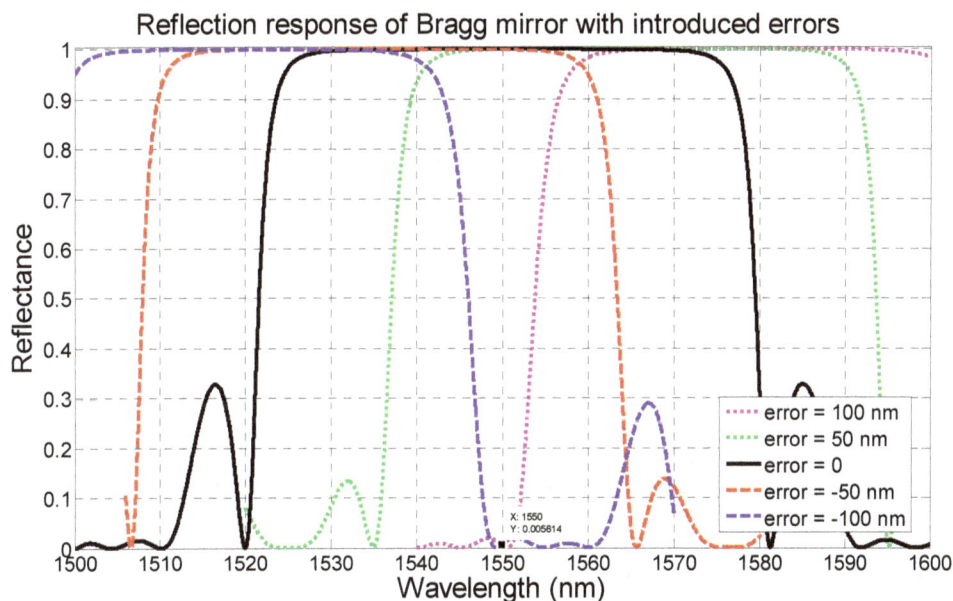

Fig. 8. Reflectance of 4 HL pairs for L thicknesses = $9\lambda_{air}/4$, H thickness = $33\lambda_{Si}/4$ and several introduced errors.

(increase or decrease) and the air layer follows the opposite trend (decreases or increases). Then, the device performance degrades. This issue is obvious in Fig. 8 where various error values are introduced into the mirror original design. We notice that the overall response shifts toward the left side as the error decreases from 100 nm to -100 nm in steps of 50 nm. Comparing the obtained responses to the error free design, we see that the mirror reflectance might turn from 99.98 % ideally to 0.6 % for an introduced error = ± 100 nm which means that the multilayered designs are not tolerant to fabrication errors exceeding 50 nm.

3.4 Modeling and simulation FP cavity based on Bragg mirrors

If instead of the stack of high reflectance mirror, we introduce a gap layer whose thickness is an integer number of half the wavelength then the characteristic matrix [M] of this layer becomes:

$$[M] = \pm \begin{bmatrix} 1 & 0 \\ 0 & 1 \end{bmatrix} \tag{45}$$

Thus, we can easily get a Fabry-Perot (FP) resonator if we combine two stacks of quarter wavelengths thick acting as high reflectance mirrors separated by a gap layer of half wavelength thick.

In the next part, we illustrate, by the help of MATLAB simulations, the properties of such FP resonators where we study the impact of several parameters on the resonator spectral response. Parameters of particular interest for this comprehensive study: the mirror reflectance controlled by the number of Bragg layers per mirror, the impact of technological errors and the cavity gap length. In what follows, unless otherwise stated, we consider that the silicon Bragg layers of thickness = 3.67 μm ($33\lambda_{Si}/4$), the air Bragg layer has a thickness of 3.49 μm ($9\lambda_{air}/4$) and the gap layer has a width = 10.075 μm ($13\lambda_{Si}/2$). The silicon refractive index n_{Si} is taken = 3.478 and all the layers are considered absorption free.

We start our study by increasing gradually the number of Bragg layers. As shown in Fig. 9 , we found that the FWHM of the resonator decreases from 7.6 nm for single Si layer/mirror, to 0.56 nm for double Si layer/mirror, to 0.046 nm for 3 Si layers/mirror and finally the FWHM becomes 0.004 nm for 4 Si layers/mirror. This is due to the increase in the mirror reflectance which goes from 71.8 % for single layer to 99.98 % for 4 Si layers. Also, the contrast improves and the minimum level goes from -10 dB up to -70 dB and the resonator sharpness improves as well.

Now, if we consider the case of 4 Si layers/mirror with introduced errors (ε) therein, we obtain the curves shown in Fig. 10 .We notice that the central wavelength λ_0 shifts from 1550 nm by ±8.5 nm as ε = ± 50 nm. For ε = ± 100 nm, λ_0 shifts by 18.15 nm. In addition, the FWHM of the peak increases from 0.004 nm for the error free case to 0.007 nm for ε = ± 50 nm and it reaches 0.029 nm for ε = ±100 nm. This might be explained by reference to previous simulations carried on Bragg mirrors with introduced errors. As mentioned earlier, the overall response of the mirror shifts right (left) as error increases (decreases) and this is the reason underlying the shift in the resonance wavelength. In addition, the maximum reflectance of the mirror decreases from 99.98 % (in the error free case) to 99.97 % (for ±50 nm error) to 99.93 % (for ±100 nm), that's why the FWHM increases.

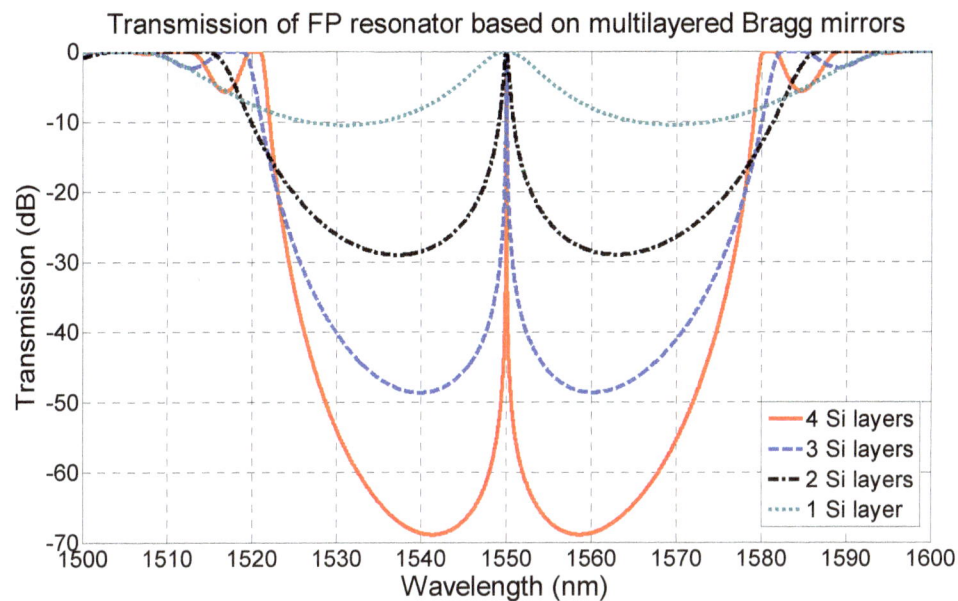

Fig. 9. Transmission of FP resonator for different number of silicon layers per mirror.

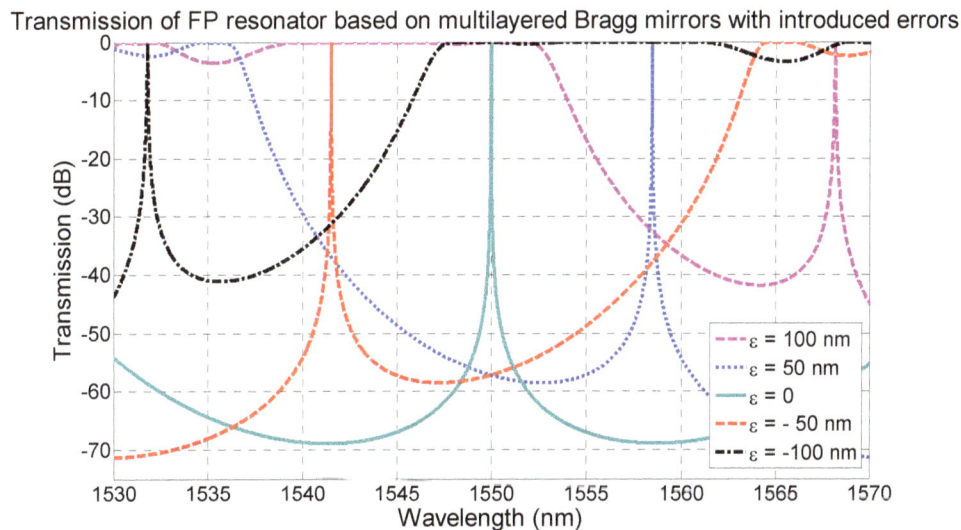

Fig. 10. Transmission of FP resonator for different errors.

By scanning over the wavelength for the cases of ε = -50 nm and ε = -100 nm, we notice that other resonance peaks, with larger FWHM and reduced contrast, appear in the spectral response of the cavity. This result seems strange and it does not go in accordance with the designed FSR for the error free cavity. In fact, the designed cavity gap length = 10.075 μm corresponding to a quasi FSR = 119.2 nm and a resonance wavelength = 1550 nm.

This issue might be explained by looking over the reflection response of the Bragg mirrors with introduced errors as shown in Fig. 11, we find that they are shifted as compared to the error free design. Moreover, they exhibit a non-negligible reflectance between 1575 nm and 1600 nm and so the design performs as a good resonator.

Analyzing the simulation results, we come out with a new definition for the cavity length named: The effective length L_{eff}. This new parameter suggests that the effective reflecting interfaces of the resonator lie inside the Bragg reflectors and not between the inner interfaces as conventionally thought and so it gives rise to unexpected resonances within the quasi FSR. Making inverse calculations for the simulation results shown in Fig. 12, we find that for $\varepsilon = -50$ nm, the FSR = 52.15 nm corresponding to $L_{eff} = 23$ μm and for $\varepsilon = -100$ nm, the FSR = 47.7 nm corresponding to $L_{eff} = 25.18$ μm.

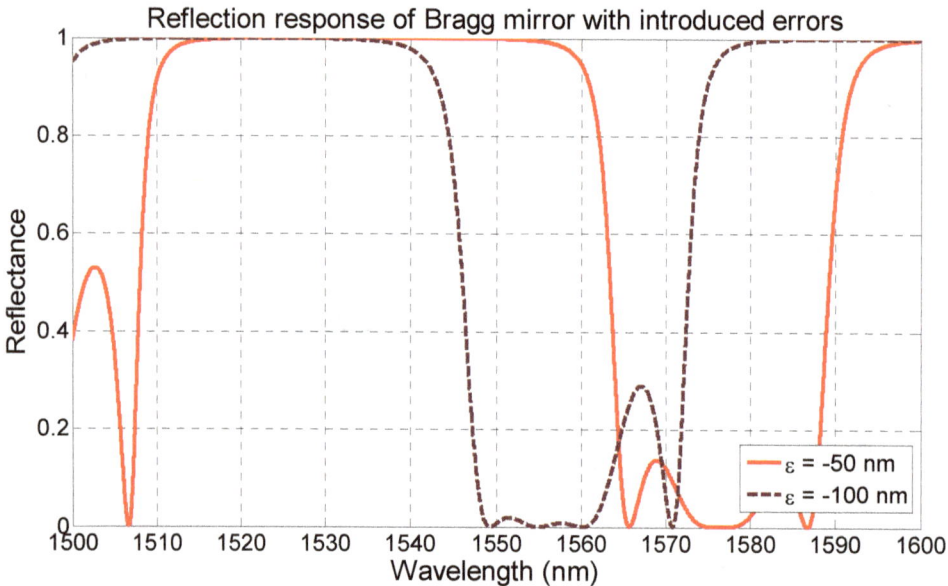

Fig. 11. Reflection response of Bragg mirrors for errors = -50 nm and -100 nm.

3.5 Multilayered Si-Air structures for anti-reflection purposes

Antireflection surfaces (usually obtained through additional material coatings) can be obtained also from silicon micromachinned Bragg structures. They can range from a simple single layer having virtually zero reflectance at just one wavelength, to a multilayer system of more than a dozen of layers, having ideally zero reflectance over a range of several decades. The type used in any particular application will depend on a variety of factors, including the substrate material, the wavelength region, the required performance and of course, the cost. There is no systematic approach for the design of antireflection coatings. Trial and error assisted by approximate techniques and by accurate computer calculation, is frequently employed. Very promising designs can be further improved by computer refinement. Several different approaches can be used in designing AR coating. In this section, we will limit our discussion to the single layer design only. Complicated analytical

Transmission of FP resonator based on multilayered Bragg mirrors with introduced errors

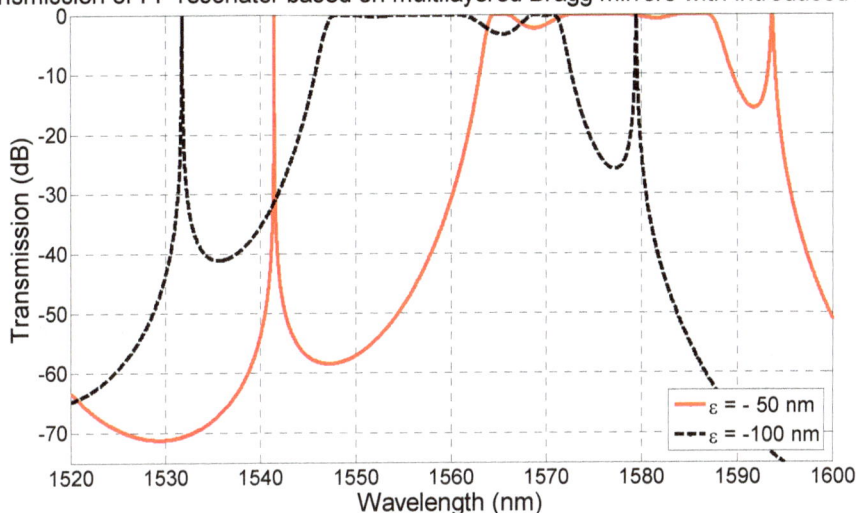

Fig. 12. Zoom out on the transmission response of FP resonator for errors = -50 nm and -100 nm.

formulas can be derived for the case of multilayer coating and they lie outside the scope of this work so they will not be presented.

The vast majority of antireflection coatings are required for matching an optical element into air. The simplest form of antireflection coating is a single layer. Consider Fig. 13. Since two interfaces are involved, we have two reflected rays, each representing the amplitude reflection coefficient at an interface. If the incident medium is air, then, provided the index of the film is lower than the index of the substrate, the reflection coefficient at each interface will be negative, denoting a phase change of 180°. The resultant minimum is at the wavelength for which the phase thickness of the layer is 90°, that is, a quarter-wave optical thickness, when the two rays are completely opposed. Complete cancellation at this wavelength, that is, zero reflectance, will occur if the rays are of equal length. This condition, in the notation of Fig. 13, is

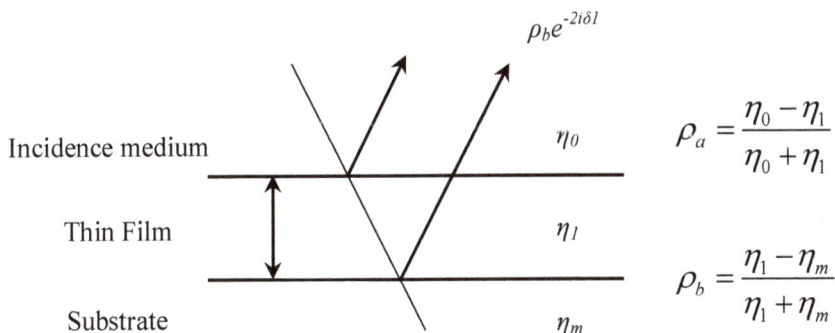

$$\rho_b e^{-2i\delta l}$$

Incidence medium $\quad \eta_0 \qquad \rho_a = \dfrac{\eta_0 - \eta_1}{\eta_0 + \eta_1}$

Thin Film $\quad \eta_1$

$$\rho_b = \dfrac{\eta_1 - \eta_m}{\eta_1 + \eta_m}$$

Substrate $\quad \eta_m$

Fig. 13. Schematic illustration of substrate coated with a single film

$$\frac{y_0 - y_1}{y_0 + y_1} = \frac{y_1 - y_m}{y_1 + y_m} \tag{46}$$

which requires:

$$\frac{y_1}{y_0} = \frac{y_m}{y_1} \tag{47}$$

Or,

$$n_1 = \sqrt{n_0 n_m} \tag{48}$$

The condition for a perfect single-layer antireflection coating is, therefore, a quarter-wave optical thickness of material with optical admittance equal to the square root of the product of the admittances of substrate and medium. It is seldom possible to find a material of exactly the optical admittance which is required. If there is a small error, ε, in y_1 such that:

$$y_1 = (1 + \varepsilon)\sqrt{y_0 y_m} \tag{49}$$

Then,

$$R = \left(\frac{-2\varepsilon - \varepsilon^2}{2 + 2\varepsilon + \varepsilon^2}\right)^2 \approx \varepsilon^2 \tag{50}$$

provided that ε is small. A 10 % error in y_1, therefore, leads to a residual reflectance of 1 %.

Zinc sulphide has an index of around 2.2 at 2 µm. It has sufficient transparency for use as a quarter-wave antireflection coating over the range 0.4–25 µm. Germanium, silicon, gallium arsenide, indium arsenide and indium antimonide can all be treated satisfactorily by a single layer of zinc sulphide. There is thus no room for manoeuvre in the design of a single-layer coating.

In practice, the refractive index is not a parameter that can be varied at will. Materials suitable for use as thin films are limited in number and the designer has to use what is available. A Better approach, therefore, is to use more layers, specifying obtainable refractive indices for all layers at the start, and to achieve zero reflectance by varying the thickness. Then, too, there is the limitation that the single-layer coating can give zero reflectance at one wavelength only and low reflectance over a narrow region. A wider region of high performance demands additional layers.

3.6 Tilted FP cavity as a notch filter

In this part, we focus on another interesting application for devices based on Bragg structures. In particular, we study FP cavity based on multilayered mirrors but under oblique incidence. The device design differs from the case of normal incidence since the rays will propagate obliquely in the layers and the optical thicknesses for both the silicon and the air layers shall be calculated differently. In this case, we must ensure that $\delta = m\pi/2$ to obtain the same matrix as in equation (41), and then we will solve the problem inversely to get the corresponding thicknesses $H(L) = d_{Si(Air)}$ which yields:

$$\delta_{Si(air)} = \frac{2\pi n_{Si(air)} d_{Si(air)}}{\lambda_0} \cos\left(\theta_{Si(air)}\right) = \frac{m\pi}{2} \tag{51}$$

$$d_{Si(air)} = \frac{m\lambda_0}{4n_{Si(air)} \cos\left(\theta_{Si(air)}\right)} \tag{52}$$

Using equation (52), we will consider $H=d_{Si}$= 3.76 μm using odd multiple m = 33 and $L=d_{Air}$ = 3.84 μm using the odd multiple m = 7. In the upcoming simulations, we will take the thickness of the HL layers as mentioned previously. For the gap thickness G under oblique incidence, we have to satisfy the condition $\delta = m\pi$. By following the same analytical treatment as before, we will get:

$$G = \frac{m\lambda_0}{2n_{air} \cos\left(\theta_{air}\right)} \tag{53}$$

So, we will consider G = 14.25 μm using odd multiple m = 13.

The studied architecture consists of two stacks of tilted Bragg mirrors separated by an air gap layer. While the FP configuration with normal incidence works only in transmission, the tilted architecture, shown in Fig. 14, allows working either in transmission or in reflection. In the case of tilted FP, it behaves as a notch filter, suitable for dropping a particular wavelength. This is due to the 45° tilt angle of the cavity with respect to incident light.

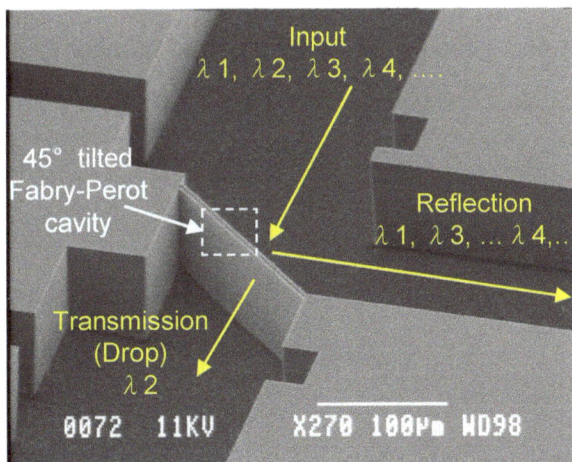

Fig. 14. 45° tilted FP filter made of two Si layers separated by an air gap

Simulating a structure based on the parameters stated above, we obtain the results shown in Fig. 15 and Fig. 16. As obvious, the FWHM of the filter reduces as the number of Silicon layers/mirror increases as it translates into higher reflectance. This device might have good potential in WDM systems where it can be used as an Add-Drop multiplexer. Also, it might be of interest for application involving tunable lasers as will be detailed in the next section of this chapter.

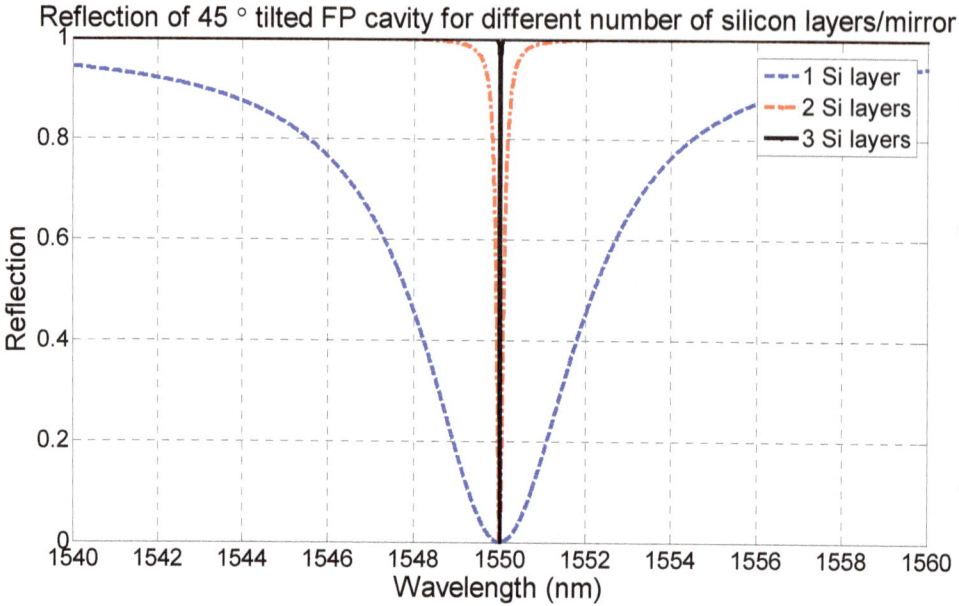

Fig. 15. Reflection response of tilted FP cavity for different number of silicon layers/mirror

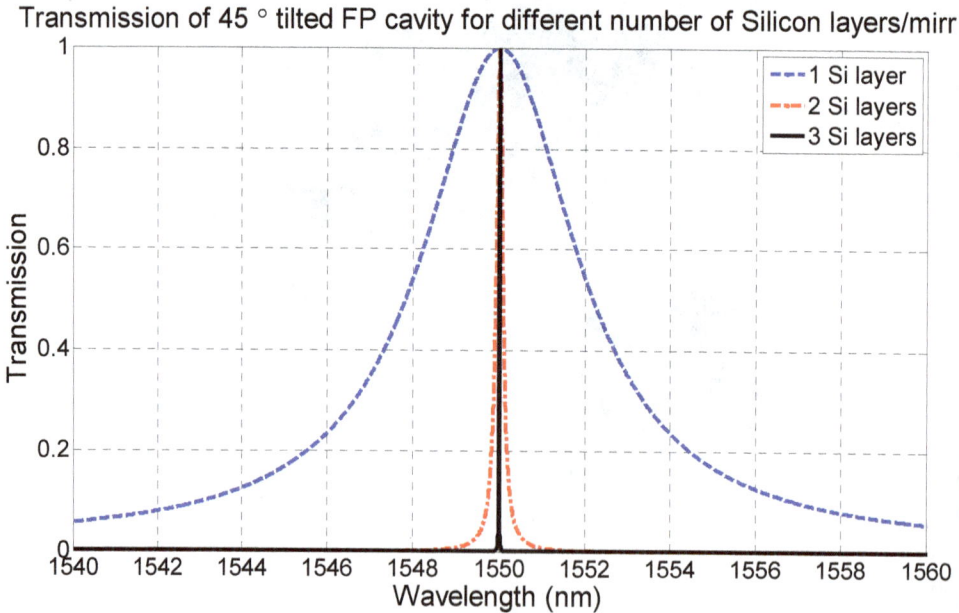

Fig. 16. Transmission response of tilted FP cavity for different number of silicon layers/mirror

Simulation results show that the FWHM decreases from 4.5 nm for the single silicon layer to 0.18 nm for the double layer design and it exhibits further decreases to 0.008 nm for the triple layer design. Now, if we consider a tilted FP cavity with mirrors of *HLH* configuration but with different angles of incidence, we obtain a spectral response with a shift in the resonance wavelength as illustrated in Fig. 17. Varying the angle of incidence by 0.5° around 45° results in 9 nm shift of the resonance wavelength. Then, proper design for rotational actuator to integrate with the tilted cavity, suggests the use of the whole package as a MEMS tunable filter. The next section highlights the potential of the tilted FP cavity in tunable laser source module.

A last point to mention about the tilted FP cavity is the sensitivity of the design to fabrication errors. Considering a HLH combination for both mirrors, and introducing errors from 100 nm down to -100 nm in steps of 50 nm, we notice from Fig. 18 that the resonance wavelength shifts by about ±7 nm for an increase of ±50 nm. Also, the FWHM increases from 0.18 nm for the error free design to 0.25 nm for an introduced error of 50 nm. It reaches 0.55 nm for an introduced error of 100 nm. Thus, the structure is not very tolerant to fabrication errors and the filter shall be designed, fabricated and tested carefully before integration into optical systems.

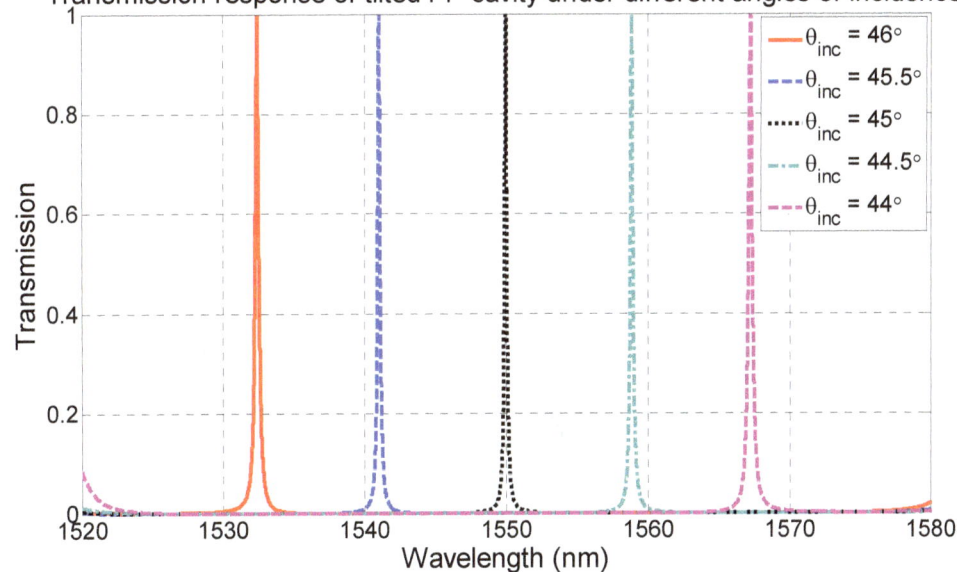

Fig. 17. Transmission response of tilted FP cavity in HLH configuration under different angles of incidence

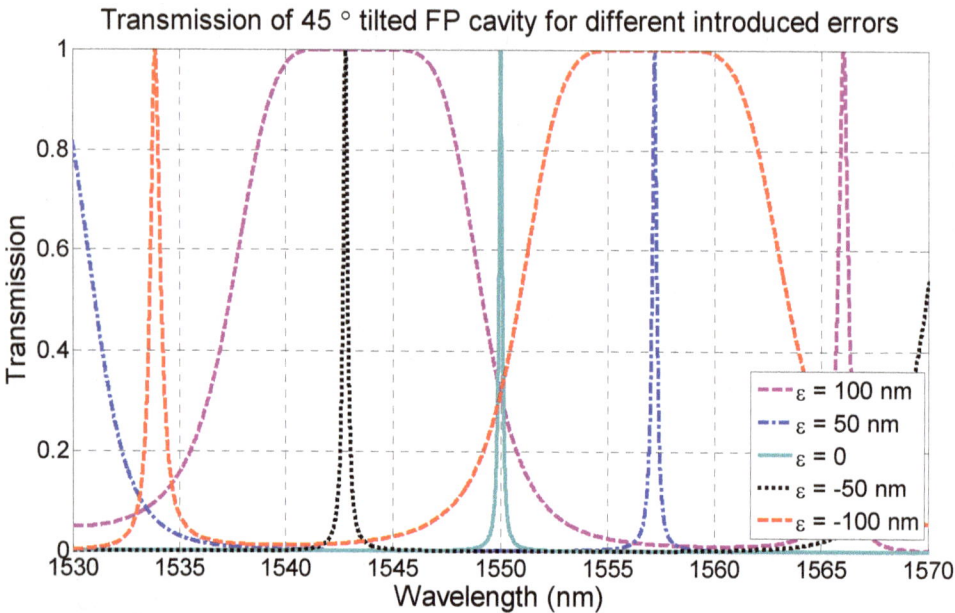

Fig. 18. Transmission response of tilted FP cavity for FP cavity with HLH mirror for different errors.

3.7 Tilted FP cavity as a mode selector

By completing the architecture surrounding the tunable tilted FP cavity with an active laser cavity and an external mirror, then we obtain a compact tunable laser by tuning the angle incident upon the tilted FP cavity. As mentioned above, the tuning might be achieved by rotating the tilted FP. Tilted FP cavities are of special interest, since they reject undesirable wavelengths off the optical axis. Therefore, they appear as interesting candidates for mode selection in external cavity tunable lasers. Indeed, as these types of lasers exhibit a competition between several longitudinal modes, there is a need for a mode selection mechanism in order to obtain single mode operation and avoid mode hopping during tuning. The main interest in using tilted FP etalon rather than a FP cavity with normal incidence is to avoid parasitic reflections due to additional FP cavities that appear when adding the mode selector. Fig. 19a illustrates the principle of the mode selector based on a 45° tilted FP cavity. The corresponding simulated transmission is shown as well, which confirms the operation principle. It is worth mentioning that the performed simulation is very basic, since it does not take into account losses. In particular, plane waves are considered here rather than Gaussian beams. Figs. 19b and Fig. 19c illustrate simulations of parts of the architecture.

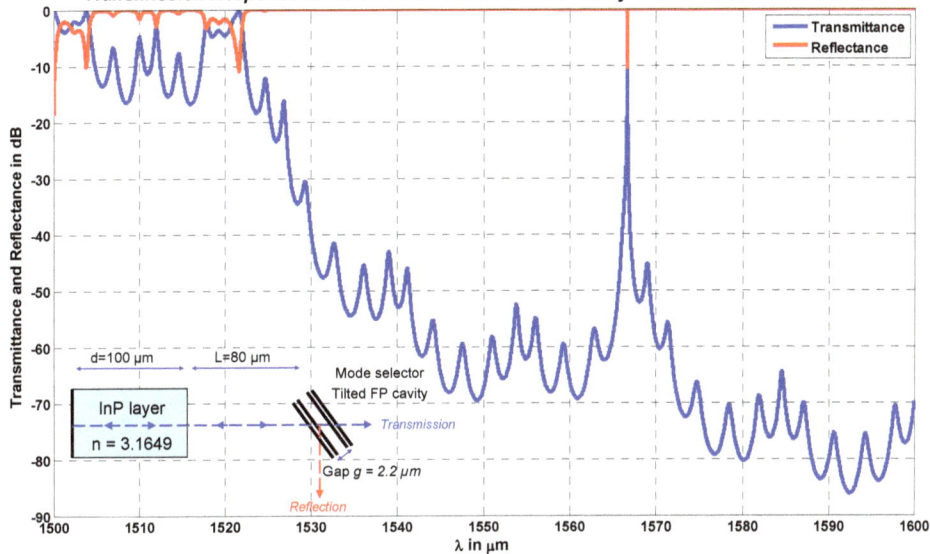

(a)

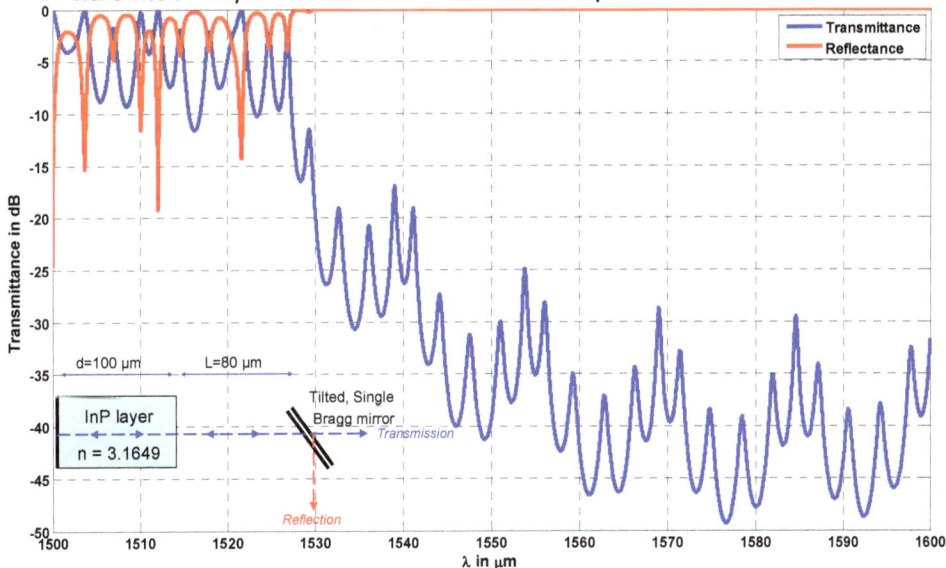

(b)

Transmission and Reflectance for the Mode Selector "tilted Fabry-Perot cavity" based on a 7 layer Bragg mirror

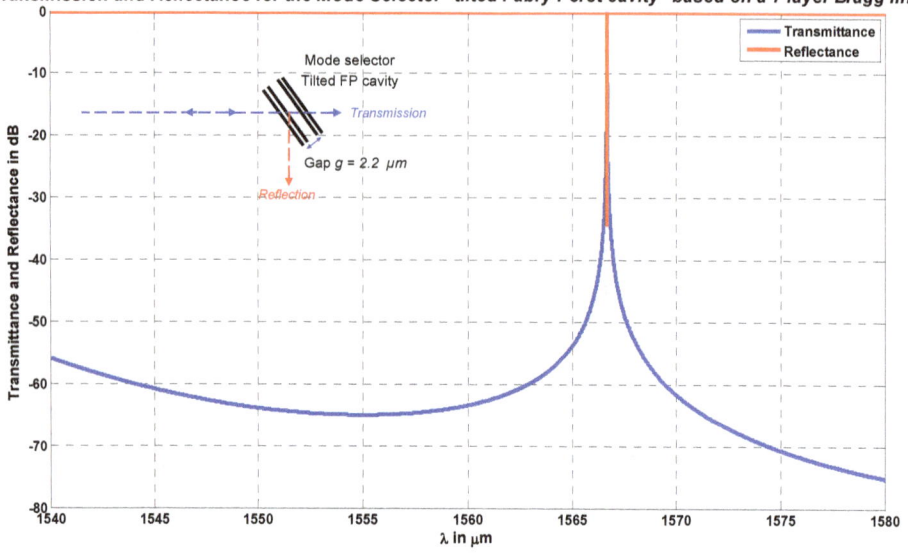

(c)

Fig. 19. Tilted FP etalon as a laser mode selector. Whole systems (a) and parts of the system (b) and (c)

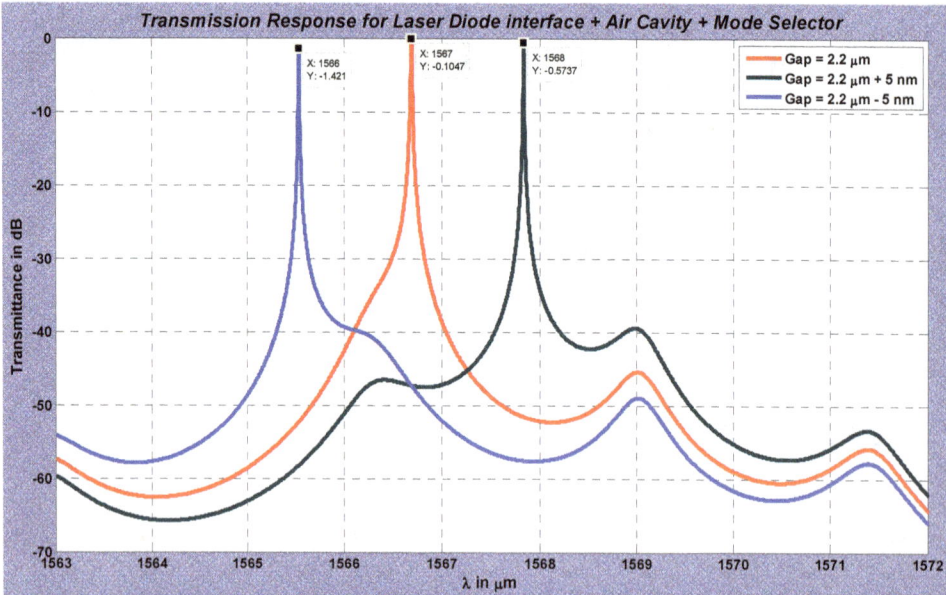

(a)

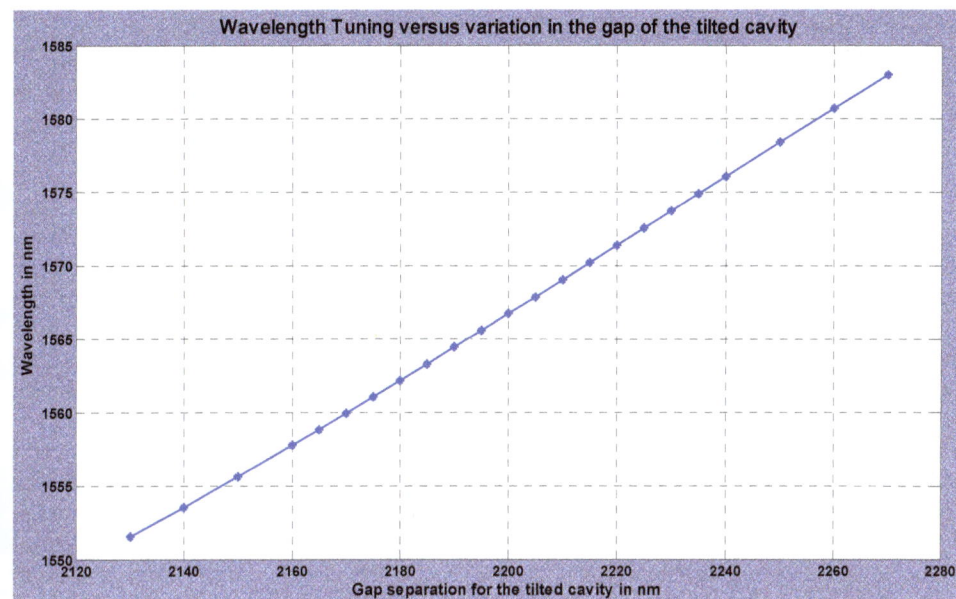

(b)

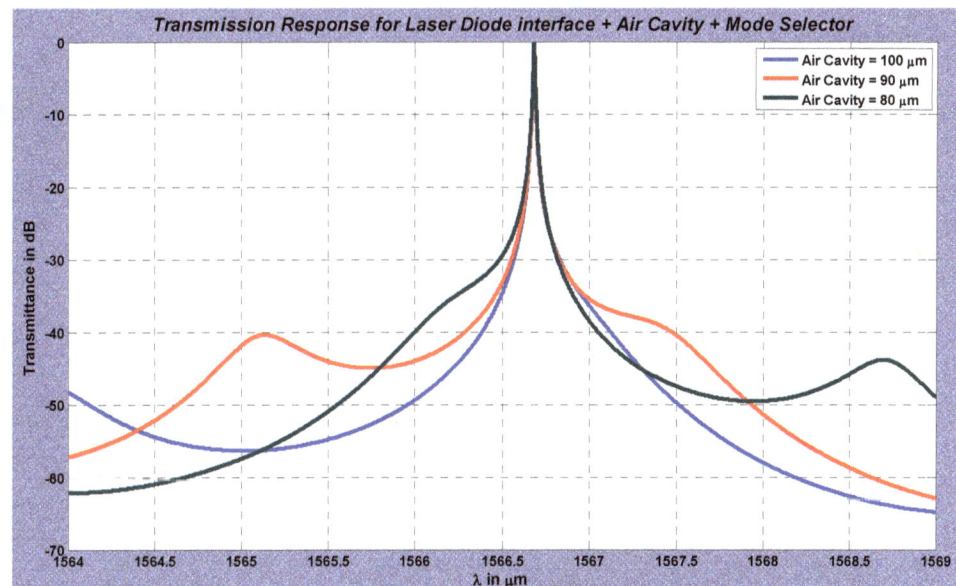

(c)

Fig. 20. (a) and (b) Simulated wavelength tuning by control of the gap g of the tilted FP cavity (c) no wavelength shift is noticed when varying the distance L between the mode selector and the InP layer.

Tuning is achieved either by rotating the cavity further or by controlling its gap g, as shown in Figs. 20a and 20b. Tuning range of 30 nm is shown as the result of gap tuning of 150 nm. The increase in the separation distance L doesn't affect the peak position as shown in Fig. 20c.

4. Advanced FP architecture

In this last section, we present two advanced architecture of FP cavity based on cylindrical 1D photonic crystal vertically etched in silicon. The first architecture is based on cylindrical Bragg mirrors to focus light beam along one transverse beam. SEM Photo of a device based on single silicon layer is presented in Fig. 21. The measured characteristic is shown in Fig. 22 pertain to three different spacing between the injection fiber and the input mirror. Numerical modeling confirms the measurements and reveals that the device exhibits selective excitation of transverse modes TEM_{20}. For more details, the interested reader may refer to [Malak et al. Transducers 2011] [Malak et al. JMEMS 2011]. The second architecture however, aims to focus the light beam in both transverse planes to reduce losses introduced by Gaussian beam expansion as well. For this purpose, the cylindrical Bragg is combined with a fiber rod lens to focus the light beam in the other transverse plane. Since the second architecture is not common, a stability model has been devised to enable the design of stable resonator [Malak el al. JMST 2011]. Photo of the realized device and corresponding response is shown in Fig. 23. This architecture provides a high quality factor (~9000) for a Bragg mirror based on four silicon layers. It has a strong potential for spectroscopic applications.

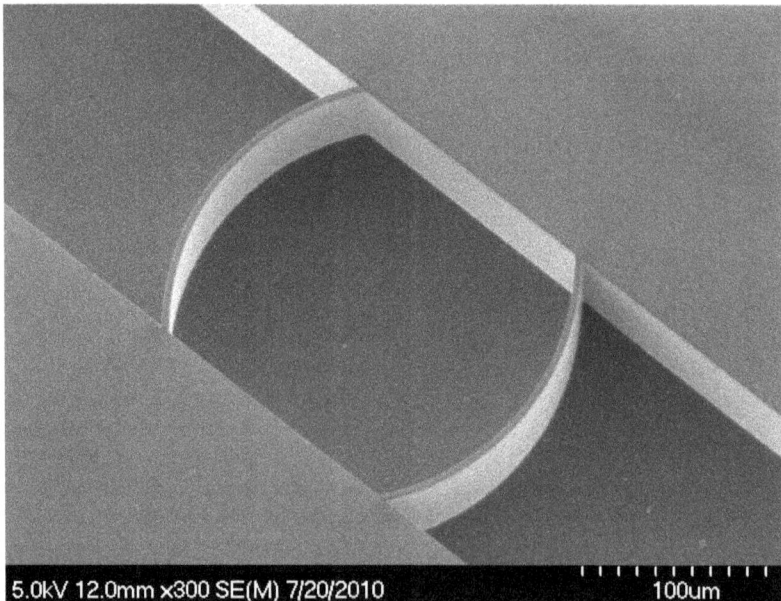

5.0kV 12.0mm x300 SE(M) 7/20/2010   100um

Fig. 21. SEM photo of the FP cavity with single cylindrical silicon layer measured with cleaved fibers.

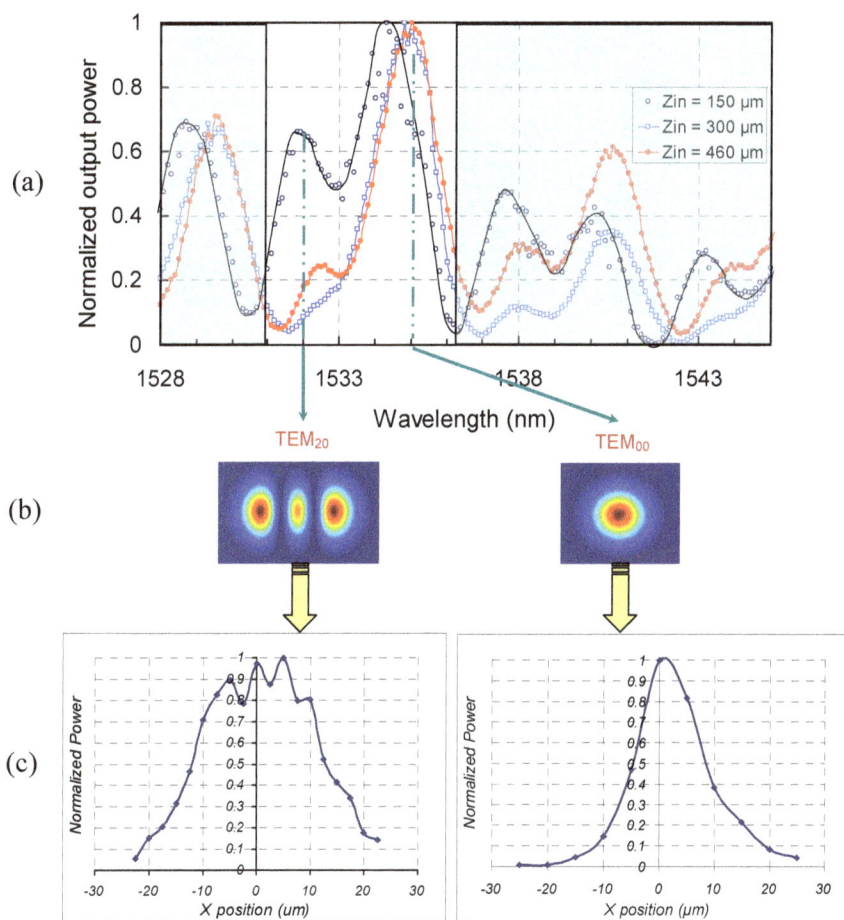

Fig. 22. Highlights on "Wavelength selective switching" and "Mode selective filtering" of the curved FP cavity (a) Recorded spectral response of the cavity, measured with lensed fiber while varying the fiber-to-cavity distance D. The quasi-periodic pattern of the curve reveals selective excitation of the resonant transverse modes TEM_{20} around 1532 nm in addition to the fundamental Gaussian mode TEM_{00}. Varying the distance D leads to different levels for mode TEM_{20} with an extinction ratio of 7:1, the maximum amplitude was at D=150 μm. (b) Ideal intensity distribution of TEM_{00} and TEM_{20} modes. (c) Measured intensity profiles (of modes TEM_{00} and TEM_{20}) obtained by lateral in-plane scanning of the detection fiber.

(a)

(b)

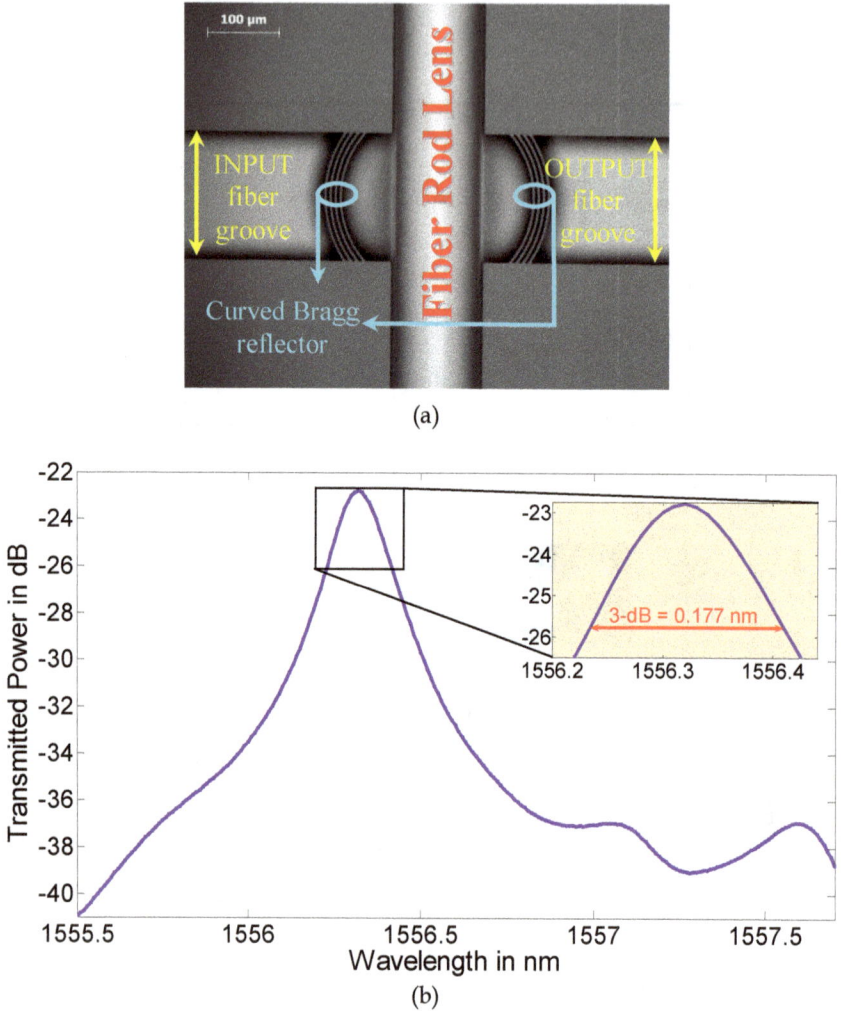

Fig. 23. (a) Top view of the curved FP cavity with the fiber rod lens (b) Typical response obtained from such device

5. Concluding remarks

1D photonic crystal structure acquired a high interest long ago due to the application domain they touch. As outlined in this chapter, they constitute a basic building block in many devices like FP resonators, multilayered coating. The attractiveness in them comes from their easy design and modeling based on multilayered stack theory and the affordable fabrication process, thanks to the advance in the fabrication processes, in particular, the advance in the DRIE process which helped producing vertical Bragg on silicon. In this context, this chapter focused on specific issues concerning 1D photonic crystal: design and modeling, fabrication technology, common applications and a brief introduction to an advanced application: The curved FP cavity.

6. References

Lipson, A. & Yeatman, E.M. (2005). Free-Space MEMS tunable optical filter on (110) silicon, *IEEE/LEOS International Conference on Optical MEMs and Their Applications*, Oulu, Finland, 1-4 August pp. 73-74

Lipson, A. & E. Yeatman, E. M. (2007). A 1-D Photonic Band Gap Tunable Optical Filter in (110) Silicon, *Journal of Microelectromechanical Systems*, vol. 16, no.3, pp. 521-527.

Macleod, H. A. (2001). *Thin Film Optical Filters*, ISBN 0 7503 0688 2, London, UK

Malak, M.; Pavy, N.; Marty, F.; Peter, Y.-A.; Liu, A. Q. & Bourouina, T. (2011). Micromachined Fabry–Perot resonator combining submillimeter cavity length and high quality factor, *Applied Physics Letters*, vol 98, 211113/1-3.

Malak, M.; Pavy, N.; Marty F. & Bourouina, T. (2011). Mode-Selective Optical Filtering And Wavelength-Selective Switching Through Fabry-Perot Cavity With Cylindrical Reflectors, *16th International Conference on Solid-State Sensors, Actuators and Microsystems (TRANSDUCERS)*, pp. 534-537.

Malak, M.; Pavy,N.; Marty, F.; Peter, Y.-A.; Liu, A. Q. & Bourouina, T. Cylindrical Surfaces Enable Wavelength-Selective Extinction and Sub-0.2 nm Linewidth in 250 μm-Gap Silicon Fabry-Pérot Cavities, submitted to the *Journal of Microelectromechanical systems*.

Malak, M.; Pavy,N.; Marty, F.; Richalot, E.; Liu, A. Q. & Bourouina, T. (2011) Design, modeling and characterization of stable, high Q-factor curved Fabry–Perot cavities, *Journal of Microsystem Technologies*, vol 17, no 4, pp. 543-552

Marty, F.; Rousseau, L.; Saadany, B.; Mercier, B.; Français, O.; Mita, Y. & Bourouina, T. (2005). Advanced Etching of Silicon Based On Deep Reactive Ion Etching For Silicon High Aspect Ratio Micro Structures And Three-Dimensional Micro- And Nano-Structures, *Microelectronics Journal*, vol. 36, pp. 673-677.

Pruessner, M.W.; Stievater, T.H. & Rabinovich, W.S. (2008). Reconfigurable Filters Using MEMS Resonators and Integrated Optical Microcavities, *IEEE MEMS conference*, pp.766-769.

Saadany, B.; Malak, M.; Kubota, M.; Marty, F.; Mita, Y.; Khalil, D. & Bourouina, T. (2006). Free-space Tunable and Drop Optical Filters Using Vertical Bragg Mirrors on Silicon, *IEEE Journal of Selected Topics in Quantum Electronics*, vol 12, no. 6, pp.1480-1488.

Yun, S. & Lee, J. (2003). A Micromachined In-Plane Tunable Optical Filter Using the Thermo-optic Effect of Crystalline Silicon, *Journal of Micromechanics and Microengineering*, vol. 13, pp.721-725.

Yun, S-S; You, S-K & Lee, J-H (2006). Fabrication of vertical optical plane using DRIE and KOH crystalline etching of (110) silicon wafer, *Sensors Actuators A*, vol. 128, pp. 387-394.

Song, I.-H. ; Peter, Y.-A. & Meunier, M. (2007). Smoothing dry-etched microstructure sidewalls using focused ion beam milling for optical applications, *Journal of Micromechanics and Microengineering*, vol. 17, pp.1593-1597.

Zandi, K.; Wong, B.; Zou, J.; Kruzelecky, R. V.; Jamroz W. & Peter, Y.-A. (2010). In-Plane Silicon-On-Insulator Optical MEMS Accelerometer Using Waveguide Fabry-Perot Microcavity With Silicon/Air Bragg Mirrors, *in Proc. 23rd IEEE Int. Conf. Micro Electro Mech. Syst.*, pp. 839-842.

Part 3

Photonic Crystal Fiber

8

Photonic Crystal Fibre Interferometer for Humidity Sensing

Jinesh Mathew, Yuliya Semenova and Gerald Farrell

Photonics Research Centre, Dublin Institute of Technology
Ireland

1. Introduction

Photonic crystal fibres (PCFs), which are also called microstructured optical fibres or holey fibres, have been extensively investigated and have considerably altered the traditional fibre optics since they appeared in the mid 1990s [Knight et al., 1996; Knight, 2003; Russell, 2003]. PCFs have a periodic array of microholes that run along the entire fibre length. They typically have two kinds of cross sections: an air–silica cladding surrounding a solid silica core or an air–silica cladding surrounding a hollow core. The light-guiding mechanism of the former is provided by means of a modified total internal reflection (index guiding), while the light-guiding mechanism of the latter is based on the photonic band gap effect (PBG guiding). The number, size, shape, and the separation between the air-holes as well as the air-hole arrangement are what confer PCFs unique guiding mechanism and modal properties [Russell, 2006]. This gives PCF many unique properties such as single mode operation over a wide wavelength range [Birks et al., 1997], very large mode area [Knight et al., 1998], and unusual dispersion [Renversez et al., 2003]. Because of their freedom in design and novel wave-guiding properties, PCFs have been used for a number of novel fibre-optic devices and fibre-sensing applications that are difficult to be realized by the use of conventional fibres.

While optical interferometers offer high resolution in metrology applications, the fibre optic technology additionally offers many degrees of freedom and some advantages such as stability, compactness, and absence of moving parts for the construction of interferometers. The two commonly followed approaches to build fibre optic interferometer are: two arm interferometer and modal interferometer. Two- arm interferometer involves splitting and recombining two monochromatic optical beams that propagate in different fibres which requires several meters of optical fibre and one or two couplers. Modal interferometer exploits the relative phase displacement between two modes of the fibre. In modal interferometers compared to their two-arm counterparts the susceptibility to environmental fluctuations is reduced because the modes propagate in the same path or fibre. Recently the unique properties of the photonic crystal fibre have attracted the sensor community. Design of PCF based interferometers in particular is interesting owing to their proven high sensitivity and wide range of applications. Photonic crystal fibre based modal interferometers include PCFs in a fibre loop mirror [Zhao et al., 2004], interferometer built with long period gratings [Lim et al., 2004], interferometers built with tapered PCFs

[Monzón-Hernández et al., 2008], and interferometers fabricated via micro-hole collapse [Choi et al., 2007; Villatoro et al., 2007a]. The latter technique is really simple since it only involves cleaving and splicing. The different configurations reported so far are a PCF with two collapsed regions separated by a few centimetres [Choi et al., 2007], a short section of a PCF longitudinally sandwiched between standard single mode fibres by fusion splicing (transmission type) [Villatoro et al., 2007] and a stub of PCF with cleaved end fusion spliced at the distal end of a single mode fibre (reflection type) [Jha et al., 2008]. The advantage of the last two configurations is that the modal properties of the PCF are exploited but the interrogation is carried out with conventional optical fibres, thus leading to more cost-effective interferometers. The interferometer with the latter configuration is demonstrated in this chapter as a relative humidity or dew sensor. The sensor presented has the unique advantages such as it does not require any special coatings to measure humidity. Also since the sensor head is made of single material (silica) it can be used in harsh and high-temperature environments to monitor humidity.

In section 2 of the chapter the operating principle of a reflection type photonic crystal fibre interferometer (PCFI), its fabrication and the dependence of the interferometer's fringe spacing on the length of the PCF are presented. Section 3 explains the water vapor adsorption/desorption phenomena on a silica surface, the working principle of a relative humidity sensor based on PCF interferometer and the humidity response of the PCF interferometer. Section 4 demonstrates the use of the PCFI as a dew sensor. The section presents the basic sensing principle of the dew sensor, the temperature dependence and the dew response of the PCF interferometer. A dew point hygrometer using PCF interferometer is also proposed in this section.

2. Photonic crystal fibre interferometer

Photonic crystal fibre interferometers based on micro-hole collapse have attained great importance in recent times due to the simple fabrication process involved and excellent sensing performance [Villatoro et al., 2007, 2009a, 2009b]. A reflection-type PCFI consists of a stub of PCF fusion spliced at the distal end of a single mode fibre [Mathew et al., 2010]. The key element of the device is the hole collapsed region close to the splice point. Some advantages of the PCF interferometers fabricated using microhole collapse are that since interferometers are fabricated by fusion splicing the splice is highly stable even at high temperatures and also its characteristics will not degrade over time.

2.1 PCFI working principle

In a PCFI the excitation and recombination of modes can be carried out by the hole collapsed region of the PCF [Choi et al., 2007; Villatoro et al., 2007]. A microscopic image of the PCFI and a schematic of the excitation and recombination of modes in the PCFI are shown in Fig. 1. The fundamental SMF mode begins to diffract when it enters the collapsed section of the PCF. Because of diffraction, the mode broadens; depending on the modal characteristics of the PCF and the hole collapsed region, the power in the input beam can be coupled to the fundamental core mode and to higher order core modes [Villatoro et al., 2007, 2009b; Barrera et al., 2010] or to cladding modes [Cárdenas-Sevilla et al., 2011; Choi et al., 2007; Jha et al., 2008] of the PCF. The modes propagate through the PCF until they reach the cleaved end from where they are reflected. Since the modes propagate at different phase velocities, thus in a certain length of PCF the modes accumulate a differential phase shift. Therefore constructive or destructive interference occurs along the length of PCF. The phase velocities and phase difference are also wavelength dependent; therefore the optical power reflected by the device will be a maximum at certain wavelengths and minimum at others [Villatoro et al., 2009b]. When the reflected modes re-enter the collapsed region they will further diffract and because the mode field of the SMF is smaller, the core acts as a spatial filter and picks up only a part of the resultant intensity distribution of the interference pattern in the PCF.

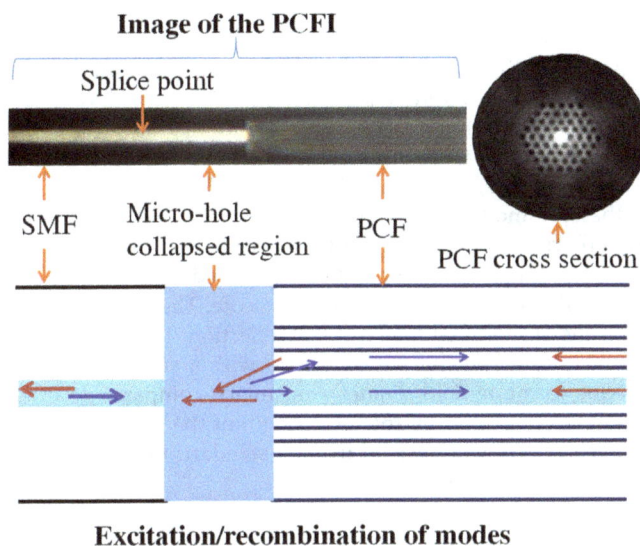

Fig. 1. Microscope image of the PCFI (upper) & a schematic of the excitation/recombination of modes in the hole collapsed region (lower).

A regular interference pattern in the reflection spectrum of the PCFI suggests that only two modes are interfering in the device. In our reported work [Mathew et al., 2010] on a PCFI using LMA 10 fibre, based on the fact that higher order modes can exist in the core of a PCF with a short length [Káčik et al., 2004; Uranus et al., 2010], the interfering modes in the PCF are considered as two core modes. However in a later experiment, which involved varying the refractive index surrounding the cladding of a PCFI, good ambient refractive index sensitivity is observed for a PCFI fabricated using the same LMA 10 fibre. This suggests that the interfering modes are a core mode and a cladding mode of the PCF, a conclusion that is supported by [Choi et al., 2007; Cárdenas-Sevilla et al., 2011] for an LMA10 fibre. Thus considering a core mode and a cladding mode as the interfering modes of the PCFI and designating the effective refractive indexes of the core mode as n_c and cladding mode as n_{cl}, the accumulated phase difference is $2\pi\Delta n(2L)/\lambda$, where $\Delta n=n_c-n_{cl}$, λ the wavelength of the optical source, and L the physical length of the PCFI [Villatoro et al., 2009a]. The power reflection spectrum of this interferometer will be proportional to $\cos(4\pi\Delta nL/\lambda)$. The wavelengths at which the reflection spectrum shows maxima are those that satisfy the condition $4\pi\Delta nL/\lambda=2m\pi$, with m being an integer. This means that a periodic constructive interference occurs when $\lambda m = (2\Delta nL/m)$. If some external stimulus changes Δn (while L is fixed) the position of each interference peak will change, a principle which allows the device to be used for sensing.

2.2 PCFI fabrication

Fusion splicing of the PCF to the SMF is undertaken using the electric arc discharge of a conventional arc fusion splicer. During the splicing process the voids of the PCF collapse through surface tension within a microscopic region close to the splice point. In fabricating such an interferometer, one critical condition for good sensor performance is achieving a regular interference pattern and good interference fringe visibility. The visibility of the interferometer depends on the power in the excited modes, which in turn depends on the length of the collapsed region [Barrera et al., 2010]. However a long collapsed region length causes activation of many cladding modes and therefore degrades the sinusoidal nature of the interference patterns and furthermore increases the splice loss. Therefore for an improved sensor performance, only one cladding mode is preferred due to its simple interference with the core mode. The collapsed region length can be controlled by the arc power and duration [Barrera et al., 2010]. In our experiments, PCF (LMA10, NKT Photonics) designed for an endless single-mode operation was used. It has four layers of air holes arranged in a hexagonal pattern around a solid silica core. The light guidance mechanism in such a fibre is by means of modified total internal reflection. The dimensions of the LMA-10 PCF simplify alignment and splicing with the SMF with a standard splicing machine and minimize the loss due to mode field diameter mismatch compared to other PCFs. For the interferometer fabricated in our study the total length of the collapsed region was 200 μm. After fusion splicing, the PCF was cleaved using a standard fibre cleaving machine so that the end surface of the PCF acts as a reflecting surface.

2.3 PCFI fringe spacing vs length of PCF

Initially to investigate the influence of the length of the PCFI on the fringe spacing thirteen PCFIs were fabricated with lengths ranging from 3.5 mm to circa 100 mm. As an example Fig. 2 shows the measured reflection spectra of three PCFIs in the 1500-1600 nm wavelength range with lengths of 92, 10.5 and 3.5 mm. The reflection spectra of the interferometers exhibit regular interference patterns with a period or fringe spacing inversely proportional to the length of the PCF section. A modulation of the expected sinusoidal pattern is observed for the spectra shown in Fig. 2, which might be due to the excitation of more than one cladding mode or possibly due to the polarization dependence of the intermodal interference [Bock et al., 2009]. Fig. 3 shows the measured fringe spacing or periods of the fabricated PCFIs as a function of length of the PCF section. The measured periods agree well with the expected ones for a two-mode interferometer given by the expression $P \approx \lambda^2/(2\Delta nL)$. The value of Δn obtained based on the experimental data is $\sim 4.2 \times 10^{-3}$.

3. Relative humidity sensor based on PCFI

Humidity refers to the water vapour content in air or other gases and its measurements can be stated in a variety of terms and units. The three commonly used terms are absolute humidity, relative humidity (RH) and dew point. Absolute humidity is the ratio of the mass of water vapour to the volume of air or gas. It is commonly expressed in grams per cubic

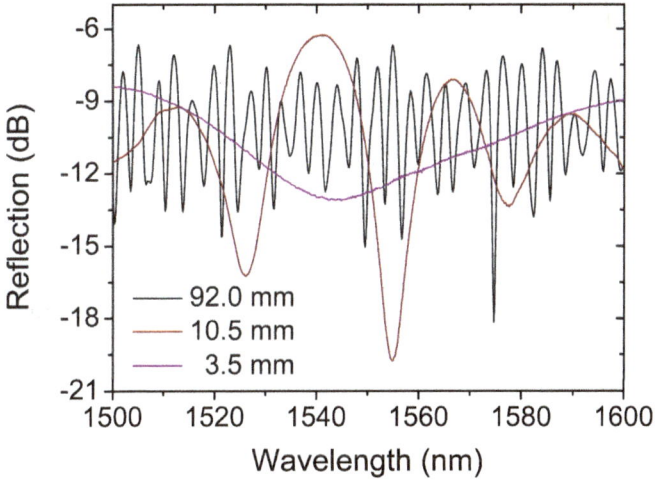

Fig. 2. The reflection spectra of interferometers with L = 92 mm, 10.5 mm and 3.5 mm in the wavelength range of 1500-1600 nm.

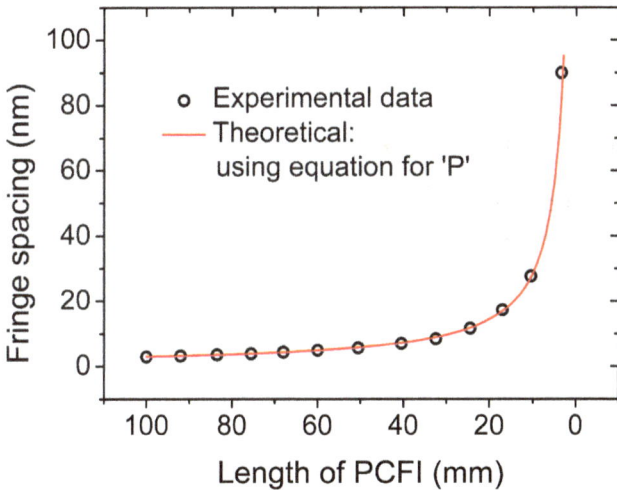

Fig. 3. The fringe spacing as a function of length of PCF observed for a reflection type interferometer.

meter. Dew point, expressed in °C or °F, is the temperature and pressure at which a gas begins to condense into a liquid. The ratio of the percentage of water vapour present in air at a particular temperature and pressure to the maximum amount of water vapour the air can hold at that temperature and pressure is the relative humidity.

The measurement of humidity is required in a range of areas, including meteorological services, the chemical and food processing industries, civil engineering, air-conditioning, horticulture and electronic processing. Compared with their conventional electronic counterparts, optical fibre humidity sensors offer specific advantages, such as small size and weight, immunity to electromagnetic interference, corrosion resistance and remote operation. A wide range of optical fibre humidity sensors have been reported in the literature. Most of these fibre optic humidity sensors work on the basis of a hygroscopic material coated over the optical fibre to modulate the light propagating through the fibre [Yeo et al., 2008; Mathew et al. 2007, 2011]. A polymer optical fibre has been adapted for humidity sensing [Zhang et al., 2010] without the use of a hygroscopic coating but the fibre is highly temperature dependent and is not suitable for high-temperature applications. An all-glass fibre-optic relative humidity sensor which does not require any special coatings to measure humidity using a reflection-type two-mode photonic crystal fibre interferometer is presented in this section. The spectrum of it exhibits good sensitivity to humidity variations.

3.1 Operating principle of the sensor

An untreated silica PCF is used for the fabrication of the PCFI, its surface is hydrophilic and there fore the adsorption of water vapour on the surface occurs when it is exposed to humid air. Two types of water-vapour adsorption mechanisms occur in sequence at the SiO_2-air interface. The chemisorption of water vapour first modifies the SiO_2 surface, resulting in a surface with silanol groups (Si-OH). The second type of adsorption, physisorption, occurs on these silanol groups. A schematic illustration of the water-vapour adsorption is given in Fig. 4. At room temperature the physisorption is a reversible function of the relative humidity of the surrounding air, while the chemisorption appears to be irreversible [Voorthuyzen et al., 1987]. So in the succeeding discussion only the physisorption is considered. Awakuni and Calderwood [Awakuni & Calderwood, 1972] investigated the adsorption of water vapour on the SiO_2 surface. They measured the amount of adsorbed water as a function of the partial vapour pressure at a constant temperature. It appeared that this so-called adsorption isotherm can be described very well by the BET (Brunauer-Emmett- Teller) adsorption theory [Brunauer et al., 1938].

The evolution of adsorbed water layer structure on silicon oxide at room temperature is demonstrated by David and Seong in [David & Seong, 2005]. They determined the molecular configuration of water adsorbed on a hydrophilic silicon oxide surface at room temperature as a function of relative humidity using attenuated total reflection (ATR)- infrared spectroscopy. A completely hydrogen-bonded ice like network of water grows up as the relative humidity increases from 0 to 30%. In the relative humidity range of 30-60%, the liquid water structure starts appearing while the ice like structure continues growing to saturation. Above 60% relative humidity, the liquid water configuration grows on top of the ice like layer. This structural evolution indicates that the outermost layer of the adsorbed water molecules undergoes transitions in equilibrium behaviour as humidity varies. Also it was shown from the adsorption isotherm that the thickness of the adsorbed layer at room temperature starts increasing exponentially above 60% RH.

Tiefenthaler and Lukosz [Tiefenthaler & Lukosz, 1985] have shown that adsorption and desorption of water vapour by the surface of a waveguide changes the effective refractive index (RI) of the guided modes, in their case for a humidity sensor based on an integrated optical grating coupler. In the case of a PCFI a similar adsorption of water vapour changes the effective refractive index (n_{cl}) of the interfering cladding mode propagating in the PCF. Since this adsorption/physisorption is a reversible process, a modulation of the n_{cl} occurs with respect to the ambient humidity values which in turn change the position of the interference pattern accordingly. An increase in humidity causes the shift of the interference pattern of a PCFI toward longer wavelengths and the value of this interference peak shift is exponential with respect to relative humidity [Mathew, 2010]. This shift of the interference peak is mainly due to the adsorption and desorption of H_2O molecules along the surface of holes within the PCF, at the interface between air and silica glass. Since the whole device is exposed to humidity the adsorption and desorption of water vapour on the PCF outer surface and on the end face also contribute to the shift of the interference pattern. But considering the field distribution of the interfering cladding mode shown in [Cárdenas-Sevilla et al., 2011; Uranus, 2010] and below the dew point temperature the main contribution to the interference shift is considered to be due to the adsorption of water molecules within the voids of the PCF. The adsorption on the end face mainly causes a shift in the overall power level of the interference pattern.

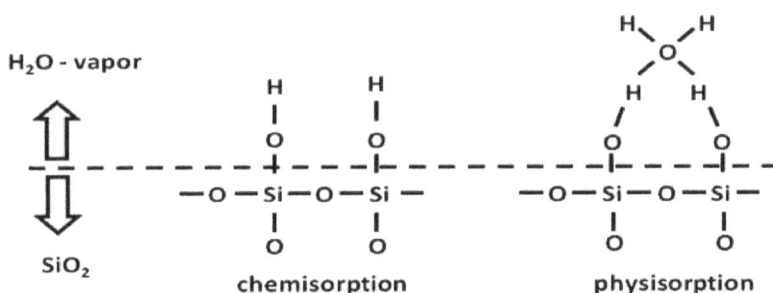

Fig. 4. Schematic representation of water vapor adsorption mechanisms on an SiO_2 surface.

3.2 Experimental characterization of the sensor

The sensor system is composed of a broadband light source (SLED), a fibre coupler/circulator (FOC), the PCF interferometer or sensor head, and an optical spectrum analyser (OSA) as shown in Fig. 5. The sensor head, as the main part of the sensor system, is composed of a small stub of PCF fusion spliced to the end of a standard SMF. The PCF in the sensor head has a microhole collapsed region near the splicing point and the free end of the PCF is exposed to ambient air. The humidity response of the device was studied at a temperature (25 °C) and at normal atmospheric pressure by placing it in a controlled environmental chamber as shown in Fig. 5. Fig. 6 shows the changes in the reflection spectrum with respect to ambient humidity for a device with L=40.5 mm. The change in the adsorption with respect to ambient humidity changes the effective refractive index of the cladding mode (n_{cl}). The resulting phase change in turn results in a shift of the interference pattern. The curves in Fig. 6 show the position of a zoomed section of the device spectrum at relative humidity values of 30, 60, 80 and 90 %RH. When humidity increases the interference pattern shifts to longer wavelengths and this shift is more significant at higher humidity values. To study the effect of reducing the length of the PCFI a second PCFI was fabricated with a shorter length of 17 mm. Fig. 7 shows the peak shift of the interferometer with respect to humidity obtained for two devices with L=17 mm and 40.5 mm.

It is observed from the Fig. 7 that the sensitivity of the device to humidity decreases as the length of the device decreases. This is due to the fact that for a small device the fibre length available for interaction between the cladding mode with the adsorbed water vapour is less so the acquired phase difference between the interfering modes will be smaller. Hence the sensitivity to humidity change is less for a device with a smaller length of PCF. It is important to note that the shift of the interference pattern is similar to the thickness variation of the adsorbed layer of water vapour on silica i.e. increases exponentially above 60 %RH [David & Seong, 2005].

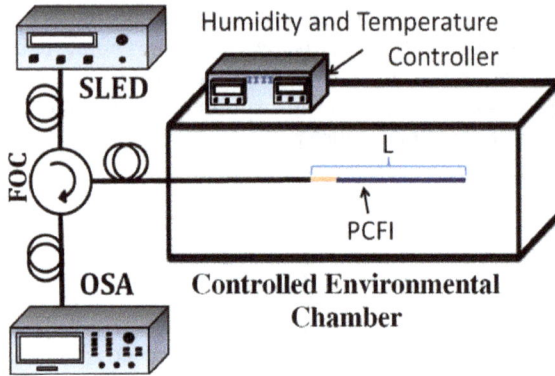

Fig. 5. Experimental arrangement for the characterisation of the PCFI with respect to relative humidity.

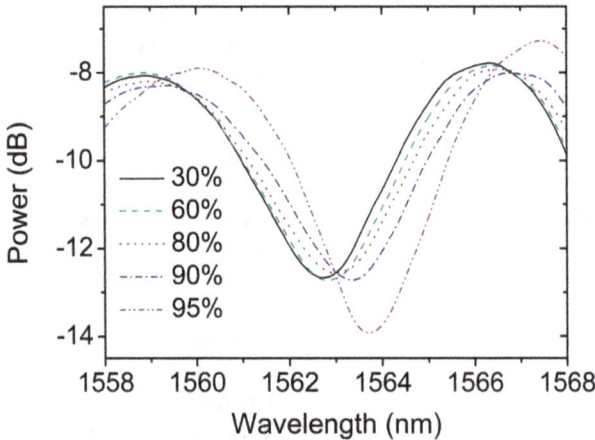

Fig. 6. Reflection spectrum of a 40.5 mm long PCFI at different humidity values.

The device sensitivity is estimated by dividing the PCFI response to humidity into three regions 27- 60 %RH, 60-80 %RH, and 80-96 %RH. The average sensitivity values observed for the PCFI with a length of 40.5 mm in these regions are 3.7, 8.5 and 64 nm/%RH respectively and for a 17 mm long PCFI they are 1.7, 3 and 23 nm/%RH respectively. Even though the PCFI with a longer length appears more sensitive, it is likely that increasing the length of the PCFI to a much longer length is not practical because in a longer device the infiltration of water molecules may take too much time. Furthermore, since the propagation loss of the interfering cladding mode is high the fringe visibility will diminish on increasing the length of PCF. Also for a longer device the fringe spacing will be shorter which limits the measurement range of the device. Decreasing the length of the PCFI to a much shorter length is also not suitable because as seen from Fig. 2 & 3 if the length is less than 3.5 mm the fringe spacing will be greater than 100 nm, the bandwidth of a typical SLED spectrum, and therefore not suitable for observing the shift in the interference spectrum. Selecting a shorter length will also result in a reduced sensitivity but that can be improved by infiltrating the microholes with suitable hygroscopic materials. Based on our experimental observations and considering the above explained factors we suggest the best lengths for an efficient humidity sensing to be in the range from 3.5 mm to 100 mm.

The response of the PCFI to humidity variations is found to be reversible and repeatable with low hysteresis. Under laboratory conditions it is reusable, but humidity is a truly analytical measurement in which the sensor must be in direct contact with the process environment. This of course has implications of contamination and degradation of the sensor to varying degrees depending on the nature of the environment. Possible contamination agents are dust particles and chemical vapours. So a further study of the sensor head contamination in different process environments and the observation of the shift in its response in those conditions are required in order to get a better understanding of the long term stability of our sensor in field applications. In the case of a PCFI based sensor this limitation can be overcome by different ways; a recalibration of the sensor head after a certain period of time and a subsequent reuse of the sensor head during another time interval, or, since the fabrication of the PCFI based sensor head is simple and cost effective, replacing the sensor head or attaching some filters to the sensor head by which it can be protected from contamination or an ultrasonic cleaning and subsequent heating (which will remove the contaminants like dust particles without damaging the sensor head) is another method to make the sensor reusable after contamination.

A study of cross sensitivity to temperature reveals that the PCFI based humidity sensor is almost temperature independent. Conventional glass fibre relative humidity sensors require coatings and thus are always temperature dependent and, furthermore, since the majority of such sensors use polymer materials as coatings, they are not suitable for use in high-temperature applications. One significant advantage of the sensor explained here is that the sensor head is made of single material silica. This suggests that apart from low and room temperature applications the PCF interferometer based humidity sensor can also be used in harsh and high-temperature environments to monitor humidity.

4. Dew sensor based on PCFI

Dew (condensed moisture) is a problem in the fields of precision electrical devices, automobiles, air conditioning systems, warehouses and domestic equipment, etc. High humidity and condensation can create an environment where the development of mould on the wooden parts can take place and it can also cause corrosion of iron parts. This is a major problem in the case of the works of art in the museums and churches [Camuffo & Valcher, 1986]. So there is a strong demand for a sensor able to accurately detect a high humidity or dew condensation state.

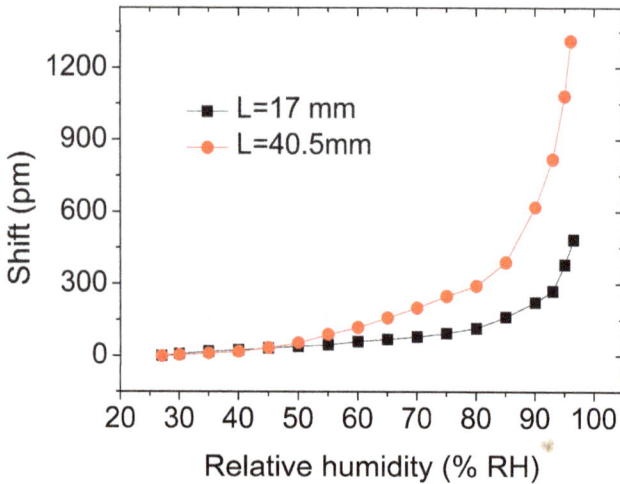

Fig. 7. Interference peak shift of the photonics crystal fibre interferometers with L= 40.5 mm and 17 mm with respect to relative humidity.

Approaches to dew detection using optical fibre have been previously reported in [Baldini et al., 2008; Kostritskii et al., 2009]. The working principle of these sensors is based on the change in the reflectivity which is observed on the surface of the fibre tip, when a water layer is formed on its distal end. The dependence on reflected power measurement scheme used in [Baldini et al., 2008; Kostritskii et al., 2009] increases the chance of measurement error due to source power fluctuations. Recently we have demonstrated a simple sensor head for dew detection based on a photonic crystal fibre interferometer (PCFI) operated in reflection mode [Mathew et al., 2011], with the advantage of good dew point measurement accuracy. The fabrication of such a sensor is very simple since it only involves cleaving and fusion splicing. Furthermore, the spectral measurement technique utilized in this work is free from errors due to source power variations. In the following section of the chapter a dew sensor based on PCFI is explained, including a study of temperature dependence of the device with different lengths of PCF. Since the sensor head is fabricated from a single material, silica, its temperature dependence is very low. From the results of the dew sensor performance with different lengths of PCF it was shown that a device with a compact length of PCF is suitable for dew sensing albeit with a reduction in the speed of response. The response of the sensor at different ambient humidity values is also included in this section.

4.1 Operating principle of the sensor

To study the response of the PCFI to dew formation it is required to set the temperature of the PCFI to dew point temperature, which is obtained from the values of ambient relative humidity and temperature. To do this let us consider a quantity of air with a constant water vapour concentration at a certain temperature, T, and relative humidity, RH < 100%. The dew point temperature, T_d, is defined as the temperature to which this quantity of air must be cooled down such that, at a constant pressure, condensation occurs (RH = 100%). In terms of relative humidity RH and temperature T, the dew point temperature is given as:

$$T_d(T,RH) = \alpha \frac{\ln\left(\dfrac{RH}{100\%}\right) + \dfrac{\beta T}{\alpha + T}}{\beta - \ln\left(\dfrac{RH}{100\%}\right) - \dfrac{\beta T}{\alpha + T}}$$

where, α=243.12 °C and β=17.62 are the so-called Magnus parameters for the temperature range -45 to 60 °C. There fore decreasing the temperature of the PCFI increases the relative humidity close to it. At a certain stage of decreasing the temperature the relative humidity becomes 100% or reaches the dew point temperature and hence the water vapour starts to condense. The condensed water vapour on the PCFI causes a large change in the effective refractive index of the interfering cladding mode (n_{cl}) which in turn causes a large phase change between the interfering modes and therefore a large wavelength shift of the interference peaks is expected.

4.2 Experimental characterization of the sensor

The dew response of the PCF interferometer was studied by placing it on a thermoelectric cooler (TEC) as shown in Fig. 8. In order to study the influence of dew on the PCFI, it was decided to limit the PCFI length used to 42 mm or less, to suit the size of the available TEC used for temperature control. The temperature of the TEC element was controlled by a temperature controller. A thermistor was used to provide temperature feedback to the controller from the TEC element. An additional handheld thermometer was used to confirm the temperature on the TEC surface. The entire setup was placed inside a controlled environmental chamber. The inside relative humidity and the temperature of the chamber can be controlled with an accuracy of ±2 %RH and ±1 °C respectively. For the purpose of this experiment the ambient temperature inside the chamber was fixed at 25 °C.

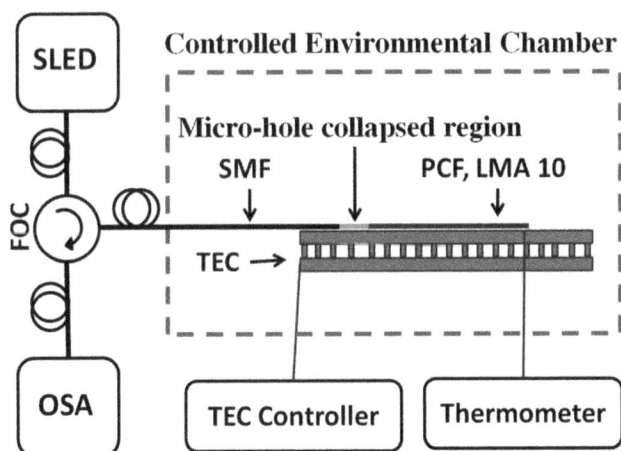

Fig. 8. Experimental arrangement for the calibration of PCFI based dew sensor.

Since the PCF is composed of only fused silica, it is expected to have minimal thermal sensitivity. The temperature dependence of the device was determined by observing the peak shift of the interference spectrum of the device for a temperature variation from 25 °C to 60 °C. The ambient humidity during the study was set to 40 % RH. When the temperature is increased from 25 °C to 60 °C the interference peak is shifted slightly to higher wavelengths. Fig. 9 shows this temperature dependence for two devices with L=17 mm and 40.5 mm. As expected the thermal sensitivity of the PCFI is very low and is further reduced for a device with the shorter length of PCF. The thermal sensitivity obtained in the experiment for a device with L= 40.5 mm is 9.5 pm/°C and that for L= 17 mm is 6.2 pm/°C.

The dew sensing experiments were carried out at an ambient temperature of 25 °C and at normal atmospheric pressure. To study the dew response of the device the temperature of the PCFI was decreased from ambient temperature (25 °C) to the dew point temperature at a fixed ambient relative humidity. It was found that the position of the interference peaks shifted to longer wavelengths with a decrease in temperature. This shift is similar to the humidity response of the PCFI as shown in Fig. 6 and 7. This occurs because the relative humidity inside the microholes and close to the PCFI increases with a decrease in temperature and causes a shift. At or below the dew point temperature (100% RH) water vapour condensation occurs, the condensed water vapour on the outer surface of the PCF also contributes to the change in the effective RI of the cladding mode, which results in a large spectral shift.

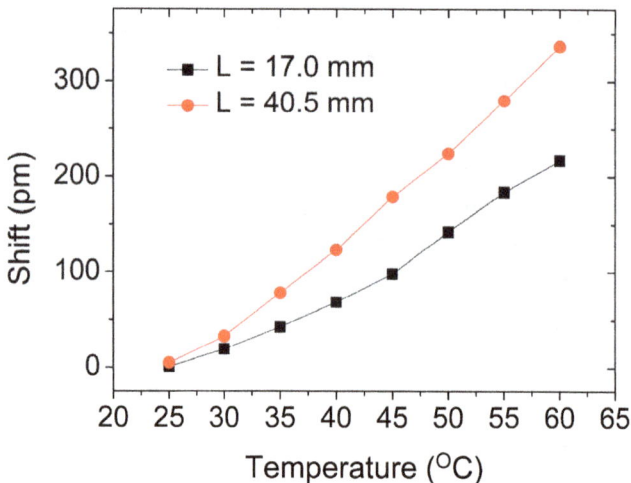

Fig. 9. Interference peak shift with respect to temperature for interferometers with PCF lengths L= 40.5 mm and 17 mm.

The spectra of two interferometers at room temperature and at the dew point temperature for devices fabricated with lengths 40.5 mm and 3.5 mm are shown in Fig. 10(a) & (b). The lengths selected are practically the largest and the smallest PCF lengths that can be studied using our experimental setup. The ambient humidity during this study was set at 60 % RH. From the Fig. 10 it is clear that relative to the period of the interferometer the shift will be larger for a longer PCFI due to a longer interaction length available for the interference between the cladding mode and the adsorbed water vapour. Hence the sensitivity to water vapour content and thus dew point temperature is high for a device with a longer length of PCF.

It is important to note that due to the large fringe spacing it is difficult to measure the peak shift accurately for a short PCFI, therefore the comparison of sensitivities for PCFIs with different lengths is not straightforward. It should also be noted that even a PCFI with a small length (3.5 mm, fringe spacing ~90 nm) when exposed to dew point temperature for a relatively long time i.e. several minutes will result in a measurable fringe shift as shown in Fig. 10(b). This is because an increasingly thicker adsorbed water layer is formed on the silica surfaces of the PCF as time progresses. Thus compared to 3.5 mm device the ~40.5 mm device is preferable for achieving a fast response time (in the order of seconds), but when a compact length is the main requirement a shorter PCFI also can be used as a dew sensor with a reduced measurement speed. The best range of lengths suitable for dew sensing is the same as the one given above for humidity sensing.

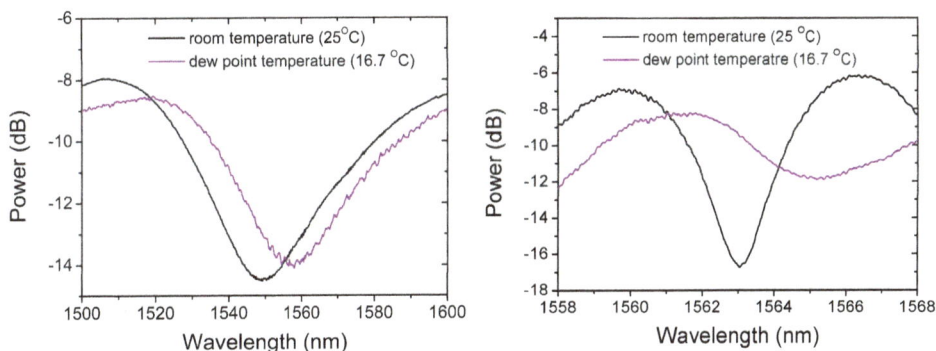

Fig. 10. (a). Interference spectra for a device with length 40.5 mm at room temperature and at dew point temperature.
(b). Interference spectra for a device with length 3.5 mm at room temperature and at dew point temperature.

The dew sensing performance of a PCFI at different environmental conditions was determined by studying the dew response of a PCFI with L= 40.5 mm at three ambient humidity values of 40, 60 and 80 %RH. At each humidity value the temperature of the PCFI is reduced from 26 °C to the corresponding dew point temperature. The peak wavelength shift of the device is plotted against temperature in Fig. 11. The three curves represent the peak shift corresponding to the ambient relative humidity values of 40, 60 and 80 %RH. The onset of the dew formation is characterized by a large shift of the interference peak which is clear in Fig. 11. The dew point temperature calculated by using equation (1) based on the corresponding ambient conditions is marked on each curve in Fig. 11. For all these three ambient humidity values the continuous spectral shift starts exactly at the dew point temperature which confirms the high dew point measurement accuracy (estimated as ±0.1 °C) of the sensor.

It is observed that at or below the dew point temperature the interference peak shifts continuously with time. This is because an increasingly thicker adsorbed water layer is formed on the silica surface of the PCF microholes as time progresses. By bringing the temperature of the PCFI back to room temperature the interference peaks also shift back to their initial position. This shows the reversibility of the sensor. Because of the small size of the sensor head and the high sensitivity to adsorbed water vapour the demonstrated sensor response time is in seconds which is relatively fast compared to existing dew point hygrometers that take several minutes for a single measurement. The simple fabrication method, small size and the all-silica nature of the demonstrated sensor head suggest that with some simple additions such as attaching a TEC element with temperature feedback on to the PCFI, the combination can be used as a dew point hygrometer.

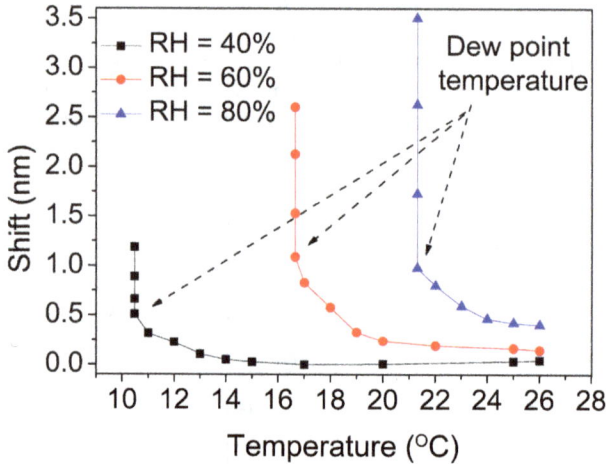

Fig. 11. Interference peak shift of PCFI with respect to temperature at three ambient humidity values of 40, 60 and 80 %RH.

5. Conclusion

A brief review of the photonic crystal fibre and the modal interferometers based on PCF are presented in this chapter. Along with the review the operating principle and the fabrication of a reflection type PCF based modal interferometer are also explained in the chapter. The dependence of the interferometer fringe spacing on the length of PCF is also explained and demonstrated experimentally. The experimental investigation and demonstration of a humidity sensor based on a PCF interferometer is presented in the chapter with a brief explanation of the operating principle of the sensor. The water vapour adsorption/desorption phenomena on silica surface are briefly addressed to explain the operating principle of the sensor. The chapter includes the experimental investigation of the relative humidity response of the sensor and the dependence of its sensitivity on the length of PCF. It is shown that a device with a longer length of the PCF section is more sensitive to relative humidity changes. A dew sensor based on PCF interferometer is presented along with the explanation of its sensing principle. The chapter presents the temperature dependence of the PCF interferometer and the dependence of its sensitivity on the length of the PCF. The dew sensing performances of PCFIs with different lengths and at different ambient relative humidity values are also presented. Based on the explained dew sensor a novel dew point hygrometer using PCF interferometer is also proposed in the chapter.

6. References

Awakuni, Y. & Calderwood, J.H. (1972). Water vapour adsorption and surface conductivity in solids. *Journal of Physics D: Applied Physics*, Vol.5, No.5, (May 1972), pp. 1038.

Baldini, F. ; Falciai, R. ; Mencaglia, A. A. ; Senesi, F. ; Camuffo, D. ; Valle, A. D. & Bergsten, C. J. (2008). Miniaturised Optical Fibre Sensor for Dew Detection Inside Organ Pipes, *Journal of Sensors*, Vol. 2008, Article ID 321065.

Barrera, D. ; Villatoro, J. ; Finazzi, V. P. ; Cardenas-Sevilla, G. A. ; Minkovich, V. P. ; Sales, S. & Pruneri, V. (2010). Low-Loss Photonic Crystal Fiber Interferometers for Sensor Networks. *Journal of Lightwave Technology*, Vol. 28, No. 24, (Dec. 2010), pp. 3542-3547.

Birks, T. A.; Knight, J. C. & Russell, P. St. J. (1997). Endlessly single-mode photonic crystal fiber, *Optics Letters*, Vol. 22, No. 13, (July 1997), pp. 961-963.

Bock, W. J.; Eftimov, T. A.; Mikulic, P. & Chen, J. (2009). An Inline Core-Cladding Intermodal Interferometer Using a Photonic Crystal Fiber, *Journal of Lightwave Technology*, Vol. 27, No. 17, (Sept. 2009), pp. 3933-3939.

Brunauer, S.; Emmett, P. H. & Teller, E. (1938). Adsorption of gases in multimolecular layers, *Journal of the American Chemical Society*, Vol.60, (February 1938), pp. 309-319.

Camuffo, D. & Valcher, S. (1986). A dew point signaller for conservation of works of art, *Environmental Monitoring and Assessment*, Vol.6, No. 2, (1986), pp. 165-170.

Cárdenas-Sevilla, G. A.; Finazzi, V. ; Villatoro, J. & Pruneri, V. (2011). Photonic crystal fiber sensor array based on modes overlapping, *Optics Express*, Vol. 19, No. 8, (April 2011), pp. 7596-7602.

Choi, H.Y. ; Kim, M. J. & Lee, B. H. (2007). All-fiber Mach-Zehnder type interferometers formed in photonic crystal fiber, *Optics Express*, Vol. 15, No. 9, (April 2007), pp. 5711-5720.

David, B. A. & Seong, H. K. (2005). Evolution of the adsorbed water layer structure on silicon oxide at room temperature, *Journal of Physical Chemistry B*, Vol. 109, No. 35, (August 2005), pp. 16760-16763.

Jha, R.; Villatoro, J. & Badenes, G. (2008). Ultrastable in reflection photonic crystal fiber modal interferometer for accurate refractive index sensing, *Applied Physics Letters*, Vol. 93, No. 19, (November 2008), pp. 191106.

Káčik, D.; Turek, I.; Martinček, I.; Canning, J.; Issa, N. & Lyytikäinen, K. (2004). Intermodal interference in a photonic crystal fibre, *Optics Express*, Vol. 12, No. 15, (July 2004), pp. 3465-3470.

Knight, J. C. (2003). Photonic crystal fibres. *Nature*, Vol. 424, No. 6950, (Aug. 2003), pp. 847-851.

Knight, J. C.; Birks, T. A.; Cregan, R. F.; Russell, P. St. J. & De Sandro, J. P. (1998). Large mode area photonic crystal fibre, *Electronics Letters*, Vol. 34, No. 13, (June 1998), pp. 1347-1348.

Knight, J. C.; Birks, T. A.; Russell, P. S. J. & Atkin, D. M. (1996). All-silica single-mode optical fiber with photonic crystal cladding. *Optics Letters*, Vol. 21, No. 19, (Oct. 1996), pp. 1547-1549.

Kostritskii, S. M.; Dikevich, A. A.; Korkishko Y. N. & Fedorov, V. A. (2009). Dew point measurement technique utilizing fiber cut reflection. *Proceedings of SPIE*, Vol. 7356, (2009), pp. 73561K.

Lim, J. H.; Jang, H. S.; Lee, K. S.; Kim, J. C. & Lee, B. H. (2004). Mach–Zehnder interferometer formed in a photonic crystal fiber based on a pair of long-period fiber gratings, *Optics Letters*, Vol. 29, No. 4, pp. 346-348.

Mathew, J.; Semenova, Y.; Rajan, G. & Farrell, G. (2010). Humidity sensor based on photonic crystal fibre interferometer, *Electronics Letters*, Vol. 46, No. 19, (September 2010), pp. 1341-1343.

Mathew, J. ; Semenova, Y. & Farrell, G. (2011). Photonic crystal fiber interferometer for dew detection, *Journal of Lightwave Technology*. DOI (identifier) 10.1109/JLT.2011.2170815.

Mathew, J. ; Semenova, Y. ; Rajan, G. & Farrell, G. (2011). Photonic crystal fiber interferometer for dew detection, *Proceedings of SPIE*, Vol. 7753, (2011), pp. 77531P.

Mathew, J.; Semenova, Y.; Rajan, G.; Wang, P. & Farrell, G. (2011). Improving the sensitivity of a humidity sensor based on fiber bend coated with a hygroscopic coating, *Optics & Laser Technology*, Vol. 43, No. 7, (October 2011) pp. 1301-1305.

Mathew, J.; Thomas, K.J.; Nampoori, V.P.N. & Radhakrishnan, P. (2007). A Comparative Study of Fiber Optic Humidity Sensors Based on Chitosan and Agarose, *Sensors & Transducers Journal*, Vol. 84, No. 10, (October 2007) pp. 1633-1640.

Monzón-Hernández, D.; Minkovich, V. P.; Villatoro, J.; Kreuzer, M. P. & Badenes, G. (2008). Photonic crystal fiber microtaper supporting two selective higher-order modes with high sensitivity to gas molecules. *Applied Physics Letters*, Vol. 93, No. 8, pp. 081106.

Renversez, G.; Kuhlmey, B. & McPhedran, R. (2003). Dispersion management with microstructured optical fibers: ultraflattened chromatic dispersion with low losses, *Optics Letters*, Vol. 28, No. 12, pp. 989-991.

Russell, P. (2003). Photonic crystal fibers. *Science*, Vol. 299, No. 5605, (January 2003), pp. 358–362.

Russell, P. (2006). Photonic-crystal fibers. *Journal of Lightwave Technology*, Vol. 24, No. 12, (December 2006), pp. 4729–4749,.

Tiefenthaler, K. & Lukosz, W. (1985). Grating couplers as integrated optical humidity and gas sensors. *Thin Solid Films*, Vol. 126, (April 1985), pp. 205–211.

Uranus, H. P. (2010). Theoretical study on the multimodeness of a commercial endlessly single-mode PCF, *Optics Communications*, Vol. 283, No. 23, (December 2010), pp. 4649–4654.

Villatoro, J. ; Kreuzer, M. P. ; Jha, R. ; Minkovich, V. P. ; Finazzi, V. ; Badenes, G. & Pruneri, V. (2009). Photonic crystal fiber interferometer for chemical vapor detection with high sensitivity, *Optics Express*, Vol. 17, No. 3, (February. 2009), pp. 1447–1453.

Villatoro, J. ; Minkovich, V. P. ; Pruneri, V. & Badenes, G. (2007). Simple all-microstructured-optical-fiber interferometer built via fusion splicing, *Optics Express*, Vol. 15, No. 4, (February 2007), pp. 1491–1496.

Villatoro, J.; Finazzi, V.; Badenes, G. & Pruneri, V. (2009). Highly Sensitive Sensors Based on Photonic Crystal Fiber Modal Interferometers, *Journal of Sensors*, Vol. 2009, Article ID 747803, 11 pages.

Villatoro, J.; Finazzi, V.; Minkovich, V. P.; Pruneri, V. & Badenes, G. (2007). Temperature-insensitive photonic crystal fiber interferometer for absolute strain sensing, *Applied Physics Letters*, Vol. 91, No. 9, (August 2007), pp. 091109.

Voorthuyzen, J.A.; Keskin, K. & Bergveld, P. (1987). Investigations of the surface conductivity of silicon dioxide and methods to reduce it, *Surface Science*, Vol.187 No.1, (August 1987), pp. 201-211.

Yeo, T.L.; Sun, T. & Grattan, K.T.V. (2008). Fibre-optic sensor technologies for humidity and moisture measurement, *Sensors and Actuators A Physical*, Vol. 144, No. 2, (June 2008), pp. 280-295.

Zhang, C. ; Zhang, W. ; Webb, D.J. & Peng, G.D. (2010). Optical fibre temperature and humidity sensor. *Electronics Letters*, Vol. 46, No. 9, (March 2010), pp. 643-644.

Zhao, C. L.; Yang, X.; Lu, C.; Jin, W. & Demokan, M. S. (2004). Temperature-insensitive interferometer using a highly birefringent photonic crystal fiber loop mirror, *IEEE Photonics Technology Letters*, Vol. 16, No. 11, (November 2004), pp. 2535-2357.

9

Optical Solitons from a Photonic Crystal Fiber and Their Applications

Naoki Karasawa and Kazuhiro Tada
Chitose Institute of Science and Technology
Japan

1. Introduction

A photonic crystal fiber (PCF) is a fiber that contains the regular (usually hexagonal) arrays of air holes in the propagation direction of an optical fiber. At the center position, the core is created by not making an air hole and the light wave propagates at the core position since the effective refractive index of the core is higher than that of the photonic crystal clad surrounding the core. Photonic crystal fibers of this type have been used to generate ultrabroadband optical pulses by propagating femtosecond optical pulses in these fibers (Ranka et al., 2000). The core diameter of a PCF for the generation of ultrabroadband optical pulses using a Ti:sapphire laser (center wavelength $\sim$800 nm) is about 1-2 μm if it is assumed that the silica core is surrounded by regular air holes. Due to the waveguide dispersion, the group velocity dispersion (GVD) becomes negative at 800 nm. Because of the small core diameter and the negative GVD, nonlinear effects are enhanced and optical solitons are generated in a PCF. Theoretical calculations for elucidating the mechanism of the ultrabroadband pulse generation in a PCF have been performed (Husakou & Herrmann, 2001) and the generation of fundamental soliton pulses by the fission of an input higher-order soliton pulse due to the third and higher order dispersion as well as the higher-order nonlinear effects including the Raman effects are found to be important for the spectral broadening. Supercontinuum generation in a PCF is reviewed in (Dudley et al., 2006). The center wavelength of the generated fundamental soliton pulse becomes longer as it propagates in a PCF due to soliton self-frequency shift and its center wavelength can be changed by the peak power or the chirp of an input pulse. Recently, it was used as a variable-wavelength light source in various applications including coherent anti-Stokes Raman scattering (CARS) spectroscopy and optical coherence tomography (OCT). The present article describes the properties of the fundamental solitons from a PCF and its applications studied in our laboratory.

2. Fundamental soliton pulse

It is well known that the soliton pulse, which does not change its shape as it propagates in a fiber, can be created when the pulse propagates in the anomalous dispersion region (Agrawal, 2007; Hasegawa, 1992). If we consider the electric field (considered to be scalar) that depends on only time t and propagation position z such that $E(z,t) = Re[A(z,t)e^{i(\beta_0 z - \omega_0 t)}]$, where Re shows the real part, ω_0 is the central angular frequency, and β_0 is the propagation constant at ω_0 ($\beta_0 = \beta(\omega_0)$) of a pulse, the slowly varying envelope approximation (SVEA) equation for the envelope $A(z,t)$ (normalized to have the unit $[W^{1/2}]$) may be obtained from the Maxwell equation after various approximations (Karasawa et al., 2001) as follows,

$$\partial_\zeta A(\zeta, T) = -\frac{i}{2}\beta_0^{(2)}\partial_T^2 A(\zeta, T) + i\gamma(\omega_0)|A(\zeta, T)|^2 A(\zeta, T). \tag{1}$$

In this equation, the coordinates are transformed $\zeta = z$, $T = t - \beta_0^{(1)}z$ such that the pulse center is always at the time origin ($\beta_0^{(1)} = \partial_\omega\beta|_{\omega_0}$ is the inverse of the group velocity of the pulse). In the right hand side of Eq. 1, the first term arises from the dispersion, where the GVD is given by $\beta_0^{(2)} = \partial_\omega^2\beta|_{\omega_0}$. The second term arises from the nonlinear self-phase modulation (SPM) with the frequency-dependent nonlinear coefficient $\gamma(\omega_0) = n(\omega_0)n_2^I(\omega_0)\omega_0^2(1 - f_R)/(c^2\beta_0 A_{\text{eff}}(\omega_0))$, where $n(\omega_0)$ and $n_2^I(\omega_0)$ are the linear and the nonlinear indices of refraction of a medium, c is the speed of light, $A_{\text{eff}}(\omega_0)$ is the effective mode area in a fiber, and f_R is the contribution of the Raman term ($f_R \simeq 0.3$ for fused silica). When the GVD is negative and the input power is chosen appropriately, the effects of these terms on the variations of the envelope cancel, and a stable soliton pulse can be propagated. By changing the variables to dimensionless ones such that $\xi = \zeta/L_D$, $\tau = T/T_0$, $u(\xi, \tau) = \sqrt{\gamma(\omega_0)L_D}A(\xi, \tau)$ (Agrawal, 2007), where T_0 is the pulse width parameter, and $L_D = T_0^2/|\beta_0^{(2)}|$ is the dispersion length, we have

$$\partial_\xi u(\xi, \tau) = \frac{i}{2}\partial_\tau^2 u(\xi, \tau) + i|u(\xi, \tau)|^2 u(\xi, \tau), \tag{2}$$

when $\beta_0^{(2)} < 0$. This equation was solved by the inverse scattering method and has a fundamental soliton solution of a form (Agrawal, 2007)

$$u(\xi, \tau) = \eta \text{ sech } \eta(\tau + \delta\xi - \tau_s)e^{-i\delta\tau+i(\eta^2-\delta^2)\xi/2+i\phi_s}, \tag{3}$$

where η and δ are determined by the eigenvalue of the inverse scattering problem, and τ_s and ϕ_s are constants. Here, ϕ_s can be included in an initial carrier wave phase, and δ and τ_s can be eliminated by shifting the carrier center frequency and the initial temporal position of the pulse. Therefore it can be written as $u(\xi, \tau) = \eta \text{ sech } (\eta\tau)e^{i\eta^2\xi/2}$, which shows that the envelope intensity does not change as it propagates in a fiber. The parameter η determines both the amplitude and the width of the soliton pulse. If the pulse width is T_0, $\eta = 1$ and the solution is given simply by $u(\xi, \tau) = \text{ sech } (\tau)e^{i\xi/2}$. If the input pulse shape of a laser is approximated as $u(0, \tau) = B \text{ sech } \tau$, the number of eigenvalues N is given by $B - 1/2 < N \le B + 1/2$ (Satsuma & Yajima, 1974). For an input pulse envelope with a peak power P_0, $A(0, T) = \sqrt{P_0} \text{ sech } (T/T_0)$ and this "soliton number" B becomes $B = \sqrt{P_0\gamma(\omega_0)L_D} = \sqrt{P_0\gamma(\omega_0)T_0^2/|\beta_0^{(2)}|}$. If B is exactly equal to N, the solution can be obtained in terms of N amplitudes $\eta_j = 1, 3, 5, ..., (2N - 1)$. For $N \ge 2$, higher-order N-soliton solutions were found (Satsuma & Yajima, 1974; Schrader, 1995). These N-soliton solutions may be considered to be consisted of N single solitons (Schrader, 1995).

In the real fiber of our interest, neglected terms deriving Eq. 1, such as higher-order dispersion terms, self-steepening terms, and the Raman term become important. If we use a slowly-evolving wave approximation (SEWA) instead of SVEA, the following propagation equation can be derived (Karasawa et al., 2001),

$$\partial_\zeta A(\zeta, T) = i(\hat{D}' + \hat{D}_{\text{corr}})A(\zeta, T) + i\gamma'(\omega_0)(1 + is\partial_T)\Big[(1 - f_R)|A(\zeta, T)|^2$$

$$+ \frac{2}{3}f_R \int_0^\infty h_R(T')|A(\zeta, T')|^2 dT'\Big]A(\zeta, T), \tag{4}$$

where

$$\hat{D}' = \sum_{n=0}^\infty \frac{i^n}{n!}\Big(\partial_\omega^n(\beta(\omega) + \frac{i\alpha(\omega)}{2})\Big)|_{\omega_0} \partial_T^n - \beta_0 - i\beta_0^{(1)}\partial_T, \tag{5}$$

is the dispersion terms that contain all higher-order terms with $\alpha(\omega)$ to be the loss constant, $\hat{D}_{\text{corr}} = (1 + i\beta_0^{(1)}\partial_T/\beta_0)^{-1}\hat{D}'^2/(2\beta_0)$ is the dispersion correction term, $s = 2/\omega_0 - \beta_0^{(1)}/\beta_0 + \partial_\omega(\log(n(\omega)n_2^I(\omega)/A_{\text{eff}}(\omega)))|_{\omega_0}$ is the self-steepening term, $\gamma'(\omega_0) = \gamma(\omega_0)/(1 - f_R)$, and $h_R(T) = (\tau_1^2 + \tau_2^2)e^{-T/\tau_2}\sin(T/\tau_1)/(\tau_1\tau_2^2)$ is the response function of the delayed Raman response with $\tau_1 = 12.2$ fs and $\tau_2 = 32$ fs for fused silica (Blow & Wood, 1989).

Because of the presence of extra terms not included in Eq. 1, the inputted pulse with $u(0, \tau) = B \operatorname{sech} \tau$ separates to multiple soliton pulses if the amplitude $B > 1.5$. The main reason of the splitting is the self-frequency shift due to the Raman effect (Gordon, 1986; Mitschke & Mollenauer, 1986), which depends on the pulse amplitude and width. If the center frequency is changed, the temporal delay of the pulse changes due to the dispersive effect ($\hat{D}'$ term). Independently, it is modified by the self-steepening effect (s term). The effects of these higher-order terms were investigated by moments equations for pulse parameters (Agrawal, 2007), where the variations of pulse parameters (pulse width, chirp, delay, and center angular frequency) were calculated as functions of propagation distance z. The variation of the center frequency of the fundamental soliton ($N = 1$) without a chirp can be approximated as

$$\omega_0(z) = -\frac{8T_R\gamma(\omega_0)P_0}{15T_0^2}z = -\frac{8T_R|\beta_0^{(2)}|}{15T_0^4}z, \tag{6}$$

where T_R is the first moment of the Raman response function, $T_R \simeq f_R \int_0^\infty t h_R(t)dt = 2.4$ fs for fused silica. This wavelength-variable fundamental soliton pulse emits a phase-matched dispersive wave, which is the important mechanism for generating supercontinuum, especially for shorter wavelength components (Husakou & Herrmann, 2001).

In Fig. 1, the model structure of a PCF used in our experiment and the dispersion properties of the PCFs with different structural parameters calculated by a multipole method (Zolla et al., 2005) using two rings of air holes surrounding a core are shown. In Fig. 2, the variations of temporal and spectral pulse intensities are shown as a function of propagation distance z, where, Eq. 4 is solved numerically for a PCF with the dispersion property shown in Fig. 1 (b) with the pitch 1.07 μm and diameter 0.7 μm (NKT Photonics NL-1.5-590). In calculations, a Gaussian input pulse, with the full width at half-maximum (FWHM) width $T_p = 50$ fs, the peak power $P_0 = 4$ kW, and the center wavelength 800 nm, where the GVD is anomalous, was used. The fiber material was assumed to be fused silica with $n_2^I = 2.48 \times 10^{-20}$ m^2/W, and $A_{\text{eff}} = 1.7671$ μm^2, which gave the soliton number $N=5.03$. The propagation loss and the derivative term in s were neglected in calculations. Also in Fig. 2, the spectrogram created by the calculation at propagation distances $z = 50$ mm and $z = 100$ mm are shown. From Fig. 2, we can see that three soliton pulses (S1, S2, and S3) are created in this case after about

$z = 25$ mm. The center frequency of the most intense fundamental soliton (S1) decreases until $z = 150$ mm, where GVD becomes zero. Because of this frequency shift, the group velocity decreases and the delay time becomes about 1000 fs at $z = 100$ mm. The pulse width, the spectral bandwidth, and the peak power of the most intense soliton are 14 to 18 fs, 60 to 65 nm, and 5 to 7 kW, respectively. Thus, this wavelength-variable fundamental soliton pulse has a shorter pulse width and a higher peak power compared with the inputted pulse. This soliton pulse emits a dispersive wave (D1) at the short wavelength near 550 nm, where the GVD is positive and its delay time increases quite rapidly. It is observed that there is a dispersive wave (D2) near 1400 nm after $z = 150$ mm. The intensities of other two soliton pulses (S2 and S3) are much weaker than the first soliton pulse (S1). The shift of the center frequency of the third soliton (S3) is small since its peak power is small. Also, it has a negative delay time since its center wavelength is shorter than 800 nm. From the spectrograms (c) and (d), we can see that three waves marked by S1, S2, and S3 are indeed localized pulses.

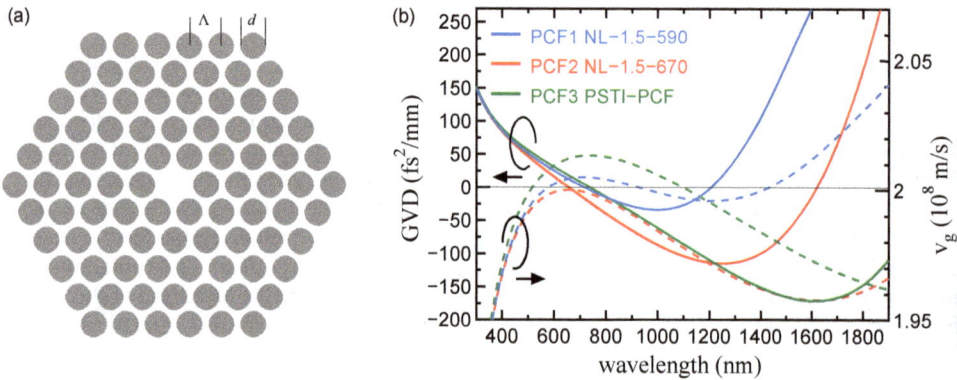

Fig. 1. (a) The structure of a five-ring PCF with a pitch Λ and a diameter d. Gray areas show air holes. (b) The GVD and the group velocity ($v_g = 1/\beta^{(1)}$) of the PCF shown in (a) calculated by a multipole method with various parameters. Blue curves show NKT Photonics NL-1.5-590 ($\Lambda = 1.07\mu m$, $d = 0.7\mu m$), red curves show NKT Photonics NL-1.5-670 ($\Lambda = 1.3\mu m$, $d = 1.105\mu m$), and green curves show PSTI-PCF ($\Lambda = 1.57\mu m$, $d = 1.31\mu m$).

3. Experiment on soliton properties

3.1 Soliton wavelength versus delay time

In the previous section, it is shown from calculations that an intense fundamental soliton pulse can be obtained by inputting a femtosecond pulse into a PCF and its center wavelength changes during propagation. However, in usual experiment, the length of a PCF is fixed and the center wavelength is controlled by modifying the input pulse parameters. In this section, the control of the center wavelength by the input pulse power and the delay property of a soliton are described.

As shown in Eq. 6, the variation of the center angular frequency due to self-frequency shift of a soliton pulse can be approximated to be proportional to the propagation distance z. This proportionality constant contains the peak power P_0 of the soliton pulse. Thus it is expected

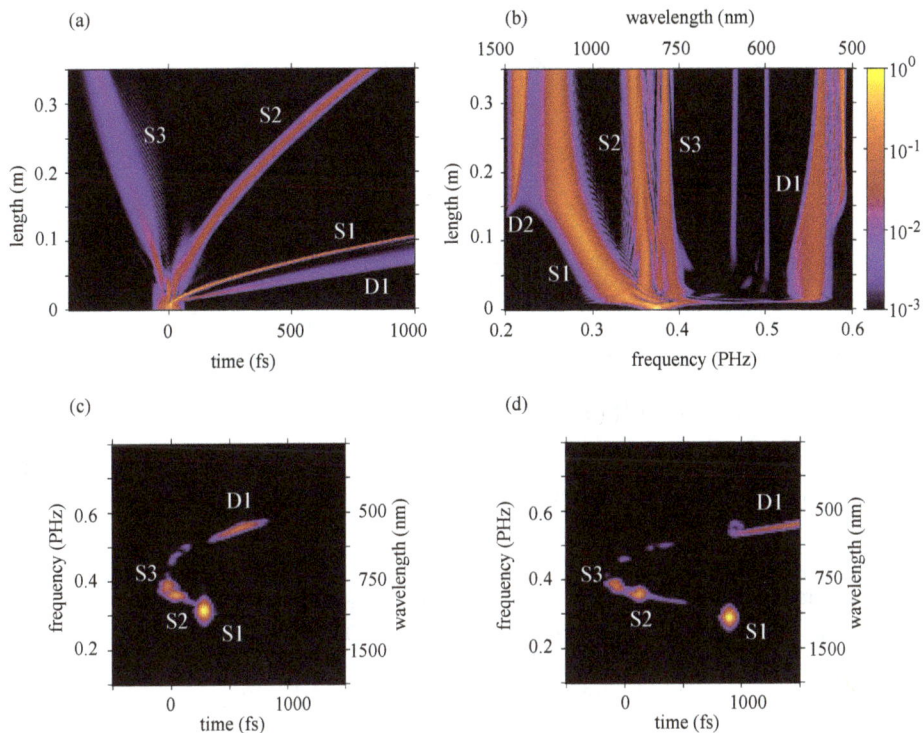

Fig. 2. The calculated temporal (a) and spectral (b) intensities of an optical pulse inputted in a PCF (NKT Photonics NL-1.5-590) versus distance. In (c) and (d), spectrograms at $z = 50$ mm (c) and $z = 100$ mm (d) are shown. S1, S2, and S3 show fundamental soliton pulses. D1 and D2 show the dispersive waves emitted from a soliton S1. The intensities are shown in a logarithmic scale.

that the amount of the shift is proportional to the power of the input pulse approximately. If the variation of the center angular frequency is proportional to the distance, the delay time of the soliton at the output end of a PCF can be estimated as follows. The propagation time $T(\omega_{0o})$ of a soliton with the angular frequency ω_{0o} at the output end of a PCF with a length L is given by

$$T(\omega_{0o}) = \int_0^L \beta^{(1)} dz = \frac{1}{K} \int_{\omega_{0i}}^{\omega_{0o}} \beta^{(1)} d\omega = \frac{1}{K}(\beta(\omega_{0o}) - \beta(\omega_{0i})), \tag{7}$$

where ω_{0i} is the center angular frequency at the input end and K is the proportional constant, which is given by $K = (\omega_{0o} - \omega_{0i})/L$. Thus, we have

$$T(\omega_{0o}) = \frac{\beta(\omega_{0o}) - \beta(\omega_{0i})}{\omega_{0o} - \omega_{0i}} L. \tag{8}$$

This equation shows that the delay time of the soliton for various center wavelengths can be estimated by the propagation constant $\beta(\omega)$ only. We have performed experiment to examine

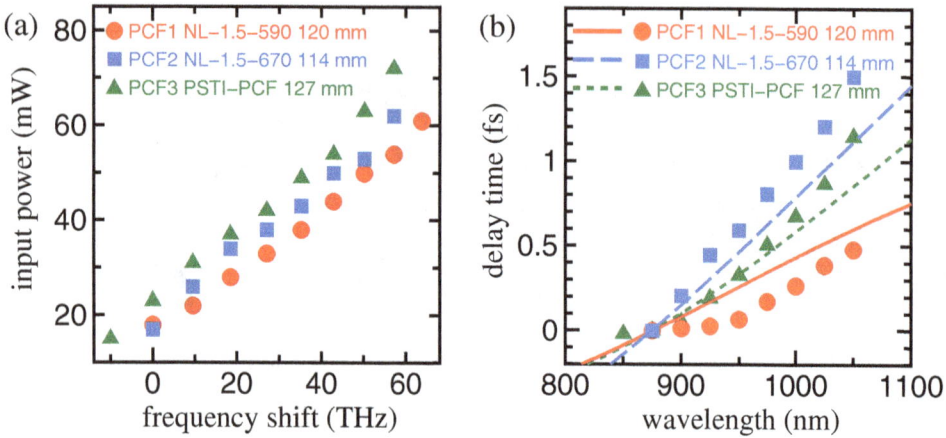

Fig. 3. (a) Frequency shift versus input power of soliton pulses for three different PCFs. (b) Wavelength versus delay time of soliton pulses for three different PCFs. Solid curves show delay times calculated by Eq. 8.

the dependence of the center wavelength and the delay time on the input power for three different PCFs, where the dispersion properties are shown in Fig. 1 (b). In experiment, a pulse from a Ti:sapphire laser oscillator (center wavelength 800 nm, pulse width 50 fs, repetition rate 78 MHz, and average power 620 mW) was propagated in PCFs, where the average power was controlled by a variable neutral density filter. In Fig. 3 (a), the frequency shifts of the most intense soliton pulses versus input pulse power are shown, where the shifts at 875 nm were set to be 0. As expected, the frequency shifts are almost proportional to the input pulse power for three different PCFs. In Fig. 3 (b), the delay times of the soliton pulses versus wavelength are shown and compared with delay times given in Eq. 8, where the delay times at 875 nm were set to be 0 and ω_{0i} was set to be $2\pi c/(800 \text{ nm})$. It is seen that the approximate delay times for the PCFs were estimated correctly. The discrepancies are presumably due to the neglect of delay times required for the soliton fission near the input ends of the PCFs in Eq. 8.

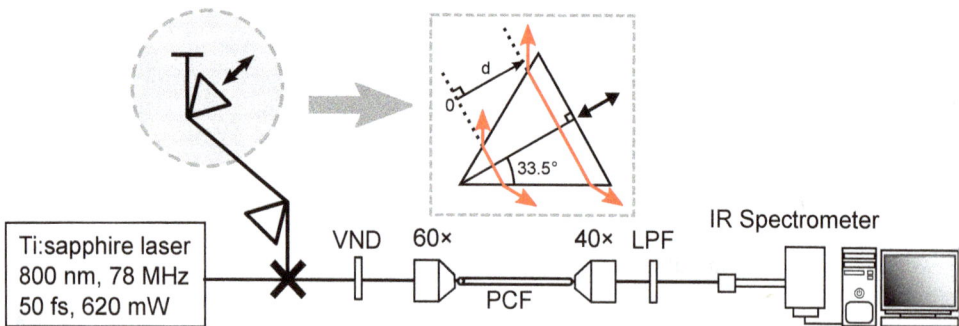

Fig. 4. Experimental setup for the control of a soliton by the chirp of an input pulse. Here, VND: variable neutral density filter, and LPF: long-wavelength pass filter.

3.2 Soliton wavelength control by chirp

We have studied the control of the soliton center wavelengths by the chirp of an input pulse (Karasawa et al., 2007; Tada & Karasawa, 2008). In experiment, the chirp of an input pulse was changed by varying the position of one of the prisms in a prism pair and the spectrum and the delay time of an output pulse from the PCF were measured. Experimental setup is shown in Fig. 4. A pulse from a Ti:sapphire laser oscillator (the center wavelength, the pulse width, and the repetition rate of the laser were 800 nm, 50 fs, and 78 MHz, respectively) was inputted into the PCF. A pair of Brewster-cut BK7 prisms was used to compensate for the dispersion of the objective and to apply the chirp for the PCF. When the insertion length was set to be d as shown in Fig. 4, the value of the group delay dispersion (GDD) was $4\beta_2 d \tan 33.5°$, where $\beta_2 = 44.7$ fs^2/mm is the GVD of BK7 at 800 nm. The chirp factor C due to this GDD was given by $C = \beta_2 d / T_0^2$. Due to this chirp, the pulse width was increased by a factor of $\sqrt{1+C^2}$ (Agrawal, 2007) and the peak power P_0 of the pulse was reduced by the same factor since the pulse energy was kept constant. Thus, the electric field envelope of the pulse may be written as a Gaussian form to be $A(T) = \sqrt{P_0}e^{-(1+iC)T^2/(2T_0^2)}$, where $T_0 = 50/1.665$ fs and this functional form was used in the calculations using Eq. 4 later in this subsection.

In the first experiment, PCFs (NKT Photonics NL-1.5-670) with three different lengths (62, 114, and 166 mm) were used. The center wavelength of a fundamental soliton pulse versus an input pulse power is shown in Fig. 5 (a) and compared with the center wavelength versus an input pulse chirp shown in Fig. 5 (b). Also, the delay time versus the center wavelength of a fundamental soliton pulse is shown in Fig. 5 (c). As shown in this figure, the center wavelength of a fundamental soliton pulse changes quadratically when a fiber length was 62 mm. Similar results are obtained for 114 mm and 166 mm, although some discrepancies are noted. From Fig. 5 (c), it is seen that for the same PCF length, the delay time is independent on the control method and its dependence on wavelength is linear. Also it is shown that the delay time decreases as the fiber length becomes shorter. In the second experiment, a

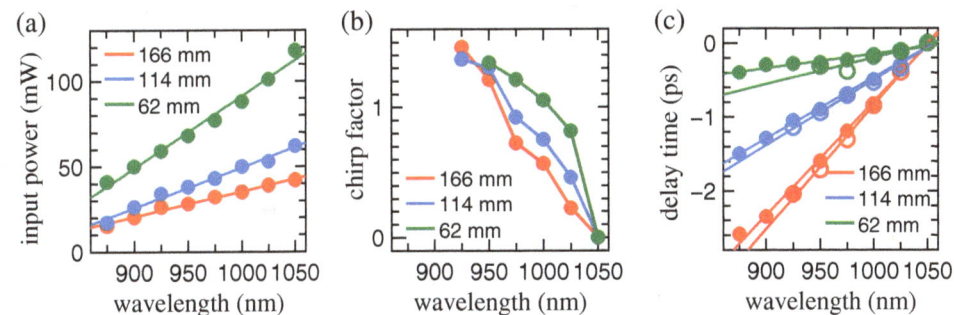

Fig. 5. Wavelength versus input power (a), wavelength versus chirp factor (b), and wavelength verses delay time (c) of soliton pulses for PCFs with different lengths.

166-mm-long PCF (NKT Photonics NL-1.5-670) whose calculated dispersion is shown in Fig. 1 (b) was used, and the results from experiment and calculations were compared. In Fig. 6 (a), experimental spectrum at 40 mW input average power is shown. There are two peaks at wavelengths near 860 nm and near 1030 nm and these correspond to fundamental soliton pulses created by the fission of an input pulse. The peak near 450 nm is a dispersive wave emitted by fundamental soliton pulses. In Fig. 6 (b), calculated spectrum at 5.74 kW input peak power, 50 fs pulse width, and $C = 0$ is shown. The experimental and calculated spectra agreed very well as shown in this figure. The temporal waveform (Fig. 6 (c)) shows two fundamental soliton pulses at $\sim$0.9 ps and at $\sim$3.6 ps. In Fig. 7, the experimental variation of the spectral peak positions of fundamental soliton pulses is shown for different prism position d. As shown in this figure, the peak position of the fundamental soliton pulse at longest wavelength changed about 70 nm according to d. In Fig. 6, the calculated spectral and temporal intensities are shown for different chirp C calculated from d in Fig. 4. The pulse energy in these calculations was set to be 0.287 nJ, which corresponded to 5.74 kW peak power at $C = 0$. The calculated spectra showed the almost identical variation of peak positions compared with experiment. As the absolute value of C was increased, the temporal position of the most intense fundamental soliton pulse became smaller and the corresponding spectral peak wavelength became shorter. It means that as the absolute value of C was increased, the timing of the fission of an input pulse was delayed more and as a result of this, the fundamental soliton pulses experienced less soliton self-frequency shifts and less delay times at the output end of the PCF.

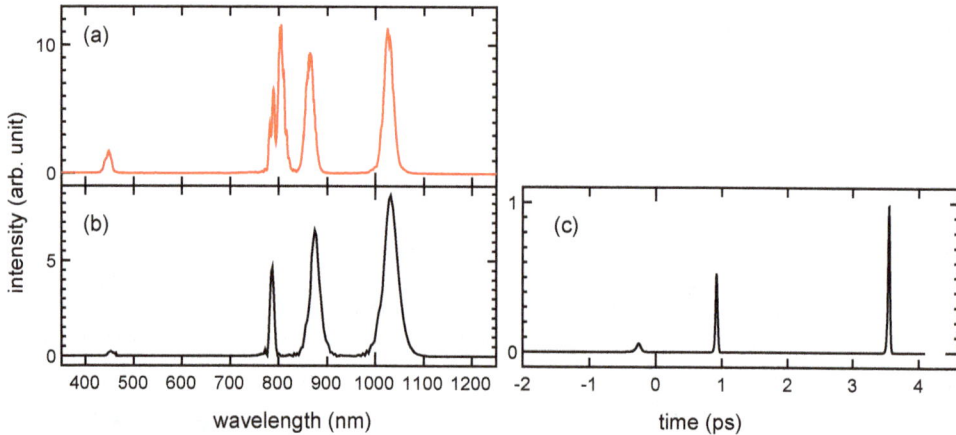

Fig. 6. (a) Experimental spectrum at 40 mW input power. Calculated spectral intensity (b) and temporal intensity (c) when a pulse with 5.74 kW peak power is inputted into a PCF.

4. Solitons for coherent anti-Stokes Raman scattering spectroscopy

Coherent anti-Stokes Raman scattering (CARS) microscopic spectroscopy is one of the nonlinear optical spectroscopy that has attracted attention recently (Cheng & Xie, 2004; Müller & Zumbusch, 2007). In CARS spectroscopy, a pump pulse (angular frequency ω_p) and a Stokes pulse (angular frequency ω_s) are used to illuminate a sample to generate an anti-Stokes signal (angular frequency $2\omega_p - \omega_s$). This signal is enhanced when the frequency difference between the pump and the Stokes pulses ($\omega_p - \omega_s$) coincides one of the Raman active vibration frequencies of the sample. The CARS signal is coherent and its intensity is much higher than that of a spontaneous Raman signal. Also, CARS signals from a fluorescent sample can be detected because the frequencies of CARS signals are higher than the frequencies of fluorescent signals. When CARS signals are detected in a microscope, the spatial resolution is expected to be high because the CARS signals are generated due to third-order nonlinear optical processes. The spectral resolution of a CARS signal is usually

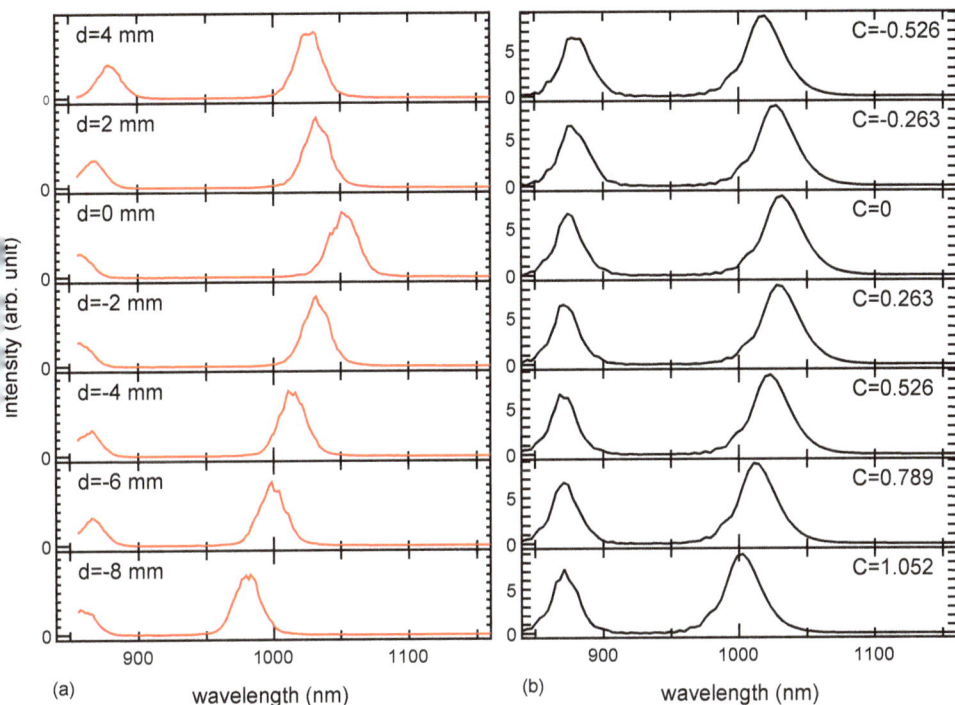

Fig. 7. (a) Experimental spectra for different prism position d. (b) Calculated spectral intensities at 0.287 nJ input pulse energy. The chirp C calculated from the position d is indicated.

determined by the spectral bandwidth of a pump pulse and its observable range is determined by the spectral bandwidth of a Stokes pulse. In multiplex CARS spectroscopy, a broadband Stokes pulse is used to detect multiple Raman vibration frequencies of a sample. To generate a broadband optical pulse, a PCF can be used and only a single laser oscillator is required to observe a broadband CARS signal by the use of a PCF. The use of a PCF to generate broadband Stokes pulses was reported in 2003 (Paulsen et al., 2003) and since then, the extent of vibration frequencies has been extended (Kano & Hamaguchi, 2005; Kee & Cicerone, 2004). However, it is difficult to generate broadband pulses with uniform spectral intensities. Also the group delays of various spectral components in a broadband pulse differ due to the dispersion property of a PCF, and thus it is difficult to generate intense broadband CARS signals. On the other hand, as shown in previous sections, the center wavelength of a soliton pulse from a PCF can be controlled easily by the power or the chirp of an input pulse, and the delay times are well known, thus it is very suitable for use in broadband or multiplex CARS spectroscopy as Stokes pluses. In this section, we show various approaches using soliton pulses for broadband CARS spectroscopy. Since the typical spectral bandwidth of a single fundamental soliton from a PCF is about 20 nm, it is necessary to change the center wavelength for broadband CARS spectroscopy. When the center wavelength is changed, the delay time changes also, so it is necessary to control the delay time at the same time. We have controlled these parameters using a pulse shaper (subsection 4.1). The other approach is to change the center wavelength of a soliton continuously by an acousto-optical modulator

(AOM) to generate quasi-supercontinuum (quasi-SC) (subsection 4.2). Also, we show results using soliton pulses for single-beam CARS spectroscopy (subsection 4.3).

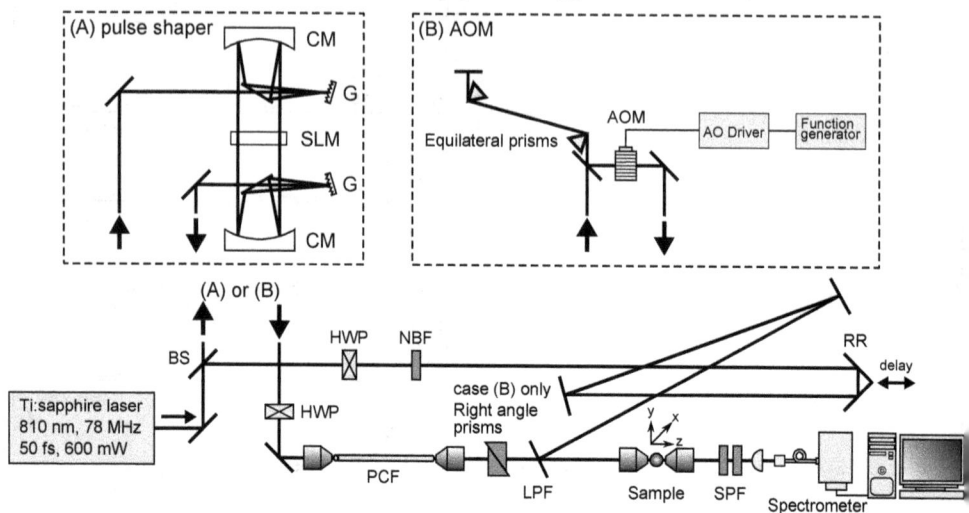

Fig. 8. Experimental setup for broadband CARS spectroscopy using a pulse shaper (A) and an AOM (B). Here, BS: beam splitter, RR: retroreflector, CM: concave mirror, G: grating, HWP: half-wave plate, NBF: narrow bandpass filter, SPF: short-wavelength pass filter, and LPF: long-wavelength pass filter.

4.1 Broadband CARS spectroscopy using a pulse shaper

Experimental setup of CARS spectroscopy using a pulse shaper is shown in Fig. 8 (A) (Tada & Karasawa, 2008; 2009). A pulse from a Ti:sapphire laser oscillator (center wavelength 810 nm, pulse width 50 fs, and repetition rate 78 MHz) was split into two pulses by a beam splitter and one of the pulse was used as a pump pulse after its spectrum was narrowed by a band pass filter (center wavelength 808 nm with a 3-nm full width at half maximum bandwidth). The other pulse was used as a Stokes pulse after it was shaped by a pulse shaper and was propagated in a 123-mm-long PCF (NKT Photonics NL-1.5-590) to generate a fundamental soliton pulse. A long-pass filter was used to eliminate the shorter-wavelength components than soliton's wavelength. Both pulses were overlapped collinearly and focused on a sample using an objective (100×, 0.5 numerical aperture). The average input powers on a sample were about 9 mW for a pump beam and about 4 mW for a Stokes beam. The signal from the sample was collected by an objective and was detected by a spectrometer (Solar TII MS-3504) with a CCD detector (Andor DV420-OE) after the spectral components of both pump and Stokes pulses were removed by the use of short-pass filters. Initially, the center wavelength of a soliton pulse from a PCF was set to be 1050 nm (which corresponds to vibration frequency ~3000 cm^{-1}) by adjusting the power of an input pulse. The center wavelength of a soliton pulse was shifted to shorter wavelength by applying a phase pattern by a spatial light modulator (SLM; JenOptik SLM-S320) in a pulse shaper. The pulse shaper consists of an SLM, two pairs of gratings, concave mirrors, and folding mirrors. In the first experiment (Tada & Karasawa, 2008), six different phase patterns were used such that the

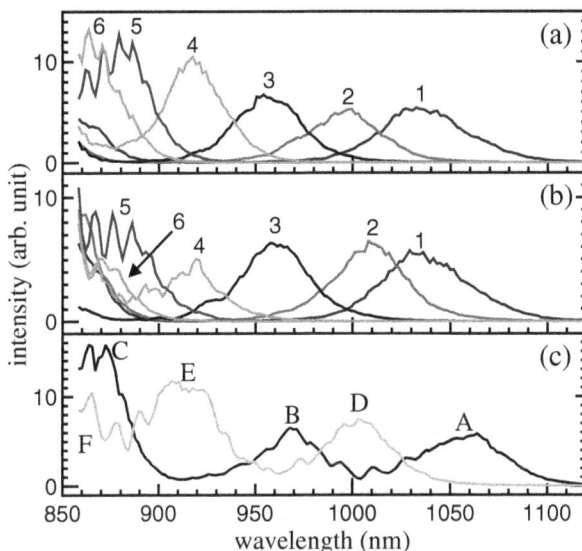

Fig. 9. Spectra of soliton pulses obtained by a pulse shaper using quadratic (a), cosine (b), and pulse train (c) phase patterns. In (c), A–F correspond to soliton pulses generated by pulses A–F in pulse trains shown in Fig. 10.

spectra of soliton pulses covered the wavelength between 850 and 1050 nm uniformly. Two different functional forms of phase patterns for varying the center wavelength of a soliton pulse were tried. One was the quadratic phase pattern of a form $(\beta_0^{(2)}/2)(\omega - \omega_0)^2$, where ω_0 was the center angular frequency of an input pulse ($\omega_0 = (2\pi c)/(800 \text{ nm})$) and $\beta_0^{(2)}$ determined the chirp of an input pulse. The other was the cosine phase pattern of a form $A \cos \omega T$. When the cosine phase pattern was used, the pulse train of an original pulse was created with a period T and the peak amplitude of the central pulse was determined by the Bessel function of zero order $J_0(A)$ (Morita & Toda, 2005). The period T was set to be 500 fs in experiment such that the timing of only the central pulse in a pulse train matched with a pump pulse. In both cases, the phase pattern of a form $\beta_0^{(1)}(\omega - \omega_0)$ was added to control the delay time of an input pulse with respect to a pump pulse, where $\beta_0^{(1)}$ is a group delay. Moreover, the phase pattern of a form $(\beta_{00}^{(2)}/2)(\omega - \omega_0)^2$ with $\beta_{00}^{(2)} = -200 \text{ fs}^2$ was added for all phase patterns to compensate for the dispersion of an objective lens in front of a PCF. To adjust the delay time of a soliton pulse with a pump pulse, the group delay ($\beta_0^{(1)}$) of an input pulse was set to be different for a soliton pulse with a different center wavelength. In Fig. 9, the spectra of soliton pulses with six different center wavelengths are shown for both the quadratic (a) and the cosine (b) phase patterns. As shown in this figure, by using six soliton pulses with different center wavelengths, the spectral regions between 850 and 1050 nm were covered almost uniformly. The exposure time to obtain CARS signals was set to be one second for each soliton pulse with a different wavelength, thus the total exposure time to obtain CARS signals between 500 and 3100 cm^{-1} was six seconds. In addition to this, the switching time of the phase pattern in a SLM was required (about 0.3 second for every switching). In the later

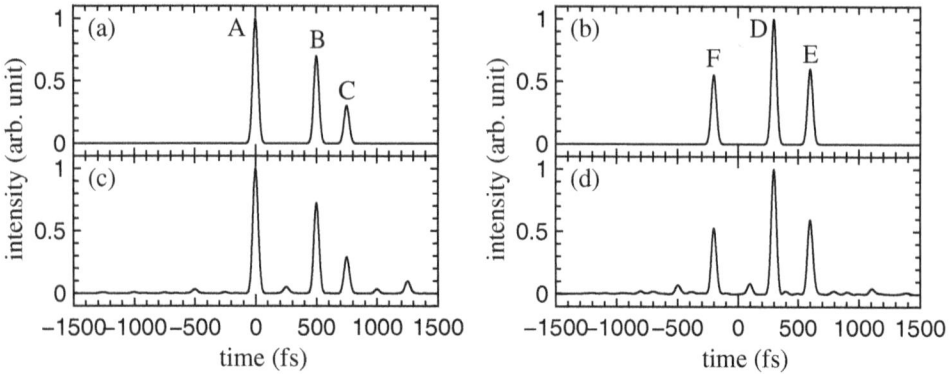

Fig. 10. Target pulse waveforms ((a) and (b)) for generating pulse train phase patterns and the optimized waveforms ((c) and (d)).

experiment (Tada & Karasawa, 2009), pulse trains were created to generate multiple solitons with different center wavelengths by a single phase pattern. Here, we prepared two different phase patterns for the SLM to generate the five fundamental soliton pulses. We adjusted the power ratios of pulses in a pulse train for varying the wavelengths of soliton pulses. Also, we adjusted the temporal delays of pulses in a pulse train such that all fundamental pulses arrived the sample at the same time. In Fig. 10, the two target pulse waveforms used in experiment for generating fundamental soliton pulses are shown. In a target pulse train 1 (Fig. 10 (a)), there were three pulses A (power ratio 1, delay time 0 fs), B (power ratio 0.7, delay time 500 fs), and C (power ratio 0.3, delay time 750 fs) for generating soliton pulses. In a target pulse train 2 (Fig. 10 (b)), there were two pulses D (power ratio 1.0, delay time 300 fs) and E (power ratio 0.6, delay time 600 fs). The pulse F (power ratio 0.55, delay -200 fs) was added in the pulse train for adjusting the total power of the pulse train 2 with respect to the pulse train 1. Since this pulse was not used in CARS spectroscopy, its delay time was shifted intentionally such that it did not arrive the sample at the same time with other pulses. All pulses in pulse trains were assumed to be Gaussian pulses with the full width at half maximum pulse widths to be 50 fs. The phase patterns for the SLM to generate these target pulses were created using genetic algorithms (Goldberg, 1989), where each phase pattern was optimized such that the difference between the shaped pulse obtained numerically and the target pulse became minimum. Here, the shaped pulse was calculated numerically by the inverse Fourier transform of the input pulse spectrum after its phase was modified according to the phase pattern, where the input pulse was assumed to be a 50-fs Gaussian pulse. The pulse waveforms obtained numerically after their phase patterns were optimized by the genetic algorithms are shown in Fig. 10 (c) and (d). As shown in Fig. 10, these numerically-obtained pulses using the optimized phase patterns agreed well with target pulses. The spectra of soliton pulses after these pulse trains were propagated in a PCF are shown in Fig. 9 (c). In this figure, two measured spectra with different phase patterns are shown and spectral peaks A–F corresponded to soliton pulses generated by pulses A–F in Fig. 10. As shown in this figure, the wavelength range between 860 and 1070 nm was covered by five fundamental soliton pulses. As mentioned above, the spectral peak F did not affect CARS measurements since its temporal timing was shifted. By switching these two phase patterns in the SLM, it was possible to measure broadband CARS signals automatically. The exposure time for each phase pattern was set to be 1 s and the additional time 0.4 s was required for switching the phase pattern in the SLM. Thus, the total measurement time was 2.4 s. The results of CARS spectroscopy of a polystyrene (PS) sample

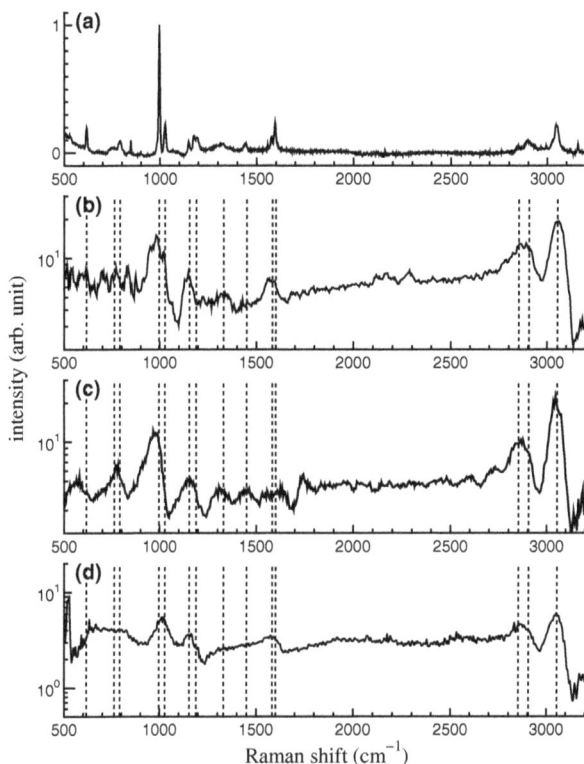

Fig. 11. Spontaneous Raman spectrum (a) and CARS spectra using a pulse shaper with a quadratic (b), cosine (c), and pulse train (d) phase patterns of a 6-μm diameter polystyrene bead sample. In (b)–(d), peak positions of spontaneous Raman spectrum are shown by dashed lines.

are shown and compared with the result of spontaneous Raman spectroscopy in Fig. 11. In this figure, the CARS signal from a sample was normalized by the signal from a glass substrate to show the resonant contribution of CARS signal clearly (Kee & Cicerone, 2004). All results using a pulse shaper (quadratic, cosine, and pulse train phase patterns) agreed well and the Raman peaks observed by spontaneous Raman spectroscopy were observed clearly in CARS signals between 500 and 3100 cm^{-1}. The spectral widths of Raman peaks were determined by the spectral width of a pump pulse and in our setup, the spectral resolution was about 50 cm^{-1}. The exposure time for the method using phase patterns for pulse trains was 2.4 s and is less than half compared with the methods using phase patterns for single soliton pulses. It was because the effective bandwidth of the Stokes pulse was multiplied by a number of pulses in a pulse train. Here, we have limited the number of pulses in a pulse train to be three due to the constraint of the available power from a laser, but it is straightforward to increase the number if enough power is available. It is demonstrated that the broadband fundamental soliton pulses can be generated by a pulse shaper and are very useful for CARS spectroscopy.

4.2 Quasi-supercontinuum broadband CARS spectroscopy using an acousto-optical modulator

Since the center wavelength of a soliton pulse can be changed by varying the input pulse power of a PCF, it is possible to generate pulse trains whose center wavelengths change continuously by modulating the input power rapidly. In this way, quasi-supercontinuum (quasi-SC) in the wavelength range from 1.56 to 1.9 μm was generated using soliton pulses from a highly nonlinear fiber by scanning the input power by an acousto-optical modulator (AOM) and its application to optical coherence tomography (OCT) was mentioned (Sumimura et al., 2008). In OCT, the adjustment of the group delays between different spectral components is not necessary since the shape of the interference signal depends on the spectrum of the light source only. On the other hand, the adjustment is very important in CARS spectroscopy to obtain strong broadband CARS signals. In this study, we have generated quasi-SC in the wavelength range from 0.85 to 1.1 μm using a PCF and applied to CARS spectroscopy, where the power modulation was performed by an AOM and the group delay adjustment was performed by simply placing a pair of prisms after the PCF, since the group delay of the soliton pulses depended on wavelength approximately linearly as mentioned in Chapter 3.

Experimental setup for quasi-SC CARS spectroscopy s shown in Fig. 8 (B) (Tada & Karasawa, 2010). A pulse from a Ti:sapphire laser oscillator (center wavelength 810 nm, pulse width 50 fs, and repetition rate 78 MHz) was split into two pulses by a beam splitter and one of the pulse was used as a pump pulse after its spectrum was narrowed by a band pass filter (center wavelength 808 nm with a 3-nm full width at half maximum bandwidth). The other pulse was used as a Stokes pulse after its power was modulated by an AOM (ISOMET M1137-SF-40L-1.5) and was propagated in a 120-mm-long PCF (NKT Photonics NL-1.5-590) to generate quasi-SC, where a pair of SFL11 equilateral prisms was used to compensate for the dispersion of the AOM. For the modulation of the AOM, a 100-kHz sinusoidal wave was used. A long-pass filter (cut-off wavelength 840 nm) was used to eliminate the shorter-wavelength components than the wavelengths of solitons. A pair of SFL11 right angle prisms was used to adjust the group delays of soliton pulses with different center wavelengths.

Both pump and Stokes pulses were overlapped collinearly and focused on a single 6-μm-diameter polystyrene bead sample using an objective (100×, 0.9 numerical aperture). The CARS signal from the sample was collected by an objective and was detected by a spectrometer (Solar TII MS-3504) with a CCD detector (Andor DV420-OE) after the spectral components of both pump and Stokes pulses were removed by the use of short-pass filters (cutoff wavelengths 785 and 850 nm). The exposure time for taking a CARS spectrum was two second. In Fig. 12 (a), the spectrum of generated quasi-SC when an AOM was modulated by a 100-kHz sinusoidal wave is shown. As shown in this figure, broadband quasi-SC, which had a sufficient spectral intensity for the CARS spectroscopy in the wavelength range from 850 to 1100 nm, was generated, which corresponded to CARS wave number between 500 and 3100 cm^{-1}. In Fig. 12 (b), the normalized CARS spectrum of a single 6-μm-diameter polystyrene bead sample using quasi-SC are shown and it is compared with the known spontaneous Raman peaks. As shown in this figure, most spectral peaks of the polystyrene sample were observed clearly between 900 and 3100 cm^{-1}. The exposure time of 2 s was shorter than the exposure time in the previous subsection (2.4 s) using the same spectrometer for the measurement with the similar signal to noise ratio. It is demonstrated that the quasi-SC is very useful for broadband CARS spectroscopy.

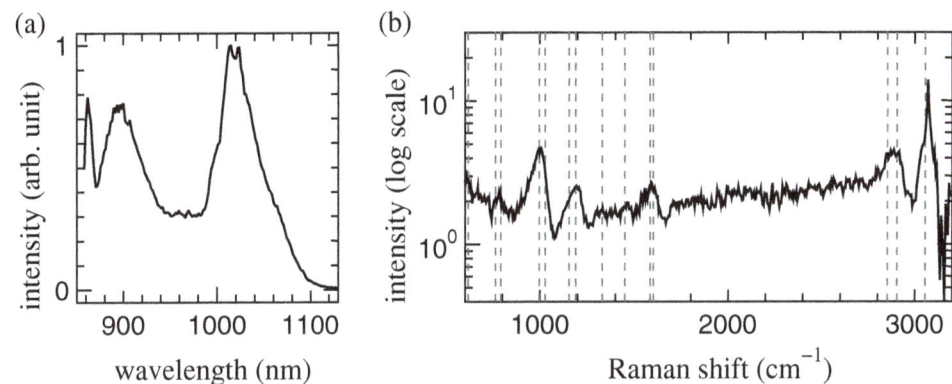

Fig. 12. (a) The spectrum of quasi-SC. (b) The normalized CARS spectrum using quasi-SC of a 6-μm diameter polystyrene bead sample. Peak positions of spontaneous Raman spectrum are shown by dashed lines.

4.3 Single-beam CARS spectroscopy

As shown in previous subsections, two separate beams, a pump beam and a Stokes beam, have to be collinearly overlapped and focused on a sample using an objective lens in CARS spectroscopy in general. However, the adjustments of these two beams, necessary to generate strong CARS signals, are sometimes difficult. Therefore, it is desirable to perform CARS spectroscopy using a single beam. It is necessary to generate a single beam that contains both pump and Stokes spectral components to perform single-beam CARS spectroscopy. In

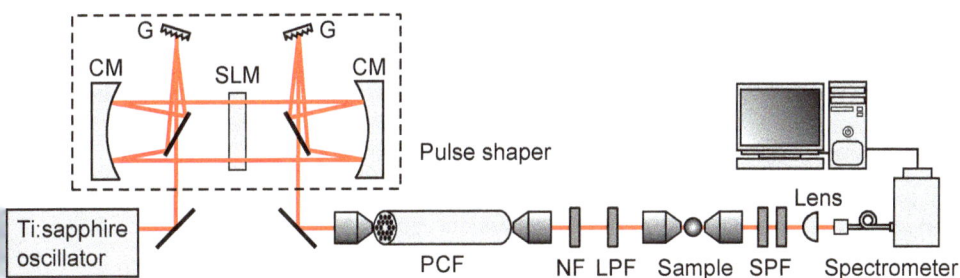

Fig. 13. Experimental setup for single-beam CARS spectroscopy. Here, NF: notch filter, LPF: long-wavelength pass filter, SPF: short-wavelength pass filter, G: grating, and CM: concave mirror.

our setup (Tada & Karasawa, 2011), two pulses, one for generating a wavelength-tunable fundamental soliton pulse and the other for generating a narrowband pump pulse, are shaped by a pulse shaper and inputted into a PCF. The fundamental soliton Stokes pulse is generated by redshifting the input pulse spectrum through the soliton self-frequency shift in a PCF and the amount of this shift is controlled by the power of an input pulse. The pulse for a pump pulse is negatively chirped by a pulse shaper for the spectral compression in a PCF

(Andresen et al., 2005), which is important for obtaining a narrowband pump pulse to achieve a high spectral resolution while retaining most of the pulse energy.

The experimental setup of single-beam CARS spectroscopy is shown in Fig. 13. Two pulses with different intensities were shaped from a single pulse from a Ti:sapphire laser oscillator (the center wavelength 797 nm, the pulse width 50 fs, the average power 570 mW, and the repetition rate 78 MHz) and propagated in a 119-mm-long and 1.5-μm-core-diameter PCF (NKT Photonics NL-1.5-590). An input pulse with a negative chirp was compressed spectrally due to the self-phase modulation until it became almost transform-limited. The amount of chirp, the input power, and the fiber length were adjusted to obtain an optimal spectral compression. In Fig. 14, the target pulse waveforms used in the experiment for generating

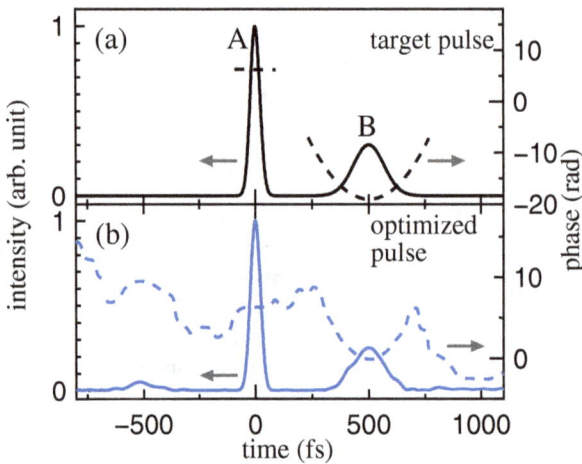

Fig. 14. (a) The intensity and the phase of a target pulse for single-beam CARS spectroscopy, where pulse A is for generating a soliton pulse and pulse B is for generating a narrowband pump pulse. (b) The intensity and the phase of an optimized pulse.

two pulses are shown by black curves. In a target pulse, there were two pulses, namely, pulse A (power ratio 1 and delay time 0 fs) for generating a Stokes soliton pulse and pulse B (power ratio 0.3 and delay time 500 fs). Pulse B was negatively chirped for spectral compression to generate a narrowband pump pulse (chirp factor -3). The phase pattern for generating these pulses were optimized by the genetic algorithms (Goldberg, 1989). By adjusting the relative delay time between these two input pulses for a PCF, it was possible to generate a single beam that contained timing-matched pump and Stokes pulses at the sample position. To determine the delay time between two pulses, a β-BaB$_2$O$_4$ (BBO) crystal was set at the sample position and the sum-frequency signals of the pump and Stokes pulses were measured. Long-pass filters (cutoff wavelengths 793 and 590 nm) and a notch filter (center wavelength 825 nm and bandwidth 40 nm) were used to limit the bandwidth of the spectrally compressed pump pulse and remove the wavelengths shorter than the pump pulse. CARS signals were generated by the output beam from a PCF, where the beam was tightly focused on a sample by an objective lens (100$\times$ and 0.9 numerical aperture). The generated CARS signals from the sample were collected by a microscope objective lens (100$\times$ and 0.7 numerical aperture), where the spectral components at the pump and longer wavelengths were removed by the

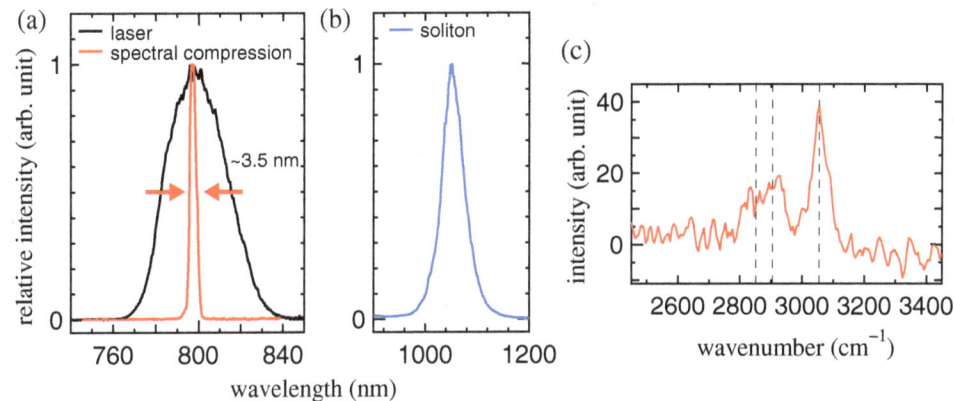

Fig. 15. In (a), the spectra from laser and a narrowband pump pulse are shown. In (b), the spectrum of a soliton pulse is shown. In (c), the CARS signal of a 6-μm-diameter polystyrene bead sample is shown, where known Raman peak positions are shown by dashed lines.

combination of short-pass filters (cutoff wavelengths 785 and 850 nm). The CARS signals from the sample were detected by a spectrometer (Solar TII MS-3504) with a CCD detector without an intensifier (Andor DV420-OE). The exposure time for taking a CARS spectrum was 5 s.

Figure 15 (a) and (b) show the spectra of the output beam from a PCF after notch and long-pass filters when a pulse with about 88 mW average power was inputted into the PCF. In Fig. 15 (a), a black curve shows the spectrum of a laser pulse with a full width at half maximum (FWHM) of about 32 nm. A red curve shows the spectrum of a pump pulse after the PCF and the combinations of filters with a width of about 3.5 nm, which was about 11% of the original bandwidth. The spectral resolution using this pump pulse was about 55 cm^{-1}. A Stokes pulse shown in Fig. 15 (b) was generated from a PCF as a fundamental soliton pulse at the center wavelength 1052 nm, which was tuned to match the Raman shift of about 3000 cm^{-1}. When the input average power for the PCF was 88 mW, the output average power was 26 mW and the average powers of the beam after a long-pass filter for pump and Stokes components were 2 and 3 mW, respectively. Figure 15 (c) shows the CARS signals generated from a single 6-μm-diameter polystyrene bead. As shown in this figure, the CARS signals of the polystyrene were observed from 2800 to 3100 cm^{-1} that corresponded to well-known C-H stretching vibration modes (the Raman shifts for symmetric aliphatic 2852 cm^{-1}, asymmetric aliphatic 2905 cm^{-1}, and aromatic 3054 cm^{-1} are shown in Fig. 15 (c) by vertical lines).

It is demonstrated that CARS spectroscopy can be performed using a single-beam setup with a pulse shaper and a fundamental soliton pulse from a PCF. Unlike previous single-beam CARS setups using a pulse shaper, this setup can be used to observe CARS signals from 2800 to 3100 cm^{-1}. The spectral resolution of the setup was determined by the bandwidth of the combinations of notch and bandpass filters. Since the wavelength of the soliton Stokes pulse can be varied by adjusting the input power, it is straightforward to perform broadband CARS measurements using this setup.

5. Solitons for optical coherence tomography

Optical coherence tomography (OCT) is a technique that images the internal structure of biological or medical samples noninvasively and nondestructively by using a low-coherent light source and a Michelson interferometer. OCT was developed in the early 1990s (Huang et al., 1991). The depth resolution of the OCT is determined by the spectral width of the light source. Thus it is important to broaden the spectral width of the light source for improving the resolution of OCT. Also it is important to make the spectral shape of the light source simple, e.g. the Gaussian shape, since in that case, the side pulses of an interference signal, which become noises in an OCT image, become small. The depth resolution of the OCT is determined by the coherence length l_c, which is the FWHM width of the interference signal and it is given by $l_c = a\lambda_0^2/\Delta\lambda$ for the intensity spectrum with the center wavelength λ_0 and the FWHM bandwidth $\Delta\lambda$. a is a constant and $a = 0.44$ for a Gaussian spectrum and $a = 0.39$ for a sech^2 spectrum. Supercontinuum from a PCF was used to improve the resolution of OCT (Hartl et al., 2001). However, since there are many peaks in the spectrum of the supercontinuum light from a PCF, there are many side pulses in the interference signal and these become noises in an OCT image. As shown in previous sections, the fundamental soliton pulses from a PCF ranges in a wavelength region between 0.85 μm and 1.05 μm when a Ti:sapphire laser is propagated and the spectral shapes of them are simple. Also, since the penetration depth of 1.0-1.3 μm light is maximum and the attenuation due to absorption and scattering is minimum in biological samples (Lim et al., 2005), it is beneficial to use a light source in this wavelength region. The center wavelength of the fundamental soliton pulse shifts to longer wavelength as it propagates in a PCF and it can be controlled by changing the power of an input pulse. When the power of an input pulse is changed continuously, the wavelength of a soliton pulse changes continuously and the soliton pulse can be used as a quasi-supercontinuum (quasi-SC) light source and its use in OCT as a light source has been studied (Sumimura et al., 2008) in wavelength ~1.5 μm. Here, we show results using a pulse train generated by a pulse shaper and quasi-SC in the wavelength range between 900 and 1000 nm generated by an AOM.

5.1 Pulse train OCT using a pulse shaper

The experimental setup is shown in Fig. 16 (A) (Takabatake et al., 2010). The pulse emitted from a Ti:sapphire laser (pulse width 50 fs, center wavelength 810 nm, repetition rate 78 MHz, average power 600 mW) was transformed into a pulse train with three pulses by a pulse shaper, where the interval times and the peak power ratios between pulses were controlled as in CARS experiment (subsection 4.1). These parameters were optimized to obtain the best interference signal in OCT. The pulse train was propagated in a 45 cm PCF (NKT Photonics NL-1.5-670) and the pulse train which had overlapped spectra of each fundamental soliton pulse was generated. The input power of a PCF was adjusted to be 123 mW and the power of a single soliton pulse generated was 3 mW. The polarization direction of the input pulse train was optimized by a half wavelength plate to maximize the output power from the PCF. Also, a polarization plate was used to select the polarization direction of an output beam from the PCF. A long-pass filter with 840-nm cut-off wavelength was used to transmit only soliton pulses. The spectrum of a soliton pulse train was measured by a spectrometer and an interference signal was measured by a Michelson interferometer with a balanced photo detector. In the experiment, a mirror was placed at a sample position to obtain an interference signal and to evaluate its width. The position of a reference mirror was varied by a piezo stage

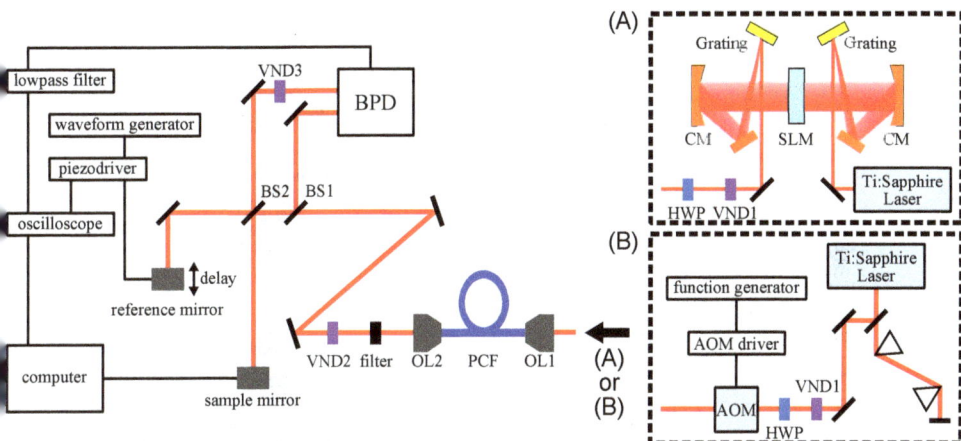

Fig. 16. Experimental setup for the soliton-based OCT using a pulse shaper (A) and an AOM (B). Here, VND: variable neutral density filter, CM: concave mirror, HWP: half-wave plate, OL: objective lens, BPD: balanced photo detector, and BS: beam splitter.

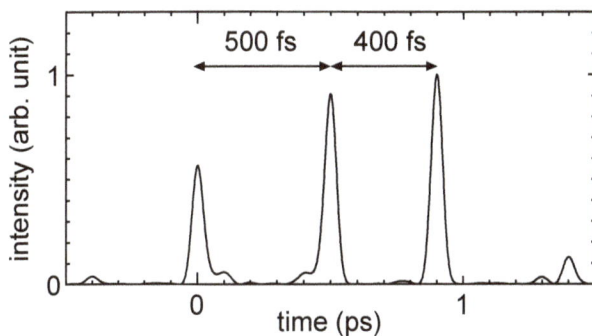

Fig. 17. Pulse waveforms for generating a pulse train of three soliton pulses obtained after optimized by a genetic algorithm.

and the modulation frequency of the delay was set to be 0.5 Hz. In experiment using a pulse shaper, a pulse train of three pulses was created, where the interval times and the power ratios of three pulses were set to be 500 fs and 400 fs, and 1.0:1.25:1.33 respectively after optimization. Since the soliton created by a pulse with a larger peak power delayed more in a fiber, it was necessary to input a pulse with a larger peak power later into a fiber to avoid overlapping with other soliton pulses. The phase pattern for generating the pulse train was optimized by a genetic algorithm (Goldberg, 1989). In Fig. 17, the temporal waveform of the pulse train, where the optimized phase pattern was used, is shown. In Fig. 18, the interference signal and the spectrum obtained by the pulse train (three soliton pulses) are compared with results without using a pulse train (a single soliton pulse). The spectral width ($\Delta\lambda$) of the single soliton was 21.4 nm and it became 64.0 nm when the pulse train was used. The coherence length (l_c) of the single soliton was 17.6 μm and it became 8.3 μm for the pulse train. From this result, it is demonstrated that the resolution of OCT can be improved by superimposing

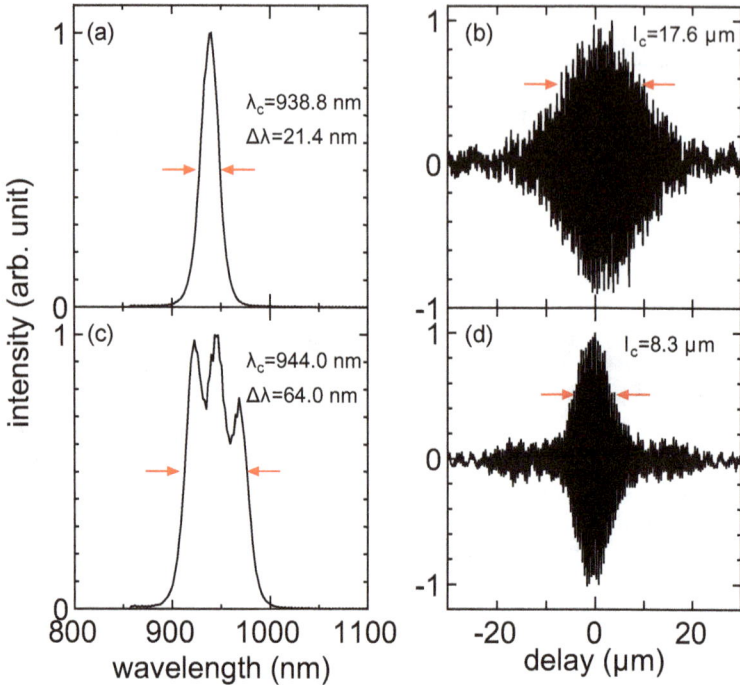

Fig. 18. The spectrum (a) and the interference signal (b) of a single soliton pulse compared with the spectrum (c) and the interference signal (d) of a pulse train composed of three soliton pulses.

soliton pulses using a pulse train. In this case, only three soliton pulses were used due to the constraint of the available power for the input power of a PCF. A pulse train that contains more pulses with the broader spectrum may be obtained if more input power is available.

5.2 Quasi-SC OCT using an acousto-optical modulator

In this experiment, a pulse shaper was replaced by an AOM and a prism pair for dispersion compensation for the AOM, as shown in Fig. 16 (B). The AOM was driven either by a sinusoidal wave or a triangle wave for modulating the input power of a PCF. The modulation frequency was set to be 5 KHz. By inputting modulated pulses (average power 64 mW) into a PCF, quasi-SC were generated, since the center wavelength of a soliton pulse changed according to the input power of a PCF due to the soliton self-frequency shift. The spectra of these quasi-SC were considered to be broadened if the measurement time was much longer than the modulation period of the AOM (0.2 ms in this case). In Fig. 19, the interference signals and the quasi-SC spectra are shown. The spectral widths ($\Delta\lambda$) of the quasi-SC for a sinusoidal wave modulation and a triangle wave modulation were 113.3 and 102.4 nm, where the exposure time of a spectrometer was 50 ms. The coherence length (l_c) of these quasi-SC were 6.1 and 6.5 μm, which are about one-third of the coherence length of a single soliton pulse shown in Fig. 18 (b). The quasi-SC spectrum obtained using a triangle wave (Fig. 19 (c)) was more uniform than the spectrum obtained using a sinusoidal wave (Fig. 19 (a)) since the

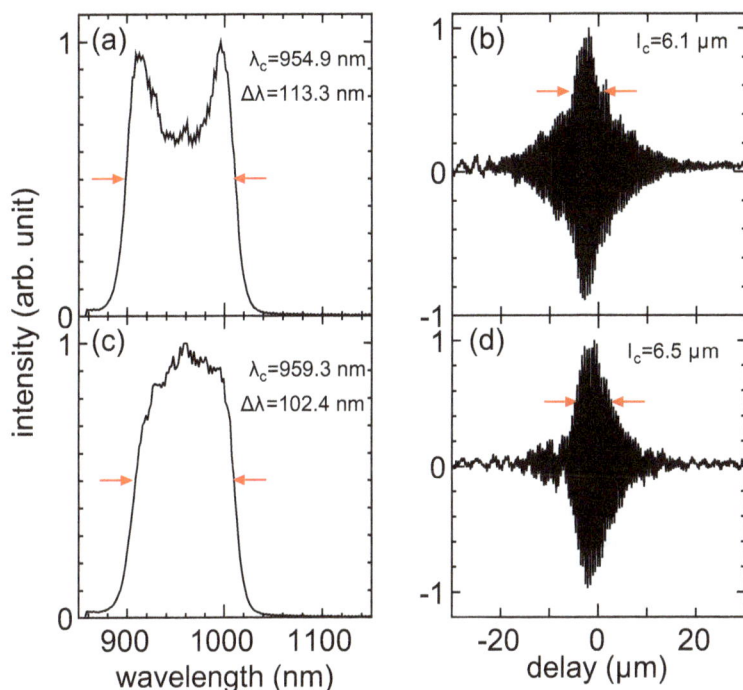

Fig. 19. The spectrum (a) and the interference signal (b) of quasi-SC using a sinusoidal wave compared with the spectrum (c) and the interference signal (d) of quasi-SC using a triangle wave.

variation of the power inputted into a PCF using a triangle wave was more uniform than that using a sinusoidal wave. From these results, it is demonstrated that the resolution of OCT can be improved by using a quasi-SC.

5.3 Comparison between methods using a pulse train and quasi-SC

By comparing the spectrum using a pulse train with three pulses shown in Fig. 18 (c) with the spectra using an AOM shown in Fig. 19 (a) and (c), we see that the spectral widths of the quasi-SC were broader than that of a pulse train and as a result, the coherent lengths of quasi-SC were smaller than that of a pulse train. Therefore it is advantageous to use quasi-SC to obtain the better resolution in OCT. This is because the throughput of an AOM was higher than that of a pulse shaper in our experimental setups. Also, it is possible to modify the spectral shape by changing the modulation waveform when using an AOM. However, the spectrum obtained using an AOM shown in Fig. 19 (a) and (c) were not stationary but the center wavelength of a soliton in the quasi-SC varied with the modulation frequency of an AOM (5 kHz). On the other hand, the spectrum obtained using a pulse train shown in Fig. 18 (c) was obtained by superimposing three soliton pulses and was stationary with the repetition rate of a laser (78 MHz). The method using quasi-SC may have a problem when the scanning speed of the OCT setup becomes comparable to the modulation speed of the AOM.

6. Conclusion

In this chapter, fundamental soliton pulses generated from a PCF were introduced and their applications in CARS spectroscopy and OCT were explained. The wavelength of the soliton pulse changes due to the self-frequency shift and it can be controlled by the power and/or the chirp of a pulse inputted into the PCF. At the same time, the delay time of the soliton pulse changes, which can be estimated relatively easily. The use of a pulse shaper is very useful to control the wavelength and the delay time simultaneously and this was used effectively in CARS spectroscopy experiment. In CARS experiment, broadband Stokes soliton pulses were generated and the broadband CARS spectroscopy setups were demonstrated. Also, a single-beam CARS spectroscopy setup was demonstrated using the fundamental soliton pulse. In OCT experiment, broadband pulses in wavelength between 900 and 1000 nm were generated using soliton pulses and the improvement of the resolution of the interference signals were demonstrated using these pulses.

7. Acknowledgements

We would like to appreciate Yoshihiko Takabatake, Minoru Bunya, Chihiro Satoh and Ryoh Shirakawa who contributed to the work presented in this book chapter. This work was supported in part by a Grant-in-Aid for Scientific Research (C) from the Japan Society for the Promotion of Science (JSPS) and a Grant-in-Aid for JSPS Fellows.

8. References

Agrawal, G. P. (2007). *Nonlinear Fiber Optics Fourth Ed.*, Elsevier, Burlington, MA, USA.

Andresen, E. R., Thøgersen, J. & Keiding, S. R. (2005). Spectral compression of femtosecond pulses in photonic crystal fibers, *Opt. Lett.* Vol. 30 pp. 2025–2027.

Blow, K. J. & Wood, D. (1989). Theoretical description of transient stimulated Raman scattering in optical fibers, *IEEE J. Quantum Electron.* Vol. 25 pp. 2665–2673.

Cheng, J. X. & Xie, X. S. (2004). Coherent anti-Stokes Raman scattering microscopy: instrumentation, theory, and applications, *J. Phys. Chem. B.* Vol. 108 pp. 827–840.

Dudley, J. M., Genty, G. & Coen, S. (2006). Supercontinuum generation in photonic crystal fiber, *Rev. Mod. Phys.* Vol. 78 pp. 1135–1184.

Goldberg, D. E. (1989). *Genetic Algorithms in Search, Optimization and Machine Learning*, Addison-Wesley, Reading, MA, USA.

Gordon, J. P. (1986). Theory of the soliton self-frequency shift, *Opt. Lett.* Vol. 11, pp. 662–664.

Hartl, I., Li, X. D., Chudoba, C., Ghanta, R. K., Ko, T. H., Fujimoto, J. G., Ranka, J. K. & Windeler, R. S. (2001). Ultrahigh-resolution optical coherence tomography using continuum generation in an air-silica microstructure optical fiber, *Opt. Lett.* Vol. 26 pp. 608–610.

Hasegawa, A. (1992). Optical solitons in fibers: theoretical review, *in* Taylor, J. R. (ed.), *Optical Solitons – Theory and Experiment* , Cambridge Univ. Press, Cambridge, pp.1–29.

Huang, D., Swanson, E. A., Lin, C. P., Schuman, J. S., Stinson, W. G., Chang, W., Hee, M. R., Flotte, T., Gregory, K., Puliafito, C. A. & Fujimoto, J. G. (1991). Optical coherence tomography, *Science* Vol. 254 pp. 1178–1181.

Husakou, A. V. & Herrmann, J. (2001). Supercontinuum generation of higher-order solitons by fission in photonic crystal fibers, *Phys. Rev. Lett.* Vol. 87 pp. 203901-1–203901-4.

Kano, H. & Hamaguchi, H. (2005). Ultrabroadband (>2500 cm^{-1}) multiplex coherent anti-Stokes Raman scattering microspectroscopy using a supercontinuum generated from a photonic crystal fiber, *Appl. Phys. Lett.* Vol. 86 pp. 121113-1–121113-3.

Karasawa, N., Nakamura, S., Nakagawa, N., Shibata, M., Morita, R., Shigekawa, H., & Yamashita, M. (2001). Comparison between theory and experiment of nonlinear propagation for a-few-cycle and ultrabroadband optical pulses in a fused-silica fiber, *IEEE J. Quantum Electron.* Vol. 37, pp. 398–404.

Karasawa, N., Tada, K., & Ohmori, H. (2007). The comparison between experiment and calculation of the chirp-controlled Raman self-frequency shift in a photonic crystal fiber, *IEEE Photon. Technol. Lett.* Vol. 19, pp. 1292–1294.

Kee, T. W. & Cicerone, M. T.. (2004). Simple approach to one-laser, broadband coherent anti-Stokes Raman scattering microscopy *Opt. Lett.* Vol. 29 pp. 2701–2703.

Lim, H., Jiang, Y., Wang, Y., Huang, Y.-C., Chen, Z. & Wise, F. W. (2005). Ultrahigh-resolution optical coherence tomography with a fiber laser source at 1 μm, *Opt. Lett.* Vol. 30, pp. 1171–1173.

Mitschke, F. M. & Mollenauer, F. (1986). Discovery of the soliton self-frequency shift, *Opt. Lett.* Vol. 11 pp. 659–661.

Morita, R. & Toda, Y. (2005). Field manipulation of ultrabroadband optical pulses, *in* Yamashita, M., Shigekawa, H. & Morita, R. eds., *Mono-Cycle Photonics and Optical Scanning Tunneling Microscopy*, Springer, Berlin, pp. 251–283.

Müller, M. & Zumbusch, A. (2007). Coherent anti-Stokes Raman scattering microscopy, *ChemPhysChem* Vol. 8 pp. 2156–2170.

Paulsen, H. N.. Hilligsøe, K. M., Thøgersen, J, Keiding, S. R. & Larsen, J. J. (2003). Coherent anti-Stokes Raman scattering microscopy with a photonic crystal fiber based light source, *Opt. Lett.* Vol. 28 pp. 1123–1125.

Ranka, J., Windeler, R. S. & Stentz, A. J. (2000). Visible continuum generation in air-silica microstructure optical fibers with anomalous dispersion at 800 nm, *Opt. Lett.* Vol. 25, pp. 25–27.

Satsuma, J. & Yajima, N. (1974). Initial value problems of one-dimensional self-modulation of nonlinear waves in dispersive media, *Suppl. Prog. Theor. Phys., Japan* Vol. 55 pp. 284–306.

Schrader, D. (1995). Explicit calculation of N-soliton solutions of the nonlinear Schroedinger equation, *IEEE J. Quantum Electron.* Vol. 31, pp. 2221–2225.

Sumimura, K., Ohta, T. & Nishizawa, N. (2008). Quasi-super-continuum generation using ultrahigh-speed wavelength-tunable soliton pulses, *Opt. Lett.* Vol. 33, pp. 2892–2894.

Tada, K. & Karasawa, N. (2008). Broadband coherent anti-Stokes Raman scattering spectroscopy using pulse-shaper-controlled variable-wavelength soliton pulses from a photonic crystal fiber, *Jpn. J. Appl. Phys.* Vol. 47 pp. 8825–8828.

Tada, K. & Karasawa, N. (2008). Coherent anti-Stokes Raman scattering microspectroscopy using a fundamental soliton pulse generated from a photonic crystal fiber, *in* Tanio, N. & Sasabe, H. eds., *Optical Materials and Devices New Stage*, PWC Publishing, Chitose, Japan, pp. 215–220.

Tada, K. & Karasawa, N. (2009). Broadband coherent anti-Stokes Raman scattering spectroscopy using soliton pulse trains from a photonic crystal fiber, *Opt. Commun.* Vol. 282 pp. 3948–3952.

Tada, K. & Karasawa, N. (2010).   Broadband coherent anti-Stokes Raman scattering spectroscopy using a quasi-supercontinuum light source, *in Conference on Lasers and Electro-Optics*, OSA Technical Digest, paper JTuD72.

Tada, K. & Karasawa, N. (2011).   Single-beam coherent anti-Stokes Raman scattering spectroscopy using both pump and soliton Stokes pulses from a photonic crystal fiber, *Appl. Phys. Express* Vol. 4 pp. 092701-1–092701-3.

Takabatake, Y., Tada, K. & Karasawa, N. (2010). Optical coherence tomography using a soliton pulse train, *in* Kawabe, Y. & Kawase, M. eds., *Polymer Photonics, and Novel Optical Technologies*, PWC Publishing, Chitose, Japan, pp. 116–119.

Zolla, F., Renversez, G., Nicolet, A., Kuhlmey, B., Guenneau, S. & Felbacq, D. (2005). *Foundations of Photonic Crystal Fibres*, Imperial College Press, London.

10

Arc Fusion Splicing of Photonic Crystal Fibres

Krzysztof Borzycki[1] and Kay Schuster[2]
[1]National Institute of Telecommunications (NIT)
[2]Institute of Photonic Technology (IPHT)
[1]Poland
[2]Germany

1. Introduction

Arc fusion splicing is an established method for joining optical fibres in communication networks, ensuring splice loss down to 0.05 dB and excellent reliability. Telecom fibres are covered by IEC 60793 and ITU-T G.651.1-G.657 standards, with common material (fused silica) and cladding size (125 µm). Splicing equipment for these fibres is widely available.

Fusion splicing of specialty fibres, like dispersion compensating fibres (DCF), polarization-maintaining fibres (PMF), rare-earth doped active fibres and photonic crystal fibres (PCF) having varying, not standardized designs, dimensions and materials is considerably more difficult. Some, like PMF and many PCFs lack axial symmetry, requiring rotational alignment before fusion. However, this functionality is not available in common splicing machines. A length of specialty fiber enclosed inside a device like optical amplifier, sensor or dispersion compensator is usually spliced at both ends to telecom single mode fibres (SMF) for connections to other components and external interfaces.

Splicing procedure must be tailored to particular fibre, often in a time-consuming trial-and-error way. Special solutions, like fiber pre-forming or insertion of intermediate fiber are sometimes needed. Dedicated splicing machines for such fibres employing both arc fusion (OFS, 2008) and hot filament methods (Vytran, 2009) exist, but are expensive.

Fusion splicing of photonic crystal or other "holey" fibres with numerous tiny, ca. 0.5-4 µm gas-filled holes is particularly hard because holes collapse quickly once glass is heated to melt; this disturbs radiation guiding, introduces loss and causes fiber shrinkage.

There is a need to splice "holey" or "microstructured" fibres for characterization and experiments using typical tools and equipment. PCF-SMF splices are most common, as PCFs need to be connected to test instruments, optical devices and circuits incorporating or designed for SMFs. Splicing PCF to a SMF requires special fiber handling and machine settings different from splicing SMF to SMF, like reduced arc power and fusion time shortened to 0.2-0.5 s. More difficult splicing of two lengths of PCF is much less common.

Fusion splicing has the advantage of gas-tight sealing a length of PCF, which is of importance in making gas-filled absorption cells or protecting the fibre against penetration of humidity, dust or vapours in hostile operating environment.

This chapter focuses on procedures and experiences with fusion splicing using equipment and tools intended for standard single mode and multimode telecom fibres, which gave acceptable results for most, but not all PCFs the authors encountered. Before this matter is presented in section 4, overview of "holey" fibres and their properties is made in section 2, and arc fusion physics and technology are summarized in section 3.

2. Microstructured optical fibres

Their common feature is substantial modification of optical characteristics by presence of multitude of longitudinal holes or concentric layer(s) of solid micro- or nano-particles around the light guiding core. There are several variants, including:

a. PCFs without doped core, e.g. suspended-core and highly birefringent fibres.
b. PCFs with doped core surrounded by layers of holes, like HAF and nonlinear fibres. The core is capable of guiding light on its own.
c. Fibres with solid nanostructured barrier around doped core.
d. Fibres with single central hole surrounded by dielectric mirror (hollow fibers and Photonic Bandgap Fibres).

Bending-tolerant Hole-Assisted Fibres (HAF) and fibres with solid nanostructured barrier found use in optical access networks (FTTH) and are covered by ITU-T G.657 recommendation, specifying their properties - but not designs. Corning ClearCurve fibre with layer of embedded solid particles around core is fusion spliced as SMF, while HAF is converted to SMF by collapse of holes on fusion splicer (Nakajima et al., 2003); both are not covered here. Properties differ: while fibres (b) and (c) are "splicing friendly", work with suspended-core PCFs and other fibres from group (a) is more difficult. Large-core "dielectric mirror" fibres for delivery of high-power laser radiation (d) are used in short lengths and not spliced. Experience of authors applies to first two groups, but recommendations are applicable to most other microstructured fibres. Key properties for fusion splicing include:

- Photonic structure - size and number of holes,
- Radiation-guiding mechanism and its expected disruption by fusion,
- Cladding diameter and protective coating.

3. Physics and technology of arc fusion splicing

This section will introduce reader to arc fusion splicing of conventional, solid fibres and techniques adopted for splicing fibres in communication networks.

3.1 Process basics and physics

Fusion splicing involves localized melting of two fibre butts pressed together, with fibre coating removed. Surface tension forces cause glass to flow when viscosity is low enough, forming a joint with continuous structure and smooth, round external surface (Figure 1).

While smoothing of edges improves splice strength, self-centring of fibres shown in Figure 2 is undesirable when lateral shift is needed to align non-concentric fibre cores.

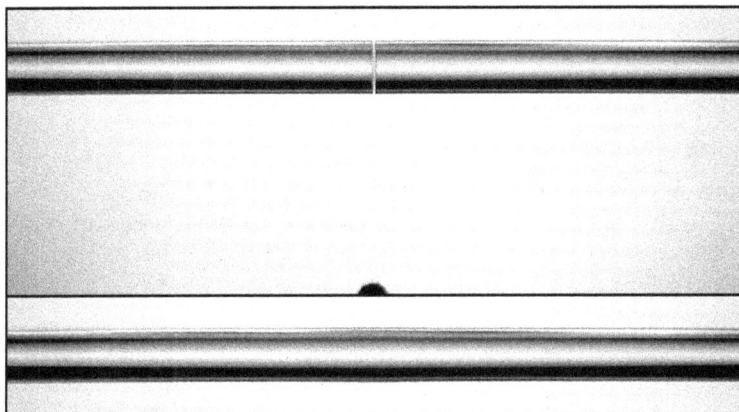

Fig. 1. Two identical 125 μm silica fibres before (top) and after (bottom) arc fusion. Electrode tip is visible as a black spot at the bottom of upper image.

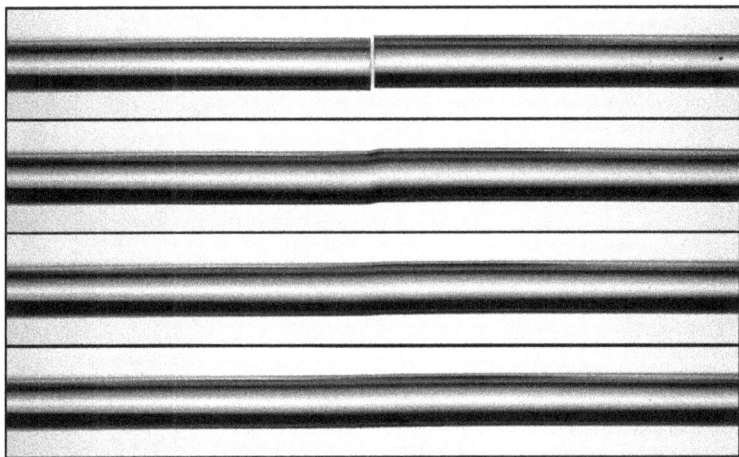

Fig. 2. Two 125 μm single-mode fibres spliced with 10 μm perpendicular offset. Top to bottom: fibres before fusion and after 0.5 s, 1 s and 2 s long fusion with 17 mA current.

Glass evaporating from the hot zone is partly deposited on fibres in the vicinity - degrading surface quality and strength, and on parts of splicing machine - contaminating them. Evaporation is compensated for by fibre overlap: a reduced volume of glass is accommodated in shorter length of fibre without change in diameter. As no air, gel, glue etc. separates fibres after fusion, strength close to one of pristine fibre, no reflection and low insertion loss are possible. The heat for fusion is provided by either:

-   external electric discharge (arc fusion splicing),
-   resistance heater located close to fibres (filament splicing),
-   hydrogen / oxygen burner (flame splicing), or
-   CO_2 laser radiation absorbed by the fibres (laser splicing).

The first method is preferred due to compact equipment, fast operation and flexible control. Filament splicing is used for specialty fibres and when high splice strength is required. Other techniques are rarely used. Descriptions below cover arc fusion splicing only.

3.2 Fused silica properties and fusion splicing

Fused silica is a glassy form of silicon dioxide (SiO_2). In comparison to most multi-component glasses, fused silica exhibits relatively slow decrease of viscosity with temperature (Figure 3). For fusion splicing, this property is desirable, as larger variations in temperature can be tolerated.

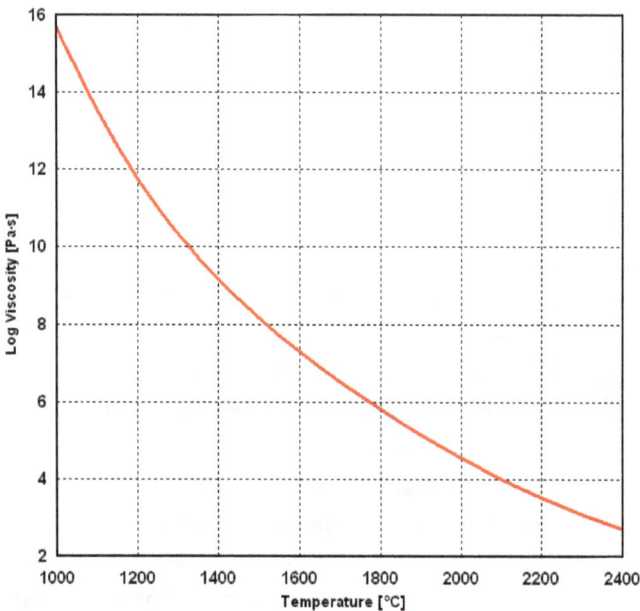

Fig. 3. Viscosity of pure fused silica as function of temperature. Approximate data compiled from sources not in full agreement (Shand, 1968, Yablon, 2005, Schott, 2007).

To ensure fusion in 0.5-2 s, glass viscosity must be reduced to about 10^4 Pa-s. Temperatures during different stages of fusion splicing range approximately from 1500°C (softening of fibres) to 2100°C (fusion), and electric power required to fuse two 125 µm solid silica fibres is about 8-10 W. As voltage drop between electrodes is almost constant, 500-600 V for 1 mm gap, discharge power is essentially proportional to arc current. The range for work with 125 µm fibres is approximately 7-20 mA.

Glass melting during fusion of 125 µm fibres takes place in zone 0.5-1 mm long, including effects of distribution of energy inside arc column, which is approx. 0.2 mm wide for 1 mm electrode gap. Because fused silica has exceptionally low linear thermal expansion coefficient and heat conductivity, as shown in Table 1, fibre strain during post-fusion cooling from the annealing temperature of approx. 1100°C to room temperature is negligible: only 0.0099% for a 1 mm hot zone and 6 mm fibre length between clamps.

Material	CTE (10^{-6}/K^{-1})	Heat conductivity (W/m•K)
Fused silica	0.55	1.38
Tungsten	4.3	173
Steel	13-18	12-45
Copper	16.6	401
Aluminium	22.2	237
PVC (hard)	50-80	0.13-0.29
PET	60-70	0.17-0.24
PMMA	50-90	0.17-0.25

Table 1. Comparison of thermal properties of fused silica and selected other materials.

Heat transfer during fusion is mostly through bulk radiation at wavelengths of 0.6-2 µm, to which the glass is transparent, and conduction along the fibre (Yablon, 2005), and:

- temperature inside fibre is essentially independent of radial position,
- such equilibrium is reached after approx. 10 ms,
- power needed to reach given temperature is proportional to fibre diameter,
- thermal time constant of 125 µm fibres is in order of few tenths of a second.

Doping with germania (GeO$_2$) or titania (TiO$_2$) considerably reduces silica viscosity, and fusing multimode fibres with 50-62.5 µm core taking 20-25% of cross-section requires lower temperature than single-mode fibres, where core is small (4-10 µm) and contains less GeO$_2$.

In a hot fibre, diffusion of dopant(s) is observed, changing refractive profile and core size. The problem is acute during splicing of "depressed cladding" or "pure silica core" single-mode fibres, where core is surrounded by a layer doped with fluorine (F), a light element easily diffusing at high temperatures; careful control of arc power and short fusion time are required.

Out-gassing of boron, phosphorus, germanium or fluorine compounds from melted glass may cause problems as well, in particular during splicing of highly-doped specialty fibres.

Silica does not burn, decompose or oxidise when heated in the air. However, exposure to humidity, alkalis or sharp objects produces surface flaws; damaged fibre breaks easily. All lengths of fibre stripped of coating must be promptly protected against humidity, abrasion, etc. Only splices intended for short-term use in laboratory conditions may be exempted.

Fusion of fibres with different dopants or doping levels can produce significant internal stress due to uneven thermal contraction after fusion, potentially affecting splice strength.

3.3 Fusion procedure

Arc fusion splicing of two single, polymer-coated, multimode or single-mode silica fibres of 125 µm cladding diameter usually includes steps listed below:

1. removal of coating from fibres, usually by mechanical stripping,
2. fibre cleaning with solvent: isopropyl alcohol, acetone, etc. or dry wiping,
3. fibre cleaving by scoring with a blade and applying controlled strain till it breaks ("scribe and load"); cleave angle shall be less than 1°,
4. clamping of fibres in supports with V-grooves,

5. cleaning of fibres by short (0.2-0.5 s), low power electric arc,
6. visual inspection of fibre tips for proper cleave and cleanliness,
7. placing fibre tips between electrodes with 10-20 μm gap (butt coupling),
8. alignment of fibres for lowest transmission loss; this may involve application of perpendicular offset with monitoring of loss or observation of cladding or cores,
9. softening of fibres by low power discharge: 6-9 mA current, 0.5-3 s duration, than pressing together with 6-15 μm overlap,
10. fusion of fibres by high power electric arc: 12-20 mA current, 0.5-2 s duration,
11. annealing / polishing of fibres with low power electric arc (optional),
12. visual inspection of splice: no distortions, slits or bubbles allowed,
13. insertion loss measurement (optional),
14. tensile strength test (optional),
15. splice protection by heat-shrinkable sleeve, re-coating, etc.

Figure 4 shows typical sequence of arc current and fibre movement.

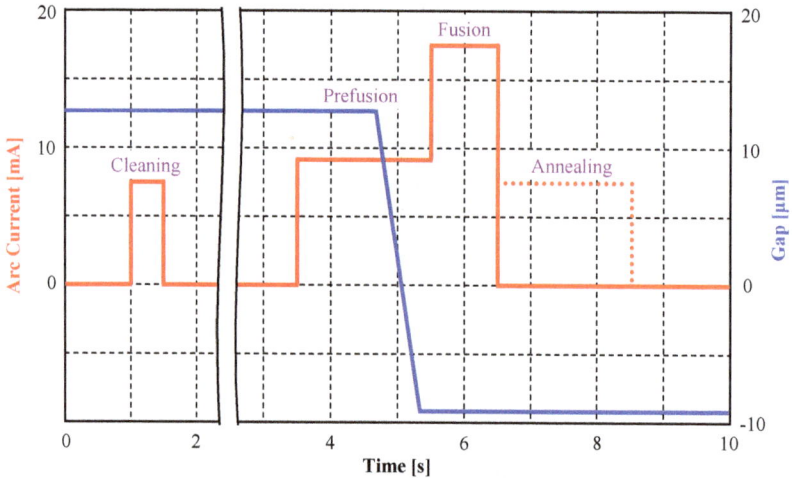

Fig. 4. Example of fibre-fibre gap and arc current variations during fusion splicing of SMF. Negative value of gap means an overlap of fibres pressed into each other.

Arc current and duration of heating in steps 9-11 and fibre overlap in step 9 depend on fibres spliced. 10% deviation from optimum arc current is usually enough to significantly rise splice loss or reduce splice strength, if current and temperature are too low. To reduce self-centring, fusion time is shorter for single mode than multimode fibres: 0.5-1 s vs. 1-2 s.

Some coatings require thermal stripping, softening with chemicals or burning for removal. When high splice strength is required, acrylate, epoxy-acrylate or polyimide coatings can be dissolved during 20-30 s immersion in hot (180-200°C) sulphuric acid (H_2SO_4) - either pure (≥95%) or with addition of approx. 5% nitric acid (HNO_3). Residual acid is removed by rinsing the fibre with water and later acetone (Matthewson et al., 1997).

3.4 Fibre alignment and splice loss

Fibre core guiding radiation is small, with effective diameter usually in the 2-15 µm range in single-mode fibres, with extreme values often encountered in PCFs, and accurate alignment of fibre cores during splicing is essential. Main factors introducing splice loss include:

-   mismatch between core sizes, characterized by mode field diameter (MFD) in single-mode fibres and physical core diameter in multimode fibres,
-   lateral offset between fibre cores,
-   angular misalignment of fibres,
-   mismatch in numerical aperture (NA) of multimode fibres,
-   distortions of splice, e.g. due to poor cleaving or incorrect fibre feed,
-   presence of gas bubbles, inclusions or un-fused gaps in the splice.

For splice between single mode fibres, insertion loss Γ resulting from core offset, MFD difference and angular misalignment is given by the following formula (Yablon, 2005), assuming a Gaussian approximation of mode fields:

$$\Gamma = -10\log\left[\frac{4w_{g1}^2w_{g2}^2}{\left(w_{g1}^2+w_{g2}^2\right)^2}\exp\left(-\frac{4\delta^2+\left(2\pi/\lambda\right)^2 n^2 w_{g1}^2 w_{g2}^2 \sin^2\theta}{2\left(w_{g1}^2+w_{g2}^2\right)}\right)\right] \quad (1)$$

where w_{g1} and w_{g2} are Gaussian radii of spliced fibres, δ is lateral offset between fibre cores, θ is the angular misalignment, n is refractive index of fibre material and λ the operating wavelength. Characteristics of loss due to each factor are shown in Figures 5, 6 and 7. Loss values calculated this way are only approximate because:

-   mode fields in real fibres like PCF often deviate from Gaussian distribution,
-   surface tension of glass produces rounded and distorted interface between fibres aligned with lateral offset (Figure 2),
-   excitation of higher order modes at splice is not included.

The mode field diameter (MFD) of fibre included in technical specifications, usually measured in accordance with Petermann II definition is roughly twice its Gaussian mode radius, but definitions of both parameters are not directly comparable.

For fibres with smaller MFD, splice loss rises faster with lateral offset, but slower with angular misalignment; the resultant loss is approximately proportional to square of given misalignment. While lateral offset between fibre claddings is easy to detect visually through splicer microscope, even large angular misalignment, most often due to debris in V-grooves holding fibres may be overlooked when field of view is small or the optical system produces image distortion. This is critical in splicing large-MFD fibres for high power applications.

Accuracy of fibre alignment depends on core size and accepted loss. For splicing single-mode fibres (w_g = 2.5-5 µm) with loss below 0.2 dB, lateral core offset must be reduced to 0.5-1 µm (Figure 6). If the alignment requires cladding offset, increase of it is needed to compensate for self-centring of fibres during fusion. Splice loss can be estimated from MFD mismatch, core misalignments and deformations measured by automated analysis of splice image. True value is obtained from bi-directional measurements with optical time domain reflectometer (OTDR), as differences between backscattering intensity in fibres can produce relative shift of fibre traces and error in OTDR measurement of splice loss in one direction, often exceeding 1 dB.

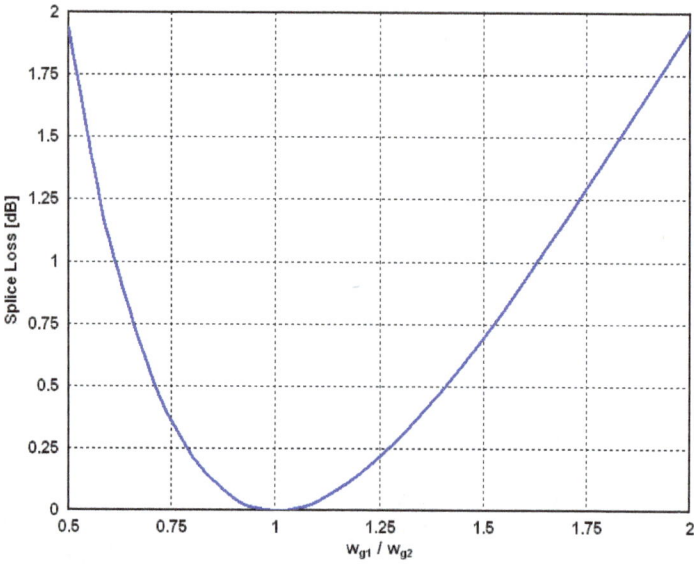

Fig. 5. Calculated loss caused by mismatch in mode field diameters of single mode fibres.

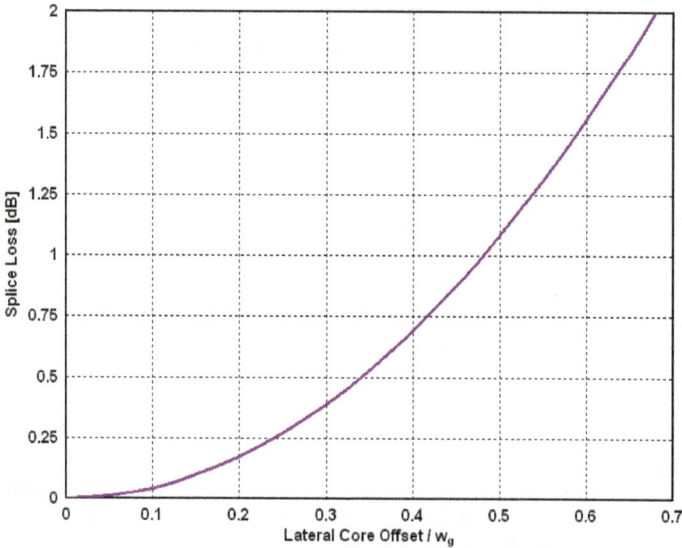

Fig. 6. Calculated loss caused by lateral offset between cores of identical single mode fibres.

As all single-mode and multimode telecom fibres exhibit radial symmetry of core and cladding, rotational alignment of fibres is not required.

Properly made splice of identical fibres has insertion loss of around 0.05 dB and return loss in excess of 70 dB. Butt-coupling with 10-15 μm gap, e.g. before fusion (Fig. 1) introduces insertion loss of approx. 0.40 dB and 15 dB return loss due to Fresnel reflection from a pair of glass/air interfaces. If a transmission loss is monitored during splicing, this difference helps to estimate splice quality. Because loss measured with laser source varies periodically with gap width due to interferometric effects (Yablon, 2005), incoherent source like LED is best. Hot fibre is a strong source of broadband thermal radiation, with emission peak close to 780 nm at 2000°C, preventing loss measurements during discharge.

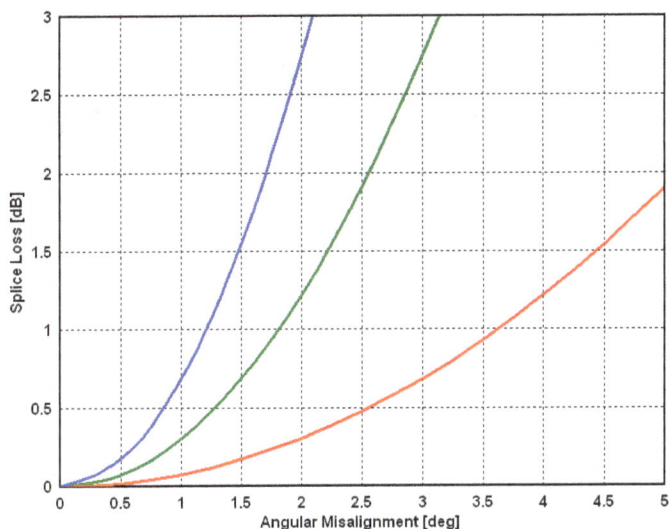

Fig. 7. Calculated loss caused by angular misalignment of identical single mode fibres. $\lambda = 1.55$ μm, n = 1.5. Blue: $w_g = 7.5$ μm, green: $w_g = 5$ μm, red: $w_g = 2.5$ μm.

Loss of splice between single-mode fibres usually shows weak wavelength dependence, because MFD increases with wavelength, typically by 10-20% between 1310 and 1550 nm (Corning, 2008). If the main source of loss is core offset, it falls with wavelength, while angular misalignment produce loss rising with wavelength, in both cases in proportion to square of MFD. For multimode fibres, splice loss is essentially wavelength-independent.

Strong increase of splice loss with wavelength shown in Figure 8 indicates excessive fibre bending due to improper handling, tight coiling or squeeze in the vicinity of splice.

For example, while single splice between fibres having 1:2 MFD ratio has best-case loss of 1.94 dB, two splices with 1 : 1.41 : 2 MFD ratio have combined loss 2 x 0.51 = 1.02 dB. This approach was adopted for splicing a small-core DCF with MFD of 2-4 μm to typical SMF having 8-11 μm MFD (Edvold & Gruner-Nielsen, 1996), and later to splice dissimilar PCFs (Xiao et al., 2007). Extra intermediate fibres can reduce loss further, e.g. to 0.56 dB for 4 splices and 1:2 MFD ratio, but difficulty with finding necessary fibres, losses due to other factors and additional labour usually make such efforts impractical.

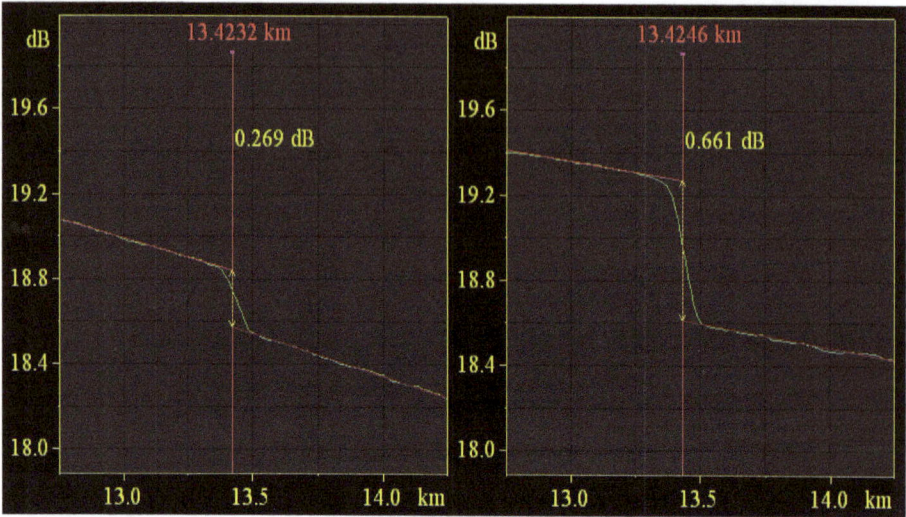

Fig. 8. Sharp bend of SMF observed with OTDR at 1310 nm (left) and 1550 nm (right).

If fibres having substantially different MFD must be spliced, loss can be reduced by introduction of short fibre with intermediate MFD, as shown in Figure 9.

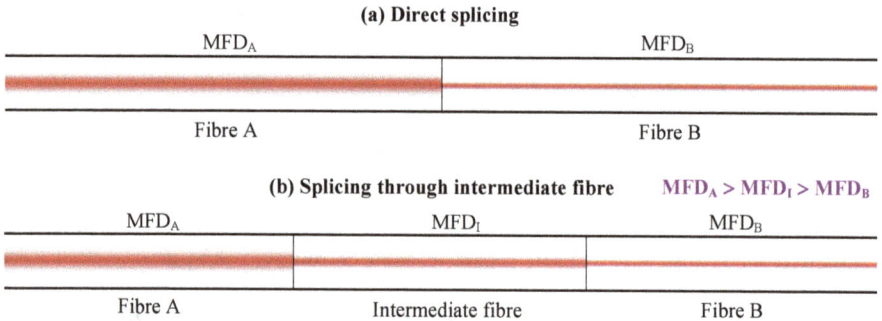

Fig. 9. Principle of splicing through intermediate fibre.

3.5 Splicing of plastic optical fibres

Fusion splicing of plastic optical fibres (POF) made of poly-methyl-methacrylate (PMMA) or TOPAS polymer is very hard. For example, the tip of microstructured fibre made of PMMA forms a cone when heated, instead of rounding typical for fused silica fibre (Figure 10).

Filament splicing and split-mould fusion of POF were reported, but failed to find acceptance. Butt-coupling is adopted for large diameter (up to 1 mm) solid POFs used in car, aircraft and industrial networks, and gluing is used as well. Microstructured plastic optical fibre (mPOF) can be glued, also to glass fibre (Bang, 2010). Polymer fibres are cleaved by cutting with razor blade at controlled speed. Blade and fibre must be heated; the temperature range for PMMA fibres is 60-90°C (Law et al., 2006).

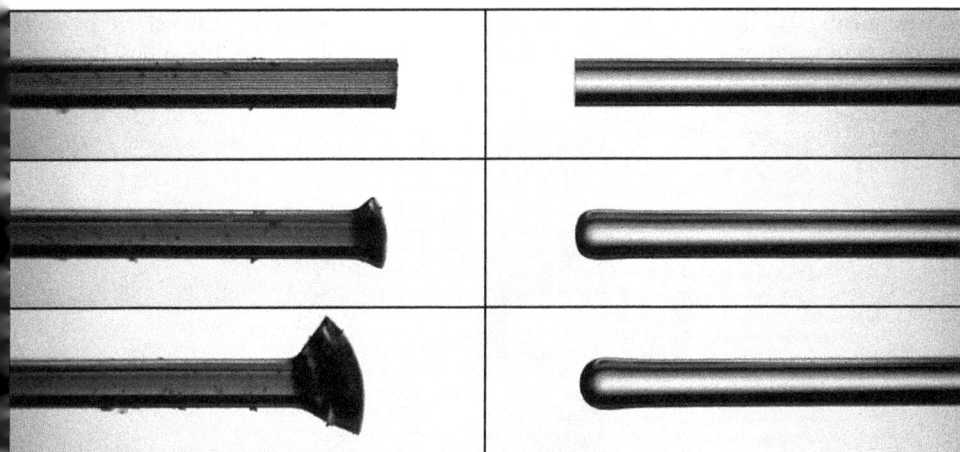

Fig. 10. Deformation of fibre tips by heat of electric arc: DTU MIK125/0.5 made of PMMA (left) and Corning SMF-28 made of fused silica (right). Top: cleaved fibres, middle: after 1st heating, bottom: after 2nd heating.

Handling and splicing of thin mPOF with standard equipment is difficult due to softness of polymers in comparison to fused silica. In experiments at NIT, mPOF of 125 μm diameter exhibited unacceptable sag and curl when clamped 4 mm away from tip, while standard V-groove clamps usually damaged the fibre.

4. Fusion splicing of PCFs

This section presents some issues specific to fusion splicing of silica "holey" fibres, primarily of single mode PCF to SMF. Due to large variety of designs, actual procedure, power, time and geometry settings must be individually tailored.

Finished splice must be protected to have adequate mechanical strength, as holes and flaws on their surfaces make PCFs inherently weaker than solid fibres. In all experiments at NIT, fusion splices were protected with commercial 60 mm heat shrinkable sleeves reinforced with stainless steel rod. Protected splices performed well during temperature cycling between -40°C and +80°C, with loss stability better than ±0.05 dB, and as grips for application of twist and tensile forces in mechanical tests (Figure 11).

Fusion splices are hermetic, keeping external contaminants out, but trapping whatever entered earlier. Exceptions include helium and hydrogen, diffusing through 60 μm thick fused silica cladding in few hours. Short suspended core PCF infiltrated with acetylene (C_2H_2) or hydrogen cyanide (HCN) and fusion spliced to SMF pigtails is used as optical frequency reference, e.g. for calibration of optical spectrum analyzers (Thapa et al., 2006).

Out-gassing of cleaved PCF is fast, but removal of liquid or dust is essentially impossible due to high pneumatic resistance of thin holes and adsorption to the surface of their walls.

Infiltration of holes with gas or liquid allows to make fibre sensors for chemical analysis, detection of pollutants or poison gas, medical diagnostics, etc. through spectral absorption measurements. Filling with liquids, including liquid crystals or suspensions of solid particles in oils allows to build tuneable, nonlinear or electrically controlled optical devices.

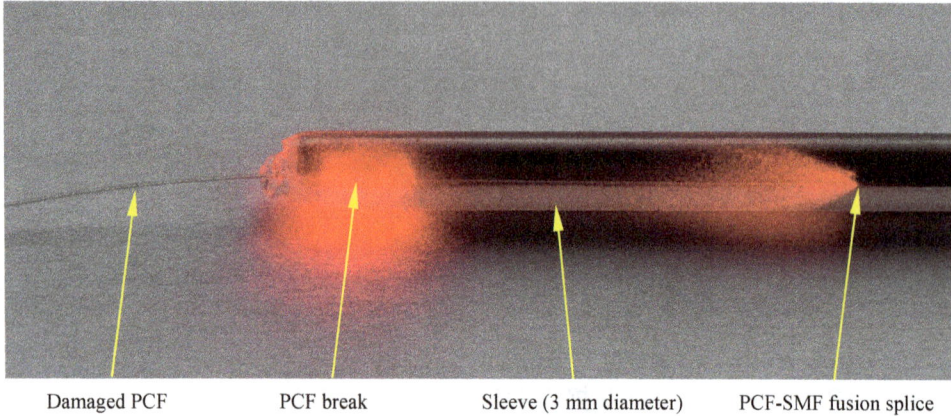

Damaged PCF   PCF break   Sleeve (3 mm diameter)   PCF-SMF fusion splice

Fig. 11. 80 μm PCF (IPHT 252b5) in acrylate coating broken by 180 rev/m twist applied using heat shrinkable sleeve as grip. 4 mm section of fibre inside sleeve was long enough to transfer a destructive force. Splice was illuminated with 650 nm laser through SMF (right).

PCFs are sometimes fused for purposes other than splicing, like:

- sealing of fibre before storage or shipment,
- consolidation of fibre tip prior to polishing.

Fusion power and duration must ensure robust collapse of all holes.

4.1 Issues and solutions specific to fusion splicing of PCFs

Besides issues presented in section 3, work with PCF brings several new challenges:

1. Solvents used for fibre cleaning infiltrate PCF holes.
2. Holes distort propagation of crack and hamper fibre cleaving.
3. Surface tension of molten silica causes collapse of holes.
4. PCF has lower fusion temperature and thermal conductivity than solid fibre.
5. Mismatch in fiber cladding diameters results in sharp edges, splice strength suffers.
6. Mode fields of PCFs do not exhibit full radial symmetry.
7. Photonic structure supports undesirable propagation of light outside of fibre core.
8. High attenuation and strong backscattering in PCF affect splice loss measurements.

It is often needed to splice PCFs having unusual cladding diameter, core design and MFD, as there are no standards for this category of fibres. Most commercially available splicing and test equipment, tools and accessories like fibre adaptors or protective sleeves are designed strictly for telecom fibres with 125 μm cladding and 250-900 μm coating.

A compromise between achieving different goals is often required, particularly between splice loss and its mechanical strength. Also, unless optimization of manufacturing process, device performance, etc. justifies labour and equipment costs involved, time and funds available for trials are limited and less-than-perfect solutions need to be accepted.

4.1.1 Fibre infiltration

Unless PCF end is sealed by fusion, low-viscosity liquid applied to it penetrates holes. Filling with colourless acetone, isopropyl alcohol, ethyl alcohol or water is invisible through coating. Removal is practically impossible, with the following consequences:

- Fall of index contrast in the infiltrated zone when solvent (isopropyl alcohol: n = 1.38, acetone, ethyl alcohol: n = 1.36, water: n = 1.33, 589 nm wavelength) replaces air (n = 1.00) in contact with fused silica (n = 1.46). Light guiding is disturbed, and loss changes with movement of solvent, interfering with alignment of fibres.
- During fusion, solvent in hot zone partly decomposes leaving thin layer of dark carbon residue on hole walls and partly evaporates, with pressure pushing the remaining liquid deeper into PCF. Resulting loss observed at NIT was up to 50 dB.

For the same reason, PCFs cannot be connectorized in conventional way because water and small (0.5-3 μm) particles of polishing materials enter holes. To avoid infiltration, fibre ends shall be protected against liquids, dust or vapours during handling and storage. In particular, water vapour degrades fibre strength by producing flaws on walls of holes. PCF is best stripped mechanically and dry wiped to remove remains of coating. When use of solvent, acid, etc. is required, fibre end must be first sealed by fusion. PCF contamination in storage or shipping can be prevented by fusing both ends. To fit a connector, PCF can be stripped, cleaved and fused to collapse holes over a 100-200 μm length and fixed in the connector ferrule for polishing. This procedure works best for fibres with doped core, whose light guiding properties are retained without photonic structure.

4.1.2 Fibre cleaving

Cleaving of glass fibres uses perpendicular propagation of indentation-initiated break at the speed of sound, approx. 5950 m/s for fused silica. Structures made of differing materials, like arrays of holes in PCF or inserts of B_2O_3-SiO_2 glass in PANDA fibre distort this propagation; these fibres are reportedly more difficult to cleave than conventional ones.

PCFs tested at NIT, with 80-200 μm cladding diameter were cleaved using a typical, simple cleaver for telecom fibres with tungsten carbide blade. Proportion of bad cleaves was around 20%, a little higher than experienced with most SMFs. It rose to some 50% for the 80 μm IPHT 252b5, presumably because tensile load was excessive for this fibre with equivalent diameter of solid glass of just 72 μm. This is consistent with literature data that best tensile load is proportional to cladding diameter raised to power of 2/3 (Yablon, 2005).

Cleaved PCFs shall be carefully inspected for perpendicular cut before further work. For non-standard fibre sizes, use of cleaver with adjustable tensile force is recommended.

4.1.3 Collapse of holes and thermal issues

In absence of differential pressure, surface tension in molten glass causes the holes to reduce their radius at constant linear speed set by the following formula (Yablon, 2005):

$$v_{collapse} = \frac{\gamma}{\eta} \tag{2}$$

where γ is surface tension, almost constant, and η is glass viscosity falling with temperature (Figure 3). If this continues long enough, holes collapse and solid fibre of reduced diameter is created. Collapse of holes can be prevented by internal gas pressure ("inflation"); equilibrium pressure $P_{critical}$ for capillary is a function of its inner (r_i) and outer (r_o) radius:

$$P_{critical} = \gamma \left(\frac{1}{r_i} + \frac{1}{r_o} \right)$$

(3)

In PCF with holes of differing sizes, the largest holes disappear last and over the shortest length. Due to longitudinal temperature gradient, only some length of PCF is subjected to collapse of holes, with gradual "thinning" in the intermediate zone - see Figures 12 and 13.

For internal fibre temperature independent of depth, all holes of given size shall collapse simultaneously, but in experiments (Xiao et al., 2007, Bourliaguet et al., 2003) holes located deeper are less affected. Example from our work is shown in Figure 13. Transfer of heat from fibre surface to its interior, predominantly of radiative type, is apparently delayed.

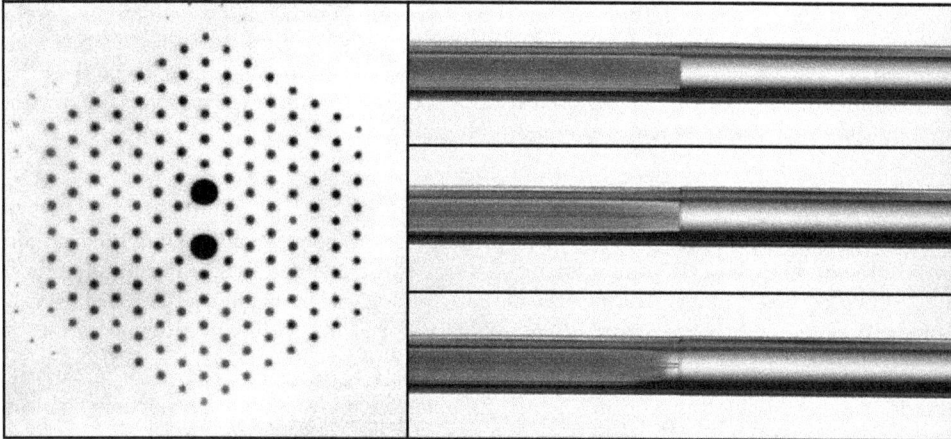

Fig. 12. Structure of PCF (UMCS 070119p2) with 3.5 µm and 1.3 µm holes and views of fusion splice to SMF. Fusion time: 0.3 s, fusion current from 13 mA (top) to 15 mA (bottom).

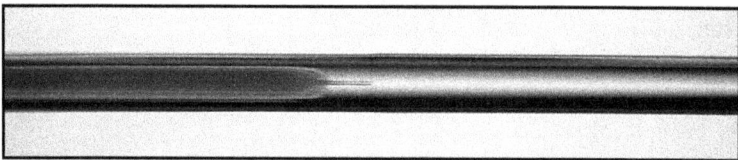

Fig. 13. Depth-dependent and diameter-dependent collapse of holes (UMCS 070119p2).

In the solidified length of fibre, light beam expands freely and proportion of power coupled to core of other fibre drops with increase of collapsed zone. Collapse of holes shall be avoided as much as possible, and if it cannot be avoided, fibre length affected must be reduced to absolute minimum. PCFs with doped core are partial exception.

Collapse is minimized by shortening fusion time to 0.2-0.5 s from 1-2 s for solid 125 μm fibres and reducing power, fusion time being more important. However, too short fusion time and too low temperature prevent full fusion of fibre-fibre boundary and proper rounding of edges if fiber diameters don't match, as the glass is too viscous and/or doesn't have enough time to flow. There is a trade-off between achieving low splice loss with little heat or good strength with more, and splice with excellent optical transmission may not be strong enough even for removal from splicing machine, as shown in Figure 14.

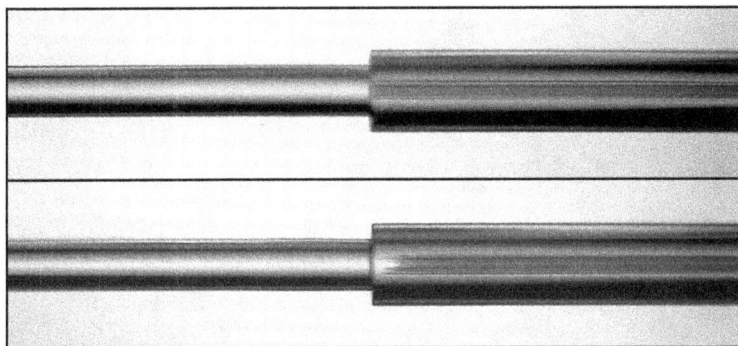

Fig. 14. Splices between 204 μm PCF (IPHT 212b1) and SMF, fused with 150 μm axial offset. Splice with intact photonic structure and lowest loss, which broke during handling (top), and splice that survived (bottom). Fusion current: 18-19 mA, fusion time: 0.5 s.

In splicing dissimilar fibres, axial offset of fibre contact point from the axis of electrodes is useful. The more heat-sensitive fibre - PCF in splice to SMF, or smaller of two PCFs, is kept away from centre of discharge column and its temperature is lower. In experiments at NIT, maximum axial offset was 1.2-1.5x fibre cladding diameter, otherwise unacceptable fibre deformation occurred in the hottest zone. Reduced fibre overlap can help.

4.1.4 Mismatch in cladding diameter

The power required to achieve given fibre temperature is approximately proportional to cladding diameter, and when fibres of different diameters are fused, the thinner fibre must receive a smaller share of arc power to obtain symmetrical temperature distribution. This is ensured by axial offset of arc centre in direction of thicker fibre. When splicing PCF to solid fibre, PCF shall be colder to prevent collapse of holes, adding second component of axial offset. In effect, even when PCF is moderately thicker than solid fibre, there is usually no offset towards PCF.

During fusion of fibres of different diameters, poor smoothing of corners at fibre-fibre transition and fragility of splice are common, as seen in Figure 14. Therefore, fusion power and duration are often selected to obtain the minimum splice strength allowing handling without break, even if collapse of holes and increased splice loss are to be accepted.

4.1.5 Non-circular mode

Typical photonic structure, e.g. of "honeycomb" type, lacks full radial symmetry and mode field distribution reflects shape of it. In PCF with circular doped core, mode field can be still more or less distorted, depending on interactions with surrounding holes. Many PCFs for signal processing, sensing, optical switching etc. have cores and/or photonic structure made deliberately non-symmetrical; mode field distribution is non-circular, but usually with lateral symmetry, and propagation is polarization-dependent.

Additionally, current PCF manufacturing technology cannot ensure perfect fibre geometry. Holes are often distorted during fibre drawing (Figure 15), affecting mode field shape.

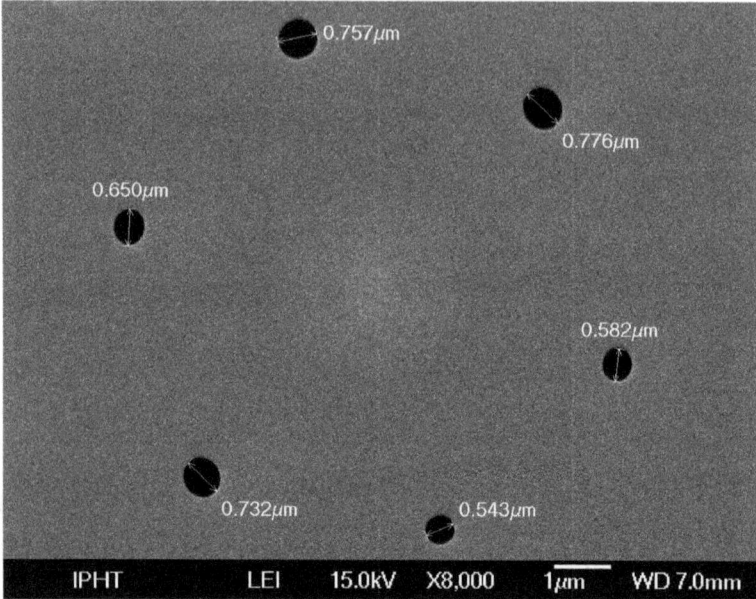

Fig. 15. Central part of doped-core PCF (IPHT 282b4) with deformed holes.

If the PCF is spliced to radially symmetrical fibre, like single-mode or multimode telecom fibre, relative rotation has no effect on splice parameters, and simpler fusion splicing machine without fibre rotation is sufficient. Rotational alignment is necessary for splicing PCF to another PCF or specialty fibre of non-circular design like Bow-Tie or PANDA. Alignment is based either on observation of fibre structure through microscope of splicing machine or monitoring of transmission through butt-coupled fibres. In the latter case, a source of linearly polarized light properly coupled to one fibre is often required.

4.1.6 MFD mismatch

This problem is not limited to PCFs - see section 3.4. Besides use of intermediate fibre(s), one can locally change MFD of one fibre to make it more compatible with another or modify propagation in the zone between fibres. Several methods were reported, including:

-   Heat-assisted diffusion of dopants to expand small core doped with GeO_2, by heating on the splicing machine. Fluorine-doped fibre can be modified as well (Yablon, 2005, Edvold & Gruner-Nielsen, 1996). Transition zone shall be at least 300 µm long.

- Slow pulling of hot (1200-1500°C) fiber to reduce its diameter and MFD before cleaving in the middle of thinned section (fiber tapering). Fusion splicing machine is used, with arc power somewhat lower than required for pre-fusion.
- Melting of fibre tip to make a ball lens. Arc power is similar to used for fusion (Wang et al., 2008, Borzycki et al., 2010a). Fibres with lenses are than fused – see 4.2.2.
- Collapsing a controlled length of small-core PCF. This results in Gaussian expansion of light beam towards interface to fibre with larger core.
- Insertion of GRIN fibre lens between two spliced fibres (Yablon & Bise, 2004). GRIN type and length must be carefully chosen to ensure proper focusing of light.

While effective, these techniques are sensitive to deviations in process parameters. Specific advice for different cases can be found in literature, but several techniques require splicing machine with precise control of fibre movements, in particular for fibre tapering.

4.1.7 Propagation of light in photonic structure and splice loss measurements

In single-mode propagation regime, insertion loss of splice is independent of transmission direction. Non-reciprocity indicates multimode propagation in one or both fibres.

Excitation of higher order modes, e.g. at splice with lateral offset is known in telecom systems, but mostly limited to short fibres, as higher order modes are strongly attenuated. In PCFs, photonic structure can support persistent propagation of own modes, especially as short lengths of such fibres, ≤ 1 m are common. This produces interference in measurements, sensing or operation of optical devices, as detectors in active instruments respond to total optical power of all modes. Examples of related measurement problems are:

- non-reciprocity of splice loss between SMF and PCF,
- noise-like interference in measurements of differential group delay (DGD) and polarization dependent loss (PDL) of PCF samples, see Fig. 17 (Borzycki et al., 2011a).

Non-reciprocity of splice loss measured with optical source and optical power meter is due to different propagation of higher-order modes. In SMF, this part of radiation escapes fibre core and is lost, leaving only the fundamental mode. In PCF, part of it reaches the end of fibre and detector of power meter (Figure 16); loss indicated by instrument connected to PCF is lower than "true" value for fundamental mode. OTDR test is less affected, as optical pulse travels through splice in both directions and the instrument shows average loss value.

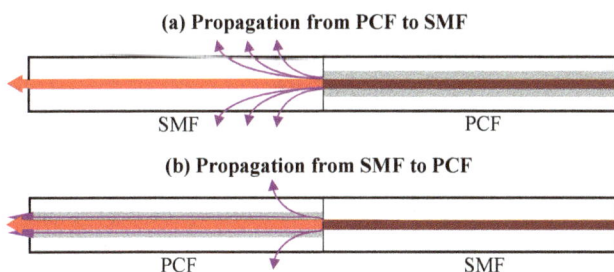

Fig. 16. Mechanism creating non-reciprocal loss in PCF-SMF splices.

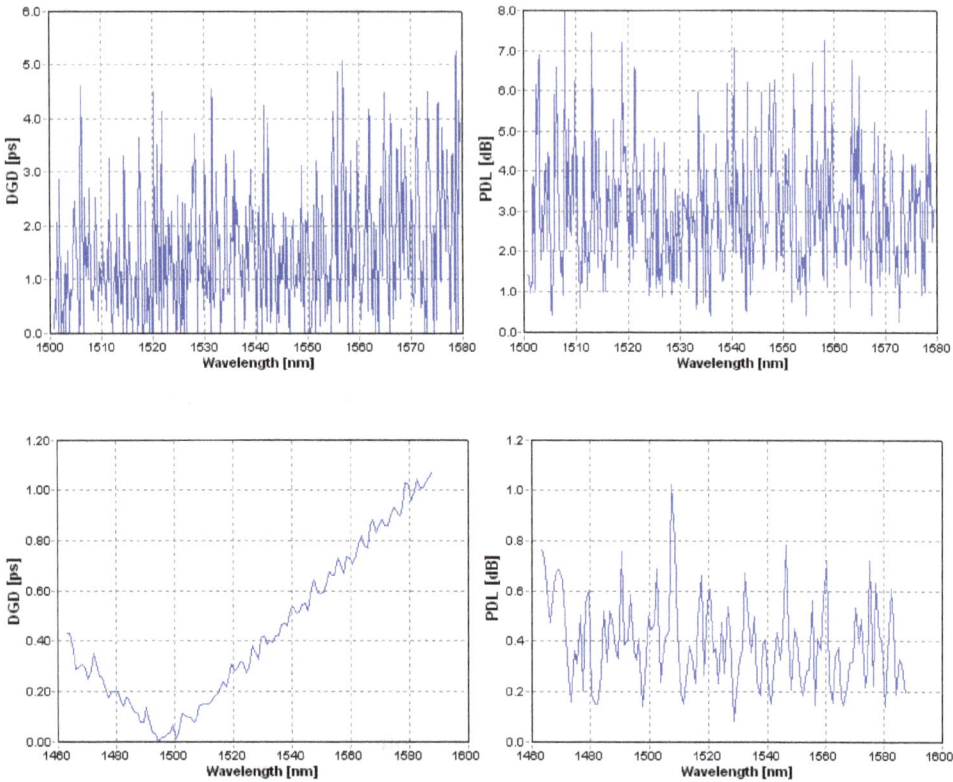

Fig. 17. DGD and PDL of 20.2 m and 18.7 m long samples of PCF prone to multimode propagation (IPHT 252b3) with poor (top) and optimized (bottom) fusion splices to SMFs.

Loss measurements are needed to align fibres, monitor fusion, and evaluate finished splice. It is often necessary to measure total loss of circuit or device incorporating PCF rather than splice(s) alone. There are two basic test methods:

- Transmitted power monitoring with optical source and optical power meter,
- OTDR, with lengths of fibres adjacent to splice required.

The first technique allows fast measurements (0.1-1 s) with high resolution (0.001 dB). As typical laser source emits linearly polarized light, apparent PCF-SMF splice loss during alignment varies with fiber rotation due to PCF non-circular structure. In experiments presented in section 4, loss was monitored during fibre alignment and after each fusion step. Test setup shown in Figure 18 was used, including HP8153A optical multimeter with HP81553SM laser source (1558 nm) and HP81532A power meter modules. For splice No. 1 it was necessary to subtract PCF loss, measured separately with OTDR and connector loss; loss of splice No. 2 was measured directly. Loss calculations must take into account high attenuation of PCF, usually 20-200 dB/km. Data in section 4.2 refer to splice No. 2.

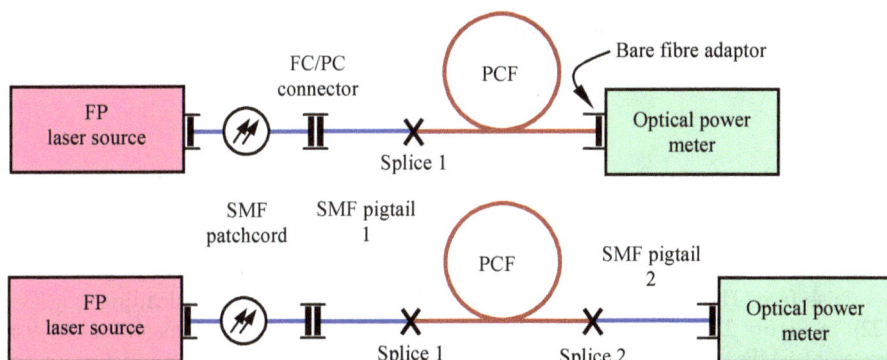

Fig. 18. Setups for loss measurements. Measurement of splice No. 1 after separate PCF attenuation measurement (top), direct measurement of splice No. 2 (bottom).

When bare fiber adapters are used, debris on PCF endface can produce errors. PCF tip could be cleaned by gentle contact with suitable sticky tape, like Scotch Magic (Figure 19).

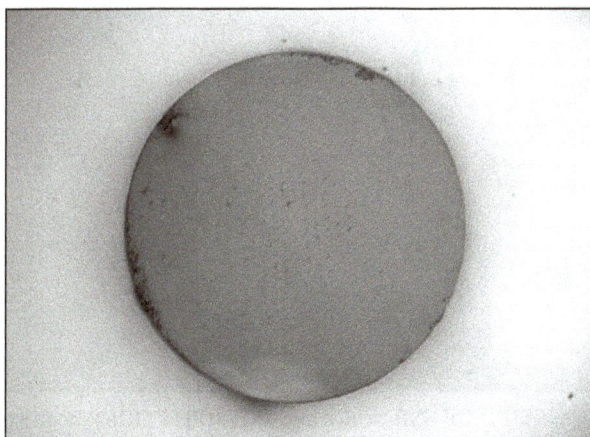

Fig. 19. Cleaved IPHT 282b4 in bare fibre adapter after cleaning of dust with sticky tape.

OTDR measurement of SMF-PCF splice(s) and PCF itself requires certain length of fibre before and after the splice, at least 50 m for instrument with 10 ns pulse width (Figure 20).

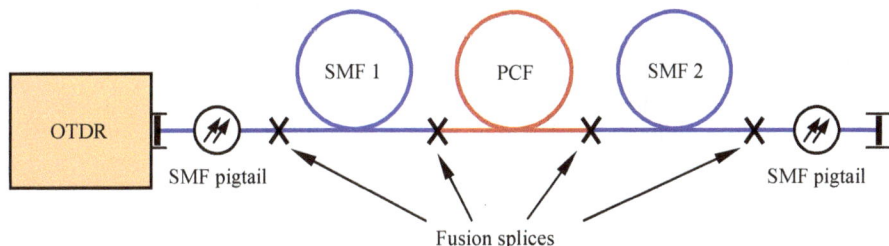

Fig. 20. Arrangement for characterization of PCF and PCF-SMF splices with OTDR.

Large differences of backscattering intensity in PCF and SMF are common, and true loss Γ of SMF-PCF splice can be established only with bi-directional measurement and averaging:

$$\Gamma = \frac{\Gamma_A + \Gamma_B}{2} \tag{4}$$

where Γ_A and Γ_B are apparent splice losses measured in A and B directions.

Many PCFs produce strong backscattering due to entrapment of scattered light by photonic structure and intense scattering in doped core, if present (Borzycki et al., 2010b, 2011a). This brings noise-free OTDR trace, but one-way OTDR measurement are misleading (Figures 21 and 22). In Figure 22, trace of highly GeO_2-doped PCF (IPHT 282b4) was shifted by 9.5 dB vs. traces of SMFs, producing "gain" in splice before PCF and exaggerated loss of splice after PCF (Borzycki et al, 2011b). Testing such samples requires high-performance OTDR.

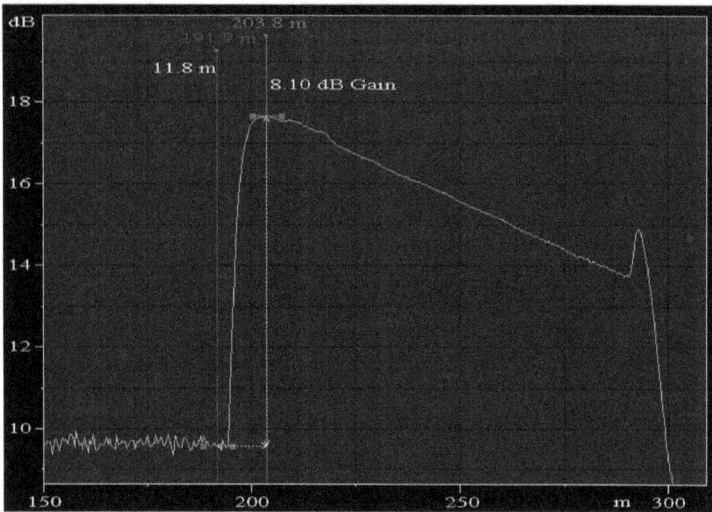

Fig. 21. OTDR trace of SMF, PCF-SMF splice and PCF (IPHT 282b3) acquired with setup shown in Figure 20. Wavelength: 1550 nm, pulse width: 10 ns, PCF length: 91 m. Instead of 8.1 dB "gain", the splice had actual loss of 2.2 dB.

In PCF characterization, improvements to quality of splices pay off with improved accuracy and fewer measurement artifacts, especially in measurements of polarization properties.

4.2 Examples

Descriptions below apply to splicing of SMF to PCFs designed and made at IPHT Jena, Germany as highly nonlinear single-mode fibres for signal processing, like wavelength conversion. For this, fibre cores were strongly doped with GeO_2, up to 36% mol. The reader is referred to separate papers on manufacturing of these fibres (Schuster et al, 2007) and their characterization (Borzycki et al., 2010b). Data of all PCFs referred to in this chapter are presented in Table 2. UMCS 070119p2 mentioned in section 4.1.3 was a "PANDA-like" birefringent fibre developed at UMCS Lublin, Poland for use in polarimetric sensors.

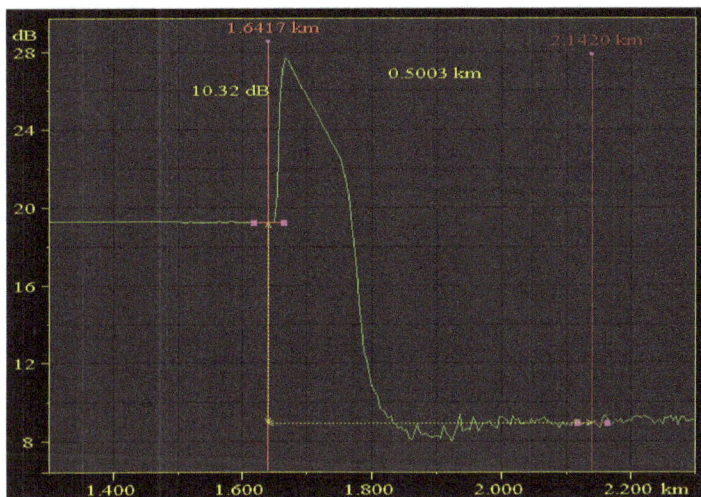

Fig. 22. OTDR trace of PCF (IPHT 282b4) spliced to SMFs as in Figure 20. Wavelength: 1550 nm, pulse width: 100 ns, PCF length: 104 m.

Parameter	Unit	IPHT 252b5	IPHT 252b3	IPHT 282b3	IPHT 282b4	UMCS 070119p2
Cladding diameter	μm	82.7	127.5	125.9	124.4	126
Hole diameter (d)	μm	3.6	5.8	0.7	0.7	3.5/1.3
Hole spacing (Λ)	μm	4.2	6.5	4.4	4.2	3.5
Diameter of holey package	μm	42.8	61.5	44.6	43.0	55
Diameter of doped core	μm	0.5/2.0/4.1*	1.4/3.3/6.6*	0.8/2.8/7.1*	1.2/3.9/7.3*	N/A **

*) Central high GeO_2 doped core / GeO_2 doped socket / total core diameter.
**) No doped core.

Table 2. Data of photonic crystal fibres

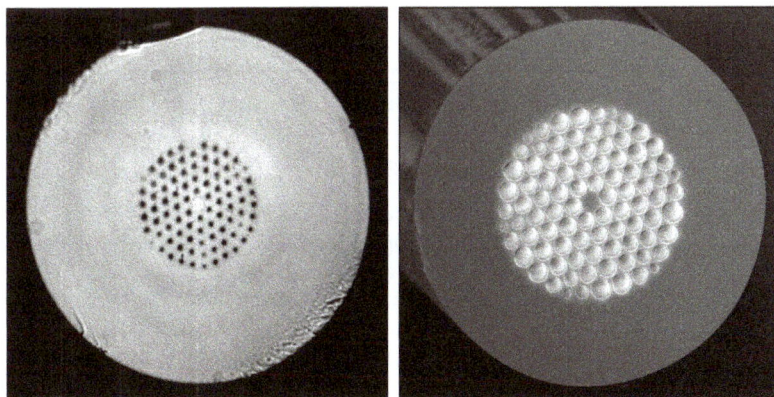

Fig. 23. Cross-sections of fibres: IPHT 282b4 (left) and IPHT 252b5 (right).

4.2.1 Splicing IPHT 282b4 to SMF

Being similar to SMF, this fibre was spliced in the same way, using the following fusion program (pre-fusion, fusion, annealing - Figure 4): 9 mA – 3 s / 17 mA – 0.5 s / 9 mA – 3 s.

PCF length, attenuation and loss at 1558 nm were 12.4 m, 61.5 dB/km and 0.76 dB respectively. The SMF fibre was Corning SMF-28 (Corning, 2008). Splice loss (Table 3) was acceptable despite destruction of photonic structure over 300 µm (Figures 24 and 25), due to guiding of light by doped core. This case was easy, as no special techniques were required besides reduction of fusion power and duration in comparison to SMF splicing. Lower loss was achievable with shorter fusion time, but at expense of reduced splice strength.

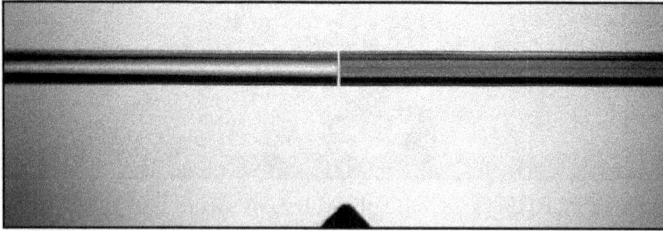

Fig. 24. SMF (left) and PCF (right) cleaved and positioned before fusion.

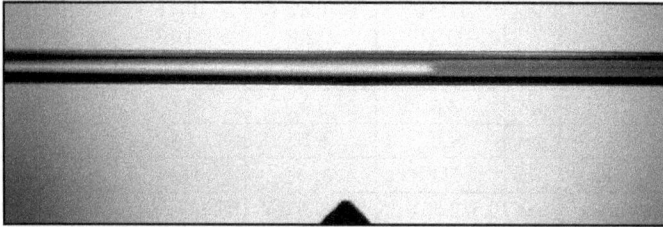

Fig. 25. Fibres fused.

Conditions	Sample loss (dB)	Splice loss (dB)
Loss with finished splice No. 1	3.19	(reference)
Fibres cleaved and aligned	4.79	2.37
Fibres spliced	3.77	1.25

Table 3. Loss of IPHT 282b4 sample measured during making of splice No. 2 (1558 nm).

4.2.2 Splicing IPHT 252b5 to SMF (Corning SMF-28)

This PCF was difficult to splice due to small diameter. Fusion procedure included melting of fiber tips into ball lenses before fusion (Wang et a., 2008). Results depended on accurate fiber movement ("feed") before fusion: too small increased necking (Figures 29-30) and loss. Fiber was 16.08 m long, with attenuation and loss at 1558 nm of 58.3 dB/km and 0.94 dB, respectively. The SMF was Corning SMF-28. Gap during alignment was 10 µm, fibre overlap 10 µm and axial offset 200 µm towards SMF. Splicing machine settings were as follows:

- Splicing (pre-fusion, fusion, annealing): 9 mA – 3 s / 18 mA – 0.5 s / 8.4 mA – 3 s.
- Melting of fibre tips: 18 mA – 0.5 s.

After measuring loss with butt coupling (Figure 26), fibres were melted to form ball lenses (Figures 26-28). Melting of SMF tip was repeated to obtain the required shape. After fusion (Figure 29), the splice was repeatedly heated at the same settings to reduce loss, but without further movement (Figures 29-30). Light transmission was monitored and work terminated after splice loss stopped to significantly decrease any further – see data in Table 4.

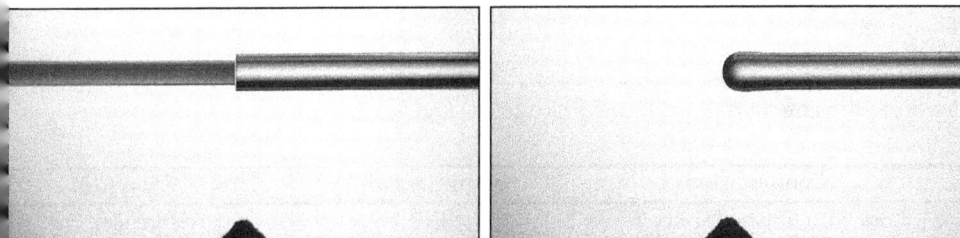

Fig. 26. Left: PCF (left) and SMF (right) aligned. Electrode tip is visible as dark triangle at the bottom of picture. Right: SMF tip melted – Phase 1.

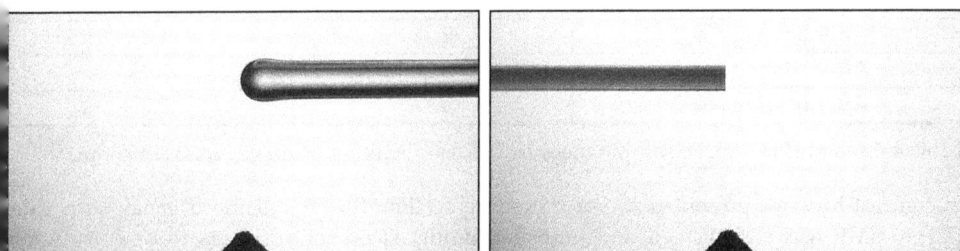

Fig. 27. Left: SMF tip melted – Phase 2. Right: PCF positioned for melting.

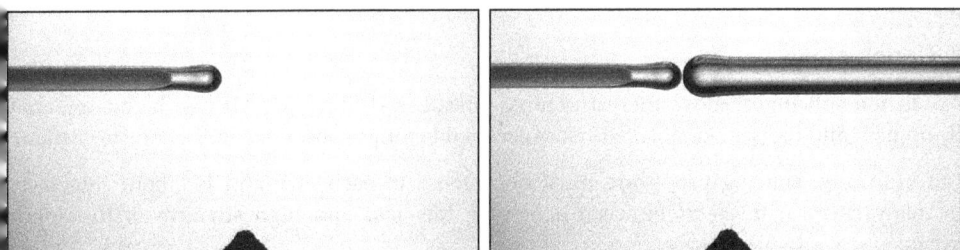

Fig. 28. Left: PCF tip melted. Right: fibres aligned for fusion with axial offset.

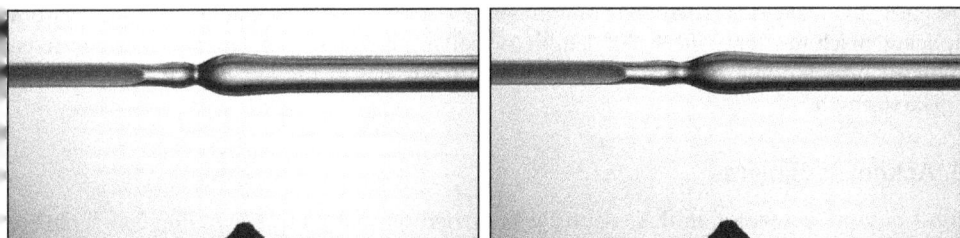

Fig. 29. Left: fibres fused. Right: splice after additional heating No. 1.

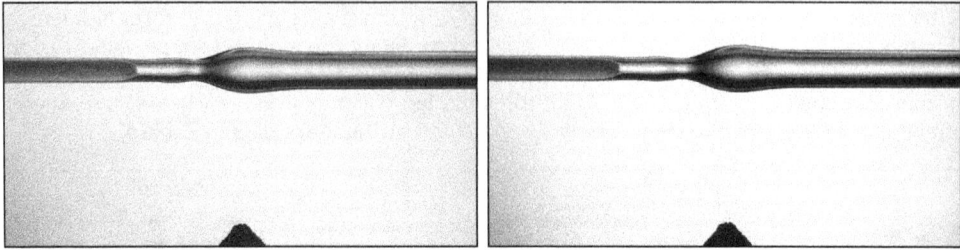

Fig. 30. Splice after additional heating No. 2 (left) and No. 3 (right).

Conditions	Sample loss (dB)	Splice loss (dB)
Loss with finished splice 1	3.19	(reference)
Fibers cleaved and aligned	8.98	5.79
Lens-tipped fibres aligned	5.37	2.18
Fibres spliced	4.61	1.42
After heating No. 1	4.38	1.19
After heating No. 2	4.28	1.09
After heating No. 3	4.17	0.98

Table 4. Loss of IPHT 252b5 sample measured during making of splice No. 2 (1558 nm).

In contrast to work presented in the preceding section, fusion splicing of small-core, thin PCF to SMF was complicated and time-consuming. However, attempts to fuse the same fibres without pre-forming resulted in very high splice loss (15.5 dB for 2 PCF-SMF splices) and frequent entrapment of small gas bubble in the centre of splice.

5. Conclusions

Arc fusion splicing of most microstructured silica-based fibres to SMFs with conventional equipment and tools is possible, with loss acceptable for purposes like PCF characterization.

Unfortunately, splicing procedure must be tailored to each PCF and is labour intensive. In many cases, a trade-off between achieving low loss and high strength of the splice exists.

PCF to PCF splicing is more demanding in term of equipment and work procedures, because of need for rotational alignment of fibres, not available in most fusion splicing machines, and increased length of collapsed holes. The latter problem can be reduced by adopting fusion time shorter than 0.3 s, but at the expense of compromised splice strength.

6. Acknowledgments

Most of work presented in this chapter was carried out within COST Action 299 "FIDES". Research at NIT was financially supported by Polish Ministry of Science and Higher Education as special research project COST/39/2007.

7. References

Bang, O. (2010). PCFs, mPOFs and THz fibers. *Proc. 2nd Workshop on Specialty Optical Fibers and their Applications (WSOF 2010)*, ISBN: 978-0-8194-8360-7, Oaxaca City, Mexico, October 13-15, 2010. Available from http://www.cio.mx/WSOF2010/archivos/Ole%20Bang.pdf

Borzycki, K., Kobelke, J., Schuster K. & Wójcik J. (2010). Arc fusion splicing of photonic crystal fibers to standard single mode fibers. *Proc. SPIE 7714-38*, ISBN 9780819481870, SPIE Photonics Europe 2010, Brussels, Belgium, April 12-16, 2010

Borzycki, K., Kobelke, J., Schuster K. & Wójcik J. (2010). Optical, thermal and mechanical characterization of photonic crystal fibers: results and comparisons. *Proc. SPIE 7714-31*, ISBN 978-0-81948-1870, SPIE Photonics Europe 2010, Brussels, Belgium, April 12-16, 2010

Borzycki, K., Kobelke, J., Mergo, P. & Schuster K. (2011). Challenges in characterization of photonic crystal fibers. *Proc. SPIE 8073B-107*, ISBN 978-0-81948-663-9, SPIE Optics + Optoelectronics 2011, Prague, Czech Republic, April 18-21, 2011.

Borzycki, K., Kobelke, J., Mergo, P. & Schuster K. (2011). Characterization of photonic crystal fibres with OTDR. *Proc. ICTON 2011*, ISBN 978-1-4577-0881-7, paper We.B4.5, Stockholm, Sweden, June 27-29, 2011

Bourliaguet, B., Paré, C., Émond, F., Croteau, A., Proulx, A. & Vallée, R. (2003). Microstructured fiber splicing. *Optics Express*, eISSN: 1094-4087, Vol. 11, No. 25, pp. 3412-3417

Corning Inc. (2008). Corning SMF-28e Optical Fiber Product Information. PI1344 (09/2008)

Edvold, B. & Gruner-Nielsen, L. (1996). New technique for reducing the splice loss to dispersion compensating fiber. *Proceedings of ECOC-1996*, Vol. 2, pp. 245-248, ISBN 82-423-0418-1, Oslo, Norway, September 15-19, 1996

IEC 60793-2-50 Ed. 3.0: Optical fibres - Part 2-50: Product specifications – Sectional specification for class B single-mode fibres, ISBN 2-8318-9824-2 (05/2008)

ITU-T Recommendation G.652: Characteristics of a single-mode optical fibre and cable (11/2009). Available from http://www.itu.int/rec/T-REC-G.652-200911-I

ITU-T Recommendation G.657: Characteristics of a bending-loss insensitive single-mode optical fibre and cable for the access network (11/2009). Available from http://www.itu.int/rec/T-REC-G.657-200911-I

Law, S.H., Harvey, J.D., Kruhlak, R.J, Song, M., Wu, E., Barton, G.W., van Eijkelenborg, M.A. & Large, M.C.J. (2006). Cleaving of microstructured polymer optical fibres. *Optics Communications*, ISSN 00304018, Vol. 258, Issue 2, pp. 193–202

Matthewson, M.J., Kurkjian, C.R. & Hamblin, J.R. (1997). Acid stripping of fused silica optical fibers without strength degradation. *Journal of Lightwave Technology*, ISSN 0733-8724, Vol. 15, Issue 3, pp. 490-497

Nakajima, K., Hogari, K., Zhou, J., Tajima, K. & Sankawa I. (2003). "Hole-Assisted Fiber Design for Small Bending and Splice Losses", *IEEE Photonics Technol. Lett.*, ISSN 1041-1135, Vol. 15, No. 12, pp. 1737-1739

OFS (2008). Fitel S183PMII - A New Standard in the Field for High-End Fusion Splicing Applications (FITEL-S183-PMII-1108)

Schott North America, Inc. (2007). Schott Technical Glasses: Physical and technical properties (11/2007). Available from: http://www.us.schott.com/english/download/technical_glass_guide_us.pdf

Schuster, K., Kobelke, J., Grimm, S., Schwuchow, A., Kirchhof, J., Bartelt, H., Gebhardt, A., Leproux, P., Couderc, V. & Urbanczyk, W. (2007). Microstructured fibers with highly nonlinear materials. *Journ. Opt. Quant. Electron.*, ISSN 0306-8919, Vol. 39, pp. 1057-1069

Shand E.B. (1968). *Engineering Glass, Modern Materials Vol. 6*, Academic Press, New York

Thapa, R., Knabe, K., Corwin, K. L. & Washburn, B. R. (2006). Arc fusion splicing of hollow-core photonic bandgap fibers for gas-filled fiber cells. *Optics Express*, eISSN 1094-4087, Vol. 14, No. 21, pp. 9576-9583

Vytran (2009). FFS-2000 Filament Fusion Splicing Workstation: An Integrated System for Production and Specialty Fiber Splicing. Available from: http://www.vytran.com/publications/datasheet_ffs_2009.pdf

Wang, Y., Bartelt, H., Brueckner, S., Kobelke, J., Rothhardt, M., Mörl, K., Ecke, W. & Willsch, R. (2008). Splicing Ge-doped photonic crystal fibers using commercial fusion splicer with default discharge parameters. *Optics Express*, eISSN 1094-4087, Vol. 16, No. 10, pp. 7258-7263. Available from http://www.opticsinfobase.org/oe/abstract.cfm?URI=oe-16-10-7258

Xiao, L., Demokan, S., Jin, W., Wang, Y. & Zhao, C. (2007). Fusion Splicing Photonic Crystal Fibers And Conventional Single-Mode Fibers: Microhole Collapse Effect. *Journal of Lightwave Technology*, ISSN 07338724, Vol. 25, No. 11, pp. 3563-3574

Yablon, A. D. & Bise R. (2004). Low-Loss High-Strength Microstructured Fiber Fusion Splices Using GRIN Fiber Lenses. *Proc. OFC 2004*, p. 41, paper MF14, ISBN 1-55752-772-5, Optical Fiber Communication Conference 2004, Los Angeles, CA, USA, 23-27 Feb. 2004

Yablon A. (2005). *Optical Fiber Fusion Splicing*, Springer Verlag, ISBN 3-540-23104-8, Berlin - Heidelberg - New York

11

Multi-Wavelength Photonic Crystal Fiber Laser

S. Shahi[1], M. R. A. Moghaddam[2] and S. W. Harun[2]

[1]Department of Electrical Engineering, Isfahan University of Technology, Isfahan
[2]Department of Electrical Engineering, University of Malaya, Kuala Lumpur
[1]Iran
[2]Malaysia

1. Introduction

The fiber lasers have some advantages compared to bulk-optics systems like compact size, high efficiency and high beam quality. The lasers in time-domain can be categorized into two groups "continuous wave fiber lasers" or "pulsed fiber lasers", and in wavelength domain as single wavelength or multi-wavelength. Such lasers were made as early as 1976 and have remained an active topic of study since then [2, 3]. Fiber lasers can be used to generate CW radiation as well as ultra-short optical pulses. The wavelength division multiplexing (WDM) techniques have shown to unlock the available fiber capacity and to increase the performances of broadband optical access networks. One of the essential components is the creation of new low-cost laser sources. Candidates for such applications are multi-wavelength fiber ring lasers as they have simple structure, are low cost, and have a multi-wavelength operation.

Recently, multi-wavelength lasers have caused considerable interests due to their potential applications such as WDM systems, fiber sensors and fiber-optics instrumentations. Requirements for multi-wavelength sources include; stable multi-wavelength operation, high signal to noise ratio and channel power flattening. Compared to a system that uses a number of discrete semiconductor diode laser [4], it is physically simpler to produce a multiple wavelength source using a single gain medium including a wavelength selective element. In order to define lasing wavelengths, wavelength selective comb filters have been included in the laser cavity. A multi-wavelength laser is highly desirable for the cost and size reduction, improvement of system integration and compatible with optical communication networks. For the past one decade or so, EDFs have been extensively studied and developed as a gain medium for the multi-wavelength laser.

In Erbium doped fiber laser (EDFL), the Erbium ions possess split Stark sublevels with multiple allowed transitions possibility of having oscillations at more than one wavelength. Therefore, the multi-transitions can be achieved in this fiber laser due to the depletion of Stark sub-levels which is selective and depends on the polarization of the wave. However, the outputs of the EDFLs are not stable at room temperature due to homogeneous broadening of lasing modes [5]. To increase the in-homogeneity one can cool Er^{+3} doped fiber at liquid nitrogen temperature [6, 7]. Generally, in order to produce the multi-wavelength, we have to employ intra-cavity filter in the EDFL cavity. In some works, a polarization controller (PC) is used in the cavity to change both the number of lasing lines and spacing of the multi-wavelength laser [8, 9].

There are also other methods to get simultaneous multi-wavelength outputs such as multi-wavelength Raman lasers [10, 11], multi-wavelength generation using semiconductor optical amplifiers (SOA) [12] and multi-wavelength Brillouin fiber lasers (BFLs) [13,14]. Special fibers such as dispersion compensating fibers (DCFs) have been used to increase the Raman gain in multi-wavelength Raman fiber lasers where the output power are limited only by the available pump sources [15]. Furthermore, the BFL is easier to be generated due to the lower threshold pump power [16].

Of the various approaches, the interest on the multi-wavelength fiber laser is increasing due to the improvements in number of lasing lines and power flatness. Furthermore, the Brillouin Erbium fiber laser (BEFL) is easier to be generated due to the lower threshold pump power for achieving the stimulated fiber laser [17]. Recently, the hybrid of EDFAs and new compact optical fibers like PCFs as a gain medium have many applications for producing amplifiers and fiber lasers.

2. Photonic crystal fiber ring laser

Photonic crystal fibers (PCFs) have generated great interest over the past few years, growing from a research-oriented field to a commercially available technology. The PCFs were first developed by Philip Russell in 1998, and can be designed to possess enhanced properties over (normal) optical fibers. They can be divided into two fundamental classes, solid-core and hollow-core as shown in Figure 1.

Fig. 1. Photonic Crystal Fibers Types, (a) Solid core PCF, (b) Hollow core PCF.

The solid core PCF is used in this report that is two dimensions (it has a periodic geometry in two directions and is homogeneous in the third) and we already introduced physical properties of that in table 1.

Figure 2 shows an electron micrograph of the cross section of this solid core PCF. Despite the hexagonal structure of the cladding, the mode field is very similar to that of the fundamental mode of a conventional fiber. The optical properties of PCFs rely on the specification of the size, shape and arrangement of the holes that surround a solid core to form a cladding. These parameters can easily be tailored to increase fiber nonlinearity, which is difficult to achieve using conventional fibers.

Fiber Type	PCF	Bi-EDF
Length(m)	20	2.15
Numerical Aperture (NA)	0.2	0.2
Core(μm)	4.8	5.4
Cladding(μm)	125	125.7
Mode field diameter(μm)	4.2	6.12
Zero dispersion wavelength(nm)	1040	1513
Cut off wavelength (nm)	1000	1180
Effective area($\mu m)^2$	27.5	29.4
V-$_{number}$	1.94	2.18
Material	Pure silica	Bi_2o_3-Er doped
Insertion loss (dB)	~2@1.06μm ~1.5@1.55μm	0.82@1.55 μm 1.18@1.48 μm
Brillouin gain, g_B(m/W)	5×10^{-7}	3.8×10^{-7}
Chromatic dispersion @1550nm (ps/nm.Km)	~70	-120
Refractive index of core/cladding at 1.55 μm	1.46/1.45	2.03/2.02
Nonlinear coefficient, γ(w.km)$^{-1}$@1550nm	~33.8	~60

Table 1. The physical parameters of PCF and Bi-EDF

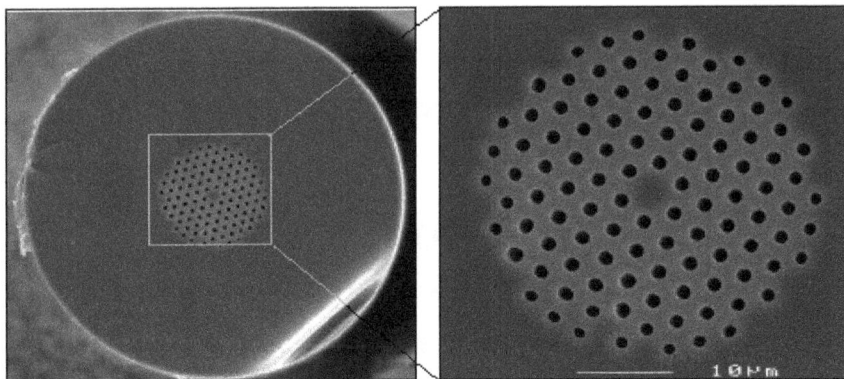

Fig. 2. The Scanning Electron Micrograph (SEM) of the PCF cross section and an enlarged view of the central "holey" cladding.

The highly nonlinear PCFs have many applications such as wavelength conversion [18] and Brillouin fiber lasers (BFLs) [13]. So far, few reports have been published on the Brillouin effects in PCFs [18, 19, 20]. The stimulated Brillouin scattering (SBS) is a nonlinear effect that results from the interaction between intense pump light and acoustic waves in a fiber, thus giving rise to backward propagating frequency shifted light [13]. In BFL applications, the required gain medium length can be substantially reduced using a holey fiber to replace the conventional SMF-28 Fiber of Corning Inc.[21]. However, most of the earlier works on PCF based BFLs are mainly on a single wavelength operation [21].

In this research, the fibre ring structure based on PCF can be used to make a very stable wavelength and narrow line-width laser. A conceptual structure of such a laser is very similar to a fibre ring resonator. In the ring configurations, a very short length of PCF (20 m) is added in the ring cavity BEFL in the proposed configurations to achieve a stable single and multi-wavelength laser generation.

3. Results and discussion

Figure 3 shows the experimental setup of the proposed PCF-based BEFL. The ring resonator consists of a circulator, a 20 m long PCF, a polarization controller (PC), two isolators and a bi-directionally pumped 215 cm long Bi-EDF. The PCF used is a polarization maintaining type with a cut-off wavelength of 1000 nm, zero dispersion wavelength of 1040 nm, nonlinear coefficient of 33.8 (W.km)$^{-1}$ at 1550 nm and a mode field diameter of 4.2 μm near zero dispersion wavelength. The Bi-EDF is pumped bi-directionally using two 1480 nm lasers. Optical isolators are used to block the Brillouin pump (BP) from oscillating in the cavity and also to ensure a unidirectional operation of the BFL. The PC is used to control the birefringence (breakage of a light ray into two different directions therefore creating two separate light rays) of the ring cavity, so that the power of the laser generated can be controlled. The experiment executed using 3 different types of couplers the 80/20, 90/10 and 95/5 and the output for BFL is tapped from the leg with the smaller coupler ratio before it is characterized using an optical spectrum analyzer (OSA).

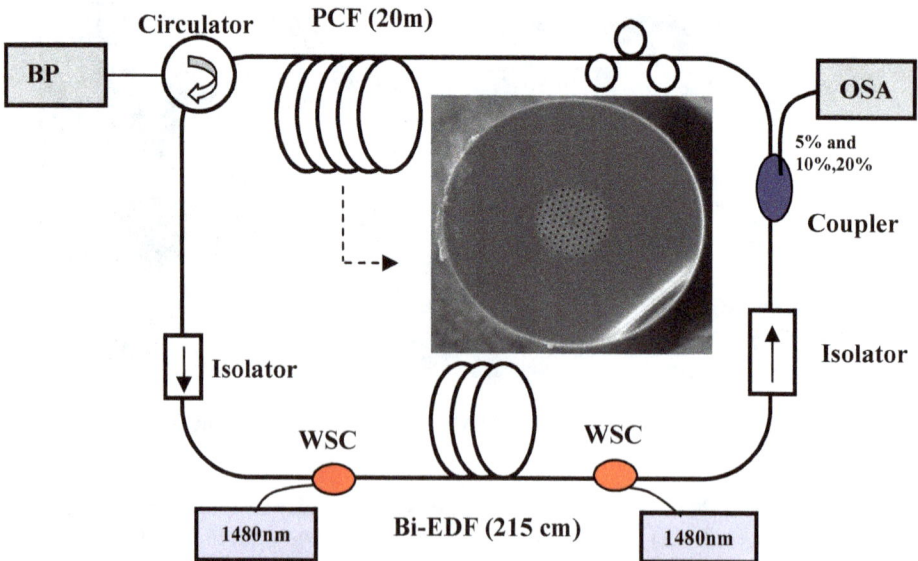

Fig. 3. Configuration of multi-wavelength BFL based PCF.

The BP is injected into the ring cavity and then PCF via the circulator to generate the backward propagating Stokes light at opposite direction. However, since the PCF length is not sufficient enough, the back-scattered light due to Rayleigh scattering is relatively higher than the Stokes light. Both back-scattered pump and the Stokes lights are amplified by the bi-directionally pumped Bi-EDF and it oscillates in the ring cavity to generate first Stokes in an anti-clockwise direction. This oscillation continues and when the intensity of the first Brillouin Stokes is higher than the threshold value for Brillouin gain, the second order SBS is generated in clockwise direction and this signal is blocked by the isolator in the cavity. However, the back-scattered light from second SBS will be amplified by the Bi-EDF. Hence, the nonlinear gain by both PCF and Bi-EDF only amplifies the Stokes light and thus the Stokes light is more dominant and laser is generated at the Stokes wavelength. The spacing between the BP and the BFL is obtained at approximately 10 GHz, which is equivalent to the Stokes shift in the single mode fiber (SMF).

The operating wavelength of the BFL is determined by the bi-directionally pumped Bi-EDF gain spectrum which covers the L-band region from 1560 nm to 1620 nm as well as the cavity loss. For comparison and the effect of different cavity resonators, three kinds of output couplers selected. Figure 4 shows the free running spectrum of the BEFL, which is obtained by turning off the BP for three different output coupler ratios; 80/20, 90/10 and 95/5. The output laser is taken from the leg with a lower portion. The peak wave generated at approximately 1574 nm with bandwidth of approximately 3 nm due to the difference between Bi-EDF's gain and cavity loss is the largest in this region. The chosen BFL operating wavelength must be within or close to the bandwidth of free running BFL. Therefore, the BP is set within 1574 nm region which is within the lasing bandwidth of the free running BFL. At the coupling ratio of 80/20, the free-running BFL exhibits the highest peak power of approximately -6 dBm with 20 dB bandwidth of approximately 1 nm. The cavity loss is the lowest with 80/20 coupler and therefore the peak power is the highest.

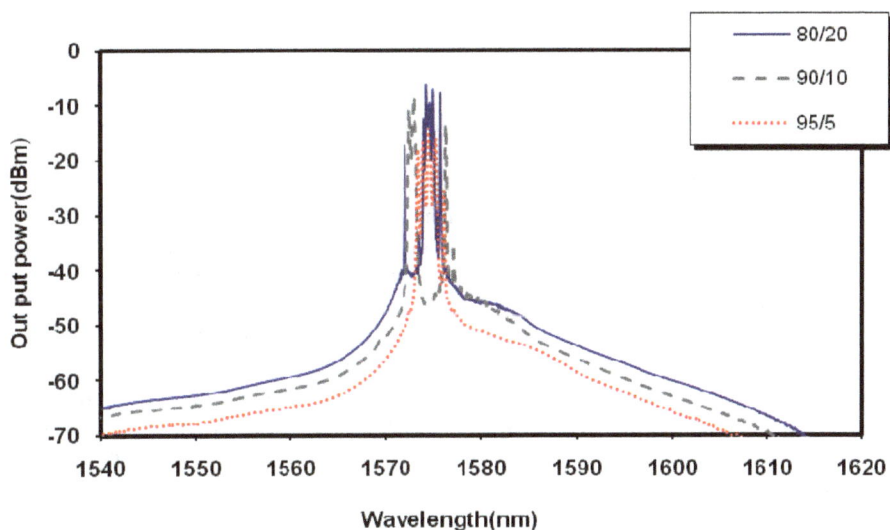

Fig. 4. Free-running spectrum of the BEFL using 80/20, 90/10 and 95/5 couplers.

Figures 5 (a), (b) and (c) show the output spectra of the BEFL at different output coupler ratios of 80/20, 90/10 and 95/5, respectively. The experiment was carried out for three different pump powers. Both the 1480 nm pumps are set at the same power and power of each pump is varied from 60 mW to 135 mW. The threshold of the BEFL is observed to be around 60 mW for all setups. At pump power below of 60 mW (threshold) the Erbium gain is very low and cannot sufficiently compensate for the loss inside the laser cavity and thus no Stokes are observed. When increasing the 1480 nm pump power the number of wavelength generated is increased and the anti-Stokes wave also surfaced, which attributed to the increment of the Erbium gain with the pump power. This situation provides sufficient signal for SBS as well as the four wave mixing (FWM) to generate Stokes and anti Stokes.

Besides that SBS, the Kerr effect or the quadratic electro-optic effect (QEO effect) was found in 1875 by John Kerr, a Scottish physicist. The Kerr effect describes a change in the refractive index of a material in response to an intense electric field. The index change is directly proportional to the square of the electric field instead of the magnitude of the field. In Kerr effect, the nonlinear phase shift induced by an intense and high power pump beam is used to change the transmission of a weak probe through a nonlinear medium [7] as such as PCF and Bi-EDF. Thus the change in the refractive index is proportional to the optical intensity and lead to nonlinear scattering and frequency shift.

However, the FWM is another nonlinear effect, which is due to the third-order electric susceptibility is called the optical Kerr effect. The FWM is a type of optical Kerr effect, and occurs when light of two or more different wavelengths is launched into a fiber. FWM is also a kind of optical parametric oscillation [22].

The FWMs in PCFs can occur at relatively low peak powers and over short propagation distances, and such processes can be possible in a much wider wavelength range (e.g. more than 120 nm). The FWM can be very efficient at the zero-dispersion wavelength. Besides the obvious advantage of shorter fiber requirement; the use of PCF would allow the operation of these nonlinear devices in the wavelength regime outside that of possible using conventional fibers. This is because, PCFs can have zero dispersion wavelength ranging from 550- 1550 nm.

In this experiment, more than 13 lines are obtained at the maximum 1480 nm pump power of 135 mW with wavelength spacing of approximately 0.08 nm for the BEFL configured with 95/5 output coupler as shown in Figure 5(c). Below of this input power, the number of lines decreased by 95/5 output couplers as such as Figures 5(a and b) which shows more restoratively in this kind coupler. However, the number of lines significantly reduced as the cavity loss increases. For instance, only two Stokes are observed with 80/20 coupler as shown in Figure 5(a). The side mode suppression ratio, which is defined as the power difference between the BFL's peak and the second highest peak (SMSR) are obtained at approximately 27.0 dB, 26.9 dB, 18.8 dB for 80/20, 90/10 and 95/5 couplers, respectively as shown in Figure 5. The multi-wavelength output of the BFL is observed to be stable at room temperature with only minor fluctuations observed coinciding with large temperature variances. The side modes are mainly due to anti Stokes and additional Stokes of the BFL, which arises due to FWM effect in the PCF.

The extremely FWM effect in PCF leads to the generation of a wave whose spectrum is the "mirror image" of the weak wave, in which the mirroring occurs about the pump

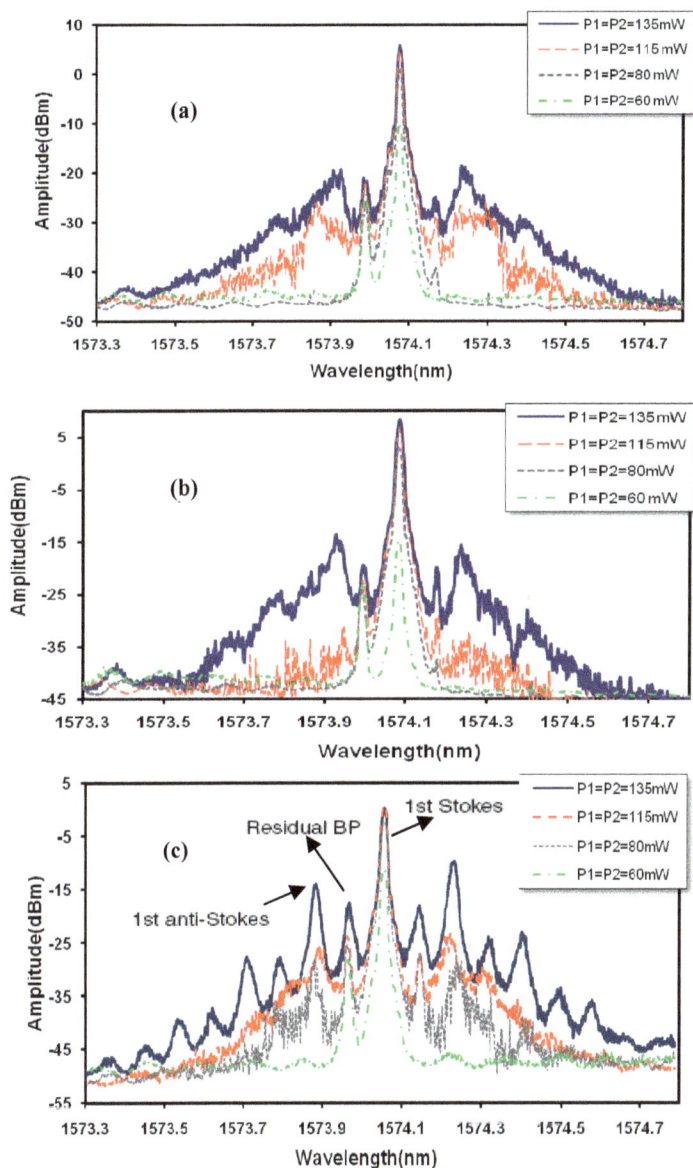

Fig. 5. The BFL output spectrum for (a) 80/20 coupler, (b) 90/10 coupler and (c) 95/5 coupler. Both pumps are at the same power for each output coupling ratio.

frequency. The representation of this image can be observed in Figures 5 where the multi-wavelength spectrum is more symmetry with the use of 95/5 output coupler.

Figure 6 shows the peak power of the first Stokes for different couplers against the input 1480 nm pump power of each pump (total pump power is double). The BP power and wavelength is fixed at 5 dBm and 1574 nm, respectively. The BEFL starts to lase at 1480 nm pump power of 60 mW which is the threshold power. Below this power, the Erbium gain is very low and cannot sufficiently compensate for the loss inside the laser cavity and thus no Stokes is observed. The output power saturates at 135 mW. As shown in the figure, the peak power is highest with 80/20 coupler and lowest with 95/5 coupler. Hence, we observed higher SMSR in Figure 5(a). Inset of Figure 6 shows the peak power of the first Stokes against the BP power at various output couplers. This figure shows that the threshold power of around 4~5dBm is required to generate the Stokes with the use of 95/5 output coupler. The threshold power reduces as smaller portion of light is allowed to oscillate in the ring cavity. For instance the threshold power is about 2 dBm with 80/20 output coupler.

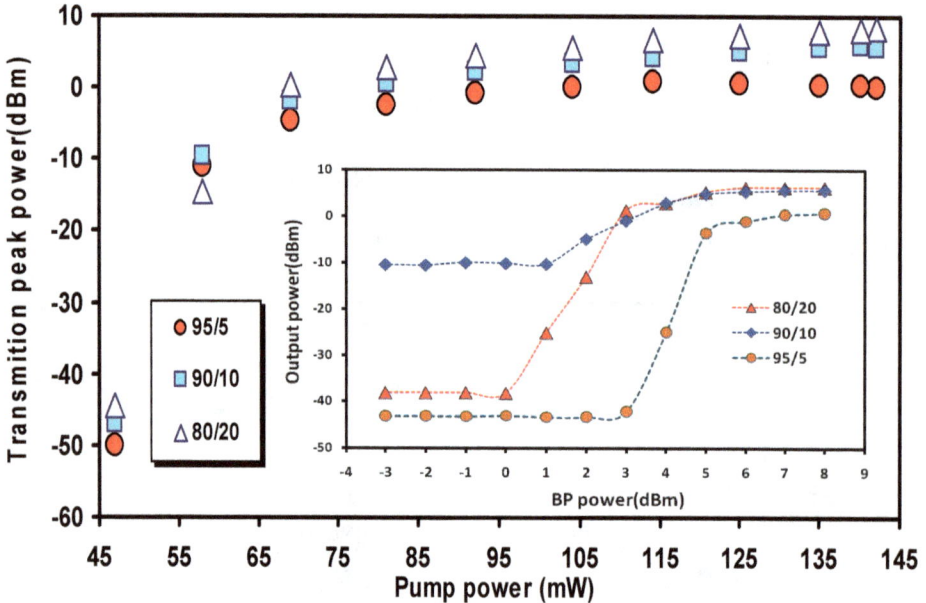

Fig. 6. Output peak power as a function of 1480 nm total pump powers. Inset shows the peak power against BP power.

In this area research, we also compared the results by another ring resonator based PCF. The second experimental setup for the proposed BFL is shown in Figure 7. The ring resonator is similar to figure 3 but consists of a forward pumped Bi-EDF and only 10 dB output coupler. The Bi-EDF is 49 cm in length and has a nonlinear coefficient of 60 W^{-1}km^{-1} at 1550 nm, an erbium concentration of 3250 ppm and a cut-off wavelength of 1440 nm as well as a pump absorption rate of 83 dB/m at 1480 nm as the same of 215 cm long of Bi-EDF.

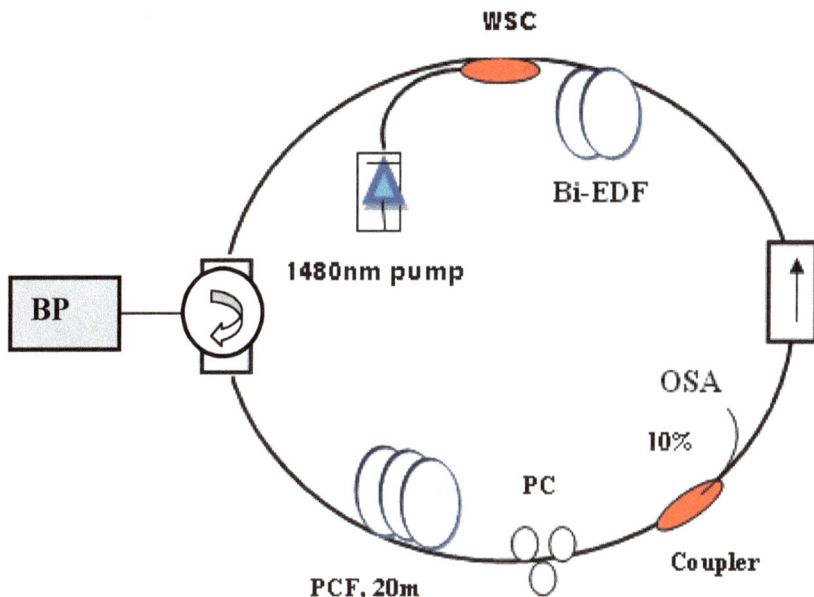

Fig. 7. Experimental setup for the proposed Bi-EDF and PCF based Brillouin fiber laser [23].

Without the BP, the larger spacing of 0.57 nm was obtained due to the incorporation of the PCF, which changes the cavity loss and dispersion parameter. The spacing is determined by the cavity length as well as the birefringence state in the cavity, which can be controlled by the PC.

The BP is injected into the ring cavity and a forward pumped Bi-EDF via the circulator to generate the backward propagating Brillouin Stokes and Rayleigh scattered light at Brillouin-shifted and BP wavelengths, respectively. The back-scattered light due to Rayleigh scattering, the reflections of the connectors and splices is relatively higher than the Brillouin Stokes light. Furthermore, the power, back scattered due to SBS, is frequency shifted to higher wavelength, because of the Doppler shift. The SBS will become significant effect, when high powers are used, because it is highly dependent on the intensity.

As we mentioned in Kerr effect, the change in the refractive index is proportional to the optical intensity and lead to nonlinear scattering and frequency shift.

However, both lights are amplified by the forward pumped Bi-EDF and oscillate in the ring cavity to generate a dual wavelength laser. The spacing between the BP and the BFL is obtained at 0.09 nm (~10 GHz) as shown in Figure 9.

The operating wavelength of this the Brillouin fiber laser set up is determined by the forward pumped Bi-EDF gain spectrum which covers the conventional band (C-band) region from 1525 to 1570 nm as well as the cavity loss. Therefore, the BP wavelength is optimized at 1559.00 nm, which is within the lasing bandwidth of the free-running erbium doped fiber laser (EDFL). The EDFL operates at around 1560 nm region due to the cavity

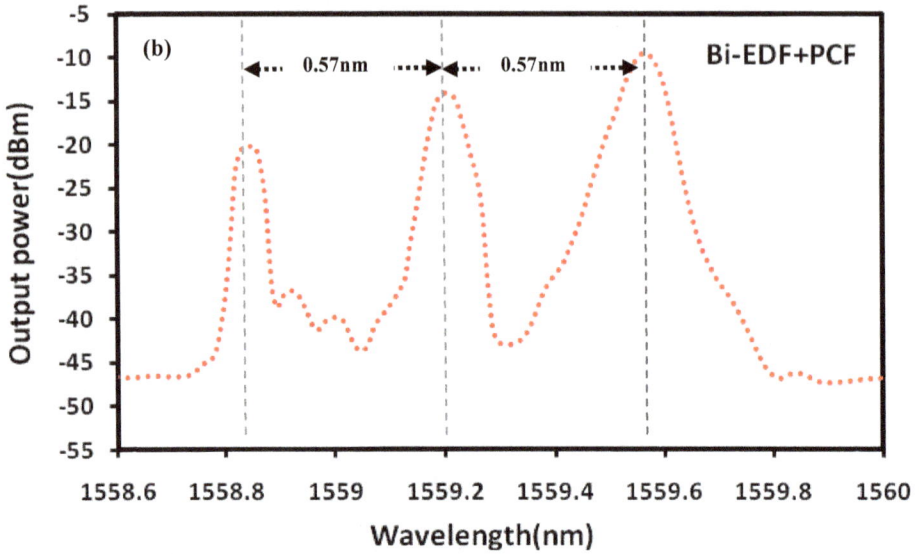

Fig. 8. Free-running spectrum of 49 cm of Bi-EDF and PCF.

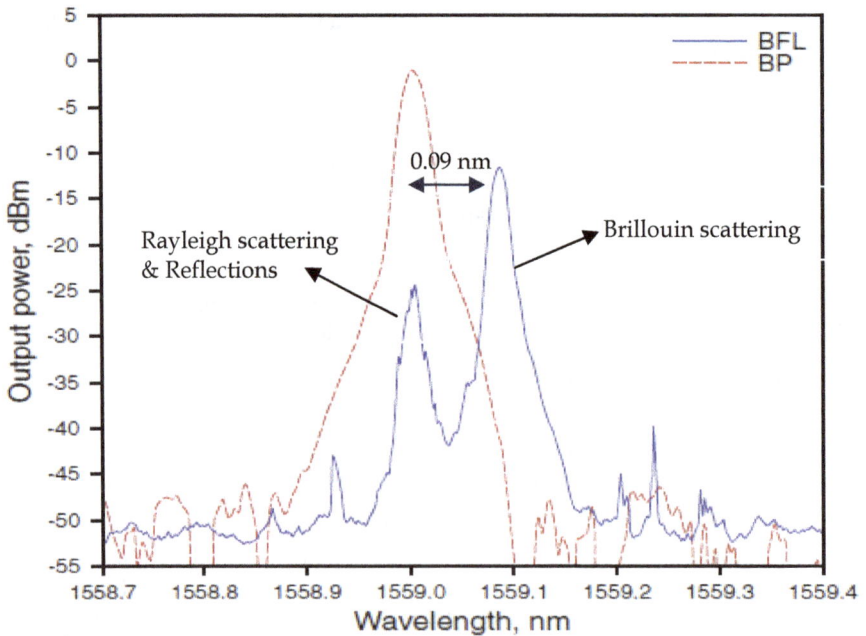

Fig. 9. Output spectrum of the proposed Bi-EDF and PCF based Brillouin fiber laser.

loss which is lower at the longer wavelength. Figure 9 compares the BP and output spectrum of the proposed BFL. The 1480 nm pump power is fixed at 150 mW. The BFL is achieved at 1559.09 nm with the peak power of approximately –12.3 dBm when the injected BP is set at 0 dBm. The 3 dB bandwidth of the BFL is measured to be approximately 0.02 nm limited by the OSA resolution. The SMSR is obtained at approximately 12 dB as shown in Figure 9. The anti-Stokes is also observed which arises due to FWM effect in the ring cavity.

However, the stimulated single wavelength BFL is obtained due to the SBS effect, which is more dominant especially in the PCF. The BFL output is observed to be stable at room temperature too.

4. Conclusion

In summary, new configurations of BFL are proposed and demonstrated using a PCF in conjunction with uni- and bi-directionally pumped Bi-EDF. By employing PCF ring configuration, more than 13 lines are obtained at the maximum 1480 nm pump power of 135 mW as shown in Figure 5(c). The wavelength spacing of setups is nearly between to 0.08-0.09 nm for the BEFL. The acceptable side mode suppression ratio (SMSR) are obtained in both configurations by 90/10 output coupler are approximately 26.9 dB and 12 dB, respectively. The side modes are mainly due to anti Stokes and additional Stokes of the BFL in the PCF. Despite of this, the 80/20 output coupler demonstrated the lowest cavity loss and the highest SBS in the peak power.

The BFL uses a ring cavity structure to generate Stokes and anti-Stokes via stimulated Brillouin scattering (SBS) and FWM processes. Hence, two nonlinear effects of SBS and FWM are extremely affective on ring fiber laser based PCF. The single- and multi-wavelength BFL are stable at room temperature with only minor fluctuations observed coinciding with large temperature variances and also is compact due to the use of only 20 m long of PCF and short long of Bi-EDF.

5. References

[1] S. Shahi, S. W. Harun, N. S. Shahabuddin and M.R. Shirazi , H. Ahmad, " Multi-wavelength generation using a Bismuth-based EDF and Brillouin effect in a linear cavity configuration," Optics & Laser Technology, Vol. 4, No. 2, pp.198-201, 2009.

[2] K. O. Hill, B. S. Kawasaki, and D. C. Johnson, "CW Brillouin laser," Appl. Phys. Lett. Vol.28, pp.608, 1976.

[3] C. Montes, D. Bahloul, I. Bongrand, J. Botineau, G. Cheval, A. Mahmhoud, E. Picholle, and A. Picozzi, Self-pulsing and dynamic bistability in CW-pumped Brillouin fiber ring lasers, J. Opt. Soc. Am. Vol. 16, No. 6, pp. 932, 1999.

[4] M. Ibsen, S. Alam, M. N. Zervas, A. B. Grudinin, and D.N. Payne, 8 and 16 Channel All-Fiber DFB Laser WDM Transmitters with Integrated Pump Redundancy, IEEE IEEE Photon. Technol. Lett. Vol. 11, No. 9, pp.1114-1116, 1999.

[5] M. Yamada, "Overview of Wideband Optical Fiber Amplification," NTT Technical, Vol.213, pp.63-66, 2002.

[6] P. W. France, "Optical fiber lasers and amplifiers," CRC Press Inc., Florida, 2000.

[7] G. P. Agrawal, " Nonlinear Fiber Optics,"Fourth edition, Academic Press, 2007.

[8] N. Park, and P.F. Wysocki, "24-line multi-wavelength operation of erbium doped fiber ring laser," IEEE Photon. Technol. Lett., Vol. 8, pp.1459- 1561, (1996).

[9] S. Yamashita and T. Baba, "Multi-wavelength fiber lasers with tunable wavelength spacing," Opt. Fiber Commun. Conf., Anaheim,CA, 3, 2001.

[10] E. Yamada, H. Takara, T. Ohara, K. Sato, T. Morioka, K. Jinguji, M. Itoh and M. Ishii, " A high SNR, 150 ch supercontinum CW optical source with precise 25 GHz spacing for 10 Gbit/s DWDM systems," Opt. Fiber Commun. Conf., Anaheim, CA, 2001.

[11] F. Koch, P. C. Reeves-Hall, S. V. Chernikov and J. R. Taylor, " CW multiple wavelength, room temperature, Raman fiber ring laser with external 19 channel, 10 GHz pulse generation in a single electro-absorption modulator," Opt. Fiber Commun. Conf., Anaheim, CA, 3, 2001.

[12] N. Pleros, C. Bintjas, M. Kalyvas, G. Theophilopoulos, K. Yiannopoulos, S. Sygletos, and H. Avamopoulos, "Multiwavelength and power equalized SOA laser sources," IEEE Photon. Technol. Lett.,Vol. 14, pp.693-695, 2002.

[13] S. W. Harun, X. S. Cheng, N. K. Saat and H. S. Ahmad, "L-band Brillouin erbium fiber laser," Electron. Lett. Vol. 4, pp.174–6, 2005.

[14] A. K. Zamzuri, M. A. Mahdi, A. Ahmad, and M. H. Al-Mansoori, "Flat amplitude multi-wavelength Brillouin–Raman comb fiber laser in Rayleigh-scattering-enhanced linear cavity," Opt. Express, Vol. 15, pp. 3000–5, 2007.

[15] M. J. Guy, S. V. Chernikov and J. R. Taylor, "Lossless transmission of 2 ps pulses over 45 km of standard fiber at 1.3 μm using distributed Raman amplification," Electron. Lett. Vol. 34, pp. 793–4, 1998.

[16] G. J. Cowle, W. H. Loh, R. J. Laming and D. Y. Stepanov, "Multi-wavelength operation of Brillouin/Erbium fiber laser with injection-locked seeding, Optical Fiber Communication Conf. OFC 97, pp. 34–5, 1997.

[17] S. Shahi, S. W. Harun, and H. Ahmad, " Multi-wavelength Brillouin fiber laser using a holey fiber and a Bismuth-oxide based erbium-doped fiber," Laser Phys. Lett. 6, No. 6, pp. 454- 457, 2009.

[18] J. H. Lee, Z. Yusoff, W. Belardi, M. Ibsen, T. M. Monro, B. Thomsen, and D. J. Richardson, Conference on Lasers and Electro-Optics, Long Beach, USA, May 19–24, (CLEO 02), Technical Digest, Vol. 2, pp. CPDB5-1–CPDB5-3, 2002.

[19] R. K. Pattnaik, S. Texier, J. Toulouse, E. J. H. Davies, P. S. J. Russell, B. J. Mangan, CLEO 2003, Paper CWJ2, Baltimore, June 2003.

[20] C. J. S. de Matos , J.R. Taylor, K.P. Hansen, "All-fibre Brillouin laser based on holey fibre yielding comb-like spectra," Optics Communications, No.238, pp.185–189, 2004.

[21] Z. Yusoff, J. H. Lee, W. Belardi, M. Ibsen, T.M. Monro, and D.J. Richardson, Conference on Lasers and Electro-Optics, Long Beach, USA, May 19–24, (CLEO 02), Technical Digest, Vol. 1, pp. 50–51, 2002.

[22] J. E. Sharping, M. Fiorentino, A. Coker, P. Kumar, and R. S. Windeler, "Four-wave mixing in microstructure fiber," Opt. Lett. Vol. 26, pp.1048-1050, 2001.

[23] S. W. Harun, S. Shahi, H. Ahmad, " Brillouin fiber laser with a 49 cm long Bismuth-based Erbium-doped fiber," Laser Physics Lett. Vol.7, No.1, pp. 60-62, 2009.

12

Long-Period Gratings Based on Photonics Crystal Fibers and Their Applications

Chun-Liu Zhao

Institute of Optoelectronic Technology, China Jiliang University, Hangzhou
China

1. Introduction

The photonic crystal fiber (PCF) is a special class of components incorporating photonic crystals with a two-dimensional (2D) periodic variation in the plane perpendicular to the fiber axis and an invariant structure along it [1-3]. Typically these fibers incorporate a number of air holes that form a so-called photonic crystal cladding and run along the length of the fibers, and the shape, size, and distribution of the holes can be designed to achieve various novel wave-guiding properties that may not be achieved readily in conventional fibers [2-19], so that they have attracted significant attention in recent ten years.

A long-period fiber grating (LPG) is a one dimension (1D) periodic structure, and is formed by introducing periodic modulation of the refractive index along an optical fiber. Since its period is about 100 to several hundreds µm and longer than that of fiber Bragg grating (FBG), LPG resonantly couples light from the fundamental core mode to some co-propagating cladding modes and leads to dips in the transmission spectrum. LPGs have been widely used in optical fiber communications and sensors. Examples of LPG-based devices include all-fiber band-rejection filters [20, 21], gain flatteners in erbium-doped fiber amplifiers [22], and sensors for strain, temperature, and external refractive index measurement [23-25]. When a LPG is formed on a PCF, a 2-D periodic structure is combined with a 1-D periodic structure. LPGs based on PCFs (PCF-LPGs) have been fabricated recently [26-36] and shown many unique properties compared with a conventional LPG (1-D periodic structure) [27, 31-34, 37, 38], which provide wide and novel applications [38-47].

In this chapter, we will first introduce the basic operation principle of LPGs, secondly, will demonstrate in detail the strain and temperature characteristics of a LPG based on an endlessly-single-mode (ESM) solid silica core PCF theoretically. To account for the effect of dispersive characteristics of the PCF, we identify a dispersion factor γ, which offers a deeper understanding into the behavior of LPGs in PCF. Following, we will move on to the fabrication of a PCF-LPG by using a CO_2 laser and demonstrate the experimental observations on the strain and temperature characteristics, which agree with the theoretical predictions very well. Finally, we will demonstrate their applications in optical sensors, including a temperature-insensitive strain sensor, and demodulation technologies for fiber Bragg grating and fiber loop mirror sensors.

2. Basic operation principle of LPGs [23, 38, 48]

A LPG is formed usually by a periodic modulation of the refractive index in a fiber core, which allows coupling from the fundamental core mode to some resonant cladding modes and leads to some dips in the transmission spectrum at wavelengths that satisfy the resonant condition. The phase matching condition of a LPG can be expressed as [23]:

$$\lambda = (n_{co} - n_{cl})\Lambda \tag{1}$$

where λ is the resonant wavelength, Λ is the index modulation period of the LPG, and n_{co} and n_{cl} are the effective indices of the fundamental core mode, and the forward-propagating cladding mode, respectively.

When an axial strain is applied on the LPG, the resonant wavelength of the LPG will shift because the Λ of the LPG will increase with stretching axially and at the same time the effective refractive index of both core and cladding modes will decrease due to the photo-elastic effect of the fiber [31]. Meanwhile, if the ambient temperature changes, the wavelength of the LPG may also be changed by linear expansion or contraction and the thermo-optic effect. From equation (1), the sensitivity of the LPG to strain or temperature is a function of the differential effective index between the core and cladding modes (or the differential propagation constant). Thus, from equation (1), the strain and temperature sensitivity can be written as [48]:

$$\frac{d\lambda}{d\varepsilon} = \lambda \cdot \gamma \cdot (1 + \frac{\eta_{co}n_{co} - \eta_{cl}n_{cl}}{n_{co} - n_{cl}}) \tag{2}$$

$$\frac{d\lambda}{dT} = \lambda \cdot \gamma \cdot (\alpha + \frac{\xi_{co}n_{co} - \xi_{cl}n_{cl}}{n_{co} - n_{cl}}) \tag{3}$$

where ε is the axial strain, T is the ambient temperature, η_{co} and η_{cl} are strain-optic coefficients of the core and cladding, ξ_{co} and ξ_{cl} are the thermo-optic coefficient of the core and cladding, respectively, and α is the linear expansion coefficient. η and ξ are defined as [23]:

$$\eta = \frac{1}{n}\frac{dn}{d\varepsilon} \tag{4}$$

$$\xi = \frac{1}{n}\frac{dn}{dT} \tag{5}$$

Different materials have different η and ξ. η and ξ may also have some difference due to the different effective index of waveguides made by the same material [49].

Since the effective index (or propagation constant) both of the fundamental mode in the fiber core and cladding modes in the fiber cladding will be affected by the waveguide change which is caused by the applied axial strain on the LPG, the dispersion factor γ is used to describe the effect of waveguide dispersion and is expressed as [48]:

$$\gamma = \frac{d\lambda/d\Lambda}{n_{co} - n_{cl}} = \frac{\Delta n_e}{\Delta n_e - \lambda \dfrac{d\Delta n_e}{d\lambda}} = \frac{\Delta n_e}{\Delta n_g} \qquad (6)$$

where $\Delta n_e = n_{co} - n_{cl}$ and $\Delta n_g = n_{co}^g - n_{cl}^g$ are respectively the differential effective index and differential group index between the core mode and the cladding mode. As will be discussed in Section 3, γ plays a significant role on the strain and temperature sensitivity of an LPG based on the PCF.

3. Theoretical properties of an LPG based on PCF [38]

In this section, we investigate in detail the strain and temperature characteristics of an LPG based on an endlessly single-mode (ESM) solid silica core PCF theoretically. To account for the effect of dispersive characteristics of the PCF, we identify a dispersion factor, which offers a deeper understanding into the behavior of PCF-LPGs. Theoretical results show that is always negative, and this causes blue-shifting of the resonant wavelength when an axial strain is applied.

3.1 Properties of the ESM-PCF

The PCF used in the work is an endlessly single-mode PCF fabricated by Crystal Fiber A/S. The fiber has a standard triangular air/silica cladding structure, as shown in Fig. 1 (a). The mode field diameter is ~6.4 μm, the center-to-center distance between the air holes (L) is ~7.78 μm, and the diameter of the air holes is ~3.55 μm. The diameter of the entire holey region is ~ 60 μm, and the outer cladding diameter of the PCF is 125 μm. A full-vector finite-element method (FEM) was used to calculate the effective index of modes of PCF. Because of the symmetric nature of the PCF, only a quarter of the cross-section as shown in Fig.1 (b) is used during calculation. A perfect electric or perfect magnetic conductor (PEC or PMC) was applied at boundaries [7]. The refractive index of pure silica was taken as 1.444.

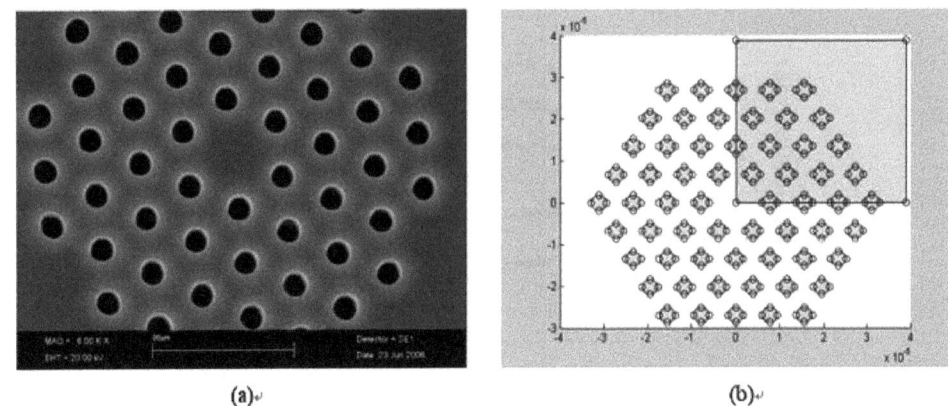

(a)   (b)

Fig. 1. (a) Micrograph of the PCF used in the experiment; (b) Schematic cross-section of a PCF, showing the quarter used in calculation

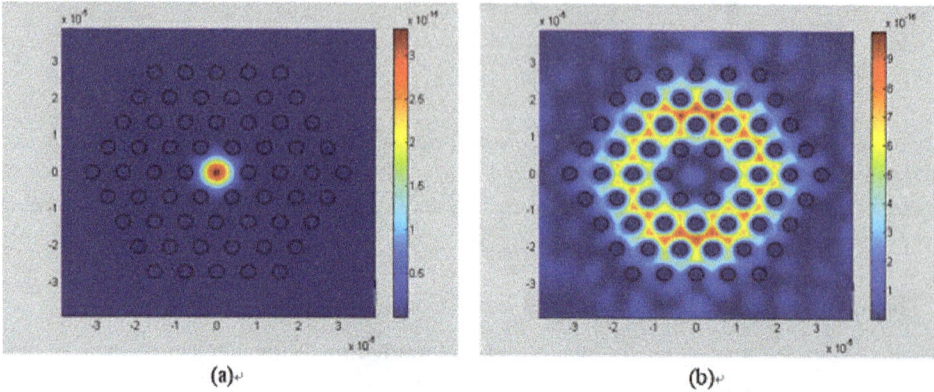

(a)   (b)

Fig. 2. Calculated intensity distribution of the PCF with L=7.78 µm and d=3.55 µm. (a) The fundamental core mode; (b) The cladding mode

Fig. 2 (a) and (b) shows the intensity distribution of the core and the cladding mode, which are considered as the two coupling modes in our PCF-LPGs. Fig. 3 shows the effective indices of the fundamental and cladding modes as functions of wavelength in the ESM-PCF. The group indices of these two modes, which were calculated by using $n_g = n_e - \lambda \frac{dn_e}{d\lambda}$, are also shown in Fig. 3. The curves of n_g are not so smooth because of the limited data available for the calculation of $\frac{dn_e}{d\lambda}$ but the trend is clear. The curve $n_{g\text{-}cl}$ shows the highly dispersive characteristics of the cladding mode. For any wavelength in the range of 1.2 ~ 1.8 µm, the group index of the cladding mode is higher than that of the core mode, which is in contrast to a conventional SMF.

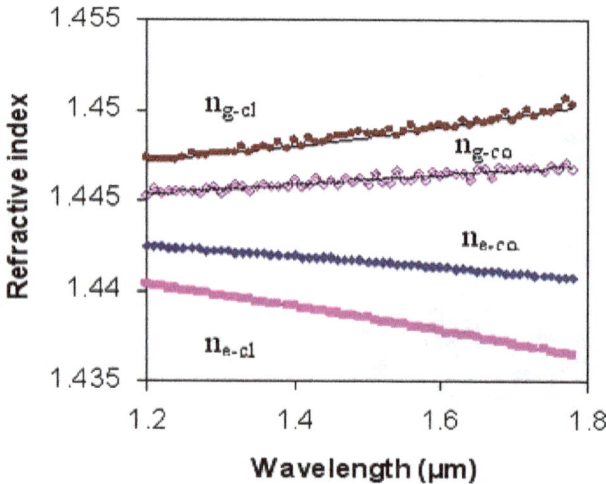

Fig. 3. Calculated dispersion curves for core and cladding modes as shown in Fig. 2.

3.2 Properties of an LPG based on the PCF in theory

Fig. 4 shows the calculated resonance wavelength as a function of the period of PCF-LPG. It is clear that the resonance wavelength of PCF-LPG decreases with increasing LPG period, which is consistent with other experimental observations [27, 33, 34]. This is in contrast to LPGs written in conventional SMFs and is because of the highly dispersive property of the cladding mode due to the existence of the air-holes. In other words, $(n_{co} - n_{cl})$ as shown in equation (1), varies significantly with wavelength.

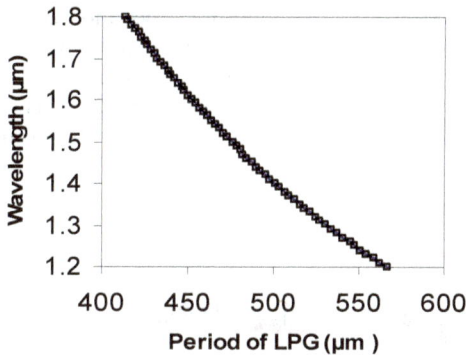

Fig. 4. Calculated resonance wavelength as a function of grating period

γ is a special factor to describe the effect of waveguide dispersion, and γ may be positive or negative. Because Δn_e is always positive, the sign of γ is determined by Δn_g. When Δn_e equals to Δn_g, the factor γ is 1. This means dispersive properties of the core and the cladding mode is similar and this is the case normally for SMFs. When the group indices of the core mode is less than that of the cladding mode, γ will be negative. This has been observed in the case of the coupling from core mode to higher-order cladding modes in B-Ge co-doped fiber [48]. In Fig. 5, we show the relationship of γ with the period of a LPG based on the theory of ESM-PCF, and γ is in the range of -1.15~ -1.35 for LPG period of from 420 to 570μm.

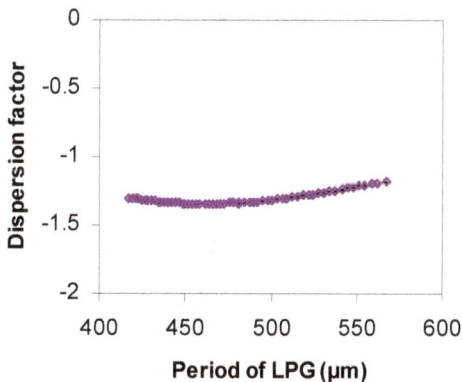

Fig. 5. Dispersion factor γ at the resonance wavelengths vs. grating period of the PCF-LPG

The strain sensitivity $d\lambda/d\varepsilon$ of a LPG based on the ESM-PCF is determined by four parameters: the elasto-optic coefficients of the core and cladding materials, waveguide properties (γ), the period of the LPG, and the mode order. Now, we choose the same coupling modes, and focus on the effect of the first three parameters on the strain sensitivity of a PCF-LPG. Fig. 6 shows the calculated strain sensitivity as a function of LPG period with different η_{cl} when we assume γ =1. In the calculation, η_{co} is assumed to be constant at a value of -0.22 for the pure silica core. For LPGs with period ranging from 400 to 600µm, the strain sensitivity is positive and relatively independent of the grating period when $|\eta_{cl}|$ is larger than 0.22. The strain sensitivity becomes negative and decreases with grating period when $|\eta_{cl}|$ is smaller than 0.218. On the other hand, when the value of γ is taken the value as shown in Fig. 5, the strain sensitivity as a function of LPG period is as shown in Fig. 7. The strain sensitivity is negative when $|\eta_{cl}|$ is larger than 0.22. This is the opposite of what is shown in Fig. 6. In ref. [49], A. Bertholds et. al. showed that the strain-optic coefficients of a bulk silica and a silica fiber are different. It's believed that, owing to the different geometry of solid core and micro-structured air-silica cladding of the ESM-PCF, η_{co} will be slightly different from η_{cl}.

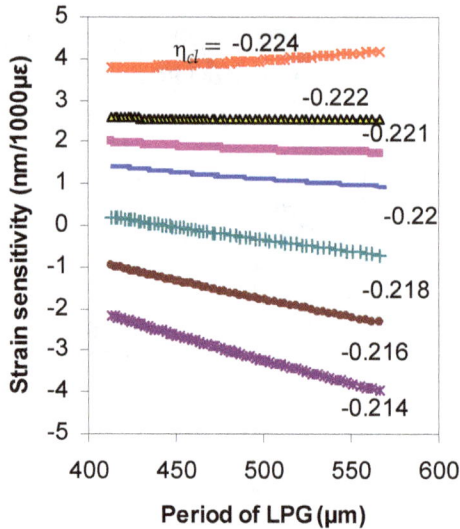

Fig. 6. Theoretical strain sensitivity at resonance wavelength vs. LPG period for various values of η_{cl} and γ = 1.

Similarly, the temperature sensitivity $d\lambda/dT$ of a LPG is determined by the thermo-optic coefficients of the core and cladding materials, waveguide properties (γ), period of LPG, and the mode order. We calculated the temperature sensitivity as a function of LPG period by assuming that the thermo-optic coefficient of the pure silica core is ξ_{co} =7.8×10^{-6} / °C and thermal expansion coefficient is α = 4.1×10^{-7}/ °C. Figs. 8 and 9 show respectively the results for the cases of γ =1 and γ taking from Fig.5. The temperature characteristics are quite different for the two cases. With ξ_{cl} less than ξ_{co} =7.8×10^{-6}, the LPG has positive temperature sensitivity for γ =1 but negative temperature sensitivity for the case of γ taking from Fig.5.

Furthermore, the dependence of temperature sensitivity on the grating period is approximately linear for $\gamma = 1$ while it is non-monotonic for the other case. Similar to the discussion for the strain coefficient, for the ESM-PCF, since the effective index n_{co} is larger than n_{cl}, from eq. (5), we expect that ξ_{co} is slightly smaller than ξ_{cl}, which has also been verified by the experiment in Section 4.

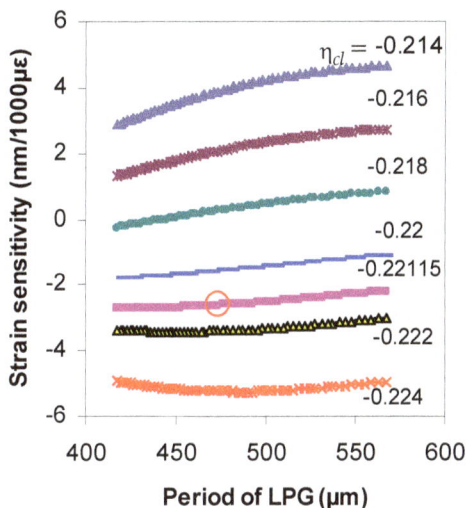

Fig. 7. Theoretical strain sensitivity at resonance wavelength vs. LPG period for various values of η_{cl} and with γ taken from Fig. 5.

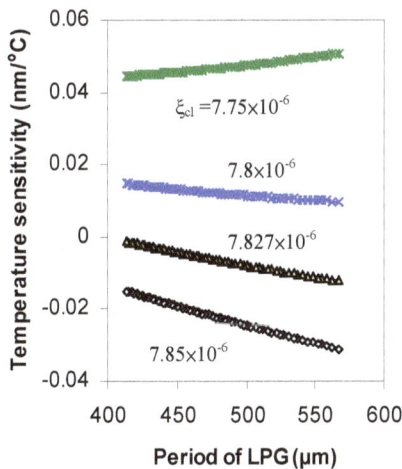

Fig. 8. Theoretical temperature sensitivity at resonance wavelength vs. LPG period for various values of ξ_{cl} and $\gamma = 1$.

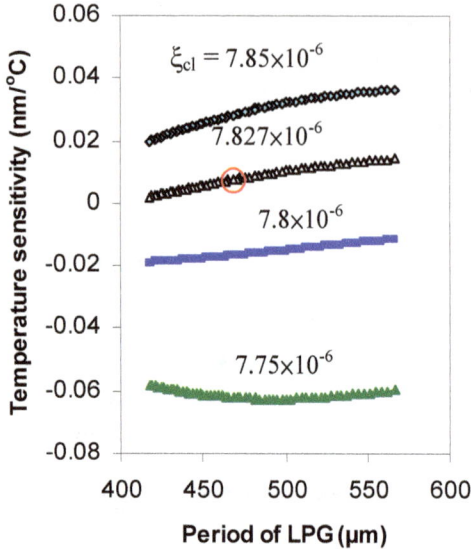

Fig. 9. Theoretical temperature sensitivity at resonance wavelength vs. LPG period for various values of ξ_{cl} and with γ taken from Fig. 5.

4. Fabrication and experimental observations on properties of a LPG based on PCF [38]

4.1 Fabrication of a PCF-LPG

Unlike conventional fibers, which contain at least two different glasses each with a different thermal expansion coefficient, thereby giving rise to a relatively high thermal expansion coefficient, PCFs are virtually insensitive to temperature because it is made of only one material (and air holes). This property can be utilized to obtain temperature-insensitive PCF-based devices, as demonstrated in [14]. However, the single material property of PCFs leads to non-photosensitivity to UV light, therefore FBGs and LPGs cannot normally be formed in PCFs by use of the conventional UV-written technique, unless a PCF with a Ge-doped photosensitive core is used [26]. Recently, several alternative methods for making LPGs in PCFs with non-photosensitive cores were introduced, including glass structure change [27], periodic structural and/ or residual stress relaxation induced by arc discharge or a CO_2 laser [28-32], refractive index modulation by periodically applied mechanical pressure [33] or by the use of an acoustic wave [34, 35], and periodic drilling micro-holes with a femto-second laser [36].

LPGs fabricated by use of a CO_2 laser are compact and stable because the perturbations are everlasting whereas in LPGs fabricated by UV light in conventional fibers, the refractive index modulation caused by UV light are prone to aging, therefore making LPGs in conventional fibers unstable over time. Here, we demonstrate the fabrication method by using a CO_2 laser, which has been used widely [30, 32, 37, 38, 40].

Fig. 10 shows the experimental setup of the PCF-LPG fabrication. The CO_2 laser operates at a frequency of 10 kHz and has a maximum power of 10 W. The laser power is controlled by the width of the laser pulses. In the experiment, the pulse-width of the CO_2 laser was chosen to be 3.8 μs. The laser beam was focused to a spot with a diameter of ~60 μm and scanned across the ESM-PCF transversely and longitudinally along the fiber by use of a two-dimensional optical scanner attached to the laser head. The scanning step of the focused beam was 1 μm and the delay time of each step was 350 μs. The LPG inscribed has a period of about 467 μm and a period number of 40. The process of the CO_2 laser scanning is repeated 9 times, which results in a LPG with a deep transmission dip and no observable deformity in the fiber structure. The spectrum measurements were performed using a broadband light source (a light-emitting diode, LED, with the wavelength range of 1200 ~1700 nm) in combination with an optical spectrum analyzer (OSA, ADVANTEST Q8384) with a resolution of 0.5 nm.

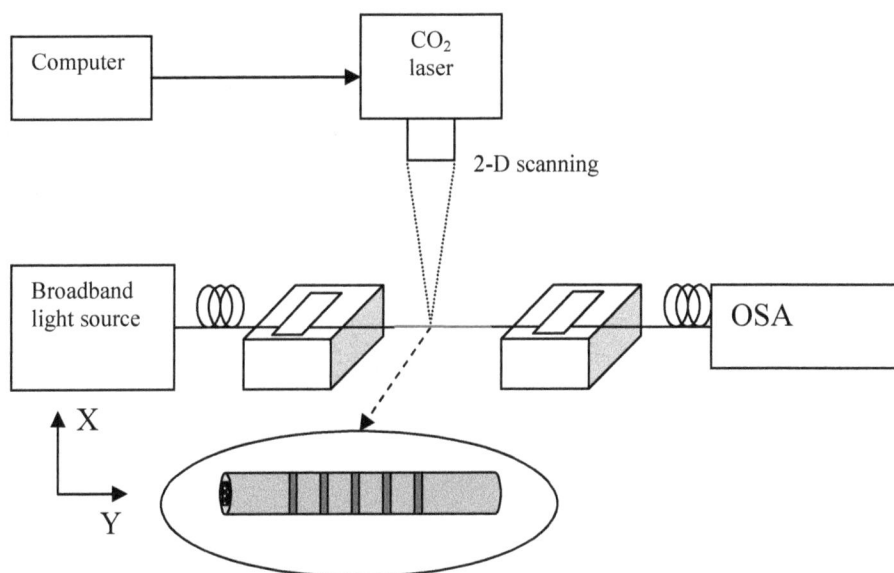

Fig. 10. Schematic of the PCF-LPG fabrication setup

Fig. 11 shows the growing process of a PCF-LPG as a function of the number of scanning procedures. The resonant wavelengths of the PCF-LPG are about 1552.45 nm and 1363.3 nm which are due to coupling of the fundamental core mode to two different cladding modes. The dip at the wavelength 1552.45 nm is nearly 20 dB. The insertion loss of the LPG is about 1.5 dB. The resonance at 1552.45 nm is due to coupling of the core mode to the cladding mode shown in Fig. 2 (b) and is in good agreement with the theoretical result (1552.45 nm resonance wavelength corresponds 467.2 μm LPG period) in Fig. 4.

4.2 Properties of a LPG based on PCF in experiment

Fig. 12 shows the experimental setup for measuring the characteristics of the PCF-LPG. The spectrum measurements were performed by using a broadband LED and an OSA with a

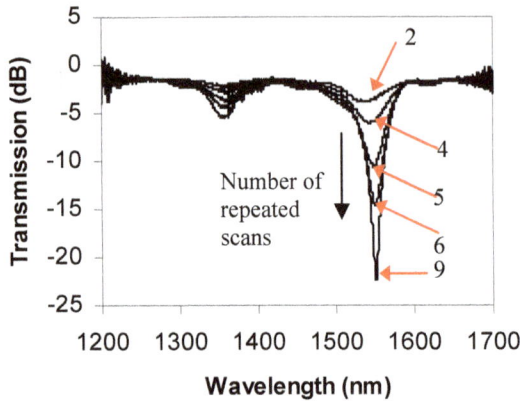

Fig. 11. Transmission spectra of the PCF-LPG with various numbers of repeated scans

resolution of 0.5 nm as mentioned in Section 4.1. The strain characteristics of the PCF-LPG was tested by stretching the PCF-LPG by moving the translation stage shown in Fig. 10 from 0 to 0.5 mm in four steps (corresponding to a strain variation of from 0 to 1604 με), the center wavelength of the transmission dip shifts toward shorter wavelength as shown in Fig. 13. This is opposite to a LPG in a conventional SMF where the transmission dip shifts toward longer wavelengths [23]. The strain dependence of the resonance wavelength 1552.45 nm on axial strain is shown in Fig. 14. The strain sensitivity, which is the slope of the curve, is estimated to be -2.68 nm/ 1000 μm and shown in Fig.7 as a small circle. By varying the strain coefficient of the cladding material η_{cl} to fit the experimental data, we found the value of η_{cl} that best fits the experimental sensitivity is η_{cl} =-0.22115. The transmitted intensity at resonance wavelength was found increases with the applied axial strain, and this is also shown in Fig. 14.

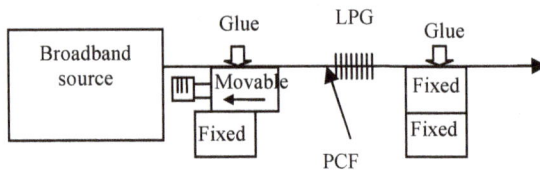

Fig. 12. Experimental setup for measuring the temperature and strain characteristics of the PCF-LPG, where a broadband source and an OSA are used.

To test the temperature characteristics of the PCF-LPG, the ambient temperature of the LPG was varied by using a temperature chamber whose temperature can be controlled within the range of 25 oC to 100 oC. As shown in Fig.15, the transmission spectrum hardly changes when temperature was raised from 25 oC to 100 oC. The estimated sensitivity of the resonant wavelength to temperature is about 0.007 nm/ oC. Again, the temperature sensitivity of the PCF-LPG is marked as a circle in Fig.9. The cladding temperature coefficient ξ_{cl} that fits this temperature insensitivity is ξ_{cl} =7.827 ×10⁻⁶, this value is slightly larger than that of the core and agrees with the prediction at the end of Section 4.2. It should be noted that the difference between ξ_{co} and ξ_{cl} for the PCF is 6 times smaller than that of conventional Ge-doped fiber and 18 times smaller than that of B-Ge co-doped fiber [48].

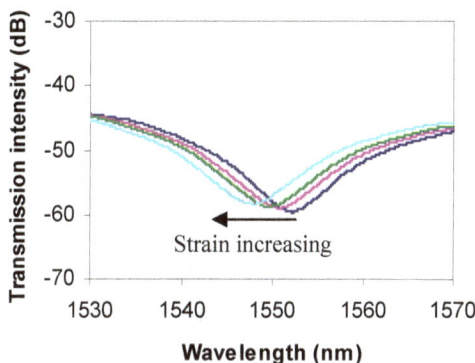

Fig. 13. Transmission spectra of the PCF-LPG for applied strain of (from right to left) 0, 535, 936, and 1604 µm

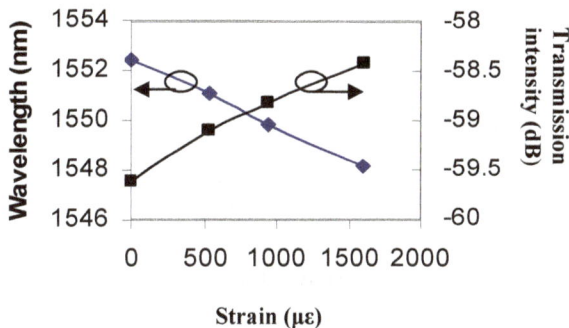

Fig. 14. Resonant wavelength and the intensity at the transmission dip as functions of applied strain

5. Applications of an LPG in PCF in optical fiber sensors

5.1 Temperature-insensitive strain sensor based on a PCF-LPG [38]

By exploiting the PCF-LPG's large sensitivity to strain and insensitivity to temperature, temperature-insensitive strain sensors are realized. These sensors can be based on either wavelength or intensity measurement. In this part, we demonstrate a simple, low cost strain sensor based on the measurement of the transmitted light intensity at a wavelength close to the LPG resonance.

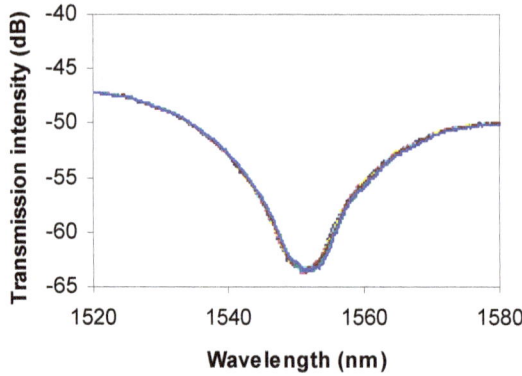

Fig. 15. Transmission spectra of the LPG in the ESM PCF at various ambient temperatures

Fig. 16 shows the proposed strain sensor that uses the PCF-LPG as the sensing element. A single wavelength light such as a DFB laser is used as a light source. The wavelength of the DFB laser is near the resonant wavelength of the LPG and hence the output light intensity from the LPG will be directly related to the LPG's transmission at the wavelength of the DFB laser. Since the LPG's transmission is insensitive to temperature, the output power will only be affected by the transmission spectrum change caused by the strain applied to the LPG. At the output, an optical power meter will be adequate to deduce the strain information and an expensive OSA would not be needed.

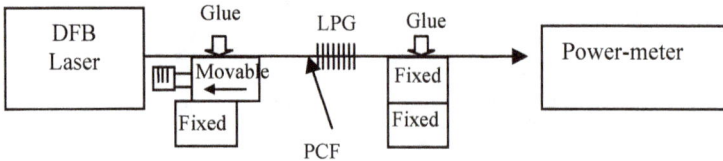

Fig. 16. The proposed temperature insensitive strain sensor with a DFB laser and an optical power meter

Fig. 17 shows the measured relationship between the output intensity of the PCF-LPG sensor and the applied axial strain for various laser wavelengths. In the experiment, a tunable laser was used for easiness of wavelength adjustment. In practice, a DFB laser with appropriate wavelength would be a better for the purpose of reducing cost. As shown in Fig.17, for laser wavelengths of 1538 nm, 1542 nm, 1545 nm, and 1547.7 nm, which are shorter than the resonant wavelength (1552.45 nm) the LPG, the output intensity decreases with applied strain and the strain sensitivity is negative and respectively -1.41, -2.17, -2.80 and -2.41 dB/1000 µε for 1538 nm, 1542 nm, 1545 nm, and 1547.7 nm. The relationship is approximately linear for strains from 0 to 1600 µε. Similarly, for laser wavelength longer than the resonant wavelength (1552.45nm), the output intensity increases with applied strain and the intensity sensitivities are positive and respectively 3.25, 3.11, 2.01 and 1.53 dB/ 1000 µε at 1553 nm, 1555.6 nm, 1560 nm, and 1563 nm. Therefore, the setup shown in Fig. 16 converts directly the strain variation to intensity variation. Assume that we choose a DFB laser at 1553nm as source, and use an optical power meter with a resolution of 0.001 dB, we may achieve a strain resolution of ~0.3µε.

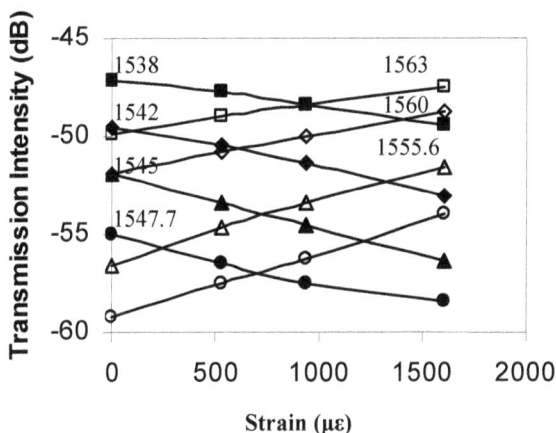

Fig. 17. Strain dependence of the transmission intensity of the PCF-LPG at different wavelengths

5.2 Temperature-insensitive demodulator based on a PCF-LPG for a fiber Bragg grating temperature sensor [45]

Fiber Bragg grating (FBG) is another kind of optical fiber grating, and is formed by a periodic modulation of the refractive index along an optical fiber with ~0.5 µm pitch Λ, which causes the coupling between the forward-propagating core mode to the backward-propagating core mode at the reflective wavelength satisfied with the phase matching condition $\lambda_B = 2n_{eff}\Lambda$, where λ_B is the reflective wavelength, n_{eff} is the effective refractive index of the core mode, respectively. By fabricating FBG with different pitch, we can get different reflective wavelength of FBG. The bandwidth of FBG is narrow compared with that of LPG, because the coupling strength decreases rapidly when the wavelength departs from λ_B. FBGs have been applied widely in sensors. Because of the wavelength-encoded nature of the FBG sensors, it is necessary to convert the wavelength-encoded signal into electronic signals for easy reading and real time monitoring.

In this part, we present the use of PCF-LPG for the purpose of making a temperature insensitive sensor interrogation based on the PCF-LPG's much lower temperature sensitivity and realize a whole fiber Bragg grating (FBG) temperature sensor, including a sensor head and a PCF-LPG readout component, in one package. Utilizing the wavelength-dependent transmission loss of the LPG, the wavelength change of the FBG due to the temperature of the environment is translated to the intensity of the output. At the output, only a power meter is required to deduce the temperature of the environment and an expensive OSA would not be needed. The experimental results show that the interrogation based on the PCF-LPG works well under different environmental temperatures.

Fig. 18 shows the schematic diagram and the experimental setup of the proposed fiber Bragg grating temperature sensor interrogation system using an LPG in a PCF. The sensor head was made of a FBG and was illuminated using a broadband LED via a 3-ports circulator. The sensor head was placed inside a temperature-controlled container in order to detect its sensitivity to temperature. The reflected light from the sensor head is returned to the circulator and enters the PCF-LPG. In the experiment, the central reflection wavelength of the FBG sensor head was chosen as shown in Fig. 19. The FBG was fabricated in a hydrogen-loaded single-mode fiber using a phase mask illuminated by UV light. The phase mask had a constant period of 1068 nm (the corresponding grating period is 534 nm). The original centre wavelength of the FBG sensor is at 1546.3 nm, which is located at the middle of the wavelength range of the transmission spectrum of the LPG with the negative slope. The reflected light from the sensor head is transmitted partially by the PCF-LPG and the amount of transmission is a function of wavelength. The wavelength shift of the FBG sensor head with temperature is almost linearly related to the transmission function of PCF-LPG filter for the range of operation. The power at the output of the PCF-LPG is directly related to the temperature applied on the FBG because as the temperature increases, the wavelength of the light reflected by the FBG shifts more to a longer wavelength, which in tune causes the light to leading to PCF-LPG to decrease in intensity. At the output, only a power meter is enough to deduce the temperature of environment. Thus it is very feasible to monitor the variation in temperature by measuring the output power of the PCF-LPG in a relatively cheap manner.

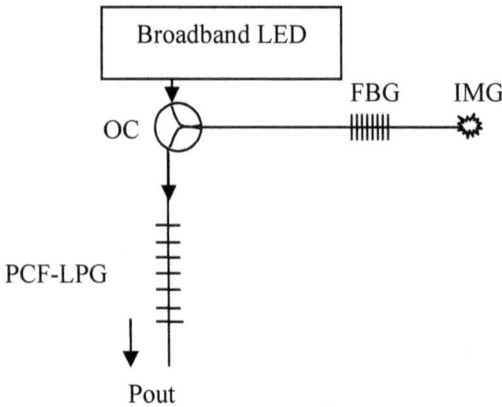

Fig. 18. Schematic diagram of the proposed temperature sensor interrogation system based on a PCF-LPG, IMG: index matched glue, OC: optical circulator.

When the FBG sensor was put into a temperature chamber and the temperature was increased from 25 to 100 °C, the Bragg reflection of the FBG would shift to longer wavelengths due to the thermo-optic effect and the thermo-expansion effect of the optical fiber. Fig. 20 shows the output spectra. The intensity of the output changes in accordance with the transmission curve of the PCF-LPG, which is as the reflection wavelength of the FBG gets longer, the output of the PCF-LPG gets smaller. Fig. 21 shows the relationship of the output intensity of the PCF-LPG with the temperature on the FBG and that this relationship is virtually linear for temperature from 25 to 100 ºC. In this temperature range (75 ºC), the change in the reflected wavelength from the FBG is 0.75 nm, and the intensity change of the output power via the LPG is 1.2 dB. Thus the sensitivity of the proposed FBG temperature sensor is about -0.0165 dB/ ºC. Because the slope of the transmission spectrum in the wavelength range from 1545 to 1548 nm is virtually constant as implied by Fig. 21, the linear dynamic range of this temperature sensor is at least 300 ºC. Therefore, this setup converts the temperature variation to wavelength variation (FBG) and then it converts the wavelength variation to intensity variation (PCF-LPG).

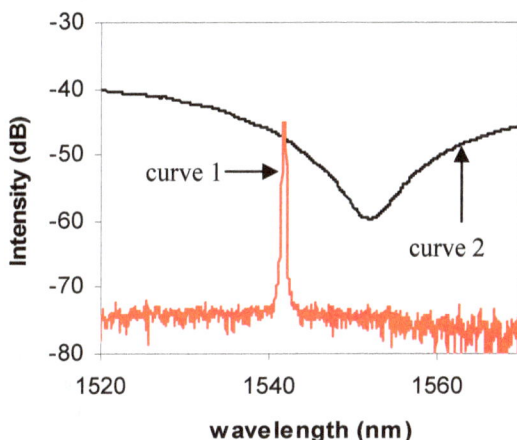

Fig. 19. The reflective spectrum of the FBG (curve 1) at room temperature and the transmission spectrum of the LPG (curve 2)

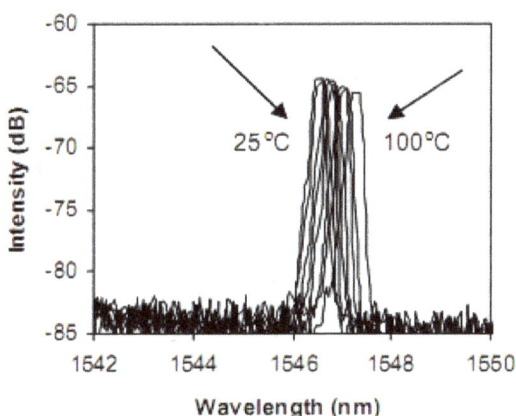

Fig. 20. The output spectra from PCF-LPG when a different temperature (25 ºC, 30, 40, 50, 60, 70, 80, 90, 100 ºC) is applied to the FBG.

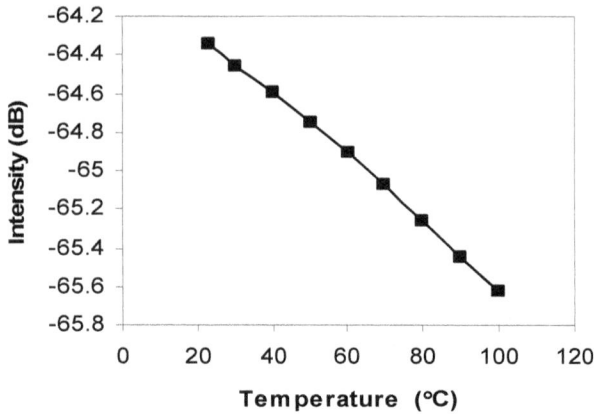

Fig. 21. The measured output power for the temperature of the grating sensor

5.3 A fiber loop mirror temperature sensor demodulation technique based on a PCF-LPG and a band-pass filter [46]

Fiber loop mirrors (FLMs) have been demonstrated for a number of applications, for example wavelength filters and sensors [14, 15, 50]. In a FLM, two interfering waves counter-propagate through the same fiber, and are exposed to the same environment. This makes it less sensitive to environmental disturbance. Various kinds of sensors based on FLMs have been realized, such as temperature sensors, strain sensors, pressure sensors, liquid level sensors, biochemical sensors, UV detection, multi-parameter measurement and refractive index sensors. However, most HiBi-FLM sensors are based on monitoring the resonant wavelength variation of FLMs [15, 51], an expensive OSA is needed.

In this part, we present a HiBi-FLM temperature sensor using a PCF-LPG and a band-pass filter as a demodulator. For the sensing principle, only the HiBi-FLM acts as a sensor head, while the PCF-LPG serves as a filter to provide wavelength dependent optical power transmission and the band-pass filter is to provide a narrow band light source. By utilizing the stable filtering function of the PCF-LPG, the resonant wavelength variation of the FLM with temperature is transferred effectively to the intensity variation of the output light. When the optical intensity of the output is monitored, temperature applied on the FLM will be deduced. The experimental results show the demodulator based on the PCF-LPG and the band-pass filter works well. By choosing a filter with an appropriate bandwidth, the temperature sensitivity of the sensor with 1.742 dB/ oC is obtained.

Fig.22 shows the experimental setup. The configuration of the proposed sensor system includes a broadband light source, a band-pass filter, a HiBi-FLM, and a PCF-LPG. A broadband superluminescent LED (SLED) source with a flatten emission intensity in the wavelength range of 1530~1570 nm was used as an input light source, whose power is ~40 mW. The SLED with a band-pass filter launches the HiBi-FLM via a 3 dB coupler. The transmitted light from the HiBi-FLM then enters the temperature-insensitive PCF-LPG. The proposed sensor demodulation was used to read in optical power and an optical power meter (NF1112) with a resolution of 0.01 dB is used to monitor the output light intensity of the sensor. For monitoring transmission spectra of the FLM and the PCF-LPG, and explaining the principle of the sensor system, an OSA is used in experiments with a spectral resolution of 0.02 nm.

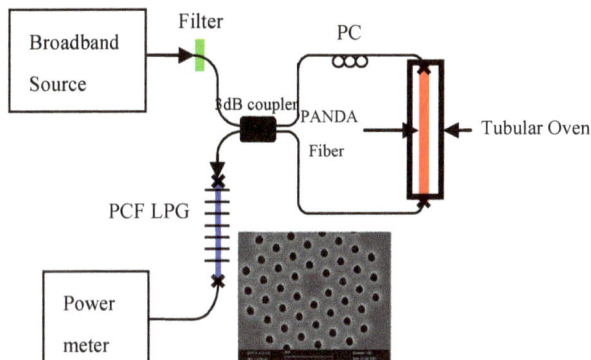

Fig. 22. Experimental setup of the proposed sensor, inset: micrograph of PCF.

The HiBi-FLM includes a 3 dB coupler, a 30.5 cm Panda PMF, and a polarization controller (PC). The Panda PMF has an attenuation of 1.0 dB/km, a measured birefringence of $\Delta n = 6.24 \times 10^{-4}$ at 1550 nm, and a core diameter~10 μm. Both ends of the Panda fiber were spliced to Corning SMF-28, and the combined loss of the two splicing points is 2.3 dB. The wavelength spacing between transmission peaks of the FLM is given by $\Delta \lambda = \lambda^2 / \Delta n \cdot L$, where λ, Δn, and L are wavelength, birefringence and length of the HiBi fiber, respectively. The wavelength spacing between the transmission peaks of the FLM is about 13 nm at 1545 nm, as shown in Fig. 23. In the experiment, this value was chosen by considering the bandwidth of the PCF-LPG. When the PCF-LPG is given, a suitable wavelength spacing of the FLM can be obtained by choosing the length of the PMF. Thus by adjusting the state of the PC, one spectral peak of the FLM is located in a negative of the linear regions with the PCF-LPG at the initial state of the sensor. The temperature-insensitive of the PCF-LPG used in the experiment is the same one as mentioned above.

The band-pass filter is made of Metal-Dielectric-Metal (MDM), which is used to provide a narrow band light source. As shown the gray region in the Fig. 23, only the intensity in the band-pass of the filter is monitored by the optical power meter. In the experiment, several filters with the same center wavelength at 1545nm and different bandwidth are used to study the effect of the bandwidth of the filter on the sensing performance. The full width at half maximum (FWHM) of the band-pass filters is 12.66nm, 8.3 nm, 6.66 nm and 3 nm, respectively, whose transmission spectra are shown in Fig.24.

Fig.25 shows transmission spectra of the PCF-LPG, the HiBi-FLM, and the output signal at different temperature when the band-pass filter is not inserted. When temperature is increased, the transmission spectrum of the FLM blue-shifts due to thermally induced refractive index change and thermal expansion of the panda fiber, and the temperature sensitivity to wavelength is -0.772 nm/ °C. Due to the filtering function of the PCF-LPG, the intensity of the resonant peak1 increases while the resonant peaks shift to shorter wavelength with temperature increasing. Therefore, when a band-pass filter is used, as shown in Fig.23, the part of the output light outside the band of the filter is attenuated owing to the function of the band-pass filter. And the intensity of the remainder will change with the transmission peak of the HiBi-FLM shifting. Therefore, converting the wavelength variation of the HiBi-FLM into intensity variation at the output directly is realized.

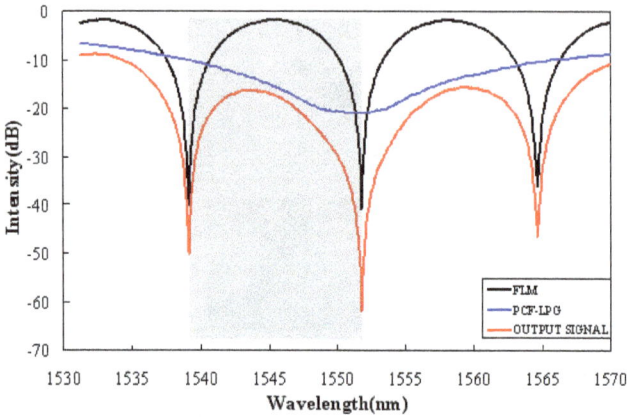

Fig. 23. Transmission spectra of the FLM, the PCF-LPG and the output signal. Gray region is transmission spectra of the sensor with band-pass filter.

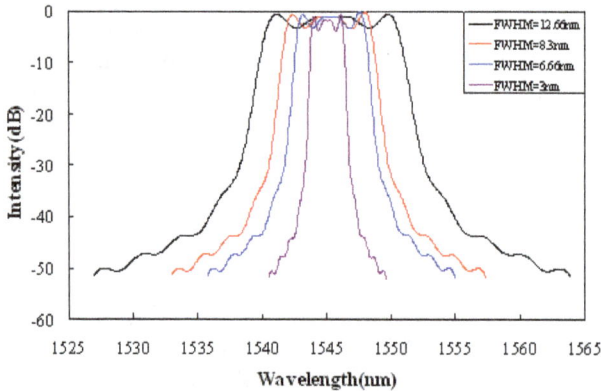

Fig. 24. Transmission spectra of the band-pass filters used in the experiment.

Fig.26 shows the relationship of the output intensity of the sensing system with temperature applied on the FLM, when the filter was chosen with a FWHM 12.66 nm and a center wavelength at 1545 nm. The output intensity is approximately a periodic function of temperature. The period is 16 °C and temperature increases from 45.5 °C to 61.5 °C in a period. This behavior can be explained by the periodic property of the FLM. Since the transmission spectrum of the FLM is approximately a periodic function of the wavelength, the peaks of the FLM enter, then shift outside the pass-band of the filter one by one when the peaks of the FLM shift to shorter wavelengths with temperature increasing. Therefore, the intensity of the output light changes periodically with temperature increasing. The change in the transmission wavelength of the FLM is 13nm corresponding to 16 °C temperature period, which is consistent with the wavelength spacing of 13 nm between the transmission peaks of the FLM. In order to certify the explaining, we take out the band-pass filter and measure the transmission spectra of the output signal at three different temperatures, which are marked in Fig.26.

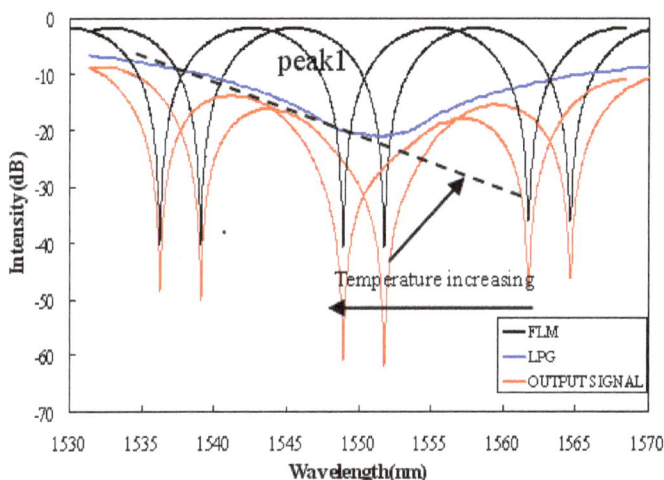

Fig. 25. Transmission spectra of the FLM, the PCF-LPG and output signal when temperature is applied to the sensing head.

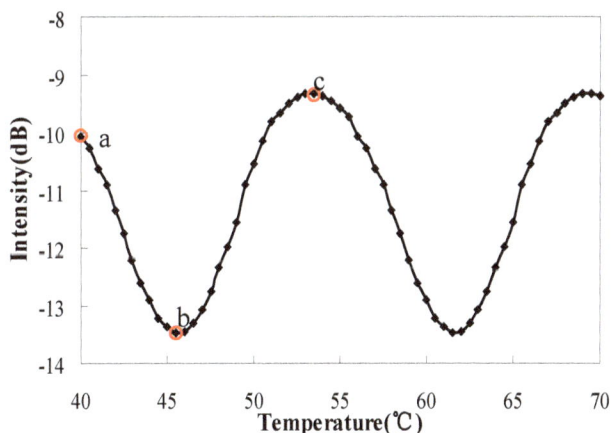

Fig. 26. The relationship of the output power and temperature when the band-pass filter is with a central a wavelength 1545 nm and FWHM 12.66nm.

Fig.27 shows the transmission spectra of the HiBi-FLM, the PCF-LPG, and the output signal without the band-pass filter when temperature is applied at 40 oC, 45.5 oC and 53.5 oC, respectively. The region between the dotted lines is the bandwidth of the band-pass filter. It is clear that the peaks of the FLM shift inside and outside the filter with temperature increasing. Due to the filtering function of the PCF-LPG, the intensity of the output is decided by the location of the FLM's peak in the band of the filter. When the peak is located within a shorter wavelength of the LPG's transmission spectrum, in which a higher transmission is provided, the intensity of the output is larger, as shown in Fig.27 (a) and (c). Obviously, Fig.27 (b) shows the opposite situation. In this case, the peak of the FLM in the filter is near to the transmission dip of the PCF-LPG, thus the intensity of the output is lowest. As shown in Fig.26, the relationship is virtually proportional for temperature from 45.5 oC to 53.5 oC. In this temperature range (8 oC), the intensity change of the output power is 4.137 dB and the change in the transmission wavelength of the FLM is ~6.52 nm which is about half period of the FLM. The fitting function can be written as $y = 0.0227x^3 + 3.3477x^2 - 163.52x + 2636.6$ with the fitting degree of $R^2 = 0.9978$. The average sensitivity of the proposed FLM temperature sensor is about 0.783 dB/ oC. When using a power meter with the resolution of 0.01dBm, the temperature resolution is obtained 0.013 oC. The measured temperature range is decided by the wavelength spacing between transmission peaks of the FLM which is given by $\lambda^2/\Delta nL$, where λ, Δn, and L are the wavelength, the birefringence and the length of the HiBi fiber, respectively. So we can widen the measured temperature range by shortening the length of the FLM. At the same time, a shorter sensor head is convenient in some applications, even though the sensitivity of the sensor will be decreased.

In order to study the effect of the band-pass filter on the sensing performance, several band-pass filters with a center wavelength 1545 nm are used, whose FWHM are 8.3 nm, 6.66 nm and 3 nm, respectively. A serial of experiments are carried out in the same way. Fig.28 shows the relationships of the output intensity and temperature when different filters are used. It is clear that all of them are periodic function and temperature period is 16 oC. The working range of the positive variation region is 8 oC, which is consistent with that of the filter with FWHM 12.66nm. This is because that the wavelength spacing between the transmission peaks of the FLM is not changed all the times. However, they have shown two marked differences: monotone interval and the output intensity range. Table I shows the properties of the proposed FLM temperature sensor when different filters are used.

From Table I, it can be concluded that the average sensitivity of the sensor is higher and the starting temperature of the monotone interval is lower when a filter with a narrower bandwidth is chosen. The reason is that, the transmission peak of the FLM shifts outside the filter band earlier and faster, the monotonicity of the relationship curve will change earlier and the output intensity range will vary larger, and the average sensitivity is also lager because of the same temperature range. When the filter is chosen a FWHM 3 nm with a center wavelength at 1545nm, the average sensitivity of the proposed FLM temperature sensor is about 1.742dB/oC. When a power meter with the resolution of 0.01dBm is used, the temperature resolution is obtained 0.006 oC.

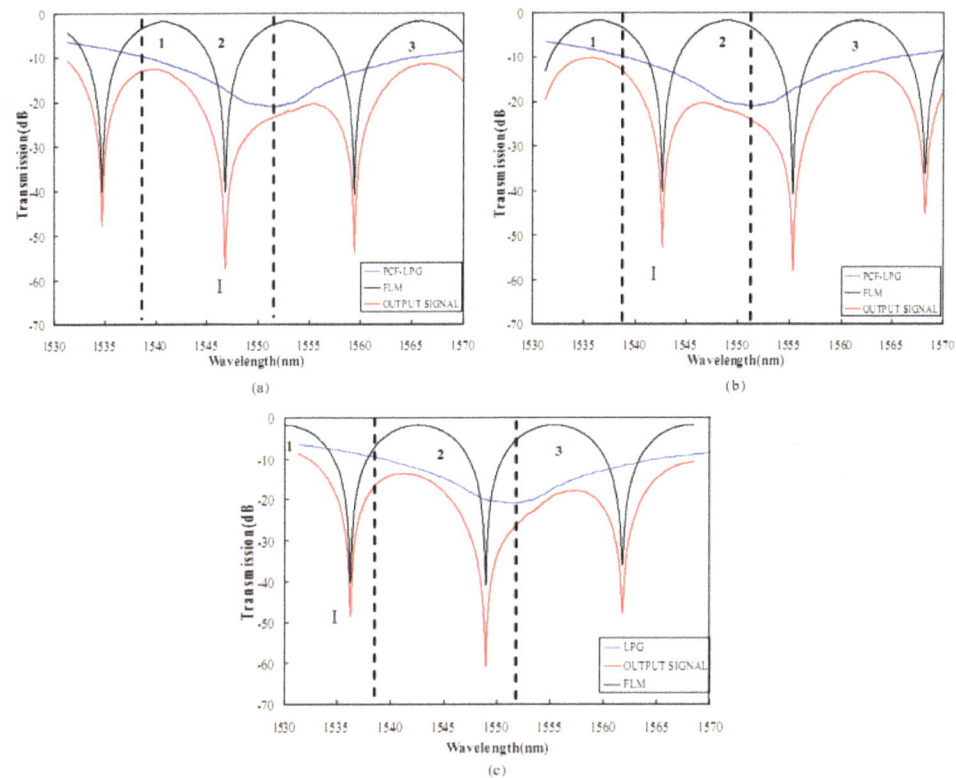

Fig. 27. Transmission spectra of the HiBi-FLM, the PCF-LPG and the output signal when temperature applied on the sensor element is (a) 40°C, (b) 45.5°C, and (c) 53.5°C.

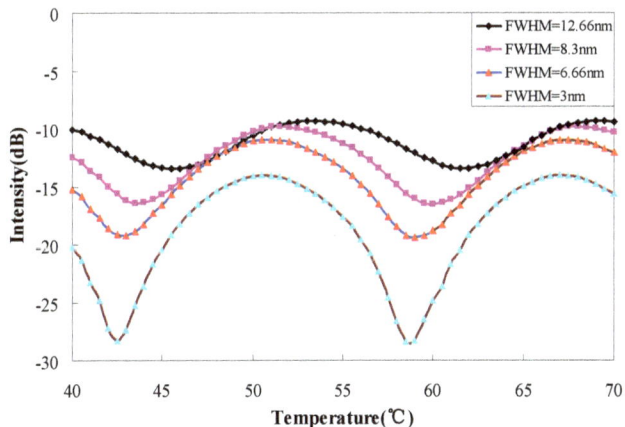

Fig. 28. The relationships of the output power and temperature when different filters are used.

TABLE 1. Properties of the proposed FLM temperature sensor when different filters are used.

FWHM	12.66 nm	8.3 nm	6.66 nm	3 nm
Properties				
Monotone interval	45.5 ~ 53.5 °C	43.5 ~ 51.5 °C	43 ~ 51 °C	42.5 ~ 50.5 °C
Output intensity range	−13.4664 dB ~ −9.3291 dB	−16.4375 dB ~ −9.8497 dB	−19.2807 dB ~ −10.9793 dB	−28.3572 dB ~ −14.0612 dB
The fitting function	$y = -0.0227x^3 - 3.3477x^2 - 163.52x - 2636.6$	$y = -0.0249x^3 - 3.5056x^2 - 163.51x - 2509.1$	$y = -0.0188x^3 - 2.5377x^2 - 112.84x - 1632.7$	$y = 0.0068x^3 - 1.2038x^2 - 69.702x - 1337.4$
R^2	0.9978	0.9988	0.9987	0.9979
Average sensitivity	0.783 dB/ °C	1.082 dB/ °C	1.222 dB/ °C	1.742 dB/ °C

Table 1. Properties of the proposed FLM temperature sensor when different filters are used

5.4 Demodulation based on a PCF-LPG with differential processing for a fiber loop mirror temperature sensor [47]

Demodulation systems based on intensity measurement are widely used in optical sensors with simple structures and low costs. However, the sensing accuracy may be decreased since fiber sensor systems often include some noises due to light source fluctuations and all of power fluctuations in the system. In this part, the demodulation based on a PCF-LPG with differential processing for FLM temperature sensor was demonstrated to eliminate the noise of the sensor and raise the sensor accuracy. Utilizing the two linear regions of the transmission spectrum of the PCF-LPG, the variations of the FLM's two resonant wavelengths, which are located respectively within the positive and negative linear region of the LPG's transmission spectrum, are transferred effectively to the intensity variations at the same time. By differential processing of the two signals separated by two band-pass filters, the noise of the sensor is eliminated effectively since the two signals (also including the noise of fiber sensors) transmit through the same path. Experimental results show that the FLM temperature sensor with demodulation based on differential processing of the PCF-LPG is stable.

Fig.29 shows the experimental setup of a HiBi-FLM temperature sensor with the proposed demodulation system based on the PCF-LPG with differential processing. A broadband SLED launches into the HiBi-FLM sensor head. The transmitted light from the HiBi-FLM enters the PCF-LPG, and the output light is split into two beams by a 3 dB coupler, then the two beams enters the signal processing units via two band-pass filters, respectively. The demodulation system is constructed using a PCF-LPG, a 3 dB coupler and two band-pass filters with the full width at half maximum (FWHM) 6.66nm whose center wavelength are 1542nm and 1561.77nm. By utilizing the stable filtering function of the PCF-LPG, the variations of the FLM's resonant wavelengths with temperature, which are located within the positive and negative linear regions of the LPG's transmission spectrum, respectively are transferred effectively to the intensity variations simultaneously. The intensity signals provided by the PCF-LPG are separated effectively by the two band-pass filters whose transmission spectra are shown as the gray regions in Fig. 30. Both signals W1 and W2 are related with the measurand, while they also may include the all of fluctuation of the sensor system. When monitoring W1 or W2, we can obtain the information of temperature applied on the FLM sensor, but the accuracy will be low due to the effect of noises. W1 and W2 are gotten at the same condition since the light from the broadband source passes through the same path and input to the monitor. By use of the differential processing $\Delta W = (W1-W2) / (W1 + W2)$, the measurement is free from the effect of power fluctuations of the light source and any other noises.

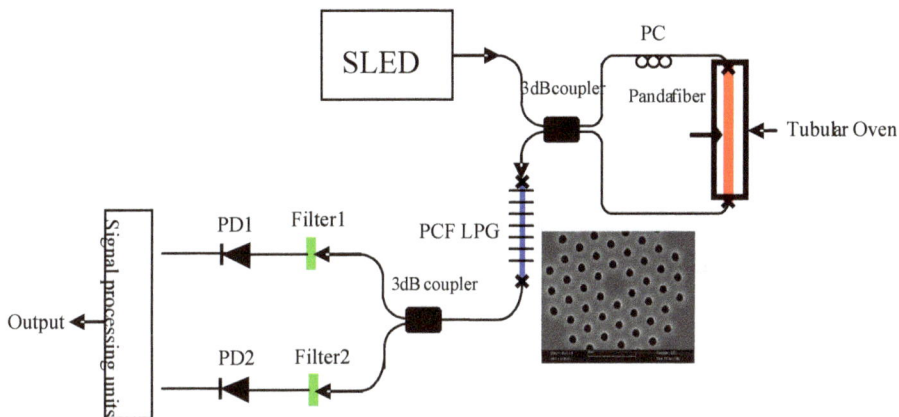

Fig. 29. Experimental setup of the proposed sensor

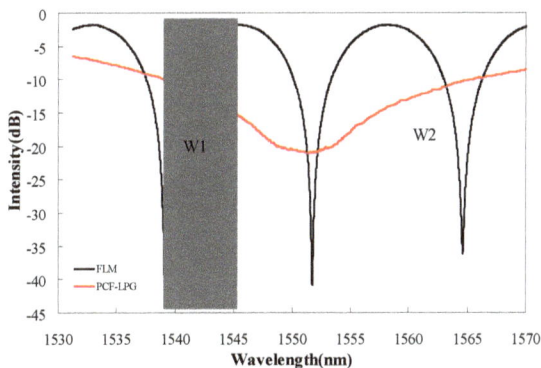

Fig. 30. Transmission spectra of the FLM, the PCF-LPG and the two band-pass filters

Fig.31 shows the relationship of the intensity signal W1 located within a negative linear region of the LPG's transmission spectrum with temperature applied on the FLM. When temperature changes from 47°C to 55°C, the intensity signal W1 will increase with temperature increasing, the sensitivity of the intensity signal W1 to temperature is about 0.01034 mw/°C, and the fitting function can be written as $Y=0.0003X^3-0.0468X^2+2.3668X-39.652$ with the fitting degree of $R2 = 0.9984$. While the intensity signal W2 will decrease when temperature increasing, the sensitivity of which is about -0.00882 mw/°C, and the fitting function can be written as $Y= -0.0004X^3 + 0.0572X - 2.8888 + 48.448$ with the fitting degree of $R2 = 0.9982$, as shown in Fig. 32. Fig. 31 and Fig. 32 also show the relationship of the intensity signal W1 and W2 with temperature when the power of the light source increases 10% and decreases 10%. It is clear that the measurement will deviate from true value and the accuracy of temperature sensor is low.

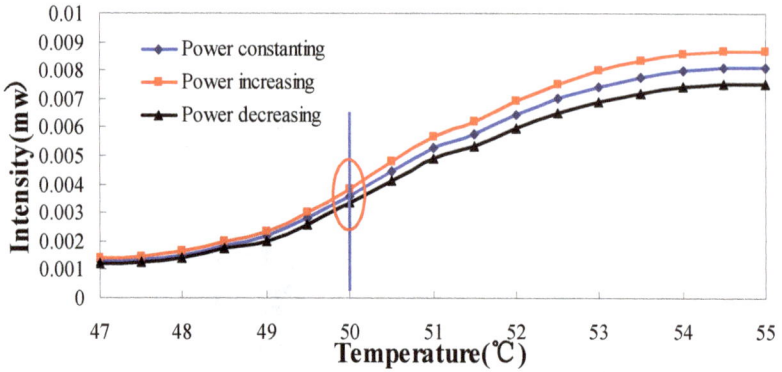

Fig. 31. Relationship of W1 and temperature when the power of light source changes

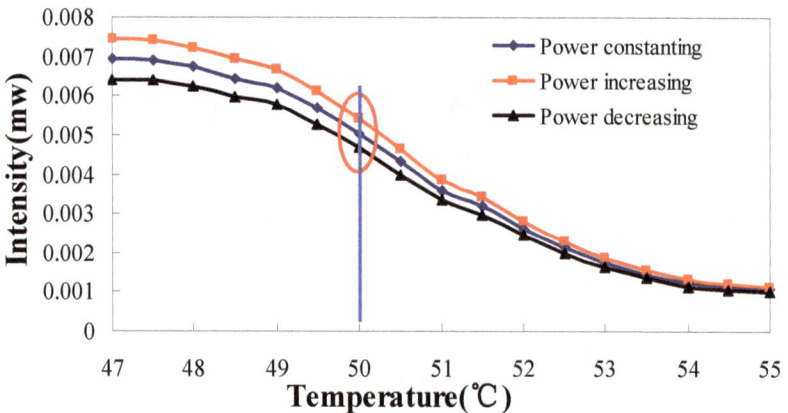

Fig. 32. Relationship of W2 and temperature when the power of light source changes.

Fig.33 shows the relationship of the differential ΔW with temperature. The fitting function can be written as $Y = -0.0075X^3 + 1.138X^2 - 57.382X + 959.39$ with the fitting degree of $R2 = 0.9976$ when temperature rises from 47°C to 55°C. When the power of the light source increases 10%, keeps constant and decreases 10%, the relationship of the differential ΔW with temperature remains stable and the differential algorithm based on a PCF-LPG eliminates the noises effectively. When the power of the light source changes ±10%, the measured temperature is 49.83 °C and the relative error is 0.34%. Compared with the result without the differential ΔW, the accuracy of the HiBi-FLM sensor rises from ~90.5% to ~99.7%.

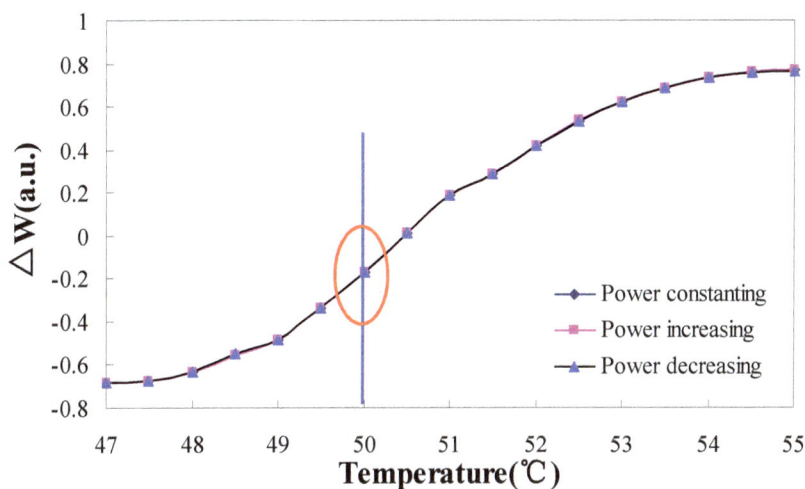

Fig. 33. Relationship of ΔW and temperature when the power of light source changes.

6. Summary

In this chapter, we have presented the basic operation principle of LPGs, and demonstrated the special properties of PCF-LPGs. By use of a dispersion factor γ, a deeper understanding of the behavior of LPG in the ESM-PCF has been achieved. Both the theoretical and experimental results clearly reveal the significant effect of the waveguide dispersive characteristics of the cladding modes on the strain and temperature characteristics of the LPG in the ESM-PCF. By selecting proper grating period, it is possible to design a LPG with specific strain and temperature properties.

We have shortly reviewed PCF-LPG fabrication methods and mainly described the fabrication method by using a CO_2 laser. The PCF-LPG fabricated under the theoretical design shows a good agreement with the theoretical predictions. Finally, applications of the PCF-LPG in optical fiber sensors have been demonstrated and discussed fully.

7. References

[1] T. A. Birks, J. C. Knight, and P. St. J. Russell, "endlessly single-mode photonic crystal fiber", *Opt. Lett.* vol. 22, no. 13, pp. 961-963, Jul. 1997.

[2] J. C. Knight, J. Broeng, and T. A. Birks, "Photonic band gap guidance in optical fibers" *Science*, vol. 282, no. 5393, pp.1476-1478, Nov. 1998.

[3] J. C. Knight. T. A. Birks, R. F. Cregan, P. St. J. Russell, And J. –P. de Sandro, " Large mode area photonic crystal fiber", *Electon. Lett.* vol. 43, no. 13, pp. 1347, Jul. 1998.

[4] Ortigosa-Blanch, J. C. Knight, W. J. Wadsworth, J. arriaga, B. J. ManganT. A. Birks, and P. St. J. Russell, "Highly birefringent photonic crystal fibers", *Optics Lett.*, vol. 25, no. 18, pp. 1325-1327, Sep. 2000.

[5]   T. A. Birks, D. Mogilevstev, J. C. Knight, and P. St. J. Russell, "Dispersion compensation using single-material fibers", *IEEE Photo. Technol. Lett.*, vol. 11, no. 6, pp. 674-676, June 1999.

[6]   J. K. Ranka, R. S. Windeler, and A. J. Stentz, "Optical properties of high-delta air silica microstructure optical fibers", *Opt. Lett.* vol. 25, no. 11, pp. 796-798, June 2000.

[7]   J. Ju, W. Jin, M. S. Demokan, "Design of single-polarization single-mode photonic crystal fiber at 1.30 and 1.55 μm", *J. Lightwave Technil.* vol. 24, no. 2, pp. 825-830, Feb. 2006.

[8]   Charles M. Jewart, Sully Mejía Quintero, Arthur M. B. Braga, and Kevin P. Chen1, "Design of a highly-birefringent microstructured photonic crystal fiber for pressure monitoring", *Optics Express*, vol.18, no. 25, pp. 25657, 2010.

[9]   V. Husakou and J. Hermann, "Supercontinuum generation, four-wave mixing, and fission of higher-order solitons in photonic-crystal fibers", *J. Opt. Soc. Am. B*, vol. 19, no. 9, pp. 2171-2182, Sep. 2002.

[10] J. E. Sharping, M. Fiorentino, A. Coker, P. Kumar, R. S. Windeler, "Four-wave mixing in microstructure fiber", *Opt. Lett.* vol. 26, no. 14, pp. 1048-1050, June 2001.

[11] W. J. Wadsworth, J. C. Knight, W. H. Reeves, P. St. J. Russell and J. Arriaga, "Yb^{3+} - doped photonic crystal fiber laser", *Electron. Lett.* Vol. 36, no. 17, pp. 1452-1454, 2000.

[12] Z. Yusoff, J. H. Lee, W. Belardi, T. M. Monro, P.C. The, and D. J. Richardson, "Raman effects in a highly nonlinear holey fiber: amplification and modulation", *Opt. Lett.*, vol. 27, no. 6, pp. 424-426, Mar. 2002.

[13] C.-L. Zhao, Z. Li, X. Yang, C. Lu, W. Jin, M.S. Demokan, "Effect of a nonlinear photonic crystal fiber on the noise characterization of a distributed Raman amplifier", *IEEE Photon. Technol. Lett.*, vol. 17, no. 3, pp. 561-563, Mar. 2005.

[14] C.-L. Zhao, X. Yang, C. Lu, W. Jin, M.S. Demokan, "Temperature- insensitive interferometer using a highly berefringent photonic crystal fiber loop mirror", *IEEE Photon. Technol. Lett.*, vol. 16, no. 11, pp. 2535-2537, Nov. 2004.

[15] W. Qian, C.-L. Zhao, S. He, X. Dong, S. Zhang, Z. Zhang, S. Jin, J. Guo and H. Wei, "High-sensitivity temperature sensor based on an alcohol-filled photonic crystal fiber loop mirror", *Opt. Lett.*, vol. 36, pp. 1548-1550, (2011).

[16] C. Kerbage, R. S. Windeler, B. J. Bggleton, P. Mach, M. Dolinski, and J. A. Rogers, "Tunable devices based on dynamic positioning of micro-fluids in micro-structured optical fiber", *Opt. Commun.*, vol. 204, no. 1-6, pp. 179-184, April 2002.

[17] W. Qian, C.-L. Zhao, Y. Wang, C. C. Chan, S. Liu, and W. Jin, "Partially liquid-filled hollow-core photonic crystal fiber polarizer", *Opt. Lett.*, vol. 36, no. 16, pp. 3296-3298, 2011.

[18] Y. L. Hoo, W. Jin, C. Shi, H. L. Ho, D. N. Wang, and S. C. Ruan, "Design and modeling of a photonic crystal fiber gas sensor", *App. Optics*, vol. 42, no. 18, pp. 3509-3515, Sep. 2003.

[19] N. Burani and J. Lægsgaard, "Perturbative modeling of Bragg-grating-based biosensors in photonic-crystal fibers", *J. Opt. Soc. Am. B*, vol. 22, no. 11, pp. 2487-2493, Nov. 2005.

[20] M. Vengsarkar, P. J. Lemaire, J. B. Judkins, V. Bhatia, T. Erdogan, and J. E. Sipe, "Long-period fiber gratings as band-rejection filters", *J. Lightwave Technol.*, vol. 14, no. 1, pp. 58-64, Jan. 1996.

[21] C.-L. Zhao, H.-Y. Tam, B.-O. Guan, X. Dong, P. K. A. Wai, X. Dong, "Optical Automatic Gain Control of EDFA Using Two Oscillating Lasers in a Single Feedback Loop", *Opt. Commun.*, vol. 225, no. 1-3, pp. 157-162, Sep. 2003

[22] M. Vengsarkar, P. J. Lemaire, G. Jacobovitz-Veselka, V. Bhatia, and J. B. Judkins, "Long-period fiber gratings as gain-flattening and laser stabilizing devices", *in Proc. IOOC'95*, June 1995, PD1-2.

[23] V. Bhatia and A. M. Vengsarkar, "Optical fiber long-period grating sensor", *Opt. Lett.*, vol. 21, no. 9, pp. 692-694, May 1996.

[24] V.Bhatia, "Applications of long-period gratings to single and multi-parameter sensing", *Optics Express*, vol. 3, no. 11, pp. 457, May 1999.

[25] H. J. Patrick, A. D. Kersey, and F. Bucholtz, "Analysis of the response of long period fiber gratings to external index of refraction," *J. Lightwave Technol.*, vol. 16, pp. 1606–1612, Sept. 1998.

[26] B. J. Eggleton, P. S. Westbrook, R. S. Windeler, S. Spalter, and T. A. Strasser, "Grating resonances in air silica microstructured optical fibers", *Opt. Lett.* vol. 24, no. 21, pp. 1460-1462, Nov. 1999.

[27] K. Morishita and Y. Miyake, "Fabrication and resonance wavelengths of long-period gratings written in a pure-silica photonic crystal fiber by the glass structure change", *J. Lightwave Technol.*, vol. 22, no. 2, pp. 625-630, Feb. 2004.

[28] J. Albert, M. Fokine and W. Margulis, "Grating formation in pure silica-core fibers," *Opt. Lett.*, vol. 27, no. 10, pp. 809- 811, May 2002.

[29] G. Kakarantzas, T. A. Birks, and P. St. J. Russell, "Structural long-period gratings in photonic crystal fibers," *Opt. Lett.*, vol. 27, no. 12, pp.1013-1015, June, 2002.

[30] Y. Wang, W. Jin, J. Ju, H. Xuan, H. L. Ho, L. Xiao, and D. Wang, "Long period gratings in air-core photonic bandgap fibers", *Optics Express*, vol. 16, no. 4, pp. 2784-2790, 2008.

[31] H. Dobb, K. Kalli and D.J. Webb, Temperature-insensitive long period grating sensors in photonic crystal fibre, *Electron. Lett.*, vol. 40, no. 11, pp. 657 – 658, May 2004.

[32] Y. Zhu, P. Shum, H. –W. Bay, M. Yan, J. Hu, J. Hao, and C. Lu, "strain-insensitive and high-temperature long-period gratings inscribed in photonic crystal fiber", *Opt. Lett.*, vol. 30, no. 4, pp. 367-369, Feb. 2005.

[33] J. H. Lim, K. S. Lee, J. C. Kim, and B. H. Lee, "Tunable fiber gratings fabricated in photonic crystal fiber by use of mechanical pressure", *Opt. Lett.*, vol. 29, no. 4, pp. 331-333, Feb. 2004.

[34] Diez, T. A. Birks, W. H. Reeves, B. J. Mangan, and P. St. J. Russell, "Excitation of cladding modes in photonic crystal fibers by flexural acoustic waves", *Opt. Lett.* vol. 25, no. 20, pp. 1499, Oct. 2000.

[35] D.-I Yeom, P. Steinvurzel, B. J. Eggleton, S. D. Lim, and B. Y. Kim, "Tunable acoustic gratings in solid-core photonic bandgap fiber", *Optics Express*, vol. 15, no. 6, pp. 3513-3518, 2007.

[36] S. Liu, L. Jin, W. Jin, D. Wang, C. Liao, Y. Wang, "Structural long period gratings made by drilling micro-holes in photonic crystal fibers with a femtosecond infrared laser", *Optics Express*, vol. 18, no. 6, pp. 5496-5503, 2010

[37] L. Jin, W. Jin, J. Ju, and Y. Wang, "Investigation of Long-Period Grating Resonances in Hollow-Core Photonic Bandgap Fibers", *J. Lightwave Technol.*vol. 29, no. 11, pp.1707-1713, 2011.

[38] C.-L. Zhao, L. Xiao, J. Ju, M.S. Demokan, W. Jin, "Strain and temperature characteristics of a long- period grating written in a photonic crystal fiber and its application as a temperature-insensitive strain sensor", *J. Lightwave Technol.* 26（2）: 220-227，2008.

[39] K. S. Hong, H. C. Park, and B. Y. Kim, I. K. Hwang, W. Jin and J. Ju, D. I. Yeom, "1000 nm tunable acousto-optic filter based on photonic crystal fiber," Appl. Phys. Lett. **92**, 031110 (2008).

[40] Y.-J. Rao, D.-W. Duan, Y.-E. Fan, T. Ke, and M. Xu, "High-Temperature Annealing Behaviors of CO2 Laser Pulse-Induced Long-Period Fiber Grating in a Photonic Crystal Fiber", *J. Lightwave Technol*, vol. 28, no.10, pp. 1531-1535, 2010.

[41] L. Jin, W. Jin, and J. Ju, "Directional Bend Sensing With a CO2 -Laser-Inscribed Long Period Grating in a Photonic Crystal Fiber", *J. Lightwave Technol*, vol. 27, no. 21, pp. 4884-4891, 2009

[42] C.-L. Zhao, J. Zhao, W. Jin, J. Ju, L. Cheng, X. Huang, "Simultaneous Strain and Temperature Measurement Using a Highly Birefringence Fiber loop Mirror and a Long Period Grating Written in a Photonic Crystal Fiber", *Opt. Communications*, 282: 4077-4080, 2009

[43] Z. He, Y. Zhu, and H. Du, "Long-period gratings inscribed in air- and water-filled photonic crystal fiber for refractometric sensing of aqueous solution", *Appl. Physics Lett.* vol. 92, pp. 044105, 2008.

[44] Y. Zhu, Z. He, H. Du, "Detection of external refractive index change with high sensitivity using long-period gratings in photonic crystal fiber", *Sensors and Actuators B*, vol. 131, pp. 265–269, 2008.

[45] C.-L. Zhao, M.S. Demokan, W. Jin, L. Xiao, "A cheap and practical FBG temperature sensor utilizing a long-period grating in a photonic crystal fiber", *Opt. Communications*, 276: 242, 2007

[46] Y. Wang, C.-L Zhao, X. Dong, J. Kang, and S. Jin. "A Fiber Loop Mirror Temperature Sensor Demodulation Technique Using a Long-period Grating in a Photonic Crystal Fiber and a Band-pass Filter", *Review of Scientific Instruments*, vol.82, pp.073101, July, 2011.

[47] Y. Wang, C.-L. Zhao, X. Dong, J. Kang, S. Jin and C. C. Chan. "Demodulation Based on a Long-period Grating in Photonic Crystal Fiber with Differential Processing for High-Birefringence Fiber Loop Mirror Temperature Sensor", International Conference on Optical Fiber Sensors (OFS), Proceedings of SPIE, Vol. 7753, no. 77539c-3, Ottawa, Canada, May 15-19, 2011.

[48] X. Shu, L. Zhang, and I. Bennion, "Sensitivity characteristics of long-period fiber gratings," *J. Lightw. Technol.*, vol. 20, no. 2, pp. 255–266, Feb. 2002.

[49] Bertholds and R. Dandliker, "Determination of the individual strain-optic coefficients in single-mode optical fibers," *J. Lightw. Technol.*, vol. 6, no. 1, pp. 17–20, Jan. 1988.

[50] D.B. Mortimore, "Fiber loop reflectors", *J. Lightwave Technol.* vol. 6, pp. 1217-1224, 1988

[51] Y. Liu, B. Liu, X. Feng, W. Zhang, G. Zhou, S. Yuan, G. Kai, and X. Dong, "High-birefringence fiber loop mirrors and their applications as sensors, " *Appl. Opt.* vol. 44, pp. 2382-2390, 2005

Part 4

Design and Modeling

13

Overview of Computational Methods for Photonic Crystals

Laurent Oyhenart and Valérie Vignéras
IMS Laboratory, CNRS, University of Bordeaux 1
France

1. Introduction

A photonic crystal (PC) is a periodic structure whose refraction index of the material is periodically modulated on the wavelength scale to affect the electromagnetic wave propagation by creating photonic band gaps. In 1887, Lord Rayleigh is the first to show a band gap in one-dimensional periodic structures *i.e.* a Bragg mirror. In 1987, Eli Yablonovitch and Sajeev John have extended the band gap concept to the two and three-dimensional structures and for the first time, they use the term "photonic crystal" (Yablonovitch, 1987; John, 1987).

Progress in computational methods for the photonic crystals is understood through an historical review (Oyhenart, 2005). At the beginning of research in the photonic crystals, the purpose was to find a structure with complete band gap by improving the computational methods. In 1988, John shows theoretically by the scalar method of Korringa-Kohn-Rostoker (KKR) that the face centered cubic lattice (FCC) has a complete band gap between the second and the third band. One year later, Yablonovitch builds this structure and finds a band gap experimentally but the W-point raises a problem. In 1990, Satpathy et al. and Leung et al. confirm the complete band gap by the scalar plane wave method (PWM). A few months later, these two teams improve their methods to obtain vectorial PWM on **D** and **E** fields. They find that FCC structure does not have complete band gap because W-point and U-point are degenerate. With these results, the editor of the journal "Nature" writes *"Photonic Crystals bite the dust"* (Maddox, 1990). Only two weeks later, Ho et al. created the vectorial PWM on **H** and they do not find the complete band gap in FCC structure but they show a complete band gap in the diamond lattice. In 1992, Sözuer et al. improve convergence of the PWM and they obtain a complete band gap for FCC lattice between 8th and 9th band. This structure that has caused many discussions has a complete band gap but not where it was expected.

To study and understand the propagation of the electromagnetic fields in the photonic crystals, computational methods were improved by using their symmetries and periodicities. We will study the classical methods for microwave devices such as the finite element method and the finite difference time domain. After some modifications of these methods, we obtain the band structure of PC which can be calculated by the methods from the solid state physics. For example the plane wave method, the tight binding method and the multiple-scattering theory will be studied. All these computational methods will be presented in this article and a PC example will be studied to compare these methods.

2. Equations, symmetries and periodicities in photonic crystals

2.1 Equations in photonic crystals

The Maxwell equations without sources control the electromagnetic wave propagation in PC.

$$\nabla \times \mathbf{E}(\mathbf{r},t) + \frac{\partial \mathbf{B}(\mathbf{r},t)}{\partial t} = 0 \quad \nabla \times \mathbf{H}(\mathbf{r},t) - \frac{\partial \mathbf{D}(\mathbf{r},t)}{\partial t} = 0$$
$$\nabla \cdot \mathbf{D}(\mathbf{r},t) = 0 \qquad \qquad \nabla \cdot \mathbf{B}(\mathbf{r},t) = 0 \tag{1}$$

In these laws, the same physical behavior is observed if we change simultaneously the wavelength and the structure dimensions in the same proportions. Therefore, it is convenient to introduce a normalized wavelength λ_0/a and a normalized frequency $af/c = a/\lambda_0$, with a the lattice constant of the photonic crystal (Joannopoulos et al., 2008).

Some methods do not solve the Maxwell equations directly but they use the Helmholtz equations, for example the E-wave equation or the H-wave equation:

$$\nabla \times [\nabla \times \mathbf{E}(\mathbf{r})] = -\nabla^2 \mathbf{E}(\mathbf{r}) = k_0^2 \varepsilon_r(\mathbf{r}) \mathbf{E}(\mathbf{r})$$
$$\nabla \times \left[\varepsilon_r^{-1}(\mathbf{r}) \nabla \times \mathbf{H}(\mathbf{r}) \right] = k_0^2 \mathbf{H}(\mathbf{r}) \quad \text{with} \quad \begin{cases} \mathbf{E}(\mathbf{r},t) = \mathbf{E}(\mathbf{r})e^{-i\omega t} \\ \mathbf{H}(\mathbf{r},t) = \mathbf{H}(\mathbf{r})e^{-i\omega t} \end{cases} \tag{2}$$

Equations number to be solved depends on dimension of the photonic crystal. In the two-dimensional case, the problem is simplified. It is assumed that the materials are uniform along z-axis. It follows that the fields are uniform and the partial derivatives with respect to the variable z vanish. The previous equation is simplified and split-up into TE-polarization and TM-polarization; we have a scalar equation for each polarization:

$$\frac{\partial^2 F(\mathbf{r})}{\partial x^2} + \frac{\partial^2 F(\mathbf{r})}{\partial y^2} + k_0^2 \varepsilon_r(\mathbf{r}) F(\mathbf{r}) = 0 \quad \text{with} \quad \begin{cases} F = E_z \text{ for TE-polarization} \\ F = H_z \text{ for TM-polarization} \end{cases} \tag{3}$$

The computation time decreases exponentially compared to the three-dimensional case. In the one-dimensional case, the problem is even more simplified. If the materials are uniform along x-axis and y-axis, we solve analytically one second-order equation:

$$\frac{d^2 F(\mathbf{r})}{dz^2} + k_0^2 \varepsilon_r(\mathbf{r}) F(\mathbf{r}) = 0 \quad \text{with} \quad F = E_x, E_y, H_x \text{ or } H_y \tag{4}$$

2.2 Symmetries of photonic crystals

Like all sets of differential equations, Maxwell's equations cannot be uniquely solved without a suitable set of boundary conditions. Photonic crystals have symmetries which define boundary conditions. Symmetries of the structure are not a sufficient condition to reduce the computational domain; the electromagnetic field must be also symmetrical. On figure 1, we study symmetries on a lattice of cylinders for different polarization of the incident wave. In the first case, we apply a TM incident wave on the lattice with a horizontal symmetry and the tangential magnetic field vanishes on the axis of symmetry. We put perfect magnetic conductor (PMC) as boundary condition on the symmetry axis and we study only half top or bottom of the problem. In the second case, we apply a TE incident wave. Similarly, we reduce the problem with a perfect electric conductor condition (PEC).

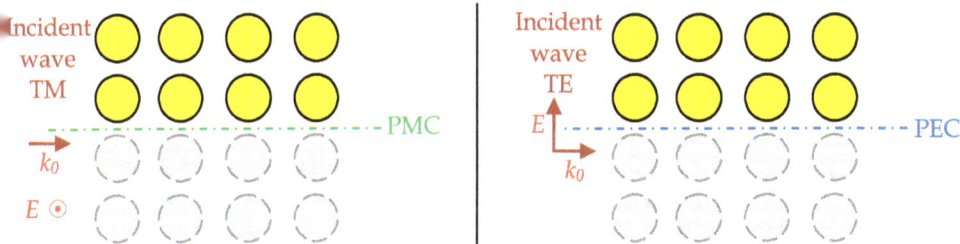

Fig. 1. Both kinds of lateral symmetries

On figure 1, the geometry has also a vertical symmetry. However, the electromagnetic field is not symmetrical because the incident wave comes only from the left side. A solution is to divide the incident wave into an even mode and an odd mode (figure 2). For the even mode, we have a vertical symmetry plane of the electric field. Only the left or right part is solved with a perfect magnetic conductor condition on the axis of symmetry. The antisymmetric problem is reduced in the same way with a perfect electric conductor condition. The sum of the symmetric and antisymmetric problem provides the solution of the total problem. The reflection of total problem is the sum $r_{even} + r_{odd}$ and the transmission is the subtraction $r_{even} - r_{odd}$. Two half-problems are solved more quickly than the whole problem because the computation time increases exponentially with the size of the problem.

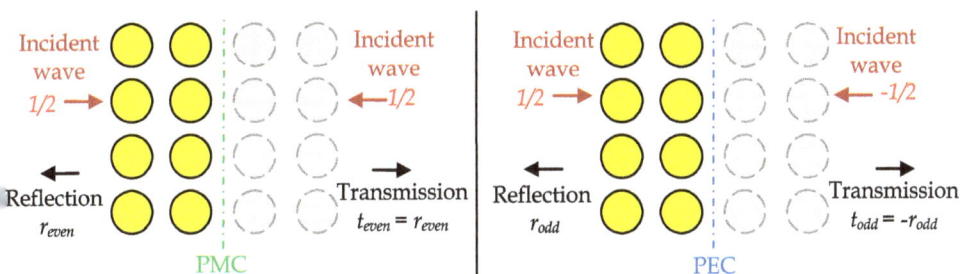

Fig. 2. Transverse symmetries of the problem

2.3 Periodicities of photonic crystals

Photonic crystals, like the familiar crystals of atoms have discrete translational symmetry $\mathbf{T}$. The dielectric function is periodic therefore the electric and magnetic fields can be written as the product of a plane wave envelope and a periodic function:

$$\varepsilon(\mathbf{r}) = \varepsilon(\mathbf{r} + n\mathbf{T})$$
$$\mathbf{F}(\mathbf{r}) = e^{i\mathbf{K}\cdot\mathbf{r}}\mathbf{u}(\mathbf{r}) \text{ with } \mathbf{u}(\mathbf{r}) = \mathbf{u}(\mathbf{r} + \mathbf{T})$$

(5)

The result above is commonly known as Bloch's theorem in solid state physics (Kittel, 2005).

If we apply these conditions to an infinite lattice of cylinders (figure 3a), the calculation of band structure is reduced to the study of a single cylinder.

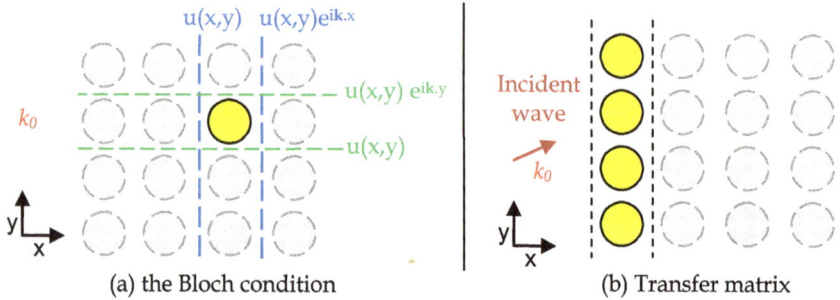

(a) the Bloch condition   (b) Transfer matrix

Fig. 3. Periodic boundary on photonic crystals

Figure 3b is a lattice of cylinders, finite along y-axis and infinite along x-axis for studying the transmission of the structure. If we know the transfer matrix of one layer, it is easy to find the transmission of the total structure thanks to the transfer matrices method (TMM). It reduces the computational domain to one layer. It will be detailed in the next sections.

By mixing TMM along y-axis and Bloch's conditions along x-axis, the structure of figure 3 can be reduced to a single cylinder. Figure 4 summarizes these techniques in the 3D case.

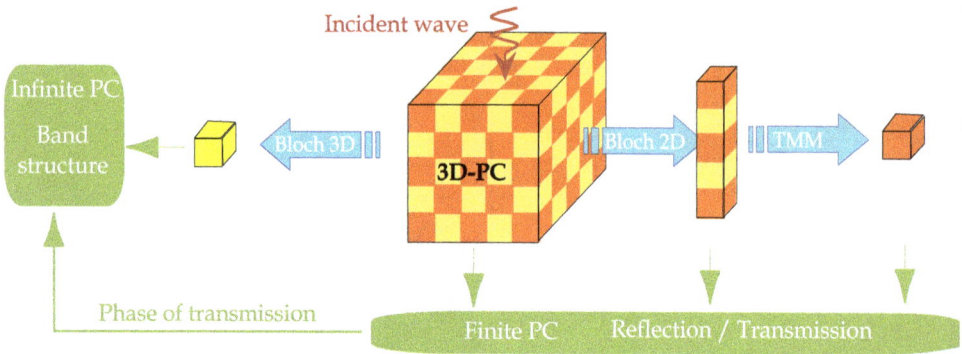

Fig. 4. Reduction of the calculation domain in PC

3. Finite element method (FEM)

The finite element method is very popular in mechanics and civil engineering. It was originally developed in the 40s to solve problems of mechanical structures. A few years later, it has been applied to electromagnetism. Since then, the finite element method extends to all branches of physics and engineering where there exists a partial differential equation (PDE) with boundary conditions. It can be formulated from the variational method or the weighted residual method. We will present in outline the second method which is simpler.

Let us consider a partial differential equation with $\Im$ the differential operator of order n applied to a function φ and a source function f:

$$\Im\varphi = f \tag{6}$$

The first step is to expand the function φ on a set of functions:

$$\varphi = \sum_{j=1}^{N} c_j v_j \tag{7}$$

v_j are the chosen expansion functions and c_j are constant coefficients to be determined. The best solution of the equation 6 is obtained when the residue $r = \Im\varphi - f$ is weakest on all points of the domain Ω. The weighted residual method requires this condition:

$$R_i = \int_{\Omega} w_i r d\Omega = \sum_{j=1}^{N} c_j \int_{\Omega} w_i \left(\Im v_j - f \right) d\Omega = 0 \tag{8}$$

If we use the Galerkin method, the weight functions w_i are the functions of previous interpolations and we write the problem in matrix form:

$$\sum_{j=1}^{N} c_j L_{ij} = b_i \text{ with } L_{ij} = \int_{\Omega} v_i \Im v_j d\Omega \text{ and } b_i = \int_{\Omega} v_i f d\Omega \tag{9}$$

So now, we have to solve a linear system with c_j unknowns where most of the entries of the matrix L are zero. Such matrices are known as sparse matrices, there are efficient solvers for such problems.

The basic idea of the finite element method is to divide the computation domain into small subdomains, which are called finite elements, and then use simple functions, such as linear and quadratic functions, to approximate the unknown solution over each element. For plane geometries, the domain is divided into finite triangular sub-domains. For three-dimensional problems, the sub-domains are tetrahedra. These two- and three-dimensional finite elements are widely used because it is a variable mesh and adapts to curved structures.

3.1 Transmission calculation of a photonic crystal

The finite element method usually solves the E-wave equation for PC (equation 5):

$$\nabla \times [\nabla \times E(r)] = k_0^2 \varepsilon_r(r) E(r) \tag{10}$$

For most of the finite elements computer programs, the frequency is fixed and the electric field is the unknown (Massaro et al., 2008). We use a commercial software, Ansys HFSS (High Frequency Structure Simulator). This three-dimensional computer program builds an adaptive tetrahedral mesh to model microwave devices, for example microstrips and antennas. These devices have a characteristic length lower than the wavelength of study. It is more difficult to study PC because it has a periodicity close to the wavelength and the matrix of calculation is large. To simplify calculations, we will study a periodic PC according to two directions of space. The directions where PC is infinitely periodic require to study only one period according to this direction. Several methods exist to model these conditions.

In the first method, the source is a plane wave and we use periodic boundary conditions (PBC) on lateral faces and absorbing boundary conditions (ABC) on bases. We study only one lateral period (figure 5). If the source has an oblique incidence, the phase shift is easily taken into account by PBC because the E-field is a complex vector in FEM.

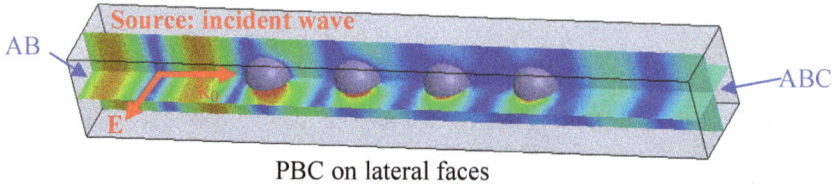

PBC on lateral faces

Fig. 5. Calculation of E-field in the PC with PBC, ABC and incident wave

The second method requires a very different source, a wave port. This source is a semi-infinite waveguide whose cross section is drawn on bases of structure. Propagative modes of this fictitious guide will be the source of the structure. For a plane wave source, we apply to lateral faces and on the fictitious guide the conditions PEC and PMC (figure 6). The source is transverse and has a normal incidence. It is not possible to change the angle of incidence without change the boundary conditions of the lateral faces. The PC studied on figure 5 can be reduced to quarter-spheres thanks to the symmetry of the fields at normal incidence.

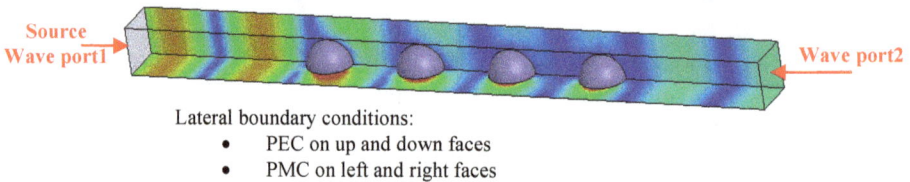

Lateral boundary conditions:
- PEC on up and down faces
- PMC on left and right faces

Fig. 6. Calculation of E-field in the PC with PEC, PMC and wave ports

3.2 Band structure calculation

The band structure shows the states which propagate in a PC. These states are differentiated by their frequency and their Bloch wave vector. FEM sets the wave vector and solves the wave equation to find the frequencies. There is no source, only boundary conditions to set because it is an eigenvalue problem. In the case of a cavity resonator, the boundary conditions of the domain are PEC. Whereas, for PC, we choose the Bloch conditions on all faces of the unit cell to set the wave vector. The phase shift of the Bloch conditions is set easily because the fields are complex vectors. The FEM calculates the band structure of dielectric with or without losses, metallic, and metallodielectric PC. Any material can be used by this method, it is the main advantage.

In this chapter, we choose a photonic crystal to study, a cubic lattice with several layers of dielectric spheres which have a permittivity equal to 5.1 and radius equal to 0.4*a (a: lattice constant). For the band structure, the number of layer is infinite, and for transmission calculation, we study four and eight layers. The structure is deliberately simple to be studied by all the methods. The band structure, the transmission coefficient in normal incidence and some modes of dielectric PC are plotted on figure 7. For normal incidence, the first two band gaps are found in the band structure and transmission curves.

The first method with incident waves takes about 7 hours and 960 Mo of working memory on a personal computer to calculate the transmission of 8 layers *i.e.* 8 spheres. Whereas if we use the second method with the wave ports, the memory is reduced to 360 Mo and the CPU time is reduced to 8 min. To obtain this optimization, we reduced the geometry to four quarter-spheres thanks to symmetries and we use the Padé interpolation on frequencies. It is necessary to make optimizations if you want to use FEM.

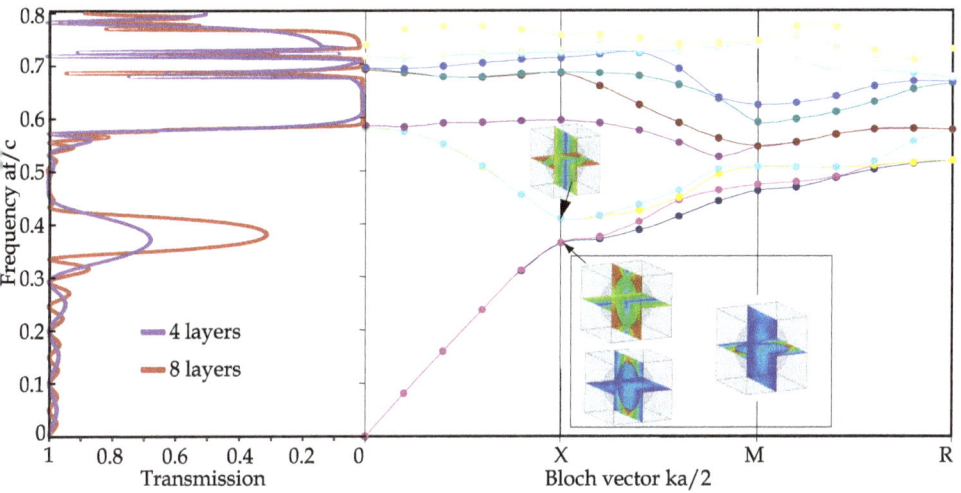

Fig. 7. Band structure and transmission coefficient of dielectric PC

4. Finite-difference time-domain method (FDTD)

Finite difference method is a numerical method for approximating differential equations using finite difference equations to approximate derivatives. This is very simple to implement but it has some limitations in the mesh geometries. In 1966, Yee proposed a finite difference scheme applied to electromagnetism. The FDTD was born (Taflove & Hagness, 2005). The success of this method is due to the scheme based on the Taylor series of second-order. For example, this method is used to model the effects of cellphones on the human body, antennas and printed circuits. The FDTD uses the two structural equations of Maxwell in a conducting, isotropic and homogeneous medium:

$$\frac{\partial \mathbf{E}}{\partial t} = \frac{1}{\varepsilon} \nabla \times \mathbf{H} - \frac{\sigma}{\varepsilon} \mathbf{E}$$

$$\frac{\partial \mathbf{H}}{\partial t} = -\frac{1}{\mu} \nabla \times \mathbf{E}$$

(11)

The FDTD uses an approximation of derivatives by centered finite differences.

$$\frac{d}{dx_i}u(x_i) = \frac{u(x_i+h)-u(x_i-h)}{2h} + o(h^2) \tag{12}$$

Let us apply this approximation to one-dimensional case in order to understand the FDTD principle. The **E** and **H**-field are stepped in time and space. By replacing curls and derivatives, we obtain one-dimensional Yee scheme:

$$\frac{E_k^{n+1/2}-E_k^{n-1/2}}{\Delta t} = -\frac{1}{\varepsilon}\frac{H_{k+1/2}^n-H_{k-1/2}^n}{\Delta x} \quad \text{with} \quad E_k^n = E(n.\Delta t, k\Delta x) = E(t,x)$$
$$\frac{E_{k+1/2}^{n+1}-E_{k+1/2}^n}{\Delta t} = -\frac{1}{\mu}\frac{H_{k+1}^{n+1/2}-H_k^{n+1/2}}{\Delta x} \qquad H_k^n = H(n.\Delta t, k\Delta x) = H(t,x) \tag{13}$$

The **E** and **H**-field are staggered and updated step by step in time. E-field updates are conducted midway during each time-step between successive H-field updates, and conversely. This explicit time-stepping scheme avoids the need to solve simultaneous equations, and furthermore it is order N *i.e.* proportional to the size of the system to model. This scheme can be generalized for two-dimensional and three-dimensional problems. The Yee scheme is stable if the wave propagates from one cell to another with a speed less than the light (the Current-Friedrichs-Lewy condition).

4.1 Calculation of transmission coefficient

FDTD is used to study PC. Calculation domain of finite PC is surrounded by absorbing boundary conditions (ABC). The periodic infinite PC uses periodic boundary conditions (PBC) on the lateral faces (Figure 8). The source is Gaussian with a spectrum which extends on the frequency range to study. The fields are calculated on the time domain and we use the Fourier transform to convert them on frequency domain.

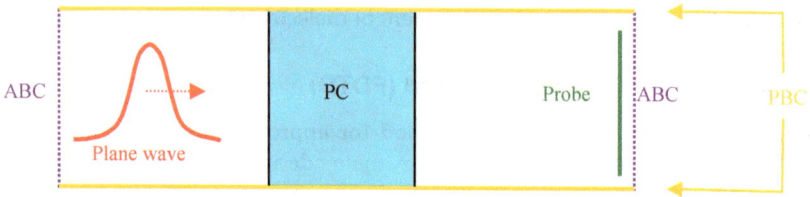

Fig. 8. Source, probe and boundary conditions for infinite PC

For the non-normal incidence, a simple periodic boundary condition cannot be applied but many methods exist to solve this problem. One method is to calculate the computing problem twice, with the sources cos(ωt) and sin(ωt). The addition of two calculations is the solution of the source exp(-iωt). Thanks to this source, we can apply the Bloch conditions on complex vector fields. This method is simplified for the study of band structure.

Figure 9 plot the transmission coefficient for the previous dielectric PC. The transmission calculation of 8 layers is calculated in 32 minutes with 58 Mo of working memory. FDTD is faster than FEM but less accurate.

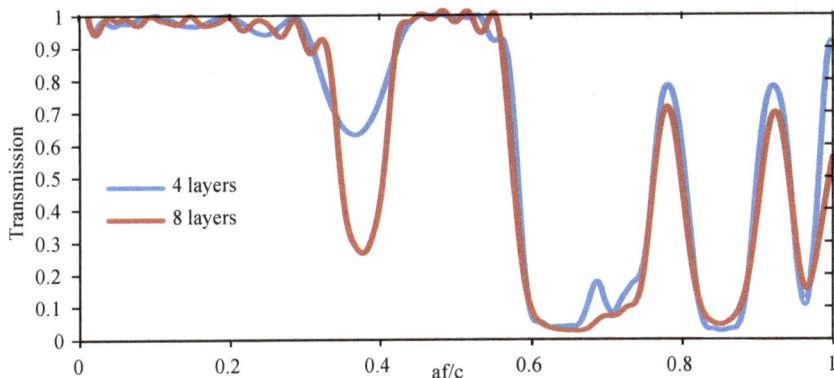

Fig. 9. Transmission coefficient of the dielectric PC

4.2 Calculation of band structure

The band structure is calculated from the eigenmodes but FDTD is not an eigenvalue problem. We use the unit cell of PC with boundary conditions of Bloch and several Gaussian functions for the source. We only need one calculation to apply the Bloch conditions. After 100000 time-steps of calculation, propagative waves are amplified and evanescent waves vanished. If the number of time steps is too small, the transmission peaks are widened, therefore imprecise. To reduce the number of time-steps without affecting the accuracy, we can use the Padé approximation or ADI-FDTD formulation (Taflove & Hagness, 2005). On Figure 10, we plot the electric field amplitude of the unit cell of PC.

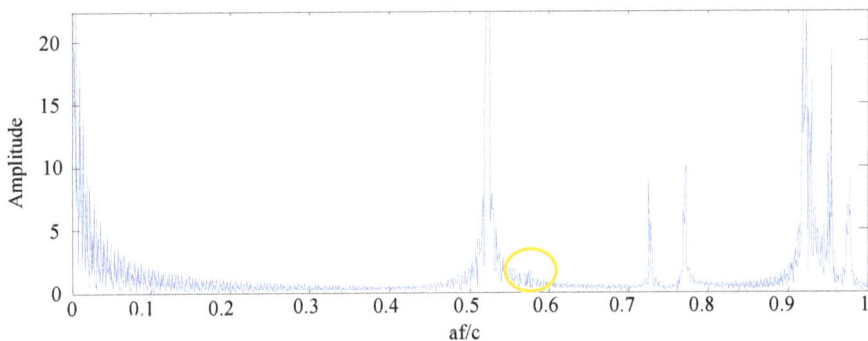

Fig. 10. Electric field amplitude of dielectric PC at the R-point of the Brillouin zone

This curve is similar to the diffraction pattern in solid state physics. It is plotted for all points of the first Brillouin zone to view the band structure. If we compare the curve of figure 10 with the band structure calculated by the FEM, the mode 0.57 is much attenuated and the following mode does not exist on figure 10. In fact, the source was not correctly selected to excite these modes, and so we must ensure that the source excites all available modes.

5. Finite-difference frequency-domain method (FDFD)

We will study a finite difference method combined with transfer matrix method and the Fourier transform. The finite difference method substitutes the derivatives in PDE to obtain finite difference schemes. In electromagnetism, FDFD uses the structural equations of Maxwell on space $(\mathbf{k}, \omega)$ and applies approximations on the wave vector:

$$\begin{cases} \mathbf{k} \times \mathbf{E} = \omega\mu_0\mathbf{H} \\ \mathbf{k} \times \mathbf{H} = -\omega\varepsilon\mathbf{E} \end{cases} \text{with} \begin{cases} k_x \approx \left(e^{ik_x a} - 1\right)/ia \\ k_y \approx \left(e^{ik_y b} - 1\right)/ib \\ k_z \approx \left(e^{ik_z c} - 1\right)/ic \end{cases} \tag{14}$$

a,b,c are lattice constants along x, y and z-axis. We obtain discrete equations which are applied to one layer of the structure on a cubic lattice. In 1992, Pendry and MacKinnon (Pendry & MacKinnon, 1992) used this method with transfer matrix method (TMM) which extended the solution of one layer to the total structure. To apply TMM, the discrete equations must be written on real space. We obtain a set of equations where z-component of the E and H-fields can be removed. The six equations are reduced to four equations which are written in a matrix form:

$$\mathbf{F}(\mathbf{r}+c) = \sum_{\mathbf{r}'} \hat{\mathbf{T}}(\mathbf{r},\mathbf{r}')\mathbf{F}(\mathbf{r}') \text{ with } \mathbf{F}(\mathbf{r}) = \begin{bmatrix} E_x(\mathbf{r}) & E_y(\mathbf{r}) & H_x(\mathbf{r}) & H_y(\mathbf{r}) \end{bmatrix}^{\mathrm{T}} \tag{15}$$

5.1 Calculation of band structure

The band structure is calculated from unit cell of structure by setting frequency and calculating wave vectors propagating in the unit cell. Bloch's theorem applies to vector $\mathbf{F}$:

$$\mathbf{F}(\mathbf{r}+a') = e^{ik_x a'}\mathbf{F}(\mathbf{r})$$
$$\mathbf{F}(\mathbf{r}+b') = e^{ik_y b'}\mathbf{F}(\mathbf{r}) \tag{16}$$
$$\mathbf{F}(\mathbf{r}+c') = e^{ik_z c'}\mathbf{F}(\mathbf{r})$$

$a' = \alpha a$, $b' = \beta b$ and $c' = \gamma c$ are the mesh size along x, y and z-axis. The transfer matrix of the unit cell is written from the unit mesh:

$$\mathbf{F}(\mathbf{r}+c') = \sum_{\mathbf{r}'} \hat{\mathbf{T}}(c',0)\mathbf{F}(\mathbf{r}') \text{ with } \hat{\mathbf{T}}(c',0) = \prod_{j=1}^{N} \hat{\mathbf{T}}(\mathbf{r}_j, \mathbf{r}_{j-1}) \tag{17}$$

If the above equations are joined together, we have an eigenvalue problem:

$$\sum_{\mathbf{r}'} \hat{\mathbf{T}}(c',0)\mathbf{F}(\mathbf{r}') = e^{ik_z c'}\mathbf{F}(\mathbf{r}) \tag{18}$$

We set the wave number $k_{//}$ and the wavelength k_0 to solve the eigenvalue problem. The eigenvalues k_z having an imaginary part are eliminated because they are not propagating waves. The remaining eigenvalues gives us the band structure $k_z(k_0)$.

5.2 Calculation of transmission coefficient

The reflection and transmission calculation is performed using the transfer matrices. It is more interesting to calculate the elements of transmission matrix than the transmitted field in some points. The incident, reflected and transmitted waves are expanded on a plane-wave basis sets thanks to the transfer matrices.

$\hat{T}_0$ is the transfer matrix of a vacuum layer. The eigenvectors of $\hat{T}_0$ define a plane-wave basis set:

$$\hat{T}_0 |r_i\rangle = e^{ik_i c} |r_i\rangle$$
$$\langle l_i| \hat{T}_0 = e^{ik_i c} \langle l_i| \tag{19}$$

The right and left eigenvectors are different because $\hat{T}_0$ operator is not self-adjoint. They are expanded on the reciprocal space k_x and k_y and translate on the direct space with the Fourier transform because the transfer matrix is known in direct space. k_z is found easily because we are in a vacuum layer. The transfer matrix $\hat{T}$ is not compatible with the new plane-wave basis set, we convert this matrix:

$$\langle l_j| \hat{T} |r_i\rangle = \langle l_j| \sum_\alpha \tilde{T}_{\alpha i} |r_\alpha\rangle = \sum_\alpha \tilde{T}_{\alpha i} \langle l_j |r_\alpha\rangle = \tilde{T}_{ji} \tag{20}$$

To calculate the transmission coefficient, this matrix is arranged to group the eigenvectors with the same propagation direction. This matrix is divided into four blocks:

$$\tilde{T} = \begin{pmatrix} \tilde{T}_{11} & \tilde{T}_{12} \\ \tilde{T}_{21} & \tilde{T}_{22} \end{pmatrix} \tag{21}$$

TMM calculates the transmission coefficient of several layers from one layer (see 1D-MST). A great number of layers can be calculated without increasing the computing power.

Figure 11 plot the band structure and the transmission coefficient calculated by FDFD, for the previous dielectric PC. If we compares with figure 7, the first bands are correct but inaccuracies on the higher band after 0.6-frequency are due to the weak mesh (7x7x7 cells). If we increase the number of cells, accuracy increases. Calculation is more difficult for high frequencies because the E- and H-field are functions that oscillate more. The transmission calculation of 8 layers is calculated in 22 seconds with 3.3 Mo of working memory for 7x7x7 cells. The computing time is low because only one sphere is actually calculated thanks TMM.

6. Finite Integration Technique (FIT)

In 1977, Weiland proposes a spatial discretization scheme to solve the integral equations of Maxwell (Weiland, 1977). This scheme called Finite Integration Technique (FIT) can be applied to many electromagnetism problems, in time and frequency domain, from static up to high frequency. The basic idea of this approach is to apply the Maxwell equations in integral form to a set of staggered grids like FDTD (figure 12).

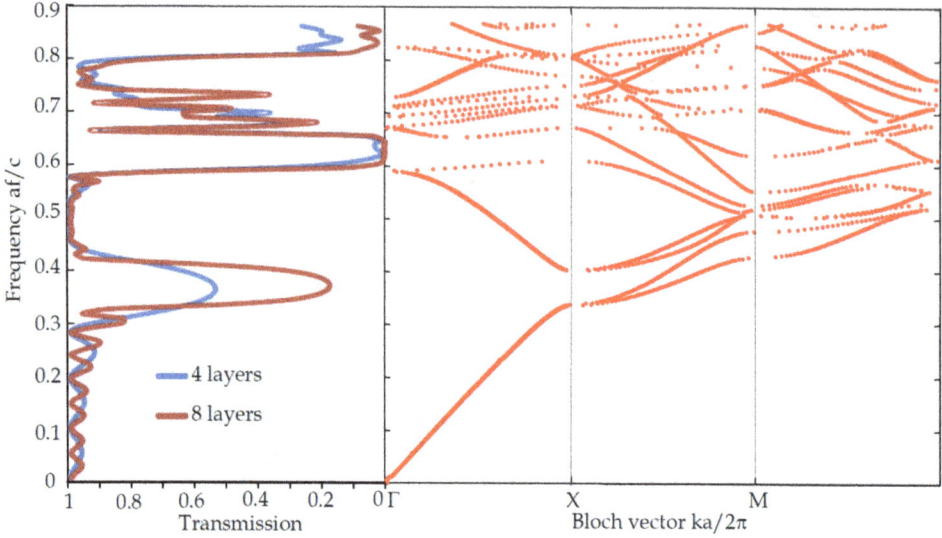

Fig. 11. Band structure and transmission coefficient of dielectric PC

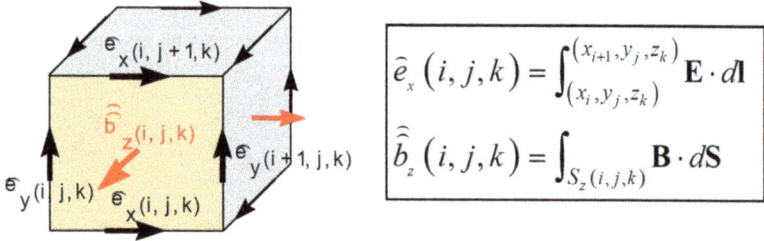

$$\hat{e}_x (i,j,k) = \int_{(x_i,y_j,z_k)}^{(x_{i+1},y_j,z_k)} \mathbf{E} \cdot d\mathbf{l}$$

$$\hat{b}_z (i,j,k) = \int_{S_z(i,j,k)} \mathbf{B} \cdot d\mathbf{S}$$

Fig. 12. Tension and flux component on the mesh

The spatial discretization process is applied to the integral form of the Faraday's law:

$$\int_{\partial S} \mathbf{E}(\mathbf{r},t) \cdot d\mathbf{l} = -\iint_S \frac{\partial}{\partial t} \mathbf{B}(\mathbf{r},t) \cdot d\mathbf{S}$$

$$\hat{e}_x (i,j,k) + \hat{e}_y (i+1,j,k) - \hat{e}_x (i,j+1,k) - \hat{e}_y (i,j,k) = -\frac{\partial}{\partial t} \hat{b}_z (i,j,k)$$

(22)

This last equation is an exact form without approximation. This process is applied to the other Maxwell equations in integral form. CST Microwave Studio is the software based on the FIT. Unlike FDTD, we use the local integral form and we can apply the technique of Perfect Boundary Approximation (PBA) which decreases the meshes on the boundaries.

7. The tight binding method (TB)

The tight binding method (Lidorikis et al., 1998) is less used than the PWM although this method is fast for the calculation of defect states in a PC. By analogy with the TB model in solid-state physics, H-field is expanded on localized wave functions. They are calculated from the Wannier functions $a_n(\mathbf{R},\mathbf{r})$ for a PC without defect:

$$a_n(\mathbf{R},\mathbf{r}) = \frac{\Omega^{\frac{1}{2}}}{(2\pi)^{\frac{3}{2}}} \int_{BZ} \mathbf{H}_n(\mathbf{k},\mathbf{r})e^{-i\mathbf{k}\cdot\mathbf{R}}d\mathbf{k} \tag{23}$$

The Wannier functions are a complete set of orthogonal functions which is defined for each band and each unit cell. The Wannier functions only depend on the quantity $(\mathbf{r}-\mathbf{R})$, $\mathbf{r}$ is the space position and $\mathbf{R}$ is any lattice vector. The reverse relation is written as follows:

$$\mathbf{H}_n(\mathbf{k},\mathbf{r}) = \frac{\Omega^{\frac{1}{2}}}{(2\pi)^{\frac{3}{2}}} \sum_G a_n(\mathbf{R},\mathbf{r})e^{i\mathbf{k}\cdot\mathbf{R}} \tag{24}$$

$\mathbf{H}_n(\mathbf{k},\mathbf{r})$ functions are the eigenvectors of H-wave equation. PWM solve this equation to get the $\mathbf{H}_n(\mathbf{k},\mathbf{r})$ functions for the PC without defect. To study a defect in PC, the H-field is expanded on the previous Wannier functions:

$$\mathbf{H}(\mathbf{r}) = \sum_n \sum_R c_n(\mathbf{R})a_n(\mathbf{R},\mathbf{r}) \tag{25}$$

Green's functions method solve the H-wave equation for the photonic crystal with defect.

8. Plane Wave Method (PWM)

Plane wave method is used to study the band structure of PC. It comes from the solid state physics where the electronic wave functions are scalar whereas the electromagnetic fields are vectors. A scalar approximation of fields is not enough to describe band structure. This method is modified to take the vectorial nature of the fields into account. Three computational methods of vectorial PWM were created quasi-simultaneously: the E-field method (Leung & Liu, 1990), the D-field method (Zhang & Satpathy, 1990) and the H-field method (Ho et al., 1990). The D and H-field are continuous in PC unlike the E-field. Moreover, only the differential operators of the wave equation on E-field and H-field are self-adjoint. We present the H-field method because of these properties. PWM expands the field and permittivity on plane-wave basis set. The permittivity is a periodic function with translational symmetry T, so Bloch's theorem can be applied to the H-field:

$$\mathbf{H}(\mathbf{r}) = e^{i\mathbf{K}\cdot\mathbf{r}}\mathbf{h}(\mathbf{r}) \quad \text{with} \quad \mathbf{h}(\mathbf{r}) = \mathbf{h}(\mathbf{r}+\mathbf{T}) \tag{26}$$

All periodic functions can be expanded on reciprocal space with Fourier series:

$$\varepsilon_r(\mathbf{r}) = \sum_G \varepsilon_G e^{i\mathbf{G}\cdot\mathbf{r}}$$
$$\mathbf{h}(\mathbf{r}) = \sum_G \mathbf{h}_G e^{i\mathbf{G}\cdot\mathbf{r}} \tag{27}$$

The H-field is written:

$$\mathbf{H}(\mathbf{r}) = \sum_{G} \mathbf{h}_{G} e^{i(\mathbf{K}+\mathbf{G}).\mathbf{r}} = \sum_{G,\lambda=1,2} h_{G}^{\lambda} e^{i(\mathbf{K}+\mathbf{G}).\mathbf{r}} \hat{\mathbf{e}}_{\perp K+G}^{\lambda} \qquad (28)$$

As the **H**-field is transverse, each plane wave is perpendicular to the propagation vector **K+G**. The transverse plane of the propagation vector is described by the unit vectors $\hat{\mathbf{e}}_{\perp K+G}^{1}$ and $\hat{\mathbf{e}}_{\perp K+G}^{2}$. The set of vectors $\left(\hat{\mathbf{e}}_{\perp K+G}^{1}, \hat{\mathbf{e}}_{\perp K+G}^{2}, \mathbf{K}+\mathbf{G}\right)$ represents an orthonormal basis. We only need to storage two vectors instead of three, consequently data storage is reduced. The Fourier series expansion is replaced in the **H**-wave equation:

$$\nabla \times \left[\varepsilon_r^{-1}(\mathbf{r}) \nabla \times \mathbf{H}(\mathbf{r}) \right] = k_0^2 \mathbf{H}(\mathbf{r})$$

$$\nabla \times \left[\sum_{G'} \varepsilon_{G'}^{-1} e^{iG'.\mathbf{r}} \nabla \times \sum_{G,\lambda} h_{G}^{\lambda} e^{i(\mathbf{K}+\mathbf{G}).\mathbf{r}} \hat{\mathbf{e}}_{\perp K+G}^{\lambda} \right] = k_0^2 \sum_{G,\lambda} h_{G}^{\lambda} e^{i(\mathbf{K}+\mathbf{G}).\mathbf{r}} \hat{\mathbf{e}}_{\perp K+G}^{\lambda} \qquad (29)$$

Some algebraic calculations simplify this equation and we get for every vector **G**, the central equation of the photonic crystals:

$$\sum_{G',\lambda'} \varepsilon_{G-G'}^{-1} \left[(\mathbf{K}+\mathbf{G}) \times \hat{\mathbf{e}}_{\perp K+G}^{\lambda} \right] \cdot \left[(\mathbf{K}+\mathbf{G'}) \times \hat{\mathbf{e}}_{\perp K+G'}^{\lambda'} \right] h_{G'}^{\lambda'} = k_0^2 h_{G}^{\lambda} \qquad (30)$$

This equation is an eigenvalue problem which is solved with classical methods. The Bloch vector **K** is set and we try to find the eigenvalues k_0. The calculation convergence depends on the N number of reciprocal lattice vectors **G**. A minimum number of vectors **G** is necessary to describe correctly the permittivity of the PC. The convergence of the problem is rather slow. As the **H**-field is transverse, the number of equations is decreased from $3N$ to $2N$. The PWM is difficult to apply to the materials whose permittivity depends on the frequency like metals. On figure 13, we plot the band structure of the dielectric PC. It is calculated in 12 seconds with 10 Mo of working memory for 343 plane-waves. Calculation is fast because the structure is simple. If we compares with figure 7, we obtain the same result.

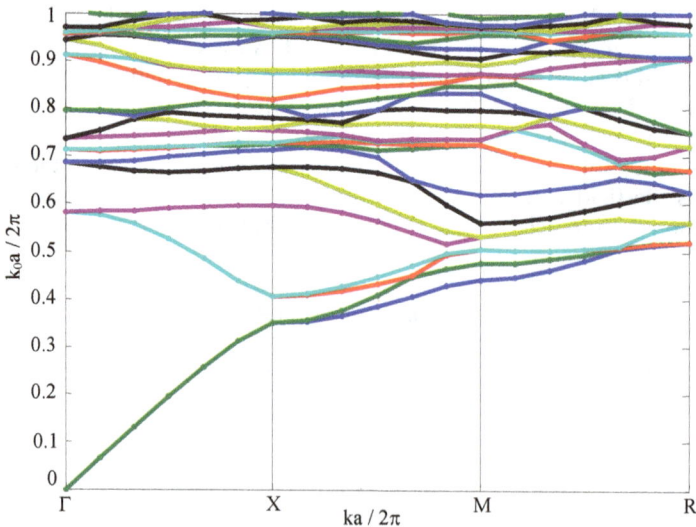

Fig. 13. Band structure of the previous dielectric PC

8.1 One and two-dimensional photonic crystals

If we study one or two-dimensional PC, the central equation is written in another form and will be simplified:

$$\sum_{G'} \varepsilon^{-1}_{G-G'} |K+G||K+G'| \begin{pmatrix} \hat{e}^2_{\perp K+G} \cdot \hat{e}^2_{\perp K+G'} & -\hat{e}^2_{\perp K+G} \cdot \hat{e}^1_{\perp K+G'} \\ -\hat{e}^1_{\perp K+G} \cdot \hat{e}^2_{\perp K+G'} & \hat{e}^1_{\perp K+G} \cdot \hat{e}^1_{\perp K+G'} \end{pmatrix} \cdot \begin{pmatrix} h^1_{G'} \\ h^2_{G'} \end{pmatrix} = k_0^2 \begin{pmatrix} h^1_{G} \\ h^2_{G} \end{pmatrix} \tag{31}$$

In the two-dimensional case, we choose the constant permittivity along z-axis. The vectors K and G are in the xy plane. The vectors of the central equation are indicated on figure 14.

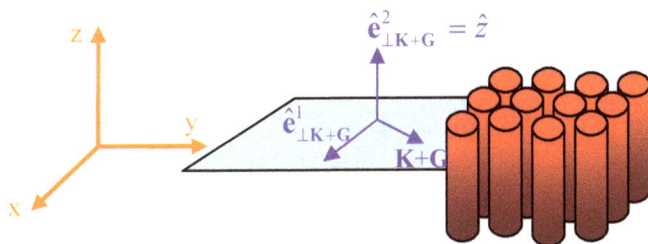

Fig. 14. Vectors definition of the central equation

In the central equation, the matrix of the scalar products is simplified:

$$\begin{pmatrix} \hat{e}^2_{\perp K+G} \cdot \hat{e}^2_{\perp K+G'} & -\hat{e}^2_{\perp K+G} \cdot \hat{e}^1_{\perp K+G'} \\ -\hat{e}^1_{\perp K+G} \cdot \hat{e}^2_{\perp K+G'} & \hat{e}^1_{\perp K+G} \cdot \hat{e}^1_{\perp K+G'} \end{pmatrix} = \begin{pmatrix} 1 & 0 \\ 0 & \cos(\theta - \theta') \end{pmatrix} \tag{32}$$

The $2N$ equations split into two parts, the TE-polarization and the TM-polarization. On the TM-polarization, the H-field vanishes on the xy-plane and the central equation is written:

$$\sum_{G'} \varepsilon^{-1}_{G-G'} |K+G||K+G'| h^1_{G'} = k_0^2 h^1_{G} \tag{33}$$

On the TE-polarization, the H-field is parallel to x-axis and the equation is written:

$$\sum_{G'} \varepsilon^{-1}_{G-G'} |K+G||K+G'| \left(\hat{e}^1_{\perp K+G} \cdot \hat{e}^1_{\perp K+G'} \right) h^2_{G'} = k_0^2 h^2_{G}$$

$$\sum_{G'} \varepsilon^{-1}_{G-G'} (K+G) \cdot (K+G') h^2_{G'} = k_0^2 h^2_{G} \tag{34}$$

In the one-dimensional case, the equations for TE and TM-polarization become similar. We solve N equations.

9. One-dimensional multiple-scattering theory (1D-MST)

The multiple-scattering studies the interaction between objects using the analytical solution for each object taken individually. In the next section, we will apply the MST to the cylinders and the spheres. Before, we will establish the analytical solution of one-dimensional PC *i.e.* multilayer structures.

9.1 Transfer-matrix method (TMM)

Transfer-matrix method is known since many years (Born & Wolf, 1999). It is essential for the study of PC. TMM reduces the computational domain. In this section, transfer matrix will be applied to one-dimensional PC. For oblique incidences, we solve the E-wave equation for TE-polarization and the H-wave equation for TM-polarization. Two polarizations are separated and we use the same steps of calculation for two polarizations. We will study only the E-wave equation:

$$\nabla^2 \mathbf{E}(\mathbf{r}) + k_0^2 \varepsilon_r(\mathbf{r}) \mathbf{E}(\mathbf{r}) = 0 \tag{35}$$

Let us suppose that the layers of 1D-PC are stacked up along e_z. The PC is uniform according to $e\perp$ and $e\parallel$ (figure 15). Because of these symmetries, the E-field and E-wave equation are simplified:

$$\mathbf{E}(\mathbf{r}) = E(r_{\parallel})E(r_\perp)E(z)\mathbf{e}_\perp = e^{i\mathbf{k}_{\parallel}\cdot\vec{r}_{\parallel}}E(z)\mathbf{e}_\perp$$

$$\frac{d^2E(z)}{dz^2} + k_z^2(z)E(z) = 0 \text{ with } k_z^2(z) = k_0^2\varepsilon_r(z) - k_{\parallel}^2 \tag{36}$$

We use the boundary conditions of each layer to solve the previous equation. The electric field E_n for a layer n is written with a forward wave and a backward wave (figure 15):

$$E_n(z) = a_n e^{ik_{z,n}\cdot z} + b_n e^{-ik_{z,n}\cdot z} \tag{37}$$

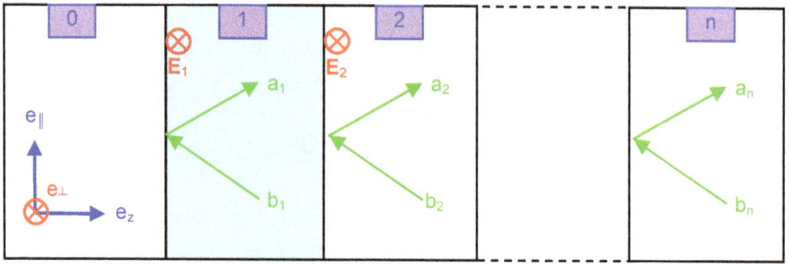

Fig. 15. Field expansion on backward wave and forward wave

We calculate the transfer matrix between layer 1 and 2. The phase shift between the beginning and the end of layer 1 can be written in a transfer matrix form:

$$\begin{pmatrix} a_{1,end} \\ b_{1,end} \end{pmatrix} = \mathbf{\Phi}_n \begin{pmatrix} a_1 \\ b_1 \end{pmatrix} \text{ with } \mathbf{\Phi}_n = \begin{pmatrix} e^{ik_{z,n}a} & 0 \\ 0 & e^{-ik_{z,n}a} \end{pmatrix} \tag{38}$$

Boundary conditions at the interface of layer 1 and layer 2 provide the following expression:

$$\begin{pmatrix} a_2 \\ b_2 \end{pmatrix} = \mathbf{\Lambda}_1^2 \begin{pmatrix} a_{1,end} \\ b_{1,end} \end{pmatrix} \text{ with } \mathbf{\Lambda}_n^{n+1} = \frac{1}{2}\begin{pmatrix} 1 + k_{z,n}/k_{z,n+1} & 1 - k_{z,n}/k_{z,n+1} \\ 1 - k_{z,n}/k_{z,n+1} & 1 + k_{z,n}/k_{z,n+1} \end{pmatrix} \tag{39}$$

The product of previous matrices provides the transfer matrix from layer 1 to layer 2:

$$\begin{pmatrix} a_2 \\ b_2 \end{pmatrix} = \mathbf{T} \begin{pmatrix} a_1 \\ b_1 \end{pmatrix} \quad \text{with} \quad \mathbf{T} = \begin{pmatrix} T_{11} & T_{12} \\ T_{21} & T_{22} \end{pmatrix} = \mathbf{\Lambda}_1^2 . \mathbf{\Phi}_1 \tag{40}$$

The transmission of the layer is calculated from inversion of transfer matrix. The inverse will be calculated directly from phase shift matrices and interface matrix to avoid inaccuracies:

$$\begin{pmatrix} a_1 \\ b_1 \end{pmatrix} = \mathbf{T}^{-1} \begin{pmatrix} a_2 \\ b_2 \end{pmatrix} = [\mathbf{\Phi}_1]^{-1} [\mathbf{\Lambda}_1^2]^{-1} \begin{pmatrix} a_2 \\ b_2 \end{pmatrix} \tag{41}$$

The following equation converts the transfer matrix to scattering matrix:

$$\begin{pmatrix} b_1 \\ a_2 \end{pmatrix} = \begin{pmatrix} S_{11} & S_{12} \\ S_{21} & S_{22} \end{pmatrix} \begin{pmatrix} a_1 \\ b_2 \end{pmatrix} = \frac{1}{T_{11}^{-1}} \begin{pmatrix} T_{21}^{-1} & T_{11}^{-1} T_{22}^{-1} - T_{12}^{-1} T_{21}^{-1} \\ 1 & -T_{12}^{-1} \end{pmatrix} \begin{pmatrix} a_1 \\ b_2 \end{pmatrix} \tag{42}$$

Reflection and transmission coefficients are equal respectively to:

$$S_{11} = \frac{T_{21}^{-1}}{T_{11}^{-1}} \quad S_{21} = \frac{1}{T_{11}^{-1}} \tag{43}$$

Several layers are calculated similarly.

9.2 Kronig-Penney model

The Kronig-Penney model evaluates the electronic levels of a crystal structure in a one-dimensional periodic potential (Kittel, 2005). This model has been modified to be used in PC and takes into account the oblique incidences (Mishra & Satpathy, 2003). We use TMM and the field expansion on backward and forward wave. The structure is periodic according to z-axis (figure 16).

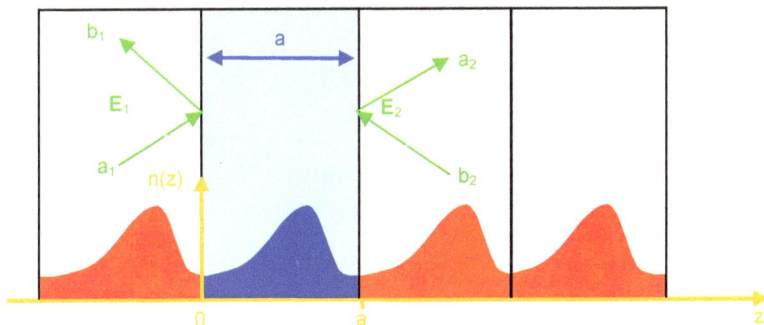

Fig. 16. Refraction index of one dimensional PC

The field $E(z)$ of the previous section is a Bloch function on z. The continuity of tangential fields and the Bloch theorem give the two following relations:

$$\begin{cases} E_2(a) = e^{iKa}E_1(0) \\ \dfrac{dE_2}{dz}\bigg|_a = e^{iKa}\dfrac{dE_1}{dz}\bigg|_0 \end{cases} \tag{44}$$

K is the wave vector of Bloch. In the unit cell, the permittivity is not uniform according to z-axis. It is splitted into sub-cells with a constant permittivity. We calculate the transfer matrix of the unit cell from the expressions of the previous section:

$$\begin{pmatrix} a_2 \\ b_2 \end{pmatrix} = \begin{pmatrix} T_{11} & T_{12} \\ T_{21} & T_{22} \end{pmatrix} \begin{pmatrix} a_1 \\ b_1 \end{pmatrix} \tag{45}$$

We expand equation 44 on forward and backward waves:

$$\begin{cases} a_2 + b_2 = e^{iKa}(a_1 + b_1) \\ a_2 - b_2 = e^{iKa}(a_1 - b_1) \end{cases} \tag{46}$$

Equation 46 is replaced in equation 45:

$$\begin{pmatrix} T_{11} & T_{12} \\ T_{21} & T_{22} \end{pmatrix} \begin{pmatrix} a_1 \\ b_1 \end{pmatrix} = e^{iKa} \begin{pmatrix} a_1 \\ b_1 \end{pmatrix} \tag{47}$$

The Bloch factor e^{iKa} and the complex conjugate value e^{-iKa} are the two eigenvalues of the transfer matrix because the determinant of this matrix is equal to one. The trace of the transfer matrix is equal to the sum of the eigenvalues:

$$T_{11} + T_{22} = e^{iKa} + e^{-iKa} \tag{48}$$

If the materials of the photonic crystal do not absorb, we have $T_{11} = T_{22}^*$. Similarly to the Kronig-Penney model, the above relation is written from the transmission coefficient:

$$S_{21} = |S_{21}|e^{i\varphi_{S_{21}}} = \left(T_{11}^*\right)^{-1} = \left(T_{22}\right)^{-1}$$
$$\frac{e^{i\varphi_{S_{21}}}}{|S_{21}|} + \frac{e^{-i\varphi_{S_{21}}}}{|S_{21}|} = e^{iKa} + e^{-iKa} \tag{49}$$

After simplification, the transcendent equation is written:

$$\frac{\cos\left(\varphi_{S_{21}}\right)}{|S_{21}|} = \cos\left(Ka\right) \tag{50}$$

We get an equation similar to electronic case. The transmission coefficient is different in the TE and TM-polarization. To plot band structure $f(k_\parallel, K, k_0) = 0$, we set the wave numbers k_0 and $k_\parallel$ and calculate the Bloch number K.

10. Two-dimensional multiple-scattering theory (2D-MST)

The multiple-scattering is an analytical theory which calculates the scattering of N objects from the scattering of each object independently (Felbacq et al., 1994). In the two-dimensional case, objects are cylinders and the theory uses the scalar wave equation:

$$\nabla^2 E(\mathbf{r}) + k^2 E(\mathbf{r}) = k^2 \left[1 - \varepsilon_r(\mathbf{r})\right] E(\mathbf{r}) \tag{51}$$

Outside the cylinders, the wave equation can be split into two equations, one for the incident field E_0 and the other one for the scattered field E_s:

$$E(\mathbf{r}) = E_0(\mathbf{r}) + E_s(\mathbf{r})$$
$$\nabla^2 E_0(\mathbf{r}) + k^2 E_0(\mathbf{r}) = 0 \tag{52}$$
$$\nabla^2 E_s(\mathbf{r}) + k^2 E_s(\mathbf{r}) = k^2 \left[1 - \varepsilon_r(\mathbf{r})\right] E(\mathbf{r})$$

Using Green's theorem, the scattered field outside the cylinder can be written as follows:

$$E_s(\mathbf{r}) = -\frac{ik^2}{4} \iint H\left(k|\mathbf{r} - \mathbf{r}'|\right)\left[1 - \varepsilon_r(\mathbf{r})\right] E(\mathbf{r}') \, dS' \tag{53}$$

The function H is a Hankel function. The surface integral can be restricted to the Cj cylinders. The scattered field is written as a sum:

$$E_s(\mathbf{r}) = \sum_j E_s^j(\mathbf{r})$$
$$E_s^j(\mathbf{r}) = \frac{ik^2\left(\varepsilon_r^j(\mathbf{r}) - 1\right)}{4} \iint_{C_j} H\left(k|\mathbf{r} - \mathbf{r}'|\right) E(\mathbf{r}') \, dS' \tag{54}$$

The cylinders interact to give the scattered field by the structure. To better understand the process of multiple-scattering between the cylinders, we will study a simple example of cylinders with circular section aligned along x-axis and excited by an incident field propagating along x-axis (figure 17).

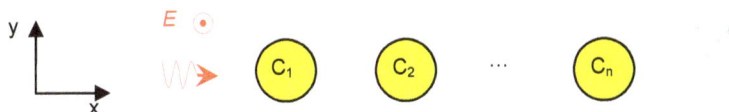

Fig. 17. Incident field on n cylinders aligned

The incident field is expanded into Bessel functions:

$$E_0(\mathbf{r}) = e^{ikx} = \sum_l i^l J_l(kr) = \sum_l a_{0l} J_l(kr) \tag{55}$$

We define the total incident field around the cylinder i which takes into account other cylinders:

$$E_i^i(\mathbf{r}) = \sum_l a_{il} J_l(kr) \tag{56}$$

In the case of the cylinders with circular section, the integral 54 is calculated analytically:

$$E_s^j(\mathbf{r}) = \sum_l b_{jl} H_l^{(1)}(kr) \text{ with } b_{jl} = S_{jl} a_{jl} \tag{57}$$

S_{jl} is the scattering coefficients (Bohren & Huffman, 1998). The field $E_s^j(\mathbf{r})$ scattered by a cylinder j is shifting on another cylinder i to become an incident field $G_{jil}.b_{jl}$ on this cylinder with G_{jil} the translation coefficients (Felbacq et al., 1994). If we shift all the scattered fields on the cylinder i, and if we add the initial incident field to it, we get the total incident field on the cylinder i:

$$a_{il} = \sum_{j \neq i} G_{jil} b_{jl} + a_{0l} \tag{58}$$

By using the equation 57, we get the multiple-scattering equation:

$$a_{il} = \sum_{j \neq i} G_{jil} S_{jl} a_{jl} + a_{0l} \tag{59}$$

G_{jil} and S_{jl} are the translation and scattering matrix coefficients. To get the total incident field and the field scattered by the structure, it is necessary to calculate G_{jil} and S_{jl} coefficients and to solve the multiple-scattering equation. This method is suitable for the study of defects in PC because it does not impose any condition on the position and the material of cylinders. On figure 18, we study a 2D-PC with 80 cylinders doped by a microcavity.

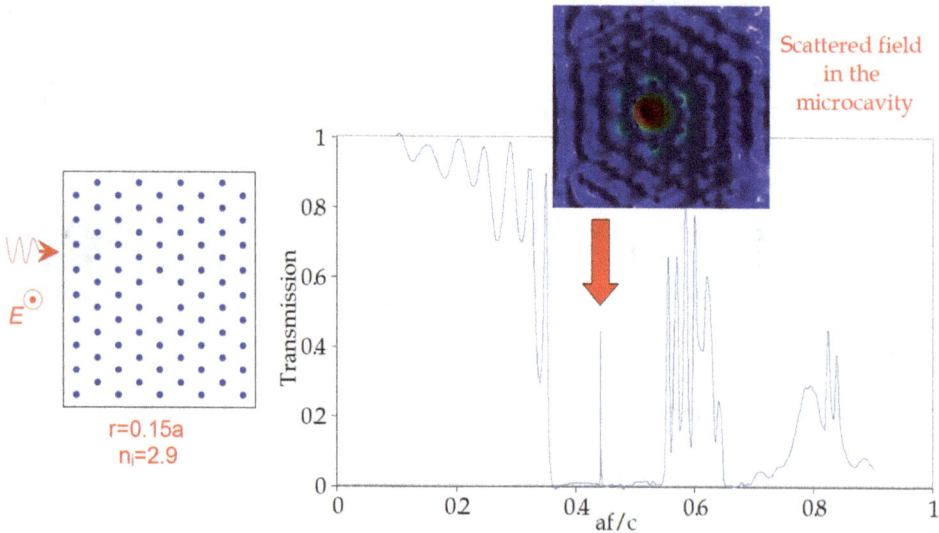

Fig. 18. Transmission coefficient of triangular 2D-PC with a defect

11. Three-dimensional multiple-scattering theory (3D-MST)

The three-dimensional multiple-scattering theory use same principles as the two-dimensional case but three-dimensional case is more complex because the Helmholtz equation is vectorial. The 3D-MST is separated into two parts, the calculation for one sphere and the generalization to N spheres (Oyhenart & Vignéras, 2007). The first part calculates the scattered wave by one sphere with Mie theory. The base of spherical harmonics is written as follows:

$$\left|\mathbf{Z}_{-1,l,m}\right\rangle = \frac{i^l\sqrt{4\pi}}{i\sqrt{l(l+1)}}z_l(kr)\mathbf{r}\times\nabla Y_{lm}\left(\theta,\phi\right)$$

$$\left|\mathbf{Z}_{1,l,m}\right\rangle = -\frac{i}{k}\nabla\times\hat{\mathbf{Z}}_{M=-1} \tag{60}$$

Z_l are the Hankel and Bessel spherical functions and $Y_{lm}(\theta,\phi)$ are scalar spherical wave functions. The incident field and the scattered field are expanded on spherical wave functions:

$$\mathbf{E}_0(\mathbf{r}) = \sum_{\sigma,l,m} E^0_{\sigma,l,m}\left|\mathbf{Z}^{(1)}_{\sigma,l,m}\right\rangle$$

$$\mathbf{E}_s = \sum_{\sigma,l,m} E^s_{\sigma,l,m}\left|\mathbf{Z}^{(3)}_{\sigma,l,m}\right\rangle \text{ with } E^s_{\sigma,l,m} = S_{\sigma l}E^0_{\sigma,l,m} \tag{61}$$

The elements of **S**-matrix are Mie's coefficients (Bohren & Huffman, 1998). Figure 19 presents an example of the incident and scattered fields by a sphere.

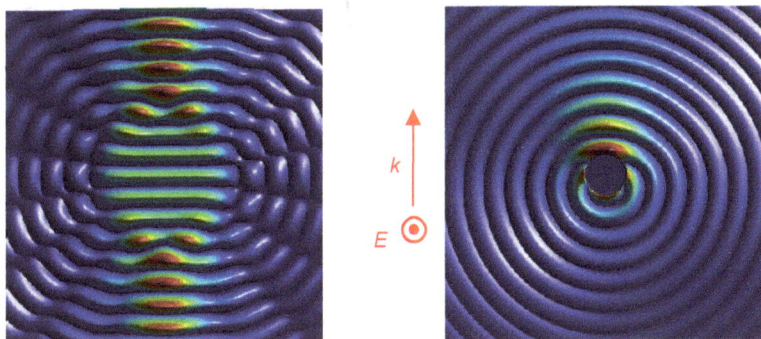

Fig. 19. Pictures of incident and scattered fields by a metallic sphere

The second part of the method is an iterative algorithm to calculate scattered field for N spheres from one sphere (figure 20). For the first order, we calculate the scattering of the incident field for each sphere. For the second order, the scattered field of first order for one sphere becomes the incident field for the N-1 other spheres. With this new incident field, the scattered field is calculated as at first order and so on, for higher orders. This iterative process stops when it is converged. The total scattered field is the contribution of all spheres and all orders. The material and the size of these N spheres can be different. Moreover, spheres may be put in a random way, without symmetry conditions on spheres positions.

These two last remarks show all the interest of this method in calculation of PC with defects and random structures.

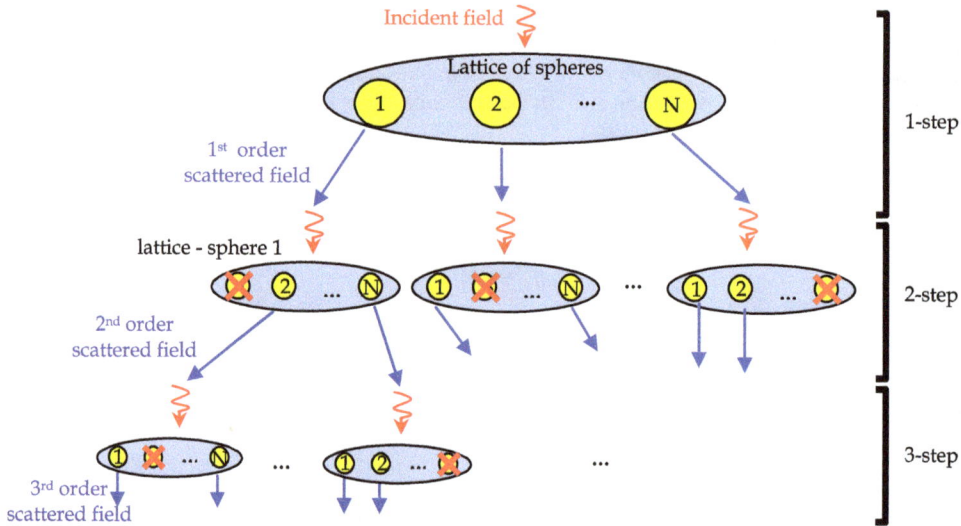

Fig. 20. Block diagram of multiple-scattering method, the total scattered field is the sum of the scattered fields for all orders.

For periodic structures, calculation is simplified. Figure 21 plot the transmission coefficient for the previous infinite dielectric PC. The transmission calculation of 8 layers is calculated in 11 minutes with 63 Mo of working memory. By using the principle of KKR-method of solid state physics, MST can also calculate the band structure (Wang and Al, 1993). MST is the fastest method for the finite structures and random structures.

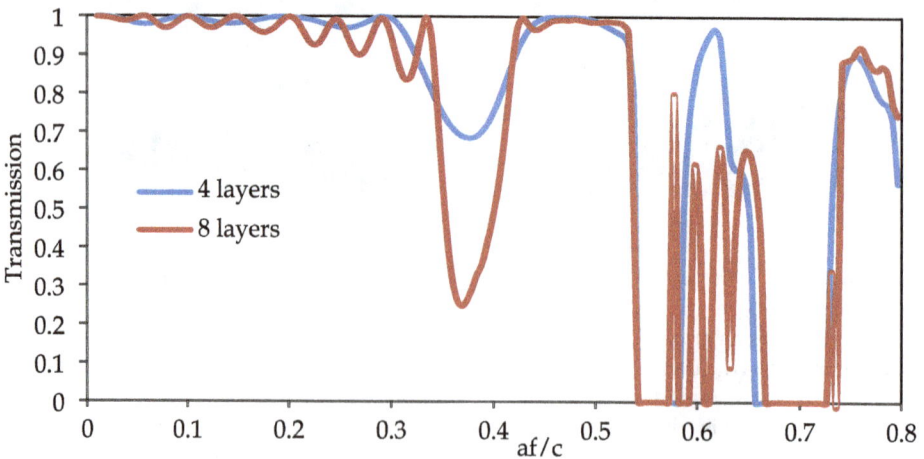

Fig. 21. Transmission coefficient of the infinite periodic dielectric PC

12. Conclusion: Comparison of computational method

The computational methods are studied and compared on table 1. The first four methods are three-dimensional numerical methods coming from electromagnetism. The FEM gives very precise results but it requires many resources systems. Another method, the FDTD is very used and converges without an excessive mesh thanks to its formulation. If you want more precise results, the mesh becomes heavy and you need to use FEM. The FDFD studies photonic crystals with a high number of layers by modeling only one layer. The FIT is a method which studies numerically PC with a high number of objects, without holding excessive resources system but the result is approximate. These methods calculate dielectric and metallic PC to obtain reflection and transmission coefficients and the band structure.

Then, we have others methods resulting from the solid state physics which require a very small computing time. They were adapted from the scalar methods of the solid state physics. The PWM calculate only the band structure. The tight binding method is less used than the PWM although this method is fast for the calculation of defect states in a PC. The multiple-scattering theory which is also used in optics studies analytically large finite PC and it easily takes the defects into account in the PC. For three-dimensional PC, there is no simple method, fast and accurate. We must remove a feature. For example, the 3D-MST is fast and accurate but the computer program is complex to write.

	FEM	FDTD	FDFD	FIT	TB	PWM	1D-MST	2D-MST	3D-MST
Origin	E.M.	E.M.	E.M.	E.M.	Q.M.	Q.M.	Q.M.	E.M.	both
Maxwell's equations	Freq.	Time	Freq.	Time	Freq.	Freq.	Freq.	Freq.	Freq.
Calculation	Num.	Num.	Num.	Num.	Num.	Num.	Analyt.	Analyt.	Analyt.
Geometries	3D	3D	3D	3D	2D/3D	3D	1D	Cyl.	Sphere
Discrete equations	X	X	X	X					
Expansion, series	X				X	X	X	X	X
Free space, finite PC	X	X		X			X	X	X
Infinite periodic PC	X	X	X	X		X	X	X	X
Band structure	X	X	X	X		X	X	X	X
Transmission/reflection	X	X	X	X			X	X	X
Metallic PC	X	X		X				X	X
Single defect in finite-PC	X	X		X			X	X	X
Periodic defect	X	X	X	X	X	X	X	X	X
computing speed	slow	med.	fast	fast	fast	med.	fast	fast	fast
data storage	large	low	low	low	low	low	low	low	low
commercial software	X	X				X			
Free software	X	X	X		—	X		X	X
Popular method in PC		X		X		X			

Abbreviations:

Q.M. : quantum mechanics and solid state physics   Analyt. : analytical method
E.M. : electromagnetism   Num. : numerical method
Freq. : frequency domain   Med. : medium
Time : time domain   Cyl. : cylinder

Table 1. Specifications of the computational methods

13. References

Bohren, C.F. & Huffman, D. R. (1998). *Absoption and scattering of light by small particles*, Wiley, ISBN: 978-0-471-29340-8, New-York, USA

Born, M. & Wolf, E. (1999). *Principles of optics* (7th edition), Cambridge Press, ISBN 0521642221, Cambridge, United Kingdom

Felbacq, D.; Tayeb, G. & Maystre, D. (1994). Scattering by a random set of parallel cylinders, *J. Opt. Soc. Am. A*, Vol.11, No. 9, (September 1994) pp. 2526-2538

Ho, K.M.; Chan, C.T. & Soukoulis, C.M. (1990). Existence of a Photonic Gap in Periodic Dielectric Structures, *Phys. Rev. Lett.*, Vol.65, No.25, (December 1990), pp. 3152-3155

Joannopoulos, J.D.; Johnson, S.J.; Winn, J.N. & Meade, R.D. (2008). *Photonic crystals: molding the flow of light*, Princeton University Press, ISBN: 9780691124568, Princeton, USA

John, S. (1987). Strong localization of photons in certain disordered dielectric superlattices, *Phys. Rev. Lett.*, Vol.58, No.23, (June 1987) pp. 2486-2489

Kittel, C. (2005). *Introduction to Solid State Physics* (8th edition), Wiley, ISBN 978-0-471-41526-8, New-York, USA

Leung, K.M. & Liu, Y.F. (1990). Full Vector Wave Calculation of Photonic Band Structures in Face-Centered-Cubic Dielectric Media, *Phys. Rev. Lett.*, Vol.65, No.21, (November 1990), pp. 2646-2649

Lidorikis, E.; Sigalas, M.M.; Economou, E.N. & Soukoulis, C.M. (1998). Tight-Binding Parametrization for Photonic Band Gap Materials, *Phys. Rev. Lett.*, Vol.81, No.7, (August 1998) pp. 1405-1408

Maddox, J. (1990). Photonic Band-Gaps Bite the Dust. *Nature*, Vol.348, (December 1990), pp. 481

Massaro, A.; Errico, V.;   Stomeo, T.; Salhi, A.; Cingolani,R.; Passaseo, A. & DeVittorio , M. (2008) 3D FEM Modeling and Fabrication of Circular Photonic Crystal Microcavity, *IEEE Journal of Lightwave Technology*, vol.26, no. 16, (August 2008), pp. 2960-2968

Mishra & S. Satpathy, S. (2003). One-dimensional photonic crystal: the Kronig-Penney model, *Phys. Rev. B*, Vol.68, No.4, (July 2003), pp. 045121.1-045121.9

Pendry, J.B. & MacKinnon, A. (1992). Calculation of Photon Dispersion Relations, *Phys. Rev. Lett.* Vol.69, No.19, (November 1992), pp. 2772-2775

Taflove, A. & Hagness, S.C. (2005). *Computational Electrodynamics: The Finite-Difference Time-Domain Method*, Artech House, ISBN 978-1-58053-832-9, Boston, USA

Oyhenart, L. (2005). *Modelling, realization and characterization of three-dimensional photonic crystal for electromagnetic compatibility applications*, thesis of Bordeaux 1 University, Bordeaux, France

Oyhenart, L. & Vignéras, V. (2007). Study of finite periodic structures using the generalized Mie theory, *Eur. Phys. J. A. P.*, Vol.39, No.2, (August 2007) pp. 95-100

Weiland, T. (1977). A Discretization Method for the Solution of Maxwell's Equations for Six-Component Fields, *Electronics and Communication*, Vol.31, (March 1977) pp. 116-120

Wang, X.; Zhang, X. G.; Yu, Q. & Harmon, B.N. (1993) Multiple-scattering theory for electromagnetic waves, *Phys. Rev. B*, Vol.47, No. 8, (February 1993) pp. 4161-4167

Yablonovitch, E. (1987). Inhibited spontaneous emission in solid-state physics and electronics, *Phys. Rev. Lett.*, Vol.58, No.20, (May 1987) pp. 2059-2062

Zhang, Z. & Satpathy, S. (1990). Electromagnetic-Wave Propagation in Periodic Structures - Bloch Wave Solution of Maxwell Equations, *Phys. Rev. Lett.*, Vol.65, No.21, (November 1990), pp. 2650-2653

14

Fourier Factorization in the Plane Wave Expansion Method in Modeling Photonic Crystals

Roman Antos[1] and Martin Veis[1,2]
[1]*Institute of Physics, Faculty of Mathematics and Physics, Charles University in Prague*
[2]*Institute of Biophysics and Informatics, First Faculty of Medicine, Charles University in Prague*
Czech Republic

1. Introduction

Photonic crystals are modern artificially designed periodical systems capable to affect the motion of photons in a similar way that the periodicity of the atomic potential in a semiconductor crystal affects the motion of electrons. The physical properties of light in a photonic crystal resemble those of electrons in atomic crystals, leading to forbidden propagation of electromagnetic modes at certain frequencies, as demonstrated by Yablonovitch (1987), Ho et al. (1990), Joannopoulos et al. (1995; 1997), etc. The existence of the optical band gap (which is the part of the spectrum for which the wave propagation is not possible) makes photonic crystals broadly interesting from many viewpoints of fundamental research and applications. Recent studies are motivated by promising applications such as purely optical integrated circuits [e.g., by Lin et al. (1998), Noda (2006), or Hugonin et al. (2007)], artificial metamaterials with high tunability [Datta et al. (1993); Genereux et al. (2001); Krokhin et al. (2002); Reyes et al. (2005)], high-sensitivity photonic biosensors [Block et al. (2008); Skivesen et al. (2007)], or devices based on phenomena not accessible in conventional media [Benisty (2009); Kosaka et al. (1998); Krokhin & Reyes (2004)]. It is also necessary for accounting for structural colors of wings of butterflies or beetles, feathers of birds, or iridescent plants [Kinoshita & Yoshioka (2005); Vukusic & Sambles (2003)].

Since the optical properties of photonic crystals strongly depend on their geometrical structure, used materials, etc., their proper design is crucial for the correct device functionality. Thorough theoretical analysis therefore takes place in the development, using various numerical methods for calculation including methods of finite difference in the time or frequency domain or finite element methods [Joannopoulos et al. (1995)]. One of the most common calculation techniques applied to photonic crystals is the plane wave expansion method, which is a frequency-domain approach based on the expansion of the fields and material parameters into the Fourier (or reciprocal) space. The components of this expansion represent the definite-frequency states. After some necessary truncation of the complete basis (plane waves with a finite cutoff), the partial differential equations are then solved as a linear-algebraic problem. However, the convergence rate of this method strongly depends on the implementation of Maxwell's equations in the truncated plane-wave basis [Meade et al. (1993); Sozuer et al. (1992)]. In the case of periodic discontinuities (typical for photonic crystals) the convergence is rather poor so that the computer calculations might become extremely time- and memory-consuming.

Because the underlying physical phenomenon in the optical behavior of photonic crystals is based on diffraction (therefore the lattice constant of a periodic structure has to be in the same length-scale as half the wavelength of the electromagnetic wave), several conclusions can be advantageously adopted from the classical coupled-wave theory, one of the most effective methods for modeling diffraction of electromagnetic waves by periodic gratings, which was developed during several past decades. This can provide a high enhancement to the plane wave expansion method, resulting in the reduction of the computation resources.

In the mid 1990s Li (1996) showed that Laurent's "direct" rule, which had always been adopted in conventional formulations to factorize the truncated Fourier series that corresponds to products of two periodic functions, presents bad convergence when the two functions of the product are simultaneously discontinuous. He suggested three Fourier factorization rules (briefly summarized in Section 2.3) and applied them to one-dimensional (1D) diffraction gratings. This major breakthrough in the grating theory (called "fast Fourier factorization") was soon applied by many authors to various grating structures with arbitrary periodic reliefs, anisotropic [Li (1998)] and slanted [Chernov et al. (2001)] periodic systems, their various combinations [Li (2003); Watanabe (2002); Watanabe et al. (2002)], and other systems [Bonod et al. (2005a;b); Boyer et al. (2004)].

Later Li (1997) applied the factorization rules to two-dimensional (2D) periodic structures treated by "zigzag" Fourier expansion, which yielded an improvement for rectangular dots or holes. However, Popov & Neviere (2000) have pointed out that the staircase approximation (of the coupled wave theory using the slicing of relief profiles) in combination with the traditional formulation of differential equation within one slice violates Li's factorization rules. This was a major complication for the analysis of the periodic systems made of rounded elements. Therefore, they applied a coordinate transform to treat individually the normal and tangential components of the electric field on 1D sinusoidal-relief gratings, which enabled the application of the correct rule for each field component and thus improved the convergence.

Later David et al. (2006) utilized the normal–tangential field separation to 2D photonic crystals composed of circular or elliptical holes. Similarly, Schuster et al. (2007) applied this method to 2D gratings, and also suggested more general distributions of polarization bases [Gotz et al. (2008)]. These approaches, always dealing with linear polarizations, enabled a significant improvement of the convergence properties, but ignored the fact that the transformation matrix between the Cartesian and the normal–tangential component bases of polarization became discontinuous at the center and along the boundaries of the periodic cell, which slowed down the resulting convergence. To overcome these discontinuities, a distribution of more complex (i.e., generally elliptic) polarization bases was recently suggested to improve optical simulations of 2D gratings and photonic crystals [Antos (2009); Antos & Veis (2010)].

Our chapter will describe in detail the application of the Fourier factorization rules to the plane wave expansion method for numerical analysis of general photonic crystals. Section 2 will introduce the principle of the plane wave expansion together with the notation of matrices and factorization theorems. Section 3 will refer to 1D photonic structures made as periodic stratified media. The consistency of the correct factorization rules with classical theory of Yeh et al. (1977) and Yariv & Yeh (1977) will be shown, pointing to the correct boundary conditions of the tangential components of the electric and magnetic field on multilayer interfaces. Section 4 will repeat our previously described methodology for 2D photonic crystals made of circular elements, and Section 5 will generalize it to elements of other shapes. Sections 6 and 7 will propose how to factorize anisotropic and three-dimensional (3D) photonic crystals, respectively.

2. Plane wave expansion

2.1 General remarks

The modes of photonic crystals are in principle the eigensolutions of the wave equation with an inhomogeneous, periodic relative permittivity $\varepsilon(r)$. One possible version of the wave equation is the equation for the unknown electric field with an unknown frequency,

$$\frac{1}{\varepsilon} \nabla \times (\nabla \times E) = \frac{\omega^2}{c^2} E, \tag{1}$$

where ω^2/c^2 is its eigenvalue (ω is the frequency and c the light velocity in vacuum) and

$$E(r) = e^{-ik \cdot r} \sum_{m,n,l} e_{mnl} e^{-ik_0(mpx+nqy+lsz)} = \sum_{m,n,l} e_{mnl} e^{-ik_0(p_m x+q_n y+s_l z)} \tag{2}$$

is its eigenfunction, which has the form of a pseudoperiodic Floquet–Bloch function. Here $p = 2\pi/\Lambda_x$, $q = 2\pi/\Lambda_y$, and $s = 2\pi/\Lambda_z$ are the normalized reciprocal lattice vectors. For simplicity we assume the periods Λ_j along the Cartesian axes throughout this chapter. For brevity we have also defined $k = k_0[p_0, q_0, s_0]$, $p_m = p_0 + mp$, $q_n = q_0 + nq$, and $s_l = s_0 + ls$. (Analogously we could write the wave equation for an unknown magnetic field H or any other field from Maxwell's equations.)

Owing to the periodicity of the problem, the plane wave expansion method is the reference method for the mode calculation. It is based on the Fourier expansion of the field such as in Equation 2 and on the Fourier expansion of a material function, either the permittivity or the impermittivity $\eta(r) = 1/\varepsilon(r)$,

$$\varepsilon(r) = \sum_{m,n,l} \varepsilon_{mnl} e^{-ik_0(mpx+nqy+lsz)} \tag{3}$$

$$\eta(r) = \sum_{m,n,l} \eta_{mnl} e^{-ik_0(mpx+nqy+lsz)} \tag{4}$$

The rules for choosing the most appropriate material parameter and the most appropriate field for the Fourier expansion are governed by various methods of Fourier factorization. In the past, the E method (η and the electric displacement D were expanded), H method (η and H were expanded), and Ho method (ε and E were expanded) were the typical choices.

2.2 Matrix notation

Now we carry out the transformation of the partial differential equations into matrix equations in order to solve the eigenproblem by linear-algebraic methods. For simplicity we limit ourselves to 1D and 2D photonic crystals, and always choose the direction of propagation in the xy plane, so that $\partial_z = 0$. With these restrictions we write

$$\varepsilon(x,y) = \sum_{m,n=-\infty}^{+\infty} \varepsilon_{mn} e^{-i(mpx+nqy)}, \tag{5}$$

$$f(x,y) = \sum_{m,n=-\infty}^{+\infty} f_{mn} e^{-i(p_m x+q_n y)}. \tag{6}$$

where f is a component of the electric field. We will now derive the matrix expressions of the fundamental relations

$$h(x,y) = \varepsilon(x,y)f(x,y), \tag{7}$$
$$g_x(x,y) = \partial_x f(x,y), \tag{8}$$
$$g_y(x,y) = \partial_y f(x,y), \tag{9}$$

i.e., the relations of multiplication by a function and applying partial derivatives. Assuming the expansions of the new functions $h = \sum_{m,n} h_{mn} e^{-i(p_m x + q_n y)}$, $g_x = \sum_{m,n} g_{x,mn} e^{-i(p_m x + q_n y)}$, and $g_y = \sum_{m,n} g_{y,mn} e^{-i(p_m x + q_n y)}$, we rewrite Equations 7–9 using the convolution rule, and applying the partial derivatives as follows:

$$h_{mn} = \sum_{k,l=-\infty}^{+\infty} \varepsilon_{m-k,n-l} f_{kl}, \tag{10}$$
$$g_{x,mn} = -ip_m f_{mn}, \tag{11}$$
$$g_{y,mn} = -iq_n f_{mn}. \tag{12}$$

Assuming furthermore a finite number of the retained Fourier coefficients, i.e., using the summation $\sum_{m=-M}^{+M} \sum_{n=-N}^{+N}$, we can renumber all the indices to replace the couple of two sets $m \in \{-M, -M+1, \ldots, M\}$ and $n \in \{-N, -N+1, \ldots, N\}$ by a single set of indices $\alpha \in \{1, 2, \ldots, \alpha_{max}\}$, with $\alpha_{max} = (2M+1)(2N+1)$, related

$$\alpha(m,n) = m + M + 1 + (n+N)(2M+1), \tag{13}$$
$$n(\alpha) = (\alpha-1)\mathbf{div}(2M+1) - N, \tag{14}$$
$$m(\alpha) = (\alpha-1)\mathbf{mod}(2M+1) - M, \tag{15}$$

where "$\mathbf{div}$" denotes the operation of integer division and "$\mathbf{mod}$" the remainder (the modulo operation). Then we can rewrite Equations 10–12 into the matrix relations

$$[h] = [\![\varepsilon]\!][f], \tag{16}$$
$$[g_x] = -i\mathbf{p}[f], \tag{17}$$
$$[g_y] = -i\mathbf{q}[f], \tag{18}$$

where $[f]$, $[h]$, $[g_x]$, and $[g_y]$ are column vectors whose αth elements are the Fourier $[m,n]$ elements of the functions f, h, g_x, and g_y, indexed by $\alpha(m,n)$ defined in Equation 13, and where $[\![\varepsilon]\!]$, $\mathbf{p}$, and $\mathbf{q}$ are matrices whose elements are defined

$$[\![\varepsilon]\!]_{\alpha\beta} = \varepsilon_{m(\alpha)-m(\beta),n(\alpha)-n(\beta)}, \tag{19}$$
$$p_{\alpha\beta} = p_{m(\alpha)}\delta_{\alpha\beta}, \tag{20}$$
$$q_{\alpha\beta} = q_{n(\alpha)}\delta_{\alpha\beta}, \tag{21}$$

where the indices on the right hand parts are defined by Equations 14 and 15 and where $\delta_{\alpha\beta}$ denotes the Kronecker delta. As a summary we can say that the multiplication by a function is in the reciprocal space represented by the matrix $[\![\varepsilon]\!]$ (in the sense of the limit $\alpha_{max} \to \infty$) and that the partial derivatives are represented by the diagonal matrices $-i\mathbf{p}$ and $-i\mathbf{q}$.

For 1D periodicity we choose the inhomogeneity along the y axis. In this case the α index is not necessary because the n index is sufficient,

$$\varepsilon(y) = \sum_{n=-\infty}^{+\infty} \varepsilon_n \, e^{-inqy}, \tag{22}$$

$$f(y) = \sum_{n=-\infty}^{+\infty} f_n \, e^{-iq_n y}, \tag{23}$$

$$[\![\varepsilon]\!]_{nk} = \varepsilon_{n-k}, \tag{24}$$

$$q_{nk} = q_n \delta_{nk}. \tag{25}$$

Here $[\![\varepsilon]\!]$ is a Toeplitz matrix (a matrix with constant diagonals).

2.3 Simplified theorems of Fourier factorization

Although the theorems were derived by Li (1996) for 1D periodic functions, we here summarize them in the matrix formalism independent of the number of dimensions. Let f, h, and ε be piecewise-continuous functions with the same periodicity related

$$h = \varepsilon f, \tag{26}$$

and let $[f]$, $[h]$, and $[\![\varepsilon]\!]$ denote their matrices as defined in Section 2.2.

Theorem 1. If ε and f have no concurrent discontinuities, then the Laurent rule applied to Equation 26 converges uniformly on the whole period and hence

$$[h] = [\![\varepsilon]\!] [f] \tag{27}$$

can be applied with fast convergence.

Theorem 2. If ε and f have one or more concurrent discontinuities but h is continuous, then Equation 26 can be transformed into the case of Theorem 1,

$$f = \frac{1}{\varepsilon} h, \tag{28}$$

and hence

$$[f] = \left[\!\!\left[\frac{1}{\varepsilon} \right]\!\!\right] [h]. \tag{29}$$

Accordingly, we can state

$$[h] = \left[\!\!\left[\frac{1}{\varepsilon} \right]\!\!\right]^{-1} [f] \tag{30}$$

referred to as the inverse rule. We say that the functions ε and f have complementary discontinuities.

Theorem 3. If none of the requirements of the first two theorems are satisfied, then none of the rules can be applied correctly because Equations 27 and 30 are no longer valid at the points of discontinuities, which considerably slows down the convergence. Therefore, we should carefully analyze the continuity of the functions and transform all the partial differential formulae to the first two cases.

3. One-dimensional photonic crystals

3.1 Geometry of the problem

Fig. 1 shows the geometrical configuration of a 1D photonic crystal made as periodic alternation of two different layers with relative permittivities ε_I and ε_{II}, thicknesses d_1 and d_2, with periodicity Λ along the y axis. The relative thickness of the first layer (with respect to the period) is denoted w; the relative thickness of the second layer is then $1 - w$. The coordinate system is chosen to get the uniform problem along the z axis. This means that the plane of incidence (plane defined by the vector of periodicity and the wave vector of incidence $k = k_0[p_0, q_0, 0]$, here only hypothetical since the photonic crystal is infinite) coincides with the xy plane.

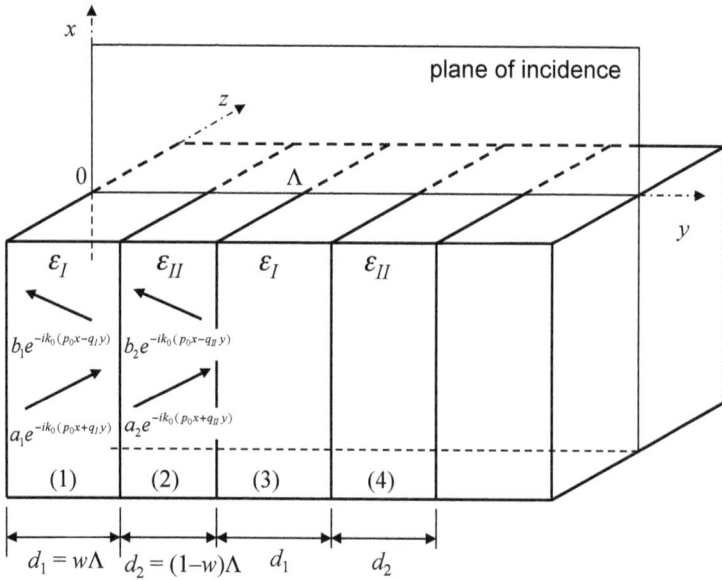

Fig. 1. Geometry of a 1D photonic crystal made as periodic alternation of two layers

Then we distinguish between two polarizations of electromagnetic fields which can be treated independently. The transverse electric (TE) polarization has E perpendicular to the plane of incidence (E_z, H_x, and H_y are nonzero). The transverse magnetic (TM) polarization has H perpendicualar to the plane of incidence (H_z, E_x, and E_y are nonzero).

3.2 Application of Fourier factorization

Propagation in a 1D periodic medium, whose inhomogeneity along the y-axis is described by the relative permittivity function $\varepsilon(y)$, is governed by Maxwell's equations (choosing the coordinate system uniform along the z-axis, i.e., $\partial_z = 0$)

$$\partial_y E_z = -i\omega\mu_0 H_x, \quad \partial_x E_z = i\omega\mu_0 H_y, \quad \partial_x H_y - \partial_y H_x = i\omega\varepsilon_0\varepsilon_{(\mathcal{L})}E_z, \tag{31}$$

$$\partial_y H_z = i\omega\varepsilon_0\varepsilon_{(\mathcal{L})}E_x, \quad \partial_x H_z = -i\omega\varepsilon_0\varepsilon_{(\mathcal{I})}E_y, \quad \partial_x E_y - \partial_y E_x = -i\omega\mu_0 H_z, \tag{32}$$

where the first and the second row corresponds to the TE and the TM polarization, respectively. The formal labels of the relative permittivity, $(\mathcal{L})$ and $(\mathcal{I})$, means that the multiplication with the corresponding component of the electric field will be (according to the factorization rules) treated by either the Laurent rule $(\mathcal{L})$ or the inverse rule $(\mathcal{I})$. This is due to the fact that E_z and E_x are continuous functions (because they are tangential to the discontinuities of ε), whereas E_y is the component normal to the discontinuities and hence discontinuous (while the product εE_z is continuous). By separating one field in each case, we obtain the wave equations

$$-\left(\partial_x^2 + \partial_y^2\right) E_z = \frac{\omega^2}{c^2}\varepsilon_{(\mathcal{L})}E_z, \quad \text{(TE)} \tag{33}$$

$$-\left(\frac{1}{\varepsilon_{(\mathcal{I})}}\partial_x^2 + \partial_y\frac{1}{\varepsilon_{(\mathcal{L})}}\partial_y\right) H_z = \frac{\omega^2}{c^2}H_z \quad \text{(TM)} \tag{34}$$

or, after expanding the permittivity or impermittivity and the fields into the Fourier series, the matrix formulae

$$[\![\varepsilon]\!]^{-1}\left(p_0^2 + \mathbf{q}^2\right)[E_z] = \frac{\omega^2}{c^2}[E_z], \quad \text{(TE)} \tag{35}$$

$$\left(\left[\!\left[\tfrac{1}{\varepsilon}\right]\!\right]p_0^2 + \mathbf{q}[\![\varepsilon]\!]^{-1}\mathbf{q}\right)[H_z] = \frac{\omega^2}{c^2}[H_z]. \quad \text{(TM)} \tag{36}$$

3.3 Consistency with Yeh's theory for the small-period limit

For the small-period limit the validity of these rules can be analytically verified by treating the periodic structure as alternation of two homogeneous layers, where we use the boundary conditions for the continuity of the tangential electric and magnetic fields on all interfaces. Now the function ε is assumed constant (ε_I or ε_{II}) within each layer of the thickness d_1 or d_2. According to the geometry in Fig. 1, the field in the jth layer has the dependence

$$E_z^{(j)}(x,y) = e^{-ik_0p_0x}[a_je^{-ik_0q_j(y-y_j)} + b_je^{ik_0q_j(y-y_j)}], \tag{37}$$

where $q_j = (\varepsilon_j - p_0^2)^{1/2}$, with $\varepsilon_j = \varepsilon_I$ for odd j and $\varepsilon_j = \varepsilon_{II}$ for even j. It is coupled with the field in the next layer by the matrix equation

$$\begin{bmatrix} a_{j+1} \\ b_{j+1} \end{bmatrix} = \mathbf{C}_j\begin{bmatrix} P_j & 0 \\ 0 & \frac{1}{P_j} \end{bmatrix}\begin{bmatrix} a_j \\ b_j \end{bmatrix}, \quad \mathbf{C}_j = \begin{bmatrix} \alpha_j & \beta_j \\ \beta_j & \alpha_j \end{bmatrix}, \tag{38}$$

where $\alpha_j = \frac{1}{2}(1 + q_j/q_{j+1})$, $\beta_j = \frac{1}{2}(1 - q_j/q_{j+1})$ for the TE polarization, or $\alpha_j = \frac{1}{2}(1 + \varepsilon_jq_{j+1}/\varepsilon_{j+1}q_j)$, $\beta_j = \frac{1}{2}(1 - \varepsilon_jq_{j+1}/\varepsilon_{j+1}q_j)$ for the TM polarization, and $P_j = e^{-ik_0q_jd_j}$ for both polarizations. Obviously, $q_j = (\varepsilon_j - p_0^2)^{1/2}$, assuming the $e^{-ik_0p_0x}$ factor of all fields.

Applying the small-period approximation $(P_j)^{\pm1} \approx 1 \mp ik_0q_jd_j$ to the problem of propagation through the whole period,

$$\begin{bmatrix} a_3 \\ b_3 \end{bmatrix} = \boldsymbol{\Omega}\begin{bmatrix} a_1 \\ b_1 \end{bmatrix}, \quad \boldsymbol{\Omega} = \mathbf{C}_2\begin{bmatrix} P_2 & 0 \\ 0 & \frac{1}{P_2} \end{bmatrix}\mathbf{C}_1\begin{bmatrix} P_1 & 0 \\ 0 & \frac{1}{P_1} \end{bmatrix}, \tag{39}$$

yields the eigenvalues of the $\boldsymbol{\Omega}$ operator

$$\Omega_\pm = 1 \pm ik_0\Lambda\sqrt{\varepsilon_0 - p_0^2}, \quad \text{(TE)} \tag{40}$$

$$\Omega_\pm = 1 \pm ik_0\Lambda\sqrt{(1 - p_0^2\eta_0)\varepsilon_0}. \quad \text{(TM)} \tag{41}$$

with

$$\varepsilon_0 = w\varepsilon_I + (1-w)\varepsilon_{II}, \tag{42}$$

$$\eta_0 = w\eta_I + (1-w)\eta_{II}, \tag{43}$$

Assuming a Floquet mode with the eigenvalues of Ω being $\Omega_\pm = e^{\pm ik_0 q_0 \Lambda} \approx 1 \pm ik_0 q_0 \Lambda$, we see that in the small-period limit the periodic structure behaves as a homogeneous anisotropic medium with the y-component of the normalized wave vector $q_0 = (\varepsilon_0 - p_0^2)^{1/2}$ for the TE polarization and $q_0 = [(1 - p_0^2 \eta_0)\varepsilon_0]^{1/2}$ for the TM polarization. These formulae are identical with Equations 35 and 36 if we retain only the 0th element of all matrices. This is a very interesting disclosure that the results obtained by Yeh and coauthors already in 1970s are consistent with the extensive, more general research carried out in 1990.

3.4 Numerical example

Example of the comparison of applying the correct Fourier factorization rules with applying the opposite ones is shown in Fig. 2 for both polarizations. The normalized eigenfrequency $\omega\Lambda/2\pi c$ of the first band is displayed according to N; the structure is made as two alternating layers of the equal thicknesses 500 nm ($\Lambda = 1000$ nm) and permittivities $\varepsilon_I = 3$ and $\varepsilon_{II} = 1$. The wave vector is chosen $\boldsymbol{k} = (0.5\pi/\Lambda)[1,1,0]$.

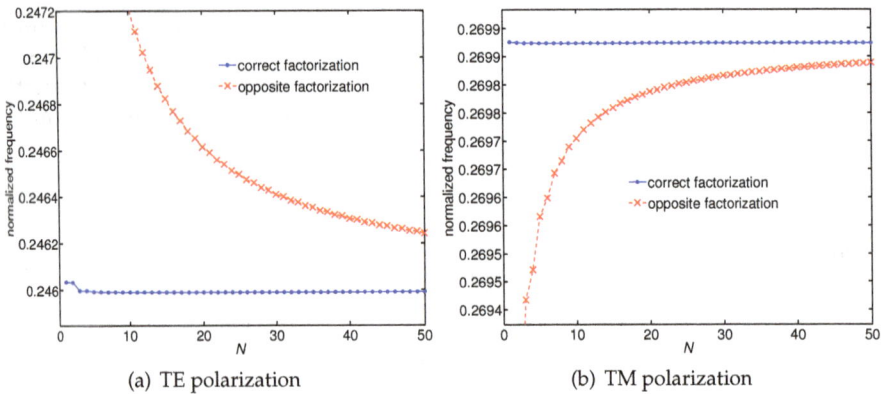

(a) TE polarization   (b) TM polarization

Fig. 2. Convergence properties of the correct factorization compared with the opposite one

4. Two-dimensional photonic crystals with circular elements

4.1 Geometry of the problem

Fig. 3 displays the geometrical arrangement of a 2D photonic crystal made as bi-periodic alternation of rods or holes with a cylindrical cross-section. Instead of a single vector of periodicity we now have the plane of periodicity (determined by two vectors of periodicities defining a unit cell), which here coincides with the xy plane. For simplicity we choose the incidence direction in the plane of periodicity, so that we can again distinguish between two independent polarizations. The plane of incidence is now determined along the propagation wave vector $\boldsymbol{k} = k_0[p_0, q_0, 0]$ and perpendicular to the plane of periodicity. The TE polarization has now $\boldsymbol{H}$ along the z axis, which is now the more difficult case (with nonzero

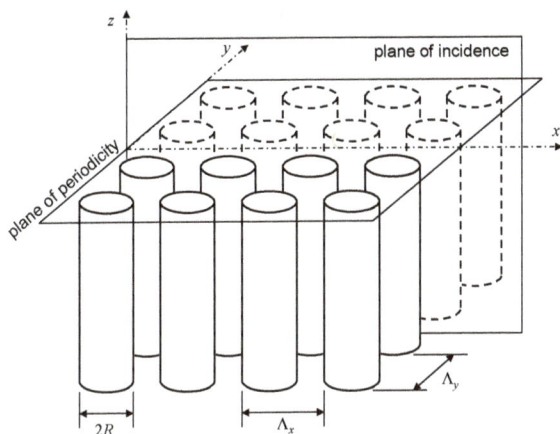

Fig. 3. Geometry of a 2D photonic crystal made as bi-periodic alternation of rods or holes

H_z, E_x, E_y), while the TM polarization has E along the z axis (nonzero E_z, H_x, H_y), which is now the simple case.

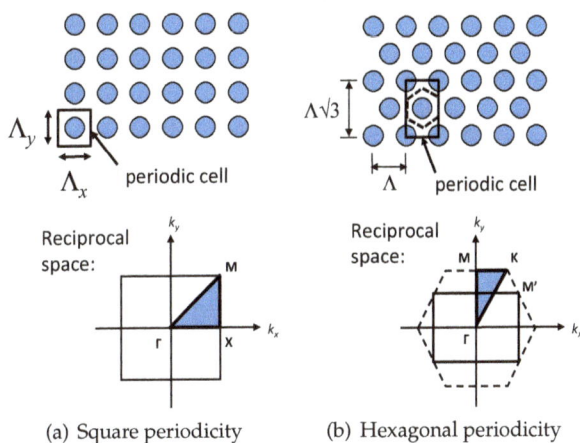

(a) Square periodicity   (b) Hexagonal periodicity

Fig. 4. Two examples of 2D periodic arrangements; below are the first Brillouin zones in the reciprocal space

As an example, in Fig. 3 we have chosen the xz plane as the plane of incidence, but we could choose any plane parallel with the z axis provided that we want to treat the TE and TM polarizations independently. In general, the plane of incidence is determined by the k vector in the reciprocal space as displayed in Fig. 4, whose particular symmetry points are denoted Γ, X, M, or K according to the corresponding periodicity.

In this section we study a 2D photonic crystal composed of infinite cylinders with a circular cross-section with either square [Fig. 4(a)] or hexagonal [Fig. 4(b)] periodicity. For the square symmetry the unit cell has the dimensions $\Lambda_x = \Lambda_y = \Lambda$, and for the hexagonal symmetry it can be chosen as an Λ-by-$\Lambda\sqrt{3}$ rectangle. The corresponding first Brillouin zones of the reciprocal space are depicted in the bottom part of Fig. 4.

Maxwel's equation are again

$$\partial_y E_z = -i\omega\mu_0 H_z, \quad \partial_x E_z = i\omega\mu_0 H_y, \quad \partial_x H_y - \partial_y H_x = i\omega\varepsilon_0\varepsilon E_z, \tag{44}$$

$$\partial_y H_z = i\omega\varepsilon_0\varepsilon E_x, \quad \partial_x H_z = -i\omega\varepsilon_0\varepsilon E_y, \quad \partial_x E_y - \partial_y E_x = -i\omega\mu_0 H_z. \tag{45}$$

However, now we cannot put there the labels $(\mathcal{L})$ or $(\mathcal{I})$ for the factorization type as easily as above, because the discontinuities of the permittivity function are now mixed among the components of E.

Assuming a hypothetical anisotropy of the relative permittivity function, we define a scaled electrical displacement $\tilde{D}$,

$$\begin{bmatrix} \tilde{D}_x \\ \tilde{D}_y \end{bmatrix} = \varepsilon \begin{bmatrix} E_x \\ E_y \end{bmatrix} = \begin{bmatrix} \varepsilon_{xx} & \varepsilon_{xy} \\ \varepsilon_{yx} & \varepsilon_{yy} \end{bmatrix} \begin{bmatrix} E_x \\ E_y \end{bmatrix}, \quad \text{(TE)} \tag{46}$$

$$\tilde{D}_z = \varepsilon_{zz} E_z = \varepsilon_{(\mathcal{L})} E_z, \quad \text{(TM)} \tag{47}$$

where ε_{zz} is obviously the only component of the permittivity tensor for which we can use the Laurent rule.

Defining also a 2-by-2 matrix of electrical impermittivity $\eta = \varepsilon^{-1}$ helps in the formulation of the wave equations

$$(-\partial_y \eta_{xx} \partial_y + \partial_x \eta_{yx} \partial_y + \partial_y \eta_{xy} \partial_x - \partial_x \eta_{yy} \partial_x) H_z = \frac{\omega^2}{c^2} H_z, \quad \text{(TE)} \tag{48}$$

$$-\left(\partial_x^2 + \partial_y^2\right) E_z = \frac{\omega^2}{c^2}\varepsilon_{(\mathcal{L})} E_z, \quad \text{(TM)} \tag{49}$$

where η_{jk} are the components of the electrical impermittivity. For the simplicity of the TM polarization case we below focus our attention only to the TE polarization.

4.2 Methods of Fourier factorization

In this section we compare several models corresponding to different factorization approaches.

4.2.1 Elementary (Cartesian) method (Model A)

First, Model A assumes the solution in the basis of the $\hat{x}$ and $\hat{y}$ polarizations uniform within the periodic cell, where in accordance with Ho et al. (1990) we choose the Laurent rules

$$[\tilde{D}_x] = [\varepsilon E_x] = [\![\varepsilon]\!][E_x], \tag{50}$$

$$[\tilde{D}_y] = [\varepsilon E_y] = [\![\varepsilon]\!][E_y]. \tag{51}$$

The components of the electric impermittivity in Equation 48 then becomes $[\![\varepsilon]\!]^{-1}$ for the cases of η_{xx}, η_{yy}, and zero for the cases of η_{xy}, η_{yx}, or

$$[\![\eta]\!]_A = \begin{bmatrix} [\![\varepsilon]\!]^{-1} & [\![0]\!] \\ [\![0]\!] & [\![\varepsilon]\!]^{-1} \end{bmatrix}. \tag{52}$$

For illustration we show the distribution of the first basis polarization vector (identical with the constant vector $\hat{x}$) in Fig. 5(a), where the black circle denotes the element boundary (the permittivity discontinuity).

4.2.2 Normal vector method (Model B)

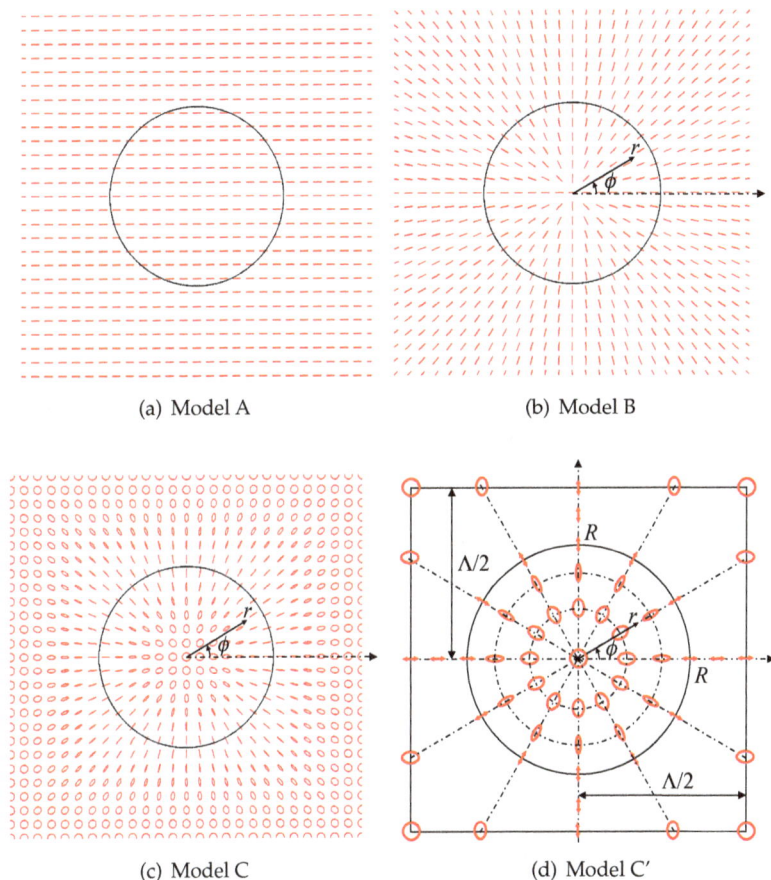

(a) Model A   (b) Model B

(c) Model C   (d) Model C'

Fig. 5. Distribution of the basis polarization vector u for the factorization models

According to the factorization theorems, neither the Laurent rule nor the inverse rule is correct for both products in Equations 50 and 51, because both pairs of functions have concurrent discontinuities and both products $\tilde{D}_x$ and $\tilde{D}_y$ are discontinuous as well. On the other hand, by an appropriate change of the polarization bases at all points (using a space-dependent Jones matrix transform $\mathbf{F}$),

$$\begin{bmatrix} E_x \\ E_y \end{bmatrix} = \mathbf{F} \begin{bmatrix} E_u \\ E_v \end{bmatrix}, \tag{53}$$

we can treat independently the normal (u) and tangential (v) components of the fields by the correct rules,

$$[\tilde{D}_u] = [\![1/\varepsilon]\!]^{-1}[E_u], \tag{54}$$

$$[\tilde{D}_v] = [\![\varepsilon]\!][E_v]. \tag{55}$$

The field components E_u, $\tilde{D}_u$ are normal to the discontinuities of the relative permittivity function, while E_v, $\tilde{D}_v$ are tangential. The factorization rules used in Equations 54 and 55 are justified simply because E_v and $\tilde{D}_u$ are continuous.

A suitable distribution of the matrix **F** within the periodic cell can obviously be the rotation

$$\mathbf{F} = \begin{bmatrix} \cos\phi & -\sin\phi \\ \sin\phi & \cos\phi \end{bmatrix}, \tag{56}$$

where the polar angle $\phi(x,y)$ is in the first cell distributed according to the polar coordinates $re^{i\phi} = x + iy$, and then periodically repeated over the entire 2D space. This enables defining the matrices $[\![c]\!]$, $[\![s]\!]$ from the corresponding 2D-periodic functions $c = \cos\phi$, $s = \sin\phi$.

Let u and v be the two columns of the matrix **F**, both being mutually orthogonal basis vectors of linear polarization. From the above definitions we see that u is a polarization vector normal to the structure discontinuities, whereas v is tangential. In Fig. 5(b) we show the distribution of u within the periodic cell. The basis polarization vectors are constant along the lines of the constant azimuth ($\phi = \text{const}$) and rotate as ϕ increases. It is obvious that the matrix function $\mathbf{F}(x,y)$ has no discontinuities concurrent with the electric field, so that we can use both Laurent and inverse rules for the transformation of polarization, e.g.,

$$\begin{bmatrix} [E_x] \\ [E_y] \end{bmatrix} = [\![\mathbf{F}]\!] \begin{bmatrix} [E_u] \\ [E_v] \end{bmatrix}, \tag{57}$$

$$[\![\mathbf{F}]\!] = \begin{bmatrix} [\![c]\!] & [\![-s]\!] \\ [\![s]\!] & [\![c]\!] \end{bmatrix}. \tag{58}$$

Combining Equations 54, 55, and 57 yields

$$\begin{bmatrix} [E_x] \\ [E_y] \end{bmatrix} = [\![\mathbf{F}]\!] \begin{bmatrix} [\![\frac{1}{\varepsilon}]\!] & [\![0]\!] \\ [\![0]\!] & [\![\varepsilon]\!]^{-1} \end{bmatrix} [\![\mathbf{F}^{-1}]\!] \begin{bmatrix} [\tilde{D}_x] \\ [\tilde{D}_y] \end{bmatrix}, \tag{59}$$

from where we derive the electric impermittivity in the reciprocal space (corresponding to Model B)

$$[\![\eta]\!]_B = [\![\mathbf{F}]\!] \begin{bmatrix} [\![\frac{1}{\varepsilon}]\!] & [\![0]\!] \\ [\![0]\!] & [\![\varepsilon]\!]^{-1} \end{bmatrix} [\![\mathbf{F}^{-1}]\!]$$

$$= \begin{bmatrix} [\![\frac{1}{\varepsilon}]\!][\![c^2]\!] + [\![\varepsilon]\!]^{-1}[\![s^2]\!], & [\![\frac{1}{\varepsilon}]\!][\![cs]\!] - [\![\varepsilon]\!]^{-1}[\![cs]\!] \\ [\![\frac{1}{\varepsilon}]\!][\![cs]\!] - [\![\varepsilon]\!]^{-1}[\![cs]\!], & [\![\frac{1}{\varepsilon}]\!][\![s^2]\!] + [\![\varepsilon]\!]^{-1}[\![c^2]\!] \end{bmatrix}, \tag{60}$$

whose components are immediately applicable to Equation 48.

4.2.3 Method with elliptical polarization bases (Model C)

The above approach (Model B) only deals with linear polarizations and thus suffers from the fact that the matrix function $\mathbf{F}(x,y)$ has a singularity at the central point of the periodic cell and other discontinuities along the cell boundaries. This slows down the convergence of the numerical implementation, as will be evidenced below. On the other hand, we can make **F** continuous by using complex functions ξ and ζ or, in other words, by defining u, v as complex vectors corresponding to generally elliptic polarizations,

$$u = \begin{bmatrix} \xi \\ \zeta \end{bmatrix}, \quad v = \begin{bmatrix} -\zeta^* \\ \xi^* \end{bmatrix} \tag{61}$$

(still orthogonal). By means of rotation θ and ellipticity ϵ we define the first basis vector

$$u = e^{i\theta} \begin{bmatrix} \cos\theta & -\sin\theta \\ \sin\theta & \cos\theta \end{bmatrix} \begin{bmatrix} \cos\epsilon \\ i\sin\epsilon \end{bmatrix}, \tag{62}$$

where

$$\theta(r,\phi) = \phi, \tag{63}$$

$$\epsilon(r,\phi) = \begin{cases} \frac{\pi}{8}\left(1 + \cos\frac{\pi r}{R}\right) & (r \leq R) \\ \frac{\pi}{8}\left\{1 + \cos\frac{\pi[r+D(\phi)-2R]}{D(\phi)-R}\right\} & (r > R). \end{cases} \tag{64}$$

Here R denotes the radius of the circular element and

$$D(\phi) = \frac{\Lambda/2}{\max(|\cos\phi|, |\sin\phi|)} \tag{65}$$

is the distance from the cell's center to its edge. In Equation 62 the Jones vector on the right represents a polarization ellipse (with ellipticity ϵ) oriented along the x coordinate, the matrix in the middle rotates this polarization by the azimuth θ, and the factor $e^{i\theta}$ preserves the continuity of the phase at the center and along the boundaries of the cell. This continuity can be easily checked by evaluating the limits

$$\lim_{r\to 0} u = \lim_{r\to D(\phi)} u = \frac{1}{\sqrt{2}}\begin{bmatrix}1 \\ i\end{bmatrix}, \tag{66}$$

which is the vector of left circular polarization (independent of ϕ).

The distribution of the basis polarization vector u within the periodic cell is shown in Fig. 5(c). Here the azimuth of the polarization ellipse is constant along the lines coming from the cell's center, which is similar to Model B. However, the ellipticity is now zero (corresponding to linear polarization) only on the boundaries of the circular element, has the maximum value ($\pi/4$ for circular polarization) at the cell's center and along its boundaries, and continuously varies (with a smooth sine dependence) in the intermediate ranges. Thus we obtain a smooth and completely continuous matrix function $\mathbf{F}(x,y)$, which is analogously used to calculate the impermittivity in the reciprocal space

$$[\![\eta]\!]_C = \begin{bmatrix} [\![\frac{1}{\varepsilon}]\!][\![\xi\xi^*]\!] + [\![\varepsilon]\!]^{-1}[\![\zeta\zeta^*]\!], & [\![\frac{1}{\varepsilon}]\!][\![\xi\zeta^*]\!] - [\![\varepsilon]\!]^{-1}[\![\xi\zeta^*]\!] \\ [\![\frac{1}{\varepsilon}]\!][\![\xi^*\zeta]\!] - [\![\varepsilon]\!]^{-1}[\![\xi^*\zeta]\!], & [\![\frac{1}{\varepsilon}]\!][\![\zeta\zeta^*]\!] + [\![\varepsilon]\!]^{-1}[\![\xi\xi^*]\!] \end{bmatrix}. \tag{67}$$

In the case of the hexagonal periodicity we define u and the other periodic quantities inside one hexagon (half the area of the rectangular unit cell) where we can use formally the same equations as above, except for

$$D(\phi) = \frac{\Lambda/2}{\max_{n=0,\dots,5}\left[\cos\left(\phi - \frac{n\pi}{3}\right)\right]}, \tag{68}$$

which is now the distance from the hexagon's center to its edge. Here Λ is the hexagon's shortest width (equal to the width Λ_x of the rectangular cell).

4.2.4 Modified method for densely arranged elements (Model C')

To analyze a more complicated situation, we consider a photonic crystal with square periodicity where circular elements are densely arranged near each other, i.e., where the radius R is almost the half width $\Lambda/2$ of the periodic cell. Then the convergence properties of $\mathbf{F}$ becomes worse, which affects all the derived quantities. For this reason we again redefine the polarization distribution. For the modified Model C' we define u to be still same inside the circle ($r < R$), but different outside. Assuming the rotation and ellipticity along the boundary of the square cell

$$\theta_b(\phi) = \theta\left(D(\phi), \phi\right) = \tfrac{\pi}{2}\mathbf{round}\left(\phi/\tfrac{\pi}{2}\right), \tag{69}$$

$$\epsilon_b(\phi) = \epsilon\left(D(\phi), \phi\right) = \tfrac{\pi}{8}(1 - \cos 4\phi) \tag{70}$$

(where "**round**" denotes rounding towards the nearest integer), we define the rotation and ellipticity outside the circle ($r > R$) as

$$\theta(r, \phi) = \tfrac{1}{2}\left\{\theta_b(\phi) + \phi + [\theta_b(\phi) - \phi]\cos\tfrac{\pi[r+D(\phi)-2R]}{D(\phi)-R}\right\}, \tag{71}$$

$$\epsilon(r, \phi) = \tfrac{\epsilon_b(\phi)}{2}\left\{1 + \cos\tfrac{\pi[r+D(\phi)-2R]}{D(\phi)-R}\right\}. \tag{72}$$

Assuming otherwise the same Equations 59, 61, 62, and 65, we obtain for $[\![\eta]\!]_{C'}$ formally the same matrix as in Equation 67, except that the functions ξ and ζ are now derived from different azimuth and ellipticity distributions of u. Note that u is again continuous along the cell's boundaries; to evaluate its precise limits [when $x \to \pm D(0)$, $y = $ const or $y \to \pm D(\phi/2)$, $x = $ const] would now be more complicated. The distribution of the basis polarization vector u within the periodic cell is depicted in Fig. 5(d) together with dimensions.

4.3 Numerical examples

We examine the numerical performances of all factorization models presented in Section 4.2 on three samples of 2D photonic crystals, for which we calculate the eigenfrequencies ω_κ (where the band number $\kappa = 1$ stands for the lowest eigenfrequency, $\kappa = 2$ for the second lowest, etc.) and the corresponding eigenvectors $[H_z]_\kappa$ of Equation 48. All convergences will be presented according to the maximum Fourier harmonics retained inside the periodic medium, which will be kept same for the x and y directions ($M = N$).

First, Sample S1 is a square array of cylindrical rods of the circular cross-section with the diameter $2R = 500$ nm, square period $\Lambda = 1000$ nm, relative permittivity of the rods $\varepsilon_1 = 9$, and relative permittivity of the surrounding medium corresponding to vacuum ($\varepsilon_2 = 1$). Its dispersion relation is displayed in Fig. 6(a). Similarly, for Sample S2 we assume exactly the same parameters except the diameter of the rods, now being $2R = 900$ nm. This corresponds to densely arranged elements (the distance between two adjacent rods is only 100 nm). Finally, for Sample H we consider a hexagonal array of cylindrical holes of the circular cross-section with the diameter $2R = 600$ nm, hexagonal periodicity $\Lambda = 1000$ nm (corresponding to the rectangular cell of the dimensions $\Lambda_x = 1 \mu m$, $\Lambda_y = \sqrt{3} \mu m$), relative permittivity of the holes corresponding to vacuum ($\varepsilon_1 = 1$), and relative permittivity of the substrate medium (surrounding holes) $\varepsilon_2 = 12$. The dispersion relation of Sample H is displayed in Fig. 6(b).

For our analysis we choose the eigenmodes Γ2 and Γ3 of Sample S1, the eigenmode X3 of Sample S2, and the eigenmode M1 of Sample H, where the letter denotes a point of symmetry

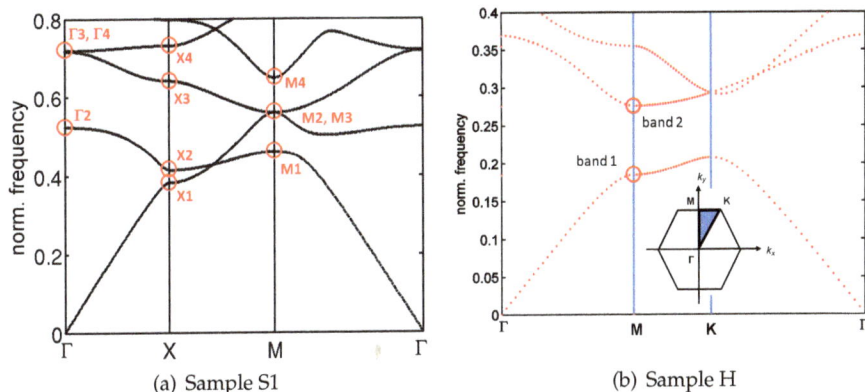

(a) Sample S1   (b) Sample H

Fig. 6. Photonic band structures of two samples

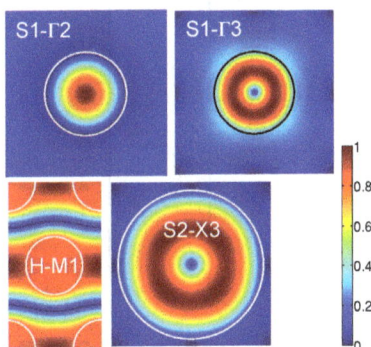

Fig. 7. Amplitude distributions of the scaled magnetic field $|H_z|$ within the unit cell of four chosen eigenmodes

and the number corresponds to the band. The amplitude distributions of the modes are displayed in Fig. 7, and the corresponding convergence properties of the eigenfrequencies calculated by the above described models are shown in Fig. 8 (Model C' is only compared for Sample S2).

The result of a careful comparison of the numerical efficiencies of all the factorization models can be summarized as follows. Models B, C, and C' always converge considerably faster than Model A. Model C converges faster (having usually one order higher precision) than Model B with two exceptions. The first exception is the case such as in Fig. 7 (S1-Γ2) and Fig. 8(a), where the discontinuities of the polarization transformation matrix **F** coincide with a nearly zero amplitude of the field, so that the discontinuities do not manifest themselves. The second exception is the case such as Fig. 7 (S2-X3) and Fig. 8(d), where the elements are densely arranged which causes rapid variations of the ellipticity between two adjacent elements (which are very close to each other); this requires more Fourier components than the weak discontinuity of the linear polarization u in Model B. The problem is solved in Model C', which obviously converges fastest among all the four models applied to Sample S2.

(a) Sample S1, point Γ2

(b) Sample S1, point Γ3

(c) Sample H, point M1

(d) Sample S2, point X3

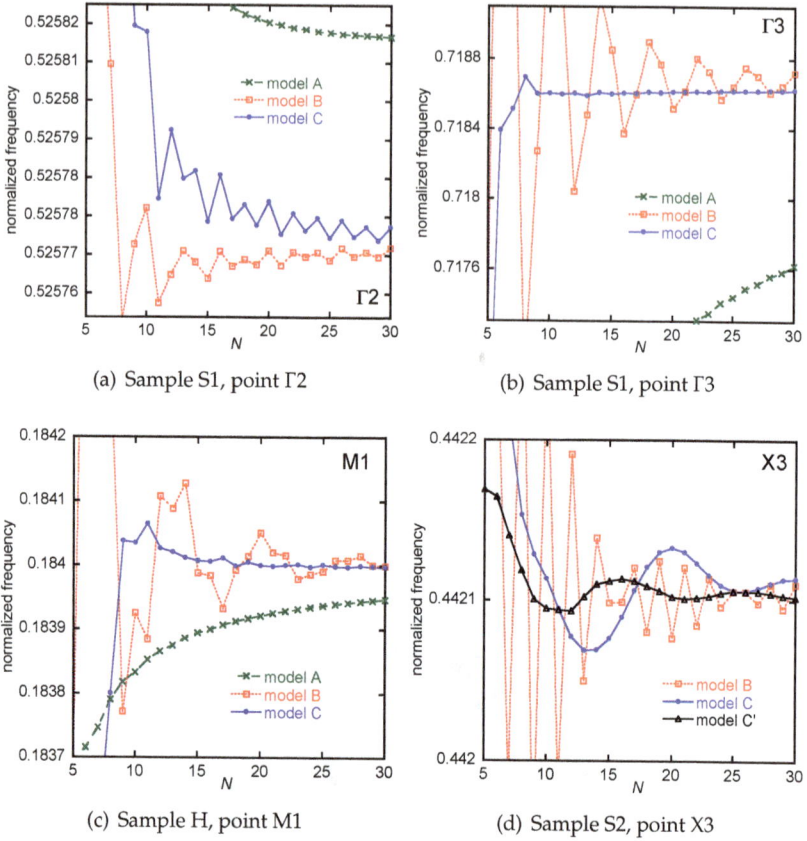

Fig. 8. Convergences of the Fourier factorization methods for three samples

5. Two-dimensional photonic crystals with non-circular elements

In this section we briefly show numerical efficiencies of three factorization models derived above (Models A, B, and C) applied to 2D photonic crystals made as long elements of other shapes, namely rods with the square cross-section and tubes with the ring and split-ring cross-sections, arranged with the square periodicity. For all the three samples we choose the permittivity of the elements $\varepsilon = 9$ and permittivity of vacuum $\varepsilon = 1$ for the surrounding medium.

5.1 Periodic rods with the square cross-section

For the photonic crystals made of square-sectioned rods, the period is chosen $\Lambda = 1000$ nm, and the width of the square $d = 600$ nm. The distribution of the basis polarization vector u, analogously to Section 4, is displayed in Fig. 9 for all the three compared models.

For Model A, of course, $u = \hat{x}$ as visible in Fig. 9(a). For Model B, we divide the unit cell into four parts by its diagonals (the lines $x = y$ and $x = -y$). The vector $u(\phi)$, depending only

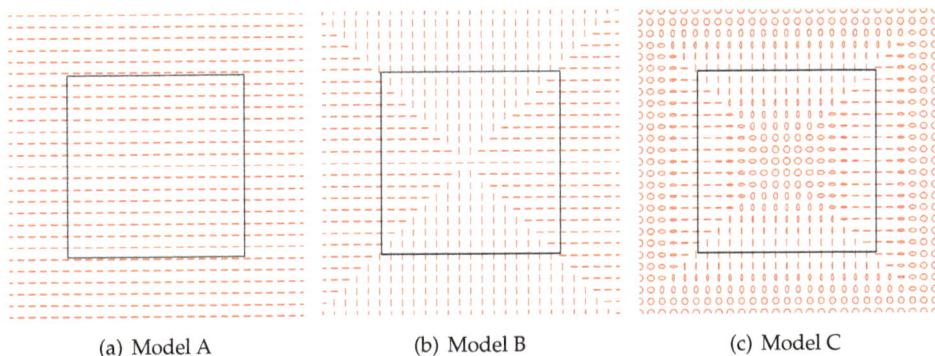

(a) Model A   (b) Model B   (c) Model C

Fig. 9. Polarization distribution of the factorization methods for periodic squares

on the azimuthal angle, is set $\hat{x}$ for $\phi \in \langle -\frac{\pi}{4}, \frac{\pi}{4} \rangle \cup \langle \frac{3\pi}{4}, \frac{5\pi}{4} \rangle$, and $\hat{y}$ for the remaining angles. This means that u is always linear, perpendicular to the permittivity discontinuities, and its discontinuities (along the lines $x = y$ and $x = -y$) have no concurrent discontinuities with the permittivity function except those at the four corners of the square. Hence, Model B here fulfills nearly the same conditions for the application of the factorization rules as demanded in Section 4 for circular elements.

For Model C we divide the unit cell into four areas in the same manner. This time, however, the basis vector $u(r, \phi)$ depends on both polar coordinates and the corresponding polarization is in general elliptic. In analogy with Section 4 we want u to be perpendicular to the permittivity discontinuities and to remove its discontinuities as much as possible. The most simple way how to do this, although the discontinuities will not be completely removed, is to set the azimuth $\theta(\phi)$ of the polarization to zero for $\phi \in \langle -\frac{\pi}{4}, \frac{\pi}{4} \rangle \cup \langle \frac{3\pi}{4}, \frac{5\pi}{4} \rangle$, and $\frac{\pi}{2}$ for the remaining angles, and to use Equation 64 for the ellipticity distribution, where $D(\phi)$ now corresponds to the square element.

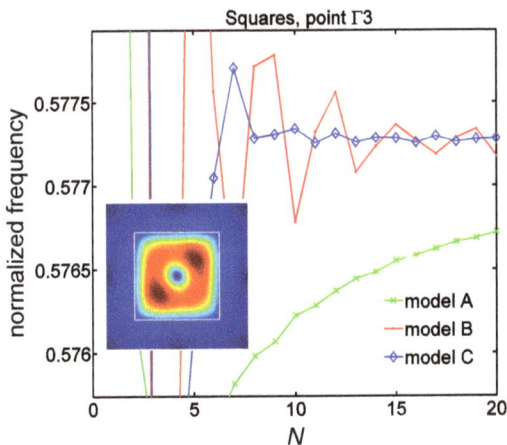

Fig. 10. Convergence of the factorization methods for rods with the square cross-section

The numerical efficiencies of all the three models are displayed in Fig. 10 for the 3rd band of the symmetry point Γ. As clearly visible, Models B and C converge considerably faster than Model A, with Model C being a little better. Although the improvement from Model B towards Model C is not so distinct, it could be improved by a more careful choice of a completely continuous distribution of the polarization basis.

5.2 Periodic hollow cylinders

For the photonic crystals made of periodic hollow cylinders (tubes with the symmetric ring cross-section) we choose the period $\Lambda = 1000$ nm, the inner diameter (the diameter of the inner circular hole) $R_1 = 400$ nm, and the outer diameter $R_2 = 680$ nm. The distribution of the basis polarization vector u for all the three compared models is displayed in Fig. 11.

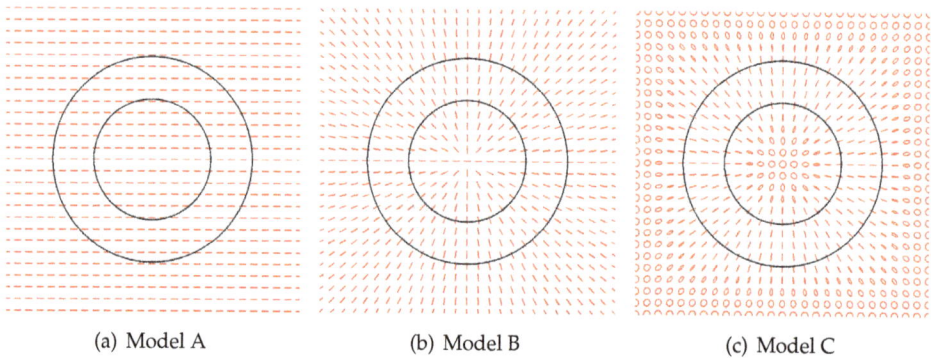

| (a) Model A | (b) Model B | (c) Model C |

Fig. 11. Polarization distribution of the factorization methods for hollow cylinders

For Model A again $u = \hat{x}$. For both Models B and C we use the same polarization distributions as in Section 4 for the inner area ($r < R_1$) and for the outer area ($r > R_2$). For the annulus area ($R_1 < r < R_2$) we simply choose the polarization distribution same as that in Model B. These distributions are obviously the most straightforward analogies of the distributions used in Section 4 for circular elements.

The numerical efficiencies of all the three models are displayed in Fig. 12 for the 3rd band of the symmetry point Γ. As clearly visible, Models B and C converge considerably faster than Model A, but now Model C exhibits no improvement against Model B. This is because the discontinuities of u at the center and along the boundaries of the periodic cell quite well coincide with the zero amplitude of the mode, as visible in the inset of Fig. 12, so that the discontinuities do not manifest themselves in the calculations.

5.3 Periodic split hollow cylinders

For the photonic crystals made of split hollow cylinders (tubes with an asymmetric, split ring cross-section) we choose the period $\Lambda = 1000$ nm, the inner diameter (the diameter of the inner semi-circular hole) $R_1 = 600$ nm, the outer diameter $R_2 = 720$ nm, and the relative azimuthal length of the ring $w_\phi = 0.9$ (where $w_\phi = 1$ means the complete, symmetric ring). The distributions of the basis polarization vector u for all the three compared models are displayed in Fig. 13.

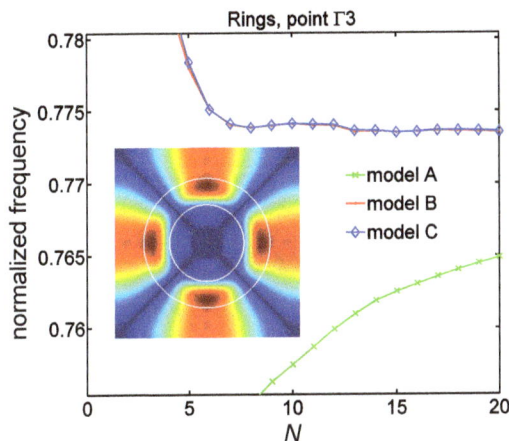

Fig. 12. Convergence of the factorization methods for hollow cylinders

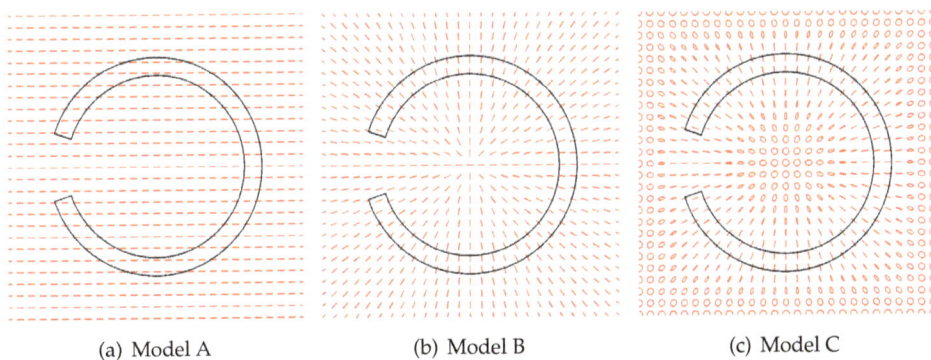

| (a) Model A | (b) Model B | (c) Model C |

Fig. 13. Polarization distribution of the factorization methods for split hollow cylinders

Since the area of splitting is quite small compared to the full area of the unit cell, we have chosen the polarization distributions of all the three models exactly same as for the symmetric rings in Section 5.2. The condition for normal u and tangential v is not satisfied on the surface proportional to the length $2 \times 120 = 240$ nm, but is satisfied on the surface proportional to $0.9 \times 2\pi(R_1 + R_2) \approx 7\,464$ nm, which is 30 times higher area, justifying this negligence. Of course, for modes with most of electromagnetic field resonating in the critical area this approximation would be insufficient.

The numerical efficiencies of all the three models are displayed in Fig. 14, again for the 3rd band of the symmetry point Γ. We can describe these performances by exactly the same conclusion as for the symmetric rings in Section 5.2. Models B and C converge similarly and both considerably faster than Model A. The discontinuities of u at the center and along the boundaries of the periodic cell again coincide with the zero amplitude of the mode, though with some deviations visible in the inset of Fig. 14.

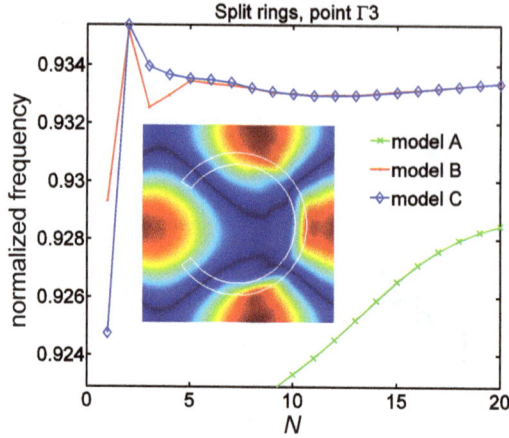

Fig. 14. Convergence of the factorization methods for split hollow cylinders

6. Anisotropic photonic crystals

In this section we will briefly demonstrate the application of the factorization rules to 2D photonic crystals made of anisotropic materials, again with the plane of incidence parallel to the z axis (with geometry of Fig. 3). Unlike the isotropic crystals, now the TE and TM polarizations are not separable. Instead of the scalar permittivity we define and expand the components of the relative permittivity tensor function

$$\varepsilon_{jk}(x,y) = \sum_{m,n=-\infty}^{+\infty} \varepsilon_{jk,mn} e^{-i(mpx+nqy)}, \tag{73}$$

where $\varepsilon_{jk,mn}$ are the Fourier coefficients. The wave equation for a generally anisotropic medium, now described by the permittivity tensor

$$\varepsilon = \begin{bmatrix} \varepsilon_{xx} & \varepsilon_{xy} & \varepsilon_{xz} \\ \varepsilon_{yx} & \varepsilon_{yy} & \varepsilon_{yz} \\ \varepsilon_{zx} & \varepsilon_{zy} & \varepsilon_{zz} \end{bmatrix}, \tag{74}$$

is the operator equation

$$\hat{\mathbf{C}}\mathbf{E} = \frac{\omega^2}{c^2}\mathbf{E}, \quad \hat{\mathbf{C}} = \varepsilon^{-1} \begin{bmatrix} -\partial_y^2 & \partial_x\partial_y & 0 \\ \partial_x\partial_y & -\partial_x^2 & 0 \\ 0 & 0 & -\partial_x^2 - \partial_y^2 \end{bmatrix}. \tag{75}$$

Similarly as above, Model A assumes all components of the permittivity tensor treated by the Laurent rule, i.e.,

$$[\tilde{D}_j] = \sum_k [\![\varepsilon_{jk}]\!][E_k]. \tag{76}$$

To apply the factorization correctly, we must again separate the normal and tangential components of all fields for which we can use the correct rules. Let us define a space-dependent matrix transform $\mathbf{F}(x, y)$ so that

$$\begin{bmatrix} E_x \\ E_y \\ E_z \end{bmatrix} = \mathbf{F} \begin{bmatrix} E_u \\ E_v \\ E_z \end{bmatrix}, \tag{77}$$

where E_u and E_v are the normal and tangential components of the vector $\mathbf{E}$ to all discontinuities of the permittivity.

Analogously to Section 4, for the case of circular elements we can choose the polarization basis distribution corresponding to Models B and C as

$$\mathbf{F} = \begin{bmatrix} \cos\phi & -\sin\phi & 0 \\ \sin\phi & \cos\phi & 0 \\ 0 & 0 & 1 \end{bmatrix}, \quad \text{(Model B)} \tag{78}$$

$$\mathbf{F} = \begin{bmatrix} \zeta & -\zeta^* & 0 \\ \zeta & \zeta^* & 0 \\ 0 & 0 & 1 \end{bmatrix}, \quad \text{(Model C)} \tag{79}$$

where the matrix elements have the same meaning as in Section 4. In the new coordinates we can write

$$\begin{bmatrix} \tilde{D}_u \\ \tilde{D}_v \\ \tilde{D}_z \end{bmatrix} = \begin{bmatrix} \varepsilon_{uu} & \varepsilon_{uv} & \varepsilon_{uz} \\ \varepsilon_{vu} & \varepsilon_{vv} & \varepsilon_{vz} \\ \varepsilon_{zu} & \varepsilon_{zv} & \varepsilon_{zz} \end{bmatrix} \begin{bmatrix} E_u \\ E_v \\ E_z \end{bmatrix}. \tag{80}$$

Now let us separate two sets of quantities, those which are continuous ($\tilde{D}_u$, E_v, E_z) and those which are not continuous ($\tilde{D}_v$, $\tilde{D}_z$, E_u) to the discontinuities of the permittivity. Expressing the second set according to the first one yields

$$\begin{bmatrix} E_u \\ \tilde{D}_v \\ \tilde{D}_z \end{bmatrix} = \mathbf{G} \begin{bmatrix} \tilde{D}_u \\ E_v \\ E_z \end{bmatrix}, \quad \mathbf{G} = \begin{bmatrix} \frac{1}{\varepsilon_{uu}} & -\frac{\varepsilon_{uv}}{\varepsilon_{uu}} & -\frac{\varepsilon_{uz}}{\varepsilon_{uu}} \\ \frac{\varepsilon_{vu}}{\varepsilon_{uu}} & \varepsilon_{vv} - \frac{\varepsilon_{vu}\varepsilon_{uv}}{\varepsilon_{uu}} & \varepsilon_{vz} - \frac{\varepsilon_{vu}\varepsilon_{uz}}{\varepsilon_{uu}} \\ \frac{\varepsilon_{zu}}{\varepsilon_{uu}} & \varepsilon_{zv} - \frac{\varepsilon_{zu}\varepsilon_{uv}}{\varepsilon_{uu}} & \varepsilon_{zz} - \frac{\varepsilon_{zu}\varepsilon_{uz}}{\varepsilon_{uu}} \end{bmatrix}, \tag{81}$$

for which we can simply use the Laurent rule,

$$\begin{bmatrix} [E_u] \\ [\tilde{D}_v] \\ [\tilde{D}_z] \end{bmatrix} = [\![\mathbf{G}]\!] \begin{bmatrix} [\tilde{D}_u] \\ [E_v] \\ [E_z] \end{bmatrix}. \tag{82}$$

From this we express $[\tilde{D}_j]$ according to $[E_j]$,

$$\begin{bmatrix} [\tilde{D}_u] \\ [\tilde{D}_v] \\ [\tilde{D}_z] \end{bmatrix} = [\![\boldsymbol{\varepsilon}_{\{uvz\}}]\!]_{B,C} \begin{bmatrix} [E_u] \\ [E_v] \\ [E_z] \end{bmatrix}, \tag{83}$$

with

$$\llbracket \varepsilon_{\{uvz\}} \rrbracket_{B,C} =$$

$$\left[\begin{array}{l} \left[\left[\frac{1}{\varepsilon_{uu}} \right] \right]^{-1}, \qquad\qquad \left[\left[\frac{1}{\varepsilon_{uu}} \right] \right]^{-1} \left[\frac{\varepsilon_{uv}}{\varepsilon_{uu}} \right], \\[2ex] \left[\left[\frac{\varepsilon_{vu}}{\varepsilon_{uu}} \right] \right] \left[\left[\frac{1}{\varepsilon_{uu}} \right] \right]^{-1}, \quad \left[\varepsilon_{vv} - \frac{\varepsilon_{vu}\varepsilon_{uv}}{\varepsilon_{uu}} \right] + \left[\frac{\varepsilon_{vu}}{\varepsilon_{uu}} \right] \left[\left[\frac{1}{\varepsilon_{uu}} \right] \right]^{-1} \left[\frac{\varepsilon_{uv}}{\varepsilon_{uu}} \right], \\[2ex] \left[\left[\frac{\varepsilon_{zu}}{\varepsilon_{uu}} \right] \right] \left[\left[\frac{1}{\varepsilon_{uu}} \right] \right]^{-1}, \quad \left[\varepsilon_{zv} - \frac{\varepsilon_{zu}\varepsilon_{uv}}{\varepsilon_{uu}} \right] + \left[\frac{\varepsilon_{zu}}{\varepsilon_{uu}} \right] \left[\left[\frac{1}{\varepsilon_{uu}} \right] \right]^{-1} \left[\frac{\varepsilon_{uv}}{\varepsilon_{uu}} \right], \end{array} \right.$$

$$\left. \begin{array}{l} \left[\left[\frac{1}{\varepsilon_{uu}} \right] \right]^{-1} \left[\frac{\varepsilon_{uz}}{\varepsilon_{uu}} \right] \\[2ex] \left[\varepsilon_{vz} - \frac{\varepsilon_{vu}\varepsilon_{uz}}{\varepsilon_{uu}} \right] + \left[\frac{\varepsilon_{vu}}{\varepsilon_{uu}} \right] \left[\left[\frac{1}{\varepsilon_{uu}} \right] \right]^{-1} \left[\frac{\varepsilon_{uz}}{\varepsilon_{uu}} \right] \\[2ex] \left[\varepsilon_{zz} - \frac{\varepsilon_{zu}\varepsilon_{uz}}{\varepsilon_{uu}} \right] + \left[\frac{\varepsilon_{zu}}{\varepsilon_{uu}} \right] \left[\left[\frac{1}{\varepsilon_{uu}} \right] \right]^{-1} \left[\frac{\varepsilon_{uz}}{\varepsilon_{uu}} \right] \end{array} \right]. \qquad (84)$$

Finally we obtain the matrix of permittivity in the Cartesian coordinates via the formula

$$\llbracket \varepsilon_{\{xyz\}} \rrbracket_{B,C} = \llbracket F \rrbracket \llbracket \varepsilon_{\{uvz\}} \rrbracket_{B,C} \llbracket F^{-1} \rrbracket, \qquad (85)$$

where the index B or C corresponds to the chosen model.

7. Distribution of polarization basis for 3D structures

In this section we will suggest how to create the polarization basis distribution for a 3D photonic crystal. Unlike the previous cases, we now have a fully vectorial wave Equation 1. The corresponding material equation $\tilde{D} = \varepsilon(x,y,z)E$ must be changed to separate the normal and tangential components of the fields to the ε discontinuities, which are now surfaces in the 3D space. For simplicity we assume a 3D photonic crystal made as spheres (with the radius R) arranged in the space with the cubic periodicity. To make a 3D analogy with Model C described in the previous sections, we must find a matrix transform $F(x,y,z)$ whose columns, denoted u, v, and w, are complex vectorial functions of space, mutually orthonormal and continuous at all points, where u is the normal vector at each point of the sphere's surface and v and w are tangential.

As the first step we choose the distribution of these vectors on the sphere. Defining the spherical coordinates

$$x = r \sin \vartheta \cos \phi, \qquad (86)$$

$$y = r \sin \vartheta \sin \phi, \qquad (87)$$

$$z = r \cos \vartheta \qquad (88)$$

and the corresponding unit vectors $\hat{r}$, $\hat{\vartheta}$, and $\hat{\phi}$ (pointing along the increase of the corresponding coordinate) helps us to define the vectors

$$u_R(\vartheta,\phi) = u(R,\vartheta,\phi) = e^{i(\vartheta+\phi)}\hat{r}, \qquad (89)$$

$$v_R(\vartheta,\phi) = v(R,\vartheta,\phi) = e^{i(\vartheta+\phi)} \frac{1}{\sqrt{1+\cos^2\vartheta}} (\hat{\vartheta} + i\hat{\phi}\cos\vartheta), \qquad (90)$$

$$w_R(\vartheta,\phi) = w(R,\vartheta,\phi) = e^{i(\vartheta+\phi)} \frac{1}{\sqrt{1+\cos^2\vartheta}} (\hat{\vartheta}\cos\vartheta - i\hat{\phi}). \qquad (91)$$

Obviously, u is normal on the whole sphere's surface. The vector v corresponds to the left circular polarization at the sphere's poles ($z = \pm R$), to the vertical linear polarization on the sphere's equator ($z = 0$), and is elliptical and continuous in the intermediate ranges. The vector w is simply chosen orthogonal to both u and v. Both v and w are tangential to the sphere's surface.

If we extend the radial dependence of the vectors defined by Equations 89–91 to the entire cubic cell (formally replacing R by r), then we obtain some analogy with Model B. The obtained matrix $\mathbf{F}$ has no concurrent discontinuities with ε, but it is discontinuous at the center and on the boundaries of the cubic cell. To remove the discontinuity at the cell's center, we will proceed differently.

As the second step, we define the basis vectors at the center of the sphere,

$$u_0 = u(0, \vartheta, \phi) = \frac{1}{\sqrt{2}}(\hat{x} - i\hat{y}), \tag{92}$$

$$v_0 = v(0, \vartheta, \phi) = \hat{z}, \tag{93}$$

$$w_0 = w(0, \vartheta, \phi) = \frac{1}{\sqrt{2}}(\hat{x} + i\hat{y}), \tag{94}$$

which are again mutually orthogonal. Then we extend u onto the whole cubic cell,

$$u_{\text{ext}}(r, \vartheta, \phi) = \begin{cases} \frac{1}{2}(u_0 + u_R) + \frac{1}{2}(u_0 - u_R)\cos\frac{\pi r}{R} & (r \le R) \\ \frac{1}{2}(u_0 + u_R) + \frac{1}{2}(u_0 - u_R)\cos\frac{\pi(r+D-2R)}{D-R} & (R < r \le D). \end{cases} \tag{95}$$

where $D(\vartheta, \phi)$ is the distance from the cell's center to its boundary along the ray determined by the spherical angles. The desired unit vector of the polarization basis is then

$$u = A_u u_{\text{ext}}, \tag{96}$$

where $A_u = [\frac{1}{2}(1 + \cos^2\frac{\pi r}{R})]^{-1/2}$ for $r < R$ and $A_u = [\frac{1}{2}(1 + \cos^2\frac{\pi(r+D-2R)}{D-R})]^{-1/2}$ for $r > R$ is a scalar function ensuring that u becomes a unit vector everywhere. We could extend v and w in a similar way,

$$v_{\text{ext}}(r, \vartheta, \phi) = \begin{cases} \frac{1}{2}(v_0 + v_R) + \frac{1}{2}(v_0 - v_R)\cos\frac{\pi r}{R} & (r \le R) \\ \frac{1}{2}(v_0 + v_R) + \frac{1}{2}(v_0 - v_R)\cos\frac{\pi(r+D-2R)}{D-R} & (R < r \le D). \end{cases} \tag{97}$$

$$w_{\text{ext}}(r, \vartheta, \phi) = \begin{cases} \frac{1}{2}(w_0 + w_R) + \frac{1}{2}(w_0 - w_R)\cos\frac{\pi r}{R} & (r \le R) \\ \frac{1}{2}(w_0 + w_R) + \frac{1}{2}(w_0 - w_R)\cos\frac{\pi(r+D-2R)}{D-R} & (R < r \le D). \end{cases} \tag{98}$$

but it is not clear whether v_{ext} or w_{ext} are both perpendicular to u and mutually. To ensure it we define

$$v = A_v(1 - \mathbf{P}_u)v_{\text{ext}}, \tag{99}$$

where $\mathbf{P}_u = uu^\dagger$ is the projector into the space of vectors proportional to u, so that $1 - \mathbf{P}_u$ is the projector to its orthogonal complement. Similarly,

$$w = A_w(1 - \mathbf{P}_u - \mathbf{P}_v)w_{\text{ext}}, \tag{100}$$

where $1 - \mathbf{P}_u - \mathbf{P}_v$ is the projector into the space of vectors perpendicular to both u and v. Here A_v and A_w are analogously chosen scalar functions ensuring the unit size of the corresponding vectors.

Finally, the matrix of permittivity in the reciprocal space with correct Fourier factorization, directly applicable to the wave Equation 1, becomes

$$[\![\varepsilon]\!]_C = [\![\mathbf{F}]\!] \begin{bmatrix} [\![\frac{1}{\varepsilon}]\!]^{-1} & [\![0]\!] & [\![0]\!] \\ [\![0]\!] & [\![\varepsilon]\!] & [\![0]\!] \\ [\![0]\!] & [\![0]\!] & [\![\varepsilon]\!] \end{bmatrix} [\![\mathbf{F}^{-1}]\!], \tag{101}$$

because the first element on the diagonal corresponds to the normal (u) components of the fields and the other two correspond to the tangential components (v, w) of the fields.

8. Conclusion

We have derived the methodology how to apply the Fourier factorization rules of Li (1996) to various photonic crystals. For the case of 1D crystals there is clear consistency of the correct rules with the classical theory of Yeh et al. (1977). For 2D crystals the convergence properties strongly depend on the chosen distribution of the polarization basis; we have shown that it is desirable to choose a distribution as smooth as possible. The method is also usable for periodic elements of any shape, where complicated shapes require complicated distributions of polarization bases. We can also use it to simulate 2D periodic elements made of anisotropic materials, as well as 3D periodic crystals. Moreover, the method can also be used to photonic devices such as photonic crystal waveguides by applying the demonstrated methodology to the device supercell. It is particularly advantageous for devices where high accuracy is required, e.g., for analyzing defect modes near photonic band edges [Dossou et al. (2009); Mahmoodian et al. (2009)], and for large devices for which the available computer memory enables calculations with only a few Fourier components (photonic crystals fibers with large cladding, or asymmetric 3D crystals).

9. Acknowledgments

This work is part of the research plan MSM 0021620834 financed by the Ministry of Education of the Czech Republic and was supported by a Marie Curie International Reintegration Grant (no. 224944) within the 7th European Community Framework Programme and by the Grant Agency of the Czech Republic (no. 202/09/P355 and P204/10/P346).

10. References

Antos, R. (2009). Fourier factorization with complex polarization bases in modeling optics of discontinuous bi-periodic structures, *Opt. Express* 17: 7269–7274.

Antos, R. & Veis, M. (2010). Fourier factorization with complex polarization bases in the plane-wave expansion method applied to two-dimensional photonic crystals, *Opt. Express* 18: 27511–27524.

Benisty, H. (2009). Dark modes, slow modes, and coupling in multimode systems, *J. Opt. Soc. Am. B* 26: 718–724.

Block, I. D., Ganesh, N., Lu, M. & Cunningham, B. T. (2008). Sensitivity model for predicting photonic crystal biosensor performance, *IEEE Sens. J.* 8: 274–280.

Bonod, N., Popov, E. & Neviere, M. (2005a). Fourier factorization of nonlinear Maxwell equations in periodic media: application to the optical Kerr effect, *Opt. Commun.* 244: 389–398.

Bonod, N., Popov, E. & Neviere, M. (2005b). Light transmission through a subwavelength microstructured aperture: electromagnetic theory and applications, *Opt. Commun.* 245: 355–361.

Boyer, P., Popov, E., Neviere, M. & Tayeb, G. (2004). Diffraction theory in TM polarization: application of the fast Fourier factorization method to cylindrical devices with arbitrary cross section, *J. Opt. Soc. Am. A* 21: 2146–2153.

Chernov, B., Neviere, M. & Popov, E. (2001). Fast Fourier factorization method applied to modal analysis of slanted lamellar diffraction gratings in conical mountings, *Opt. Commun.* 194: 289–297.

Datta, S., Chan, C. T., Ho, K. M. & Soukoulis, C. M. (1993). Effective dielectric constant of periodic composite structures, *Phys. Rev. B* 48: 14936–14943.

David, A., Benisty, H. & Weisbuch, C. (2006). Fast factorization rule and plane-wave expansion method for two-dimensional photonic crystals with arbitrary hole-shape, *Phys. Rev. B* 73: 075107.

Dossou, K. B., Poulton, C. G., Botten, L. C., Mahmoodian, S., McPhedran, R. C. & de Sterke, C. M. (2009). Modes of symmetric composite defects in two-dimensional photonic crystals, *Phys. Rev. A* 80: 013826.

Genereux, F., Leonard, S. W., van Driel, H. M., Birner, A. & Gosele, U. (2001). Large birefringence in two-dimensional silicon photonic crystals, *Phys. Rev. B* 63: 161101(R).

Gotz, P., Schuster, T., Frenner, K., Rafler, S. & Osten, W. (2008). Normal vector method for the RCWA with automated vector field generation, *Opt. Express* 16: 17295–17301.

Ho, K. M., Chan, C. T. & Soukoulis, C. M. (1990). Existence of a photonic gap in periodic dielectric structures, *Phys. Rev. Lett.* 65: 3152–3155.

Hugonin, J. P., Lalanne, P., White, T. P. & Krauss, T. F. (2007). Coupling into slow-mode photonic crystal waveguides, *Opt. Lett.* 32: 2638–2640.

Joannopoulos, J. D., Meade, R. D. & Winn, J. N. (1995). *Photonic Crystals: Molding the Flow of Light*, Princeton Univ.

Joannopoulos, J. D., Villeneuve, P. R. & Fan, S. (1997). Photonic crystals: Putting a new twist on light, *Nature (London)* 386: 143–149.

Kinoshita, S. & Yoshioka, S. (2005). Structural colors in nature: The role of regularity and irregularity in the structure, *ChemPhysChem* 6: 1442–1459.

Kosaka, H., Kawashima, T., Tomita, A., Notomi, M., Tamamura, T., Sato, T. & Kawakami, S. (1998). Superprism phenomena in photonic crystals, *Phys. Rev. B* 58: R10096–R10099.

Krokhin, A. A., Halevi, P. & Arriaga, J. (2002). Long-wavelength limit (homogenization) for two-dimensional photonic crystals, *Phys. Rev. B* 65: 115208.

Krokhin, A. A. & Reyes, E. (2004). Homogenization of magnetodielectric photonic crystals, *Phys. Rev. Lett.* 93: 023904.

Li, L. (1996). Use of Fourier series in the analysis of discontinuous periodic structures, *J. Opt. Soc. Am. A* 13: 1870–1876.

Li, L. (1997). New formulation of the Fourier modal method for crossed surface-relief gratings, *J. Opt. Soc. Am. A* 14: 2758–2767.

Li, L. (1998). Reformulation of the Fourier modal method for surface-relief gratings made with anisotropic materials, *J. Mod. Opt.* 45: 1313–1334.

Li, L. (2003). Fourier modal method for crossed anisotropic gratings with arbitrary permittivity and permeability tensors, *J. Opt. A* 5: 345–355.

Lin, S.-Y., Chow, E., Hietala, V., Villeneuve, P. R. & Joannopoulos, J. D. (1998). Experimental demonstration of guiding and bending of electromagnetic waves in a photonic crystal, *Science* 282: 274–276.

Mahmoodian, S., McPhedran, R. C., de Sterke, C. M., Dossou, K. B., Poulton, C. G. & Botten, L. C. (2009). Single and coupled degenerate defect modes in two-dimensional photonic crystal band gaps, *Phys. Rev. A* 79: 013814.

Meade, R. D., Rappe, A. M., Brommer, K. D., Joannopoulos, J. D. & Alerhand, O. L. (1993). Accurate theoretical analysis of photonic band-gap materials, *Phys. Rev. B* 48: 8434–8437.

Noda, S. (2006). Recent progresses and future prospects of two- and three-dimensional photonic crystals, *J. Lightwave Technol.* 24: 4554–4567.

Popov, E. & Neviere, M. (2000). Grating theory: new equations in Fourier space leading to fast converging results for TM polarization, *J. Opt. Soc. Am. A* 17: 1773–1784.

Reyes, E., Krokhin, A. A. & Roberts, J. (2005). Effective dielectric constants of photonic crystal of aligned anisotropic cylinders and the optical response of a periodic array of carbon nanotubes, *Phys. Rev. B* 72: 155118.

Schuster, T., Ruoff, J., Kerwien, N., Rafler, S. & Osten, W. (2007). Normal vector method for convergence improvement using the RCWA for crossed gratings, *J. Opt. Soc. Am. A* 24: 2880–2890.

Skivesen, N., Tetu, A., Kristensen, M., Kjems, J., Frandsen, L. H. & Borel, P. I. (2007). Photonic-crystal waveguide biosensor, *Opt. Express* 15: 3169–3176.

Sozuer, H. S., Haus, J. W. & Inguva, R. (1992). Photonic bands: Convergence problems with the plane-wave method, *Phys. Rev. B* 45: 13962–13972.

Vukusic, P. & Sambles, J. R. (2003). Photonic structures in biology, *Nature (London)* 424: 852–855.

Watanabe, K. (2002). Numerical integration schemes used on the differential theory for anisotropic gratings, *J. Opt. Soc. Am. A* 19: 2245–2252.

Watanabe, K., Petit, R. & Neviere, M. (2002). Differential theory of gratings made of anisotropic materials, *J. Opt. Soc. Am. A* 19: 325–334.

Yablonovitch, E. (1987). Inhibited spontaneous emission in solid-state physics and electronics, *Phys. Rev. Lett.* 58: 2059–2062.

Yariv, A. & Yeh, P. (1977). Electromagnetic propagation in periodic stratified media. II. Birefringence, phase matching, and x-ray lasers, *J. Opt. Soc. Am.* 67: 438–447.

Yeh, P., Yariv, A. & Hong, C.-S. (1977). Electromagnetic propagation in periodic stratified media. I. General theory, *J. Opt. Soc. Am.* 67: 423–438.

15

Coupled Mode Theory of Photonic Crystal Lasers

Marcin Koba[1] and Pawel Szczepanski[2]
[1]*University of Warsaw, National Institute of Telecommunications*
[2]*Warsaw University of Technology, National Institute of Telecommunications*
Poland

1. Introduction

Photonic crystals (PC) are structures with periodic variation of the refractive index in one, two or three spatial dimensions. The dynamic development of experimental and theoretical work on photonic crystals has been launched by Yablonovitch (1987; 1993) and Sajeev John (1987) publications, although the idea of periodic structures had been known since Strutt (1887).

The main property of photonic crystal is the existence of a frequency range, for which the propagation of electromagnetic waves in the medium is not permitted. These frequency ranges are commonly known as photonic band gaps, giving the ability to modify the structure parameters, e.g. group velocity, coherence length, gain, and spontaneous emission. This type of periodic structures is used in both passive and active devices.

1.1 Two-dimensional photonic crystal lasers

Much of the research on active structures is devoted to efficient photonic sources of coherent radiation. Photonic crystals are one of these structures, and they are used in lasers as mirrors (Dunbar et al. (2005); Scherer et al. (2005)), active waveguides (Watanabe & Baba (2006)), coupled cavities (Steinberg & Boag (2006)), defect microcavities (Asano et al. (2006); Lee et al. (2004)), and the laser active region (Cojocaru et al. (2005)).

Lasers with defect two-dimensional photonic crystals are known for their high finesse (Monat et al. (2001)) and very low threshold (Nomura et al. (2008)).

Photonic crystal band-edge lasers allow to obtain edge (Cojocaru et al. (2005)) and surface emission (Turnbull et al. (2003); Vurgaftman & Meyer (2003)) of coherent light from large cavity area. They also allow to control the output beam pattern by manipulation of the primitive cell geometry (Iwahashi et al. (2010); Miyai et al. (2006)), provide low threshold (Susa (2001)), and beams which can be focused to a size less than the wavelength (Matsubara et al. (2008)).

The photonic crystal structures lasing wavelengths span from terahertz (Chassagneux et al. (2009); Sirigu et al. (2008)), through infrared (Kim et al. (2006)) to visible (Lu et al. (2008); Zhang et al. (2006)).

1.2 Radiation generation modeling in photonic crystal lasers

Laser action in photonic crystal structures has been theoretically studied and centered on the estimation of the output parameters (Czuma & Szczepanski (2005); Lesniewska-Matys et al. (2005)) and models describing light generation processes e.g. (Florescu et al. (2002); Koba, Szczepanski & Kossek (2011); Sakai et al. (2010)). The most sophisticated and general (it describes one-, two-, and three-dimensional structures) semiclassical model of light generation in photonic structures is presented in (Florescu et al. (2002)). Theoretical analysis of photonic crystal lasers based on two-dimensional plane wave expansion method (PWEM) (Imada et al. (2002); Sakai et al. (2005)) and finite difference time domain method (FDTD) (Imada et al. (2002); Noda & Yokoyama (2005)) confirm experimental results. Nevertheless these methods suffer from important disadvantages, i.e. plane wave method gives a good approximation for infinite structures, whereas finite difference time domain method is suited for structures with only a few periods and consumes huge computer resources for the analysis of real photonic structures. Therefore these methods are not very convenient for design and optimization of actual photonic crystal lasers. Hence, different, less complicated methods of analysis of two-dimensional photonic crystal lasers are developed. These methods are meant to effectively support the design process of such lasers. They are based on a coupled mode theory (Sakai et al. (2006); Vurgaftman & Meyer (2003)) and focused on square and triangular lattice photonic crystals (Koba & Szczepanski (2010); Koba, Szczepanski & Kossek (2011); Koba, Szczepanski & Osuch (2011); Sakai et al. (2007; 2010; 2008)).

The Sakai et al. (2007; 2010) works contain a mathematical description and numerical results of the threshold analysis of two-dimensional (2-D) square lattice photonic crystal laser with TM and TE polarization. They introduce general coupled mode relations for a threshold gain, a Bragg frequency deviation and field distributions, and give calculation results for some specific values of coupling coefficients. Additionally, in (Sakai et al. (2007)) the effect of boundary reflections has been investigated, and it has been shown that the mode properties can be adjusted by changing refractive index or boundary conditions.

In Sakai et al. (2008) paper, the analytical description of triangular lattice photonic crystal cavity for TE polarization has been given. In this work the analysis was focused on the coupled wave equations and the dependence of the resonant frequencies on the coupling coefficients.

In Sakai et al. (2007; 2010; 2008) works threshold analysis has been conducted for specific values of coupling coefficient and TM polarization for triangular has not been considered.

The equations for triangular lattice photonic crystal laser with TM polarization has been shown in Koba, Szczepanski & Kossek (2011), and the evaluation of these is shown in this chapter.

The mentioned semiclassical model, presented by Florescu et al. (2002) describing an above threshold analysis is complicated and difficult to implement. To overcome this drawback, this chapter also includes an overview of our works (Koba & Szczepanski (2010); Koba, Szczepanski & Kossek (2011); Koba, Szczepanski & Osuch (2011)), where we introduced easy to implement models for an above threshold analysis of a two-dimensional photonic crystal laser.

Therefore, in the subsequent parts of this chapter we addressed the issues of the laser threshold characteristics in the wide range of the coupling coefficient and described all four cases of square and triangular lattice photonic crystal structures with TE and TM polarization. We also describe an above threshold analysis for these structures.

Thus, in this chapter we will summon the analytical models of the threshold and above threshold light generation in photonic crystal band-edge lasers considering square and triangular lattice structures with TE and TM polarization. Theoretical evaluation in this chapter is based on coupled wave model and energy theorem.

2. Structure definition

This paper describes the two-dimensional photonic crystals which properties can be described by the complex relative electrical permittivity ε. The cross sections of these structures are shown schematically in Fig. 1

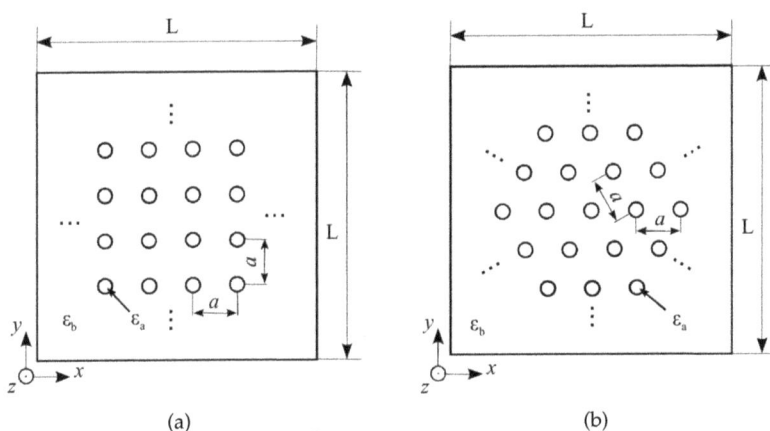

Fig. 1. a) Square and b) triangular lattice photonic structure cross section. (ε_a and ε_b are relative permittivities of rods and background material, respectively, a - lattice constant, L - cavity length)

In crystallography the ideal crystal is described by the elementary cell. The shape of the cell is defined by the basic vectors which linear combination allows to specify the location of all nodes of the structure. Each node is connected to the base which may be constituted by an atom, a group of atoms, molecules, etc. The photonic structures perfectly resemble the microscopic nature of the crystal lattice in the mesoscopic scale. This allows using the terminology adopted in the solid state physics to describe the photonic crystal.

In this chapter, only 2-D photonic crystals will be discussed. In the two-dimensional space, there are five basic types of crystal lattice. This comprises a square, hexagonal, rectangular, oblique, and rhombic lattice (Kittel (1995)). The square and hexagonal (also known as triangular) lattices are the most common types of symmetry used in the practical realizations of photonic cavities. The role of the base in such systems is often played by cylinders called rods or holes depending on the relative difference between the refractive index of the cylinders and the surrounding material.

The structures in Fig. 1 a) and b) are constrained in the xy plane by the square region of length L, and are assumed to be uniform and much larger than the wavelength in the the z direction. The permittivity of the holes and background material is ε_a and ε_b, respectively. The number of periods in the xy plane is finite, but large enough to be expanded in Fourier series with small error. Schemes in Fig. 1 a) and 1 b) illustrate two spatial distributions of rods for two-dimensional photonic crystal, respectively, with square and triangular lattice.

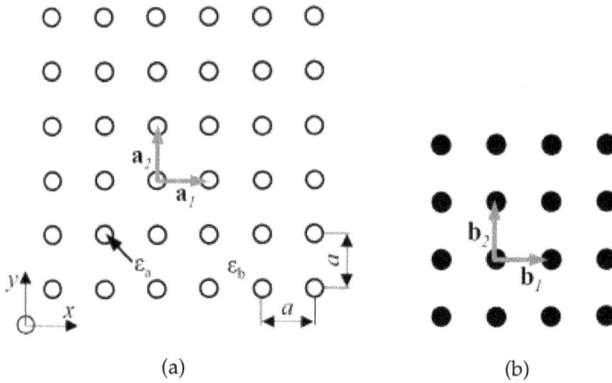

(a)   (b)

Fig. 2. The scheme of a) a square lattice photonic crystal with primitive vectors; and b) its representation in reciprocal space with reciprocal primitive vectors.

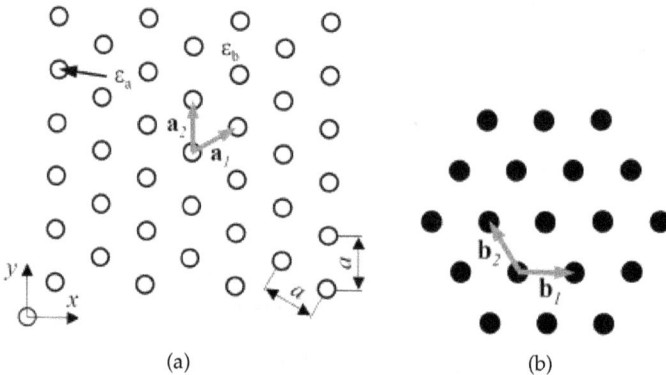

(a)   (b)

Fig. 3. The scheme of a) a triangular lattice photonic crystal with primitive vectors; and b) its representation in reciprocal space with reciprocal primitive vectors.

Fig. 2 a) and 3 a) show photonic crystal cross sections in xy plane with cylinders arranged in square or triangular lattice with period a, and with depicted primitive vectors $\mathbf{a}_1$ and $\mathbf{a}_2$.

Fig. 2 b) and 3 b) show the reciprocal lattices corresponding, respectively, to the real square and triangular lattice. In the described case, the nodes of a two-dimensional structure can be expressed by

$$x_{\parallel}(l) = l_1 a_1 + l_2 a_2 \tag{1}$$

where a_1 and a_2 are primitive vectors (Kittel (1995)), l_1 and l_2 are arbitrary integers, $x_{\parallel}$ specifies the placement on the plane, $x_{\parallel} = \hat{x}x + \hat{y}y$, where $\hat{x}$ and $\hat{y}$ are unit vectors along x and y axis, respectively. The area of primitive cell is $a_c = |a_1 \times a_2| = a^2$ in case of square lattice, and $a_c = |a_1 \times a_2| = \sqrt{3}a^2/2$ in case of triangular lattice. Primitive vectors for square lattice are described by the expressions: $a_1 = (a,0)$, $a_2 = (0,a)$, and for the triangular lattice: $a_1 = \left(\sqrt{3}a/2, a/2\right)$, $a_2 = (0,a)$.

In general, the reciprocal vecotrs can be written in the following form:

$$G(h) = h_1 b_1 + h_2 b_2 \tag{2}$$

where h_1 and h_2 are arbitrary integers, b_1 and b_2 are the primitive vectors of the two-dimensional reciprocal lattice:

$$b_1 = \frac{2\pi}{a_c}\left(a_y^{(2)}, -a_x^{(2)}\right), \qquad b_2 = \frac{2\pi}{a_c}\left(-a_y^{(1)}, a_x^{(1)}\right), \tag{3}$$

where $a_j^{(i)}$ is the j-th cartesian component (x or y) of the a_i vector (i = 1 lub 2) (Sakai et al. (2010)).

Using Equation 3 and the expressions for square and triangular lattice primitive vectors the reciprocal primitive vectors are described by the following formulas:

$$b_1 = (2\pi/a, 0), \qquad b_2 = (0, 2\pi/a) - \text{square lattice} \tag{4}$$

and

$$b_1 = \left(4\pi/\sqrt{3}a, 0\right), \qquad b_2 = \left(-2\pi/\sqrt{3}a, 2\pi/a\right) - \text{triangular lattice.} \tag{5}$$

The infinite square or triangular photonic crystal can be described in terms of relative permittivity by the functions:

$$\varepsilon^{-1}\left(x_{\parallel}\right) = \varepsilon_b^{-1} + \left(\varepsilon_a^{-1} - \varepsilon_b^{-1}\right)\sum_l S\left(x_{\parallel} - x_{\parallel}(l)\right) \tag{6}$$

in case of TE polarization, where it is more convenient to use the inverse of relative permittivity, and

$$\varepsilon\left(x_{\parallel}\right) = \varepsilon_b + (\varepsilon_a - \varepsilon_b)\sum_l S\left(x_{\parallel} - x_{\parallel}(l)\right) \tag{7}$$

for TM polarization. In previous Equations, function S

$$S\left(x_{\parallel}\right) = \begin{cases} 1 & \text{dla } x_{\parallel} \in O \\ 0 & \text{dla } x_{\parallel} \notin O \end{cases} \tag{8}$$

specifies the location of rods in the structure, O is the area of the xy plane defined by the cross section of the rod, which symmetry axis intersects the plane at the point $x_{\parallel} = 0$.

The functions describing the structure need to be transformed to the frequency domain in order to solve the wave equations. To do so, the crystal geometry is expressed in terms of reciprocal lattice vector by the Fourier transformation of functions 6 and 7 (M. Plihal & Maradudin (1991); M. Plihal et al. (1991)).

For TE polarization function $\epsilon^{-1}(\mathbf{G})$ is written in the following form:

$$\epsilon^{-1}(\mathbf{G}) = \begin{cases} \varepsilon_a^{-1} f + \varepsilon_b^{-1} (1-f), & \mathbf{G}_{\parallel} = 0 \\ \left(\varepsilon_a^{-1} - \varepsilon_b^{-1} \right) f \dfrac{2J_1(\mathbf{G}_{\parallel} R)}{(\mathbf{G}_{\parallel} R)}, & \mathbf{G}_{\parallel} \neq 0 \end{cases} \tag{9}$$

and for the TM polarization function $\epsilon(\mathbf{G})$:

$$\epsilon(\mathbf{G}) = \begin{cases} \varepsilon_a f + \varepsilon_b (1-f), & \mathbf{G}_{\parallel} = 0 \\ (\varepsilon_a - \varepsilon_b) f \dfrac{2J_1(\mathbf{G}_{\parallel} R)}{(\mathbf{G}_{\parallel} R)}, & \mathbf{G}_{\parallel} \neq 0 \end{cases} \tag{10}$$

where $f = \pi r^2 / a^2$ – square lattice filling factor, $f = \left(2\pi/\sqrt{3} \right) r^2 / a^2$ – triangular lattice filling factor, r – rod radius, J_1 – Bessel function of the first kind.

In further parts of this chapter four different cases have been analyzed. Two of them are dedicated to square lattice cavities with TE and TM polarization, and two remaining to triangular lattice structures also with TE and TM polarization.

In the next parts of this chapter the threshold and above threshold analysis of the photonic crystal laser operation has been shown for the defined structures.

3. A threshold analysis

3.1 Coupled-wave equations

In general, the scalar wave equations for the electric and magnetic fields E_z and H_z, respectively, are written in the following form (M. Plihal & Maradudin (1991); M. Plihal et al. (1991)):

$$\frac{\partial^2 E_z}{\partial x^2} + \frac{\partial^2 E_z}{\partial y^2} + k^2 E_z = 0 \tag{11}$$

and

$$\frac{\partial}{\partial x} \left\{ \frac{1}{k^2} \frac{\partial}{\partial x} H_z \right\} + \frac{\partial}{\partial y} \left\{ \frac{1}{k^2} \frac{\partial}{\partial y} H_z \right\} + H_z = 0 \tag{12}$$

where the constant k is given by (Sakai et al. (2007))

$$k^2 = \beta^2 + 2i (\alpha - \alpha_L) \beta + 2\beta \sum_{\mathbf{G} \neq 0} \kappa (\mathbf{G}) \exp (i (\mathbf{G} \cdot \mathbf{r})) \tag{13}$$

in case of TM modes, and (Sakai et al. (2010))

$$\frac{1}{k^2} = \frac{1}{\beta^4}\left(\beta^2 - i2\left(\alpha - \alpha_L\right)\beta + 2\beta \sum_{G \neq 0} \kappa(G)\exp\left(i\left(G \cdot r\right)\right)\right) \tag{14}$$

in case of TE modes. In Equations 13 and 14 $\beta = 2\pi\varepsilon_0^{1/2}/\lambda$ where $\varepsilon_0 = \varepsilon\left(G = 0\right)$ is the averaged dielectric permittivity ($\varepsilon_0^{1/2}$ corresponds to averaged refractive index n), α is an averaged gain in the medium, $\kappa(G)$ is the coupling constant, λ is the Bragg wavelength, and $G = (mb_1, nb_2)$ is the reciprocal lattice vector, m and n are arbitrary integers, b_1 and b_2 vary depending on the structure symmetry. Therefore, these vectors are expressed in the following forms $b_1 = (\beta_0^s, 0)$ and $b_2 = (0, \beta_0^s)$ for square lattice, and $b_1 = (\beta_0^t, 0)$ and $b_2 = \left(-\beta_0^t/2, \sqrt{3}\beta_0^t/2\right)$ for triangular lattice structure, where $\beta_0^s = 2\pi/a$ and $\beta_0^t = 4\pi/\sqrt{3}a$. In the derivation of Equations 13 and 14 following e.g. (Sakai et al. (2007)), we set $\alpha \ll \beta \equiv \frac{2\pi\varepsilon_0^{1/2}}{\lambda}$, $\varepsilon_{G\neq0} \ll \varepsilon_0$, and $\alpha_G \ll \beta$. In these equations the periodic variation in the refractive index is included as a small perturbation and appears in the third term through the coupling constant $\kappa(G)$ of the form:

$$\kappa(G) = -\frac{\pi}{\lambda\varepsilon_0^{1/2}}\varepsilon(G) \pm i\frac{\alpha(G)}{2}. \tag{15}$$

In Equation 15, plus sign refers to TM polarization (Equation 13), while minus sign refers to TE polarization (Equation 14). Furthermore, we set $\alpha(G)|_{G\neq0} = 0$ neglecting spatial periodicity of gain. In the vicinity of the Bragg wavelength only some of the diffraction orders contribute in a significant way, where in general, a periodic perturbation produces an infinite set of diffraction orders. Therefore the Bragg frequency orders have to be cautiously chosen. The Bragg frequency corresponding to the Γ point in the photonic band structure, e.g. (Sakai et al. (2007)) is chosen for the purpose of this paper, and the most significantly contributing coupling constants are expressed as follows:

$$\kappa_1 = \kappa(G)|_{|G|=\beta_0^{s,t}} \quad \kappa_2 = \kappa(G)|_{|G|=\sqrt{3}\beta_0^{s,t}} \quad \kappa_3 = \kappa(G)|_{|G|=2\beta_0^{s,t}} \tag{16}$$

In Equations 11 and 12 electric and magnetic fields for the infinite periodic structure are given by the Bloch modes, (M. Plihal & Maradudin (1991); Vurgaftman & Meyer (2003)):

$$E_z(r) = \sum_G e(G)\exp\left(i\left(k + G\right)\cdot r\right) \tag{17}$$

and

$$H_z(r) = \sum_G h(G)\exp\left(i\left(k + G\right)\cdot r\right) \tag{18}$$

where the functions $e(G)$ and $h(G)$ correspond to plane wave amplitudes, and the wave vector is denoted by k. In the first Brillouin zone at the Γ point the wave vector vanishes $k = 0$, see e.g. (Sakai et al. (2006)). For a finite structure, the amplitude of each plane wave is not constant, so $e(G)$ and $h(G)$ become functions of space. At the Γ point we consider only the amplitudes ($e(G)$, $h(G)$) which are meant to be significant, i.e. in most cases with $|G| = \beta_0^{s,t}$, except for square lattice with TE polarization where additional $h(G)$ amplitudes with $|\mathbf{G}| = \sqrt{2}\beta_0^s$ have to be included (Sakai et al. (2006)). The contributions of other waves of higher order in the Bloch mode are considered to be negligible.

3.1.1 Square lattice - TM polarization

Considering square lattice photonic crystal with TM polarization, it is assumed that at the Γ point the most significant contribution to coupling is given by the electric waves which fulfill the condition $\left(|\mathbf{G}| = \beta_0^s\right)$. Thus, all higher order electric wave expansion coefficients $\left(|\mathbf{G}| \geq \sqrt{2}\beta_0\right)$ are negligible. Four basic waves most significantly contributing to coupling are depicted in Fig. 4.

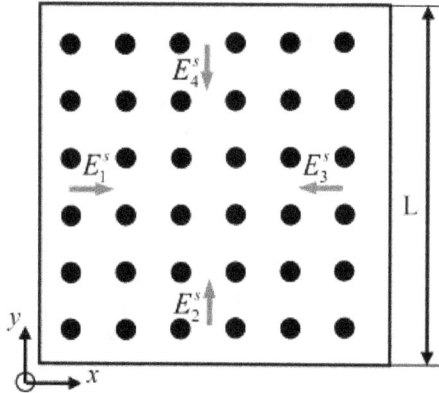

Fig. 4. Schematic cross section of square lattice photonic crystal laser active region, where the four basic waves involved in coupling for TM polarization are shown.

Equation 17 describes infinite structures. It is possible to take into account the fact that the structure is finite by using the space dependent amplitudes, e.g. (Sakai et al. (2007)). Thus, the electric field given by Equation 17 in the finite periodic structure can be expressed in the following way:

$$E_z = E_1^s(x,y)e^{-i\beta_0^s x} + E_2^s(x,y)e^{i\beta_0^s x} + E_3^s(x,y)e^{-i\beta_0^s y} + E_4^s(x,y)e^{i\beta_0^s y} \qquad (19)$$

In Equation 19 E_i^s, $i = 1..4$ are the four basic electric field amplitudes propagating in four directions $+x, -x, +y, y$. These amplitudes correspond to $e(\mathbf{G})$ in Equation 17. In the further analysis, we will drop the space dependence notation.

Knowing the reciprocal lattice vectors for the square lattice PC, the coupling coefficients $\kappa(\mathbf{G})$ 16 can be written as:

$$\kappa_1 = \frac{\pi\left(\varepsilon_a - \varepsilon_b\right)}{a\left(\varepsilon_a f + \varepsilon_b\left(1 - f\right)\right)} \frac{2f J_1\left(2\sqrt{\pi f}\right)}{\left(2\sqrt{\pi f}\right)} \qquad (20)$$

$$\kappa_2 = \frac{\pi\left(\varepsilon_a - \varepsilon_b\right)}{a\left(\varepsilon_a f + \varepsilon_b\left(1 - f\right)\right)} \frac{2f J_1\left(2\sqrt{2\pi f}\right)}{\left(2\sqrt{2\pi f}\right)} \qquad (21)$$

$$\kappa_3 = \frac{\pi\left(\varepsilon_a - \varepsilon_b\right)}{a\left(\varepsilon_a f + \varepsilon_b\left(1 - f\right)\right)} \frac{2f J_1\left(4\sqrt{\pi f}\right)}{\left(4\sqrt{\pi f}\right)} \qquad (22)$$

Putting Equations 13 and 19 into Equation 11, and assuming the slow varying electromagnetic field, one can get the set of coupled mode equations (Sakai et al. (2007)):

$$-\frac{\partial}{\partial x} E_1^s + (\alpha - \alpha_L - \kappa_0 - i\delta) E_1^s = (i\kappa_3 - \kappa_0) E_3^s + i\kappa_2 (E_2^s + E_4^s) \tag{23}$$

$$\frac{\partial}{\partial x} E_3^s + (\alpha - \alpha_L - \kappa_0 - i\delta) E_3^s = (i\kappa_3 - \kappa_0) E_1^s + i\kappa_2 (E_2^s + E_4^s) \tag{24}$$

$$-\frac{\partial}{\partial y} E_2^s + (\alpha - \alpha_L - \kappa_0 - i\delta) E_2^s = (i\kappa_3 - \kappa_0) E_4^s + i\kappa_2 (E_1^s + E_3^s) \tag{25}$$

$$\frac{\partial}{\partial y} E_4^s + (\alpha - \alpha_L - \kappa_0 - i\delta) E_4^s = (i\kappa_3 - \kappa_0) E_2^s + i\kappa_2 (E_1^s + E_3^s) \tag{26}$$

where

$$\delta = (\beta^2 - \beta_0^{s2})/2\beta \approx \beta - \beta_0^s \tag{27}$$

is the Bragg frequency deviation, κ_2 and κ_3 are coupling coefficients expressed by Equations 21 and 22 (Sakai et al. (2007)). The κ_2 coefficient is responsible for orthogonal coupling (e.g. the coupling of E_1^s to E_2^s and E_4^s), and κ_2 corresponds to backward coupling (e.g. the coupling of E_1^s to E_3^s). The additional coefficient κ_0 denotes surface emission losses, and it is proportional to κ_1 (Sakai et al. (2007; 2010)). Solution of Equations 23-26 for the boundary conditions:

$$E_1^s(-\frac{L}{2}, y) = E_3^s(\frac{L}{2}, y) = 0, E_2^s(x, -\frac{L}{2}) = E_4^s(x, \frac{L}{2}) = 0 \tag{28}$$

defines eigenmodes of the photonic structure. The analysis of this solution will be shown in section 3.2.

3.1.2 Square lattice - TE polarization

In the square lattice photonic crystal cavity with TE polarization, as mentioned before, the coupling process involves magnetic waves satisfying following conditions: ($|\mathbf{G}| = \beta_0$) and $\left(|\mathbf{G}| = \sqrt{2}\beta_0\right)$, (Sakai et al. (2010)), neglecting higher order Bloch modes. Eight basic waves most significantly contributing to coupling are depicted in Fig. 5.

Similarly as in the case of TM polarization, the equation for magnetic field (Equation 18) describes modes for infinite structure. Thus, the finite dimensions of the structure are described by spatial dependence of magnetic field amplitudes (Sakai et al. (2010)), and the magnetic field 18 is written in the following form:

$$H_z(\mathbf{r}) = H_1^s(x, y)e^{-i\beta_0^s x} + H_5^s(x, y)e^{i\beta_0^s x} + H_3^s(x, y)e^{-i\beta_0^s y} + H_7^s(x, y)e^{i\beta_0^s y} + H_2^s(x, y)e^{-i\beta_0^s x - i\beta_0^s y}$$

$$+ H_4^s(x, y)e^{i\beta_0^s x - i\beta_0^s y} + H_6^s(x, y)e^{i\beta_0^s x + i\beta_0^s y} + H_8^s(x, y)e^{-i\beta_0^s x + i\beta_0^s y} \tag{29}$$

In Equation 29 H_i^s, $i = 1..8$ are the eight basic magnetic field amplitudes of waves propagating in directions schematically shown in Fig. 5. These amplitudes correspond to $h(\mathbf{G})$ in Equation 18. Joining Equations 14, 29, and 12, and assuming slowly varying amplitudes, the coupled

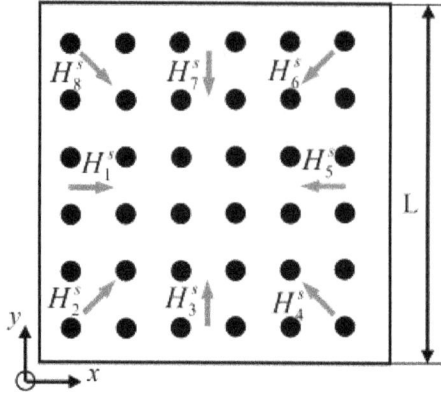

Fig. 5. Schematic cross section of square lattice photonic crystal laser active region, where the eight basic waves involved in coupling for TE polarization are shown.

wave equations for TE modes in square lattice PC are obtained (Sakai et al. (2010)):

$$-\frac{\partial}{\partial x}H_1^s + (\alpha - \alpha_L - \kappa_0 - i\delta)\,H_1^s = (i\kappa_3 - \kappa_0)\,H_5^s + i\frac{2\kappa_1^2}{\beta_0^s}\left(2H_1^s + H_3^s + H_7^s\right) \tag{30}$$

$$\frac{\partial}{\partial x}H_5^s + (\alpha - \alpha_L - \kappa_0 - i\delta)\,H_5^s = (i\kappa_3 - \kappa_0)\,H_1^s + i\frac{2\kappa_1^2}{\beta_0^s}\left(2H_5^s + H_3^s + H_7^s\right) \tag{31}$$

$$-\frac{\partial}{\partial x}H_3^s + (\alpha - \alpha_L - \kappa_0 - i\delta)\,H_3^s = (i\kappa_3 - \kappa_0)\,H_7^s + i\frac{2\kappa_1^2}{\beta_0^s}\left(2H_3^s + H_1^s + H_5^s\right) \tag{32}$$

$$\frac{\partial}{\partial x}H_7^s + (\alpha - \alpha_L - \kappa_0 - i\delta)\,H_7^s = (i\kappa_3 - \kappa_0)\,H_3^s + i\frac{2\kappa_1^2}{\beta_0^s}\left(2H_7^s + H_1^s + H_5^s\right) \tag{33}$$

In Equations 30-33, the spatial dependence of H_i^s, $i = 2, 4, 6, 8$ amplitudes was neglected, and it was assumed that $\alpha \ll \delta$. In Equations 30-33, δ is the Bragg frequency deviation, given by 27. The coupling coefficients κ_1, κ_2, and κ_3, defined by Equations 16 are expressed by (Sakai et al. (2010; 2008)):

$$\kappa_1 = \frac{\pi\left(\varepsilon_a^{-1} - \varepsilon_b^{-1}\right)}{a\left(\varepsilon_a^{-1}f + \varepsilon_b^{-1}(1-f)\right)}\frac{2fJ_1\left(2\sqrt{\pi f}\right)}{\left(2\sqrt{\pi f}\right)} \tag{34}$$

$$\kappa_2 = \frac{\pi\left(\varepsilon_a^{-1} - \varepsilon_b^{-1}\right)}{a\left(\varepsilon_a^{-1}f + \varepsilon_b^{-1}(1-f)\right)}\frac{2fJ_1\left(2\sqrt{2\pi f}\right)}{\left(2\sqrt{2\pi f}\right)} \tag{35}$$

$$\kappa_3 = \frac{\pi\left(\varepsilon_a^{-1} - \varepsilon_b^{-1}\right)}{a\left(\varepsilon_a^{-1}f + \varepsilon_b^{-1}(1-f)\right)}\frac{2fJ_1\left(4\sqrt{\pi f}\right)}{\left(4\sqrt{\pi f}\right)} \tag{36}$$

In contrast to TM polarization, in Equations 30-33, the coupling coefficient responsible for coupling in perpendicular direction κ_2 vanishes. The coupling coefficient κ_3 has the same meaning as described in the previous (TM) case, whereas the coupling coefficient κ_1 describes the coupling of e.g. waves H_1^s, H_2^s, and H_8^s. Solution of Equations 30-33 for the following boundary conditions:

$$H_7^s\left(-\frac{L}{2},y\right) = H_5^s\left(\frac{L}{2},y\right) = 0, H_3^s\left(x,-\frac{L}{2}\right) = H_7^s\left(x,\frac{L}{2}\right) = 0 \tag{37}$$

defines structure eigenmodes at lasing threshold i.e. in the linear case.

3.1.3 Triangular lattice - TM polarization

In the triangular lattice photonic crystal cavity with TM polarization, the coupling process involves waves satisfying following conditions ($|\mathbf{G}| = \beta_0$), neglecting higher order Bloch modes (Koba, Szczepanski & Kossek (2011); Sakai et al. (2008)). Six basic waves most significantly contributing to coupling are depicted in Fig. 6.

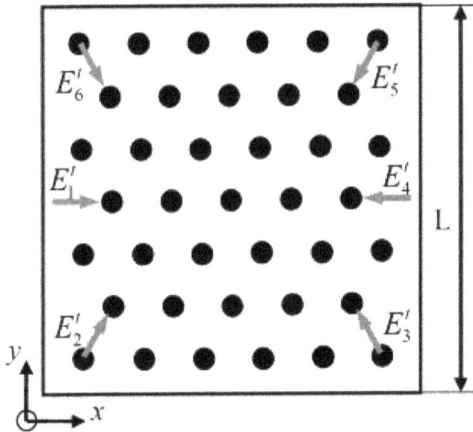

Fig. 6. A schematic cross section of a triangular lattice photonic crystal laser active region, where the six basic waves involved in the coupling for TM polarization are shown.

The space dependent amplitudes for electric field $e(\mathbf{G})$ (Equation 17) in triangular lattice photonic crystal cavity are written in the following form (Koba, Szczepanski & Kossek (2011)):

$$E_z = E_1^t(x,y)e^{-i\beta_0^t x} + E_2^t(x,y)e^{-i\frac{\beta_0^t}{2}x-i\frac{\sqrt{3}\beta_0^t}{2}y} + E_3^t(x,y)e^{i\frac{\beta_0^t}{2}x-i\frac{\sqrt{3}\beta_0^t}{2}y}$$

$$+E_4^t(x,y)e^{i\beta_0^t x} + E_5^t(x,y)e^{i\frac{\beta_0^t}{2}x+i\frac{\sqrt{3}\beta_0^t}{2}y} + E_6^t(x,y)e^{-i\frac{\beta_0^t}{2}x+i\frac{\sqrt{3}\beta_0^t}{2}y} \tag{38}$$

In Equation 38, E_i^t, $i = 1..6$, are the six electric field amplitudes propagating in the symmetry directions, Fig. 6. Combining Equations 13, 38 and 11, and assuming slowly varying amplitudes, the coupled wave equations for TM modes in triangular lattice PC are obtained:

$$-\frac{\partial}{\partial x}E_1^t + (\alpha - \alpha_L - \kappa_0 - i\delta)E_1^t = i\kappa_1\left(E_2^t + E_6^t\right) + i\kappa_2\left(E_3^t + E_5^t\right) + (i\kappa_3 - \kappa_0)E_4^t \tag{39}$$

$$-\frac{1}{2}\frac{\partial}{\partial x}E_2^t - \frac{\sqrt{3}}{2}\frac{\partial}{\partial y}E_2^t + (\alpha - \alpha_L - \kappa_0 - i\delta) E_2^t =$$

$$= i\kappa_1 \left(E_1^t + E_3^t\right) + i\kappa_2 \left(E_4^t + E_6^t\right) + (i\kappa_3 - \kappa_0) E_5^t \qquad (40)$$

$$\frac{1}{2}\frac{\partial}{\partial x}E_3^t - \frac{\sqrt{3}}{2}\frac{\partial}{\partial y}E_3^t + (\alpha - \alpha_L - \kappa_0 - i\delta) E_3^t =$$

$$= i\kappa_1 \left(E_2^t + E_4^t\right) + i\kappa_2 \left(E_1^t + E_5^t\right) + (i\kappa_3 - \kappa_0) E_6^t \qquad (41)$$

$$\frac{\partial}{\partial x}E_4^t + (\alpha - \alpha_L - \kappa_0 - i\delta)E_4^t = i\kappa_1 \left(E_3^t + E_5^t\right) + i\kappa_2 \left(E_2^t + E_6^t\right) + (i\kappa_3 - \kappa_0) E_1^t \qquad (42)$$

$$\frac{1}{2}\frac{\partial}{\partial x}E_5^t + \frac{\sqrt{3}}{2}\frac{\partial}{\partial y}E_5^t + (\alpha - \alpha_L - \kappa_0 - i\delta) E_5^t =$$

$$= i\kappa_1 \left(E_4^t + E_6^t\right) + i\kappa_2 \left(E_1^t + E_3^t\right) + (i\kappa_3 - \kappa_0) E_2^t \qquad (43)$$

$$-\frac{1}{2}\frac{\partial}{\partial x}E_6^t + \frac{\sqrt{3}}{2}\frac{\partial}{\partial y}E_6^t + (\alpha - \alpha_L - \kappa_0 - i\delta) E_6^t =$$

$$= i\kappa_1 \left(E_1^t + E_5^t\right) + i\kappa_2 \left(E_2^t + E_4^t\right) + (i\kappa_3 - \kappa_0) E_3^t \qquad (44)$$

In Equations 39-44, like in the case of square lattice, δ is the Bragg frequency deviation, given by Equation 27, while κ_1, κ_2, and κ_3 are the coupling coefficients, which are defined by 16 and as follows (Koba, Szczepanski & Kossek (2011)):

$$\kappa_1 = \frac{\pi (\varepsilon_a - \varepsilon_b)}{a (f\varepsilon_a + (1 - f)\varepsilon_b)} \frac{2fJ_1(\sqrt{8\pi f/\sqrt{3}})}{\sqrt{8\pi f/\sqrt{3}}} \qquad (45)$$

$$\kappa_2 = \frac{\pi (\varepsilon_a - \varepsilon_b)}{a (f\varepsilon_a + (1 - f)\varepsilon_b)} \frac{2fJ_1(\sqrt{\sqrt{3}8\pi f})}{\sqrt{\sqrt{3}8\pi f}} \qquad (46)$$

$$\kappa_3 = \frac{\pi (\varepsilon_a - \varepsilon_b)}{a (f\varepsilon_a + (1 - f)\varepsilon_b)} \frac{fJ_1(2\sqrt{8\pi f/\sqrt{3}})}{\sqrt{8\pi f/\sqrt{3}}} \qquad (47)$$

These coefficients describe strength and direction of the coupling of the waves, e.g. the coupling of E_1^t and E_4^t is described by κ_3, the coupling of E_1^t, E_2^t, and E_6^t by κ_1, and the coupling of E_1^t, E_3^t, and E_5^t by κ_2. In Equations 39-44, there is an additional coefficient κ_0 which, like in the square lattice case, is responsible for surface emission losses (Kazarinov & Henry (1985); Vurgaftman & Meyer (2003)). Solution of Equations 39-44 for the boundary conditions:

$$E_1^t(-\frac{L}{2},y) = 0, E_2^t(-\frac{L}{2},y) = E_2^t(x,-\frac{L}{2}) = 0, E_3^t(\frac{L}{2},y) = E_3^t(x,-\frac{L}{2}) = 0,$$

$$E_4^t(\frac{L}{2},y) = 0, E_5^t(\frac{L}{2},y) = E_5^t(x,\frac{L}{2}) = 0, E_6^t(-\frac{L}{2},y) = E_6^t(x,\frac{L}{2}) = 0 \qquad (48)$$

defines structure eigenmodes at lasing threshold.

3.1.4 Triangular lattice - TE polarization

In the triangular lattice photonic crystal cavity with TE polarization, the coupling process involves waves satisfying the same condition as it was stated in TM polarization case, i.e. $(|G| = \beta_0)$, (Sakai et al. (2008)), neglecting higher order Bloch modes. Six basic waves most significantly contributing to coupling are depicted in Fig. 7.

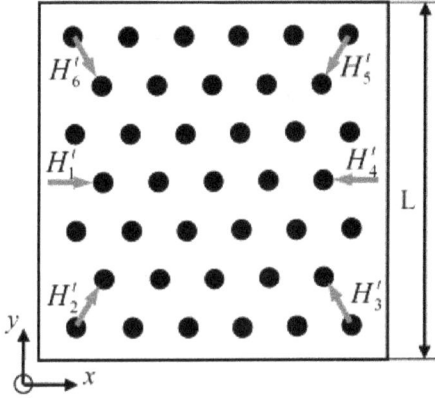

Fig. 7. A schematic cross section of a triangular lattice photonic crystal laser active region, where the six basic waves involved in the coupling for TE polarization are shown.

The magnetic field amplitudes $h(G)$ (Equation 18) in the triangular lattice photonic crystal cavity are written as follows (Sakai et al. (2008)):

$$H_z = H_1^t(x,y)e^{-i\beta_0 x} + H_2^t(x,y)e^{-i\frac{\beta_0}{2}x - i\frac{\sqrt{3}\beta_0}{2}y} + H_3^t(x,y)e^{i\frac{\beta_0}{2}x - i\frac{\sqrt{3}\beta_0}{2}y}$$

$$+ H_4^t(x,y)e^{i\beta_0 x} + H_5^t(x,y)e^{i\frac{\beta_0}{2}x + i\frac{\sqrt{3}\beta_0}{2}y} + H_6^t(x,y)e^{-i\frac{\beta_0}{2}x + i\frac{\sqrt{3}\beta_0}{2}y} \qquad (49)$$

In Equation 49, $H_i^t, i = 1..6$, are the six magnetic field amplitudes propagating in the symmetry directions, Fig. 7. Combining Equations 14, 49 and 12, and assuming slowly varying magnetic field amplitudes, the coupled wave equations for TE modes in triangular lattice PC are obtained:

$$-\frac{\partial}{\partial x}H_1^t + (\alpha - \alpha_L - \kappa_0 - i\delta)H_1^t = -i\frac{\kappa_1}{2}\left(H_2^t + H_6^t\right) + i\frac{\kappa_2}{2}\left(H_3^t + H_5^t\right) + (i\kappa_3 - \kappa_0)H_4^t \quad (50)$$

$$-\frac{1}{2}\frac{\partial}{\partial x}H_2^t - \frac{\sqrt{3}}{2}\frac{\partial}{\partial y}H_2^t + (\alpha - \alpha_L - \kappa_0 - i\delta)H_2^t =$$
$$= -i\frac{\kappa_1}{2}\left(H_1^t + H_3^t\right) + i\frac{\kappa_2}{2}\left(H_4^t + H_6^t\right) + (i\kappa_3 - \kappa_0)H_5^t \qquad (51)$$

$$\frac{1}{2}\frac{\partial}{\partial x}H_3^t - \frac{\sqrt{3}}{2}\frac{\partial}{\partial y}H_3^t + (\alpha - \alpha_L - \kappa_0 - i\delta)H_3^t =$$
$$= -i\frac{\kappa_1}{2}\left(H_2^t + H_4^t\right) + i\frac{\kappa_2}{2}\left(H_1^t + H_5^t\right) + (i\kappa_3 - \kappa_0)H_6^t \qquad (52)$$

$$\frac{\partial}{\partial x}H_4^t + (\alpha - \alpha_L - \kappa_0 - i\delta)H_4^t = -i\frac{\kappa_1}{2}\left(H_3^t + H_5^t\right) + i\frac{\kappa_2}{2}\left(H_2^t + H_6^t\right) + (i\kappa_3 - \kappa_0)H_1^t \quad (53)$$

$$\frac{1}{2}\frac{\partial}{\partial x}H_5^t + \frac{\sqrt{3}}{2}\frac{\partial}{\partial y}H_5^t + (\alpha - \alpha_L - \kappa_0 - i\delta) H_5^t =$$
$$= -i\frac{\kappa_1}{2}\left(H_4^t + H_6^t\right) + i\frac{\kappa_2}{2}\left(H_1^t + H_3^t\right) + (i\kappa_3 - \kappa_0) H_2^t \qquad (54)$$

$$-\frac{1}{2}\frac{\partial}{\partial x}H_6^t + \frac{\sqrt{3}}{2}\frac{\partial}{\partial y}H_6^t + (\alpha - \alpha_L - \kappa_0 - i\delta) H_6^t =$$
$$= -i\frac{\kappa_1}{2}\left(H_1^t + H_5^t\right) + i\frac{\kappa_2}{2}\left(H_2^t + H_4^t\right) + (i\kappa_3 - \kappa_0) H_3^t \qquad (55)$$

where the coupling coefficients κ_1, κ_2, and κ_3 are described by

$$\kappa_1 = \frac{-\pi \left(\varepsilon_a^{-1} - \varepsilon_b^{-1}\right)}{a \left(f\varepsilon_a^{-1} + (1-f)\,\varepsilon_b^{-1}\right)} \frac{2fJ_1(\sqrt{8\pi f/\sqrt{3}})}{\sqrt{8\pi f/\sqrt{3}}} \qquad (56)$$

$$\kappa_2 = \frac{-\pi \left(\varepsilon_a^{-1} - \varepsilon_b^{-1}\right)}{a \left(f\varepsilon_a^{-1} + (1-f)\,\varepsilon_b^{-1}\right)} \frac{2fJ_1(\sqrt{8\pi f\sqrt{3}})}{\sqrt{8\pi f\sqrt{3}}} \qquad (57)$$

$$\kappa_3 = \frac{-\pi \left(\varepsilon_a^{-1} - \varepsilon_b^{-1}\right)}{a \left(f\varepsilon_a^{-1} + (1-f)\,\varepsilon_b^{-1}\right)} \frac{fJ_1(2\sqrt{8\pi f/\sqrt{3}})}{\sqrt{8\pi f/\sqrt{3}}} \qquad (58)$$

and have the same physical meaning like it was described in the TM polarization case. The boundary conditions for the square region of PC with triangular symmetry are written as:

$$H_1^t(-\frac{L}{2},y) = 0, H_2^t(-\frac{L}{2},y) = H_2^t(x,-\frac{L}{2}) = 0, H_3^t(\frac{L}{2},y) = H_3^t(x,-\frac{L}{2}) = 0,$$
$$H_4^t(\frac{L}{2},y) = 0, H_5^t(\frac{L}{2},y) = H_5^t(x,\frac{L}{2}) = 0, H_6^t(-\frac{L}{2},y) = H_6^t(x,\frac{L}{2}) = 0. \qquad (59)$$

3.2 Numerical analysis of the PC laser threshold operation

3.2.1 Square lattice - TM and TE polarization

In Fig. 8 enlarged areas of a square lattice photonic crystal dispersion characteristics for the first four modes (A,B,C,D) in the vicinity of Γ point are shown. At the photonic band edge, i.e. at the Γ point, the cavity finesse increases, hence the active medium is used more efficiently. The dispersion curves are plotted for a) TM polarization and b) TE polarization. The plane wave method (Johnson & Joannopoulos (2001)) was used to plot the dispersion characteristic for the infinite two-dimensional PC structure with circular holes $\varepsilon_b = 9.8$ arranged in square lattice with background material $\varepsilon_a = 12.0$. The rods radius to lattice constant ratio was set to 0.24. In each plot, i.e. Fig. 8a) and Fig. 8b), one can observe two degenerate modes: B,C for TM polarization and C,D for TE polarization. They have the same frequency at Γ point. Modes A have the lowest frequency.

In Fig. 8 each of the marked points (A,B,C,D) represents a mode, which is characterized by: Bragg frequency deviation δ, threshold gain α, and threshold field distribution. These characteristic values were calculated by the numerical solution of Equations 23-26 for TM

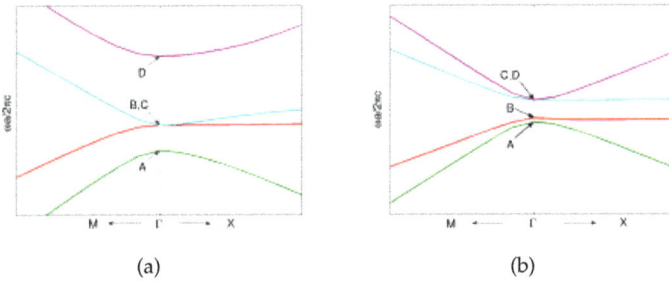

(a) (b)

Fig. 8. An enlarged area of a square lattice photonic crystal dispersion curves for the first four modes in the vicinity of Γ point. Square lattice, a) TM polarization, and b) TE polarization.

polarization and Equations 30-33 for TE polarization. In order to assign appropriate points A,B,C,D to the obtained numerical values, it was necessary to use the analytic expressions for the Bragg frequency deviation (Sakai et al. (2006)):

$$\delta_A = -2\kappa_2 - \kappa_3, \quad \delta_{B,C} = \kappa_3, \quad \delta_D = 2\kappa_2 - \kappa_3 \tag{60}$$

in case of TM polarization, and

$$\delta_A = -8\kappa_1^2/\beta_0 - \kappa_3, \quad \delta_B = -\kappa_3, \quad \delta_{C,D} = -4\kappa_1^2/\beta_0 + \kappa_3 \tag{61}$$

in case of TE polarization. These expressions were obtained from Equations 23-26 and 30-33 where no gain ($\alpha = 0$), no loss ($\kappa_0 = 0$, $\alpha_L = 0$), and no spatial dependence of electric or magnetic field amplitude were assumed. Sets of Equations 23-26 and 30-33 were solved numerically for the wide range of coupling coefficients (κ_1, κ_2, κ_3). We grouped obtained solutions: $\left((\delta, \alpha, E_m^s)^j\right)_{\kappa_{3i}}$ or $\left((\delta, \alpha, H_m^s)^j\right)_{\kappa_{3i}}$, where κ_{3i} corresponds to subsequent values of coupling coefficient for different modes $j = A, B, C, D$; $m = 1..4$, s-denotes square lattice. Assigning numerical values of δ_j to analytical solutions 60 and 61 ($\delta_A, \delta_{B,C}, \delta_D$), we obtained the mode structure of 2-D square lattice PC laser with TM and TE polarization.

Fig. 9 and 10 show the field distributions $|\sum_m |E_m^s|^2|$ and $|\sum_m |H_m^s|^2|$, respectively, corresponding to the modes: A - Fig. 9a), D - Fig. 9b), B,C - Fig. 9c), d) for TM modes, and A - Fig. 10a), B - Fig. 10b), C, D - Fig. 10c), d) for TE modes. The plots were made for the normalized coupling coefficients $|\kappa_1 L| = 10.96$, $|\kappa_2 L| = 8$, $|\kappa_3 L| = 4$ and filling factor $f = 0.16$. In each case (TM and TE polarization), the doubly degenerate modes are orthogonal and show saddle-shaped patterns. All non-degenerate modes are similar and exhibit Gaussian-like pattern, and this suggests that these modes should more efficiently use the photonic cavity. These modes also have lower threshold , Fig. 11.

In Fig. 11a) and 11b), the normalized threshold gain αL was plotted as a function of Bragg frequency deviation δL, for various values of the normalized coupling coefficient $|\kappa_3 L|$ (it takes values from 0.01 to 50).

Fig. 11a) and 11b) show that by increasing the value of coupling coefficient the Bragg frequency deviation increases and the threshold gain decreases. Simultaneously, for larger

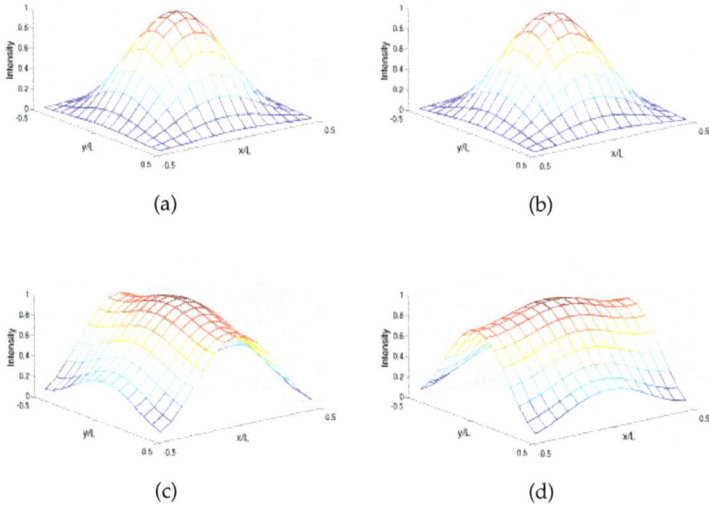

(a)   (b)

(c)   (d)

Fig. 9. Electromagnetic field distributions corresponding to a)A, b)D, c)B, and d) C points from Fig. 8a), respectively. Square lattice, TM polarization.

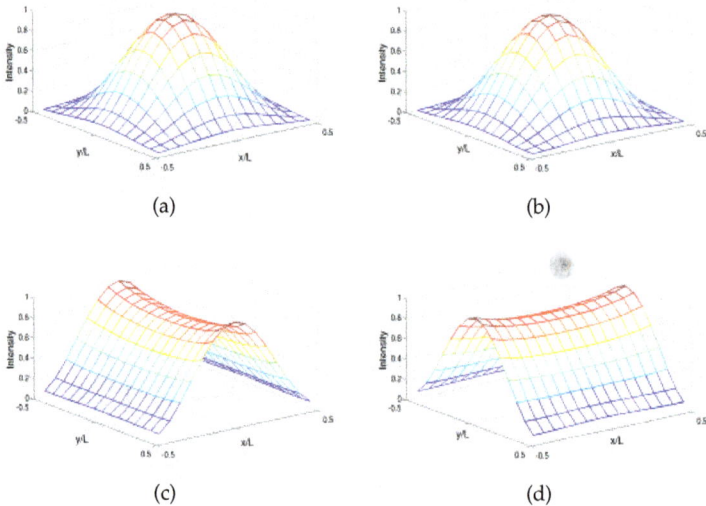

(a)   (b)

(c)   (d)

Fig. 10. Electromagnetic field distributions corresponding to a) A, b) B, c) C, and d) D points from Fig. 8b), respectively. Square lattice, TE polarization.

values of coupling coefficient the threshold gain tends to similar values. This tendency is due to growing field confinement in the cavity (all modes become Gaussian-like). In this case the

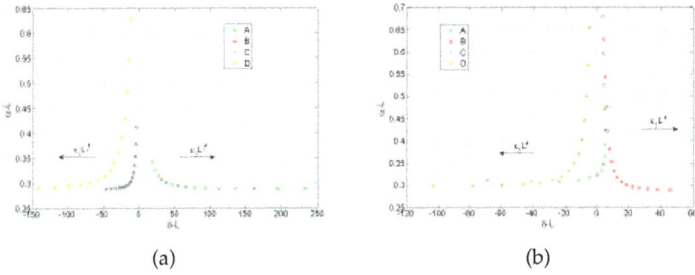

(a)   (b)

Fig. 11. The dependence of threshold gain versus Bragg frequency deviation. Square lattice, a) TM polarization and b) TE polarization.

mode designation is only possible by the frequency deviation δ. It is also worth noting that the threshold gain values for mode A are the lowest in wide range of coupling coefficient. These modes (A for TM and TE polarization) by having the lowest threshold and by using the active medium in the most efficient way, are favored for lasing.

3.2.2 Triangular lattice - TM and TE polarization

Repeating all the calculations shown for square lattice structures, we obtained threshold characteristics for triangular lattice structures. In Fig. 12 enlarged areas of triangular lattice photonic crystals dispersion curves for the first six modes (A,B,C,D,E,F) in the vicinity of Γ point are shown. Fig. 12a) coresponds to TM polarization, and Fig. 12b) refers to TE polarization. The circular holes $\varepsilon_b = 9.8$ arranged in triangular lattice with background material $\varepsilon_a = 12.0$ were assumed. The rods radius to lattice constant ratio was set to 0.24. In each plot, i.e. Fig. 12a) and Fig. 12b), there can be two pairs of doubly degenerate modes observed: B,C and D,E for TM polarization, and B,C and E,F for TE polarization (they have the same frequency at the Γ point). Modes A have the lowest frequency.

Bragg frequency deviation (for points marked as A,B,C,D,E,F in Fig. 12) depending on coupling coefficient is analytically expressed in the following form for the TM polarization:

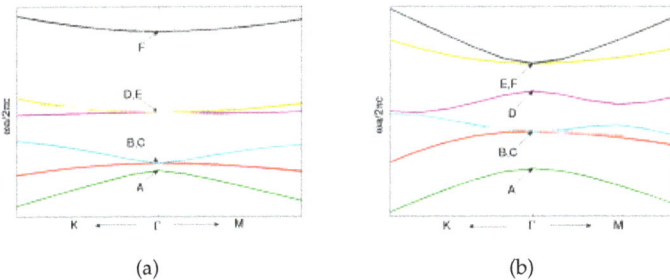

(a)   (b)

Fig. 12. An enlarged area of dispersion curves of photonic crystal for the first four modes in the vicinity of Γ point. Triangular lattice, a) TM polarization, and b) TE polarization.

$$\delta_A = -2\kappa_1 - 2\kappa_2 - \kappa_3, \quad \delta_{B,C} = -\kappa_1 + \kappa_2 + \kappa_3,$$
$$\delta_{D,E} = \kappa_1 + \kappa_2 - \kappa_3, \quad \delta_F = 2\kappa_1 - 2\kappa_2 + \kappa_3 \tag{62}$$

and for TE polarization:

$$\delta_A = -2\kappa_1 - 2\kappa_2 - \kappa_3, \quad \delta_{B,C} = -\kappa_1 + \kappa_2 + \kappa_3,$$
$$\delta_{D,E} = \kappa_1 + \kappa_2 - \kappa_3, \quad \delta_F = 2\kappa_1 - 2\kappa_2 + \kappa_3. \tag{63}$$

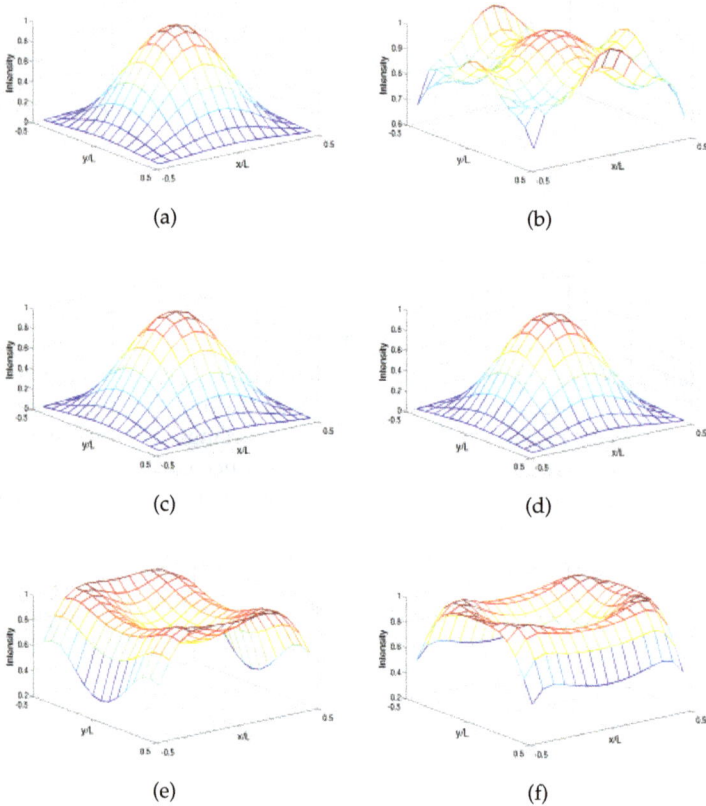

(a)   (b)

(c)   (d)

(e)   (f)

Fig. 13. Electromagnetic field distributions corresponding to a)A, b)F, c)B, d)C, e)D, and f)E points from Fig. 12a), respectively. Triangular lattice, TM polarization.

Fig. 13 shows the field distributions $|\sum_m |E_m^t|^2|$, $m = 1..6$ corresponding to the modes: A - Fig. 13a), F - Fig. 13b), B,C - Fig. 13c), d), D,E - Fig. 13e), f). Fig. 14 shows the field distributions $|\sum_m |H_m^t|^2|$, $m = 1..6$ corresponding to the modes: A - Fig. 14a), D - Fig. 14b), B,C - Fig. 14c), d), E,F - Fig. 14e), f). We set the values of the normalized coupling coefficients for TM and TE polarization as follows $|\kappa_1 L| = 13.96$, $|\kappa_2 L| = 6.6$, $|\kappa_3 L| = 4$, and the value of the filling factor $f = 0.16$. In case of TM and TE polarization, all degenerate modes are

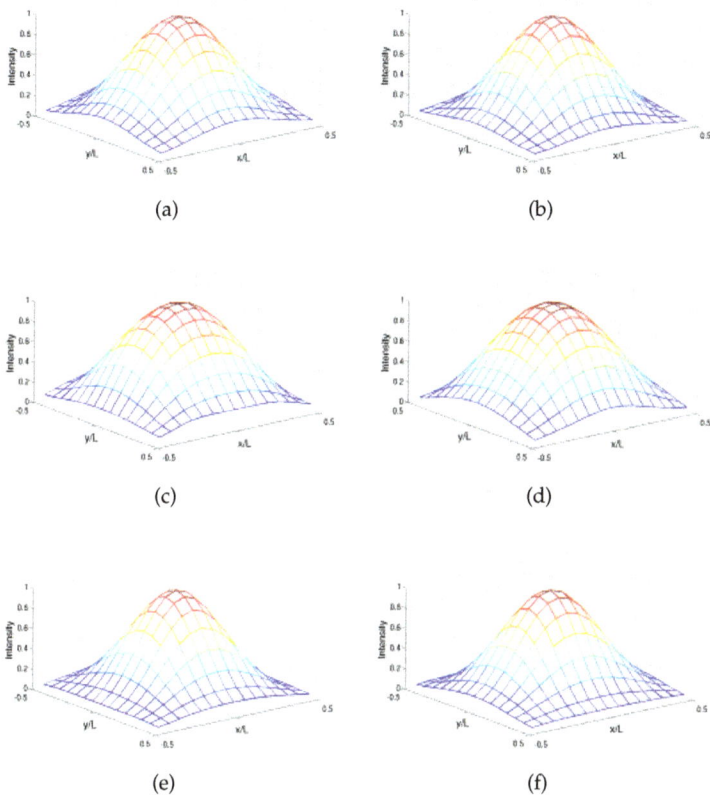

(a)　　　　　　　　　　　　　　(b)

(c)　　　　　　　　　　　　　　(d)

(e)　　　　　　　　　　　　　　(f)

Fig. 14. Electromagnetic field distributions corresponding to a)A, b)D, c)B, d)C, e)E, and f)F points from Fig. 12b), respectively. Triangular lattice, TE polarization.

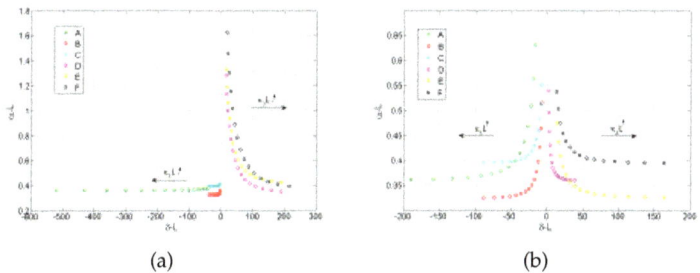

(a)　　　　　　　　　　　　　　(b)

Fig. 15. The dependence of threshold gain versus Bragg frequency deviation. Triangular lattice, a) TM polarization, and b) TE polarization.

orthogonal and show similar patterns. For TM polarization, Fig. 13, modes B,C are very similar to the non-degenerate mode A. This means that the coupling coefficients have values for which the modes tend to converge. Similarly for TM polarization, Fig. 14, where two pairs of doubly-degenerate modes are similar to non-degenerate mode A. Likewise, it is due to high values of coupling coefficients and mode convergence.

In Fig. 15a), and 15b) the normalized threshold gain αL was plotted as a function of Bragg frequency deviation δ, for various values of the normalized coupling coefficient $|\kappa_3 L| \in (0.01; 50)$.

Fig. 15 shows similar tendency as in square lattice examples, i.e. by increasing the values of coupling coefficient the Bragg frequency deviation increases and the threshold gain decreases. Simultaneously, for larger values of coupling coefficient the threshold gain tends to similar values. This fact is due to the growing field confinement in the cavity (all modes become Gaussian-like, e.g. Fig. 13 and 14). The mode designation is only possible by obtaining the frequency deviation δ values. The difference in the threshold gain values of degenerate modes stems from numerical inaccuracy, and the threshold gain values should be averaged.

4. An above threshold analysis

The above threshold analysis of light generation in square and triangular lattice two-dimensional photonic crystal laser is based on the energy theorem, presented in e.g. (Koba & Szczepanski (2010)). The introduction of the energy theorem into previously presented coupled wave equations is straightforward but requires laborious calculations. This section presents the results of these calculations, while accurate derivations can be found in (Koba & Szczepanski (2010); Koba, Szczepanski & Kossek (2011); Koba, Szczepanski & Osuch (2011)).

At the basis of the described analysis lies a statement that the energy generated in the structure is equal to the energy leaving the structure and the energy lost in it. In general, the gain coefficient is a function of a small signal gain coefficient α_0, saturation intensity I_S, electric field intensity in the laser structure I, and the shape of gain bandwidth. In the case of a homogenous broadening and the laser action near resonance the gain coefficient is expressed in the following form:

$$\alpha = \frac{\alpha_0}{1 + (I_{in} + \eta I_{coh})/I_S}. \tag{64}$$

In this equation $I_{in} = \sum_i |E_i|^2$ denotes noncoherent component of the electric field, whereas $I_{coh} = \sum_{i \neq j} E_i E_j^*$ is the coherent component, and is responsible for the spatial hole burning effect. The strength of this effect is described by the phenomenological coefficient $\eta \in (0, 1)$.

Equations presented in this section describe the relations between normalized small signal gain coefficient and the laser output power, structure losses, and structure coupling coefficient.

4.1 Square lattice - TM and TE polarization

In order to obtain the expressions describing the small signal gain coefficient in square lattice photonic crystal laser for TM and TE polarization we used the sets of coupled wave Equations 23 - 26 and 30 - 33, (Koba & Szczepanski (2010); Koba, Szczepanski & Osuch (2011)). We added these sets of equations respectively with their complex conjugates and into each obtained equation we introduced the expression for the nonlinear gain Equation 64. These steps led us to the equations for small signal gain with above threshold field distributions. We replaced the above threshold distributions with the threshold field distributions which we found by numerical solutions of the sets of Equations 23-26 and 30-33. The accuracy of this threshold approximation has been discussed in (Szczepanski (1985)). The final expressions for the small signal gain coefficient of square lattice photonic crystal laser are:

$$\alpha_0 = \left\{ \iint (\alpha_L + \kappa_0) M_{th} - 2\kappa_0 \Re (T_{th}) \, dxdy + \frac{W_{th}}{2} \right\} \left\{ \iint \frac{M_{th}}{1 + \frac{P_{out}}{P_S} \frac{M_{th} + \eta 2 \Re T_{th}}{W_q}} dxdy \right\}^{-1} \quad (65)$$

where

$$M_{th} = \sum_{m=1}^{4} |E_m^s|^2, \quad T_{th} = E_3^s E_1^{s*} + E_4^s E_2^{s*},$$

and

$$W_{th} = \int_{-L/2}^{L/2} \left| E_1^s \left(\tfrac{L}{2}, y \right) \right|^2 + \left| E_3^s \left(-\tfrac{L}{2}, y \right) \right|^2 dy + \int_{-L/2}^{L/2} \left| E_2^s \left(x, \tfrac{L}{2} \right) \right|^2 + \left| E_4^s \left(x, -\tfrac{L}{2} \right) \right|^2 dx$$

in case of TM polarization, and

$$\alpha_0 = \left\{ \iint (\alpha_L + \kappa_0) M_{th} - 2\kappa_0 \Re (T_{th}) \, dxdy + \frac{W_{th}}{2} \right\}$$

$$\cdot \left\{ 2 \iint \frac{M_{th}}{1 + \frac{P_{out}}{P_S} \frac{c^2}{(\omega \varepsilon_{av})^2} \frac{(M_{th}^{TE} + \eta T_{th}^{TE})}{W_{th}}} dxdy \right\}^{-1} \quad (66)$$

where

$$M_{th} = \sum_{m=1}^{4} \left| H_{2m-1}^s \right|^2, \quad T_{th} = H_5^s H_1^{s*} + H_7^s H_3^{s*}, \quad M_{th}^{TE} = \sum_{m=1}^{4} \left| \frac{\partial}{\partial x} H_{2m-1}^s \right|^2 + \left| \frac{\partial}{\partial y} H_{2m-1}^s \right|^2,$$

$$W_{th} = \int_{-L/2}^{L/2} \left| H_1^s \left(\tfrac{L}{2}, y \right) \right|^2 + \left| H_5^s \left(-\tfrac{L}{2}, y \right) \right|^2 dy + \int_{-L/2}^{L/2} \left| H_3^s \left(x, \tfrac{L}{2} \right) \right|^2 + \left| H_7^s \left(x, -\tfrac{L}{2} \right) \right|^2 dx,$$

and

$$T_{th}^{TE} = \sum_{\substack{n,m=1 \\ m \neq n}}^{4} \frac{\partial}{\partial x} H_{2m-1}^s \frac{\partial}{\partial x} H_{2n-1}^s + \frac{\partial}{\partial y} H_{2m-1}^s \frac{\partial}{\partial y} H_{2n-1}^s$$

in case of TE polarization. In these equations E_i^t, $i = 1..4$ and H_i^t, $i = 1, 3, 5, 7$ are the electric and magnetic field amplitudes at the lasing threshold (Koba & Szczepanski (2010); Koba, Szczepanski & Osuch (2011)).

4.2 Triangular lattice - TM and TE polarization

Expressions describing the small signal gain coefficients for triangular lattice photonic crystal laser are obtained in the analogical way as we have done for square lattice structure. All necessary calculations can be found in (Koba, Szczepanski & Kossek (2011); Koba, Szczepanski & Osuch (2011)). The starting points for these calculations are Equations 39-44 and 50-55 for TM and TE polarization, respectively. The small signal gain coefficient in triangular lattice photonic crystal laser with TM polarization is described as follows:

$$
\alpha_0 = \left\{ \iint (\alpha_L + \kappa_0) M_{th} - 2\kappa_0 \Re \left(E_1^t E_4^{t*} + E_2^t E_5^{t*} + E_3^t E_6^{t*} \right) dx dy + \frac{W_{th}}{2} \right\}
$$

$$
\cdot \left\{ \iint \frac{M_{th}}{1 + \frac{P_{out}}{P_S} \frac{M_{th} + \eta T_{th}}{W_{th}}} dx dy \right\}^{-1}
\tag{67}
$$

where

$$
M_{th} = \sum_{m=1}^{6} |E_m^t|^2, \quad T_{th} = \sum_{\substack{m,n=1 \\ m \neq n}}^{6} E_m^t E_n^{t*},
$$

and

$$
W_{th} = \int_{-L/2}^{L/2} \left[\left| E_1^t \left(\tfrac{L}{2}, y \right) \right|^2 + \tfrac{1}{2} \left| E_2^t \left(\tfrac{L}{2}, y \right) \right|^2 + \tfrac{1}{2} \left| E_3^t \left(-\tfrac{L}{2}, y \right) \right|^2 + \left| E_4^t \left(-\tfrac{L}{2}, y \right) \right|^2 \right.
$$

$$
\left. + \tfrac{1}{2} \left| E_5^t \left(-\tfrac{L}{2}, y \right) \right|^2 + \tfrac{1}{2} \left| E_6^t \left(\tfrac{L}{2}, y \right) \right|^2 \right] dy + \frac{\sqrt{3}}{2} \int_{-L/2}^{L/2} \left[\left| E_2^t \left(x, \tfrac{L}{2} \right) \right|^2 \right.
$$

$$
\left. + \left| E_3^t \left(x, \tfrac{L}{2} \right) \right|^2 + \left| E_5^t \left(x, -\tfrac{L}{2} \right) \right|^2 + \left| E_6^t \left(x, -\tfrac{L}{2} \right) \right|^2 \right] dx,
$$

and for the TE polarization:

$$
\alpha_0 = \left\{ \iint (\alpha_L + \kappa_0) M_{th} - 2\kappa_0 \Re \left(H_1^t H_4^{t*} + H_2^t H_5^{t*} + H_3^t H_6^{t*} \right) dx dy + \frac{W_{th}}{2} \right\}
$$

$$
\cdot \left\{ \iint \frac{M_{th}}{1 + \frac{P_{out}}{P_S} \frac{(c/\omega \varepsilon_{av})^2 \left(M_{th}^{TE} + \eta T_{th}^{TE} \right)}{W_{th}}} dx dy \right\}^{-1}
\tag{68}
$$

where

$$
M_{thq} = \sum_{m=1}^{6} |H_m^t|^2, \quad M_{thq}^{TE} = \sum_{m=1}^{6} \left| \frac{\partial}{\partial x} H_m^t \right|^2 + \left| \frac{\partial}{\partial y} H_m^t \right|^2,
$$

$$
T_{th}^{TE} = \sum_{\substack{m,n=1 \\ m \neq n}}^{6} \frac{\partial}{\partial x} H_m^t \frac{\partial}{\partial x} H_n^{t*} + \frac{\partial}{\partial y} H_m^t \frac{\partial}{\partial y} H_n^{t*},
$$

and

$$
W_{th} = \int_{-L/2}^{L/2} \left[\left| H_1^t \left(\tfrac{L}{2}, y \right) \right|^2 + \tfrac{1}{2} \left| H_2^t \left(\tfrac{L}{2}, y \right) \right|^2 + \tfrac{1}{2} \left| H_3^t \left(-\tfrac{L}{2}, y \right) \right|^2 + \left| H_4^t \left(-\tfrac{L}{2}, y \right) \right|^2 \right.
$$
$$
\left. + \tfrac{1}{2} \left| H_5^t \left(-\tfrac{L}{2}, y \right) \right|^2 + \tfrac{1}{2} \left| H_6^t \left(\tfrac{L}{2}, y \right) \right|^2 \right] dy + \tfrac{\sqrt{3}}{2} \int_{-L/2}^{L/2} \left[\left| H_2^t \left(x, \tfrac{L}{2} \right) \right|^2 + \left| H_3^t \left(x, \tfrac{L}{2} \right) \right|^2 \right.
$$
$$
\left. + \left| H_5^t \left(x, -\tfrac{L}{2} \right) \right|^2 + \left| H_6^t \left(x, -\tfrac{L}{2} \right) \right|^2 \right] dx
$$

for TE polarization. In Equations 67 and 68 E_i^t, $i = 1..6$ and H_i^t, $i = 1, 3, 5, 7$ are the electric and magnetic field components at the lasing threshold, respectively.

In Equations 65-68, the distinguished factors M_{th}, W_{th}, and T_{th} are associated with total power in the structure, outgoing power, and the spatial hole burning effect. Moreover, in case of TE polarization, an additional factors T_{th}^{TE} and M_{th}^{TE} are included to take into account the electric dipole interaction in terms of magnetic field.

Equations 65-68 allow us to plot the characteristics showing the behavior of small signal gain for different structure parameters.

4.3 Numerical analysis

This section is devoted to the analysis of numerical solutions of Equations 65-68. As mentioned earlier, the field distributions in Equations 65, 66, 67, and 68 are those which exist at lasing threshold. We obtained these threshold field distributions by numerically solving the sets of the coupled equations 23-26, 30-33, 39-44, and 50-55. The presented results describe above threshold operation of square and triangular lattice photonic crystal laser with TM and TE polarization. These results include nonlinear gain, structure imperfections losses, surface emission losses and spatial hole burning effect. In this section we discus modes which are marked as A in Fig. 8 and 12, section 3.

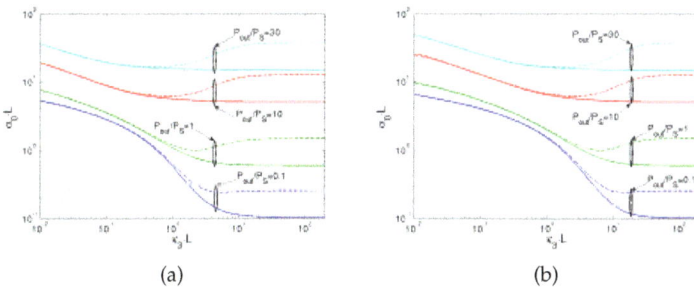

(a)   (b)

Fig. 16. Normalized small signal gain $\alpha_0 L$ vs. the normalized coupling constant $\kappa_3 L$ with the normalized output power level P_{out}/P_S as a parameter, for two values of the normalized losses in the structure, $\alpha_L L = 0$ (solid line) and $\alpha_L L = 0.05$ (dashed line). Surface emission loss $\kappa_0 = 0$. Square lattice photonic crystal structures with a)TM, and b)TE polarization.

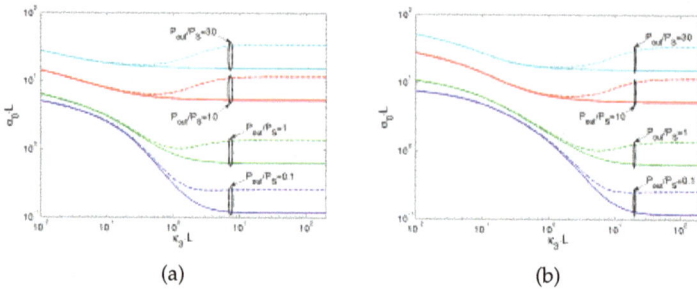

(a)   (b)

Fig. 17. Normalized small signal gain $\alpha_0 L$ vs. the normalized coupling constant $\kappa_3 L$ with the normalized output power level P_{out}/P_S as a parameter, for two values of the normalized losses in the structure, $\alpha_L L = 0$ (solid line) and $\alpha_L L = 0.05$ (dashed line). Surface emission loss $\kappa_0 = 0$. Triangular lattice photonic crystal structures with a)TM, and b)TE polarization.

Fig. 16 and 17 represent normalized small signal gain coefficient $\alpha_0 L$ as a function of the normalized coupling constant $\kappa_3 L$ with the normalized output power level P_{out}/P_S as a parameter, for two values of the normalized losses in the structure, $\alpha_L L = 0$ (solid line) and $\alpha_L L = 0.05$ (dashed line), respectively.

In case of square lattice, we set the coupling coefficients ratios constant, and they are $\kappa_2/\kappa_3 = 2$ and $\kappa_1/\kappa_3 = 2.74$ (this corresponds to the filling factor $f = 0.16$). Whereas, for triangular lattice we set $\kappa_1/\kappa_3 = 3.49$ and $\kappa_2/\kappa_3 = 1.65$, which is related to the same filling factor as in square lattice case, i.e. $f = 0.16$. Constant ratio of the coupling coefficients corresponds to the situation in which the relative refractive indexes difference vary, but the filling factor remains the same, e.g. Equations 20-22 or 45-47. In the lossless structure with an increasing coupling strength (i.e., increasing Q-factor of the cavity), the small signal gain required to maintain given output power monotonically decreases. This tendency changes, when we introduce losses. In this situation (depicted by dashed lines in Fig. 16 and 17) plotted curves have minima within the considered values of the coupling coefficient $\kappa_3 L$. The minima are caused by nonlinear gain, i.e. the gain saturation effect. Their depth and curve shape depends on the output power P_{out}/P_S, refractive index difference, and filling factor. The minima represent the lowest value of small signal gain for considered system parameters. Thus, for each power level and given other structure parameters, there exists an optimal coupling strength that results in the minimal small signal gain required to maintain that output level. The small signal gain is related to the active medium pumping rate, thus we expect that the pumping level of the laser structure is also minimal. Therefore, we can say that for the optimal coupling strength the laser structure operates at the maximal power efficiency. Moreover, with an increasing output power level, the optimal coupling strength is shifted towards lower values (Koba & Szczepanski (2010); Koba, Szczepanski & Kossek (2011); Koba, Szczepanski & Osuch (2011)).

5. Perspectives

Here, we point out a interesting path for further investigation of photonic crystal lasers. In this chapter we discussed 2-D PC lasers, but since a lot of publications on three-dimensional (3-D) coupled mode theory are issued e.g.(Hamam et al. (2007)) and 3-D photonic crystal lasers are developed e.g. (Tandaechanurat et al. (2011)) it would be interesting to introduce this 3-D theory to PC lasers. This formulation would have to face some important issues, e.g. the estimation of the number of coupling waves, and increasing number of coupled equations, but it would give a crucial insight into 3-D photonic cavities.

6. Conclusions

In our work we have presented the systematic studies on the threshold and above threshold two-dimensional photonic crystal laser operation. We have shown the comprehensive coupled mode description of photonic crystal laser threshold operation, completing the works of Sakai et al. by presenting the threshold model for triangular lattice structure with TM polarization. Moreover, we conducted our calculations in the wide range of coupling coefficient for all four cases (square and triangular lattice with TM and TE polarization), which also has not yet been done. In addition, we have presented an approximate method of the above threshold analysis of a 2-D photonic crystal laser operation. We showed the approximate formulas for the small signal gain coefficients as a function of system parameters. Furthermore, we made necessary calculations to obtain above threshold characteristics, which depicted that it is possible to attain the optimal coupling strength providing maximal power efficiency of a given 2-D photonic laser structure. We believe that our analysis and methods could be useful in supporting the design process of a laser structure and help understand the principles of photonic crystal band-edge laser operation.

7. References

Asano, T., Song, B.-S. & Noda, S. (2006). Analysis of the experimental Q factors ($\sim$1 million) of photonic crystal nanocavities, *Opt. Express* 14(5): 1996–2002.

Chassagneux, Y., Colombelli, R., Maineult, W., Barbieri, S., Beere, H., Ritchie, D., Khanna, S., Linfield, E. & Davies, A. (2009). Electrically pumped photonic-crystal terahertz lasers controlled by boundary conditions, *Nature* 457: 174–178.

Cojocaru, C., Raineri, F., Raj, R., Monnier, P., Drisse, O., Legouezigou, L., Chandouineau, J.-P., Pommereau, F., Duan, G.-H. & Levenson, A. (2005). Room-temperature simultaneous in-plane and vertical laser operation in a deep-etched inp-based two-dimensional (2d) photonic crystal, *Optoelectronics, IEE Proceedings* - 152(2): 86–89.

Czuma, P. & Szczepanski, P. (2005). Analytical model of one-dimensional SiO_2:Er-doped photonic crystal Fabry-Perot laser: semiclassical approach, Vol. 5723, SPIE, pp. 307–315.

Dunbar, L., Moreau, V., Ferrini, R., Houdré, R., Sirigu, L., Scalari, G., Giovannini, M., Hoyler, N. & Faist, J. (2005). Design, fabrication and optical characterization of quantum cascade lasers at terahertz frequencies using photonic crystal reflectors, *Opt. Express* 13(22): 8960–8968.

Florescu, L., Busch, K. & John, S. (2002). Semiclassical theory of lasing in photonic crystals, *J. Opt. Soc. Am. B* 19(9): 2215–2223.

Hamam, R., Karalis, A., Joannopoulos, J. & Soljacic, M. (2007). Coupled-mode theory for general free-space resonant scattering of waves, *Phys. Rev. A* 75: 053801.

Imada, M., Chutinan, A., Noda, S. & Mochizuki, M. (2002). Multidirectionally distributed feedback photonic crystal lasers, *Phys. Rev. B* 65(19): 195306.

Iwahashi, S., Liang, Y., Sakai, K., Miyai, E., Kunishi, W., Ohnishi, D., Noda, S. & Kurosaka, Y. (2010). On-chip beam-steering photonic-crystal lasers, *Nature Photonics* 4(7): 447–450.

John, S. (1987). Strong localization of photons in certain disordered dielectric superlattices, *Physical Review Letters* 58(23): 2486–2489.

Johnson, S. & Joannopoulos, J. (2001). Block-iterative frequency-domain methods for maxwell's equations in a planewave basis, *Opt. Express* 8(3): 173–190.

Kazarinov, R. & Henry, C. (1985). Second-order distributed feedback lasers with mode selection provided by first-order radiation losses, *Quantum Electronics, IEEE Journal of* 21(2): 144–150.

Kim, M., Kim, C., Bewley, W., Lindle, J., Canedy, C., Vurgaftman, I. & Meyer, J. (2006). Surface-emitting photonic-crystal distributed-feedback laser for the midinfrared, *Applied Physics Letters* 88(19): 191105 –191105–3.

Kittel, C. (1995). *Introduction to Solid State Physics*, Vol. 2011, John Wiley & Sons.

Koba, M. & Szczepanski, P. (2010). Approximate analysis of nonlinear operation of square lattice photonic crystal laser, *Quantum Electronics, IEEE Journal of* 46(6): 1003–1008.

Koba, M., Szczepanski, P. & Kossek, T. (2011). Nonlinear operation of a 2-D triangular lattice photonic crystal laser, *Quantum Electronics, IEEE Journal of* 47(1): 13–19.

Koba, M., Szczepanski, P. & Osuch, T. (2011). Nonlinear analysis of photonic crystal laser, *Journal of Modern Optics* 0(0): 1–13. to be published.

Lee, K.-H., Baek, J.-H., Hwang, I.-K., Lee, Y.-H., Lee, G.-H., Ser, J.-H., Kim, H.-D. & Shin, H.-E. (2004). Square-lattice photonic-crystal vertical-cavity surface-emitting lasers, *Opt. Express* 12(17): 4136–4143.

Lesniewska-Matys, K., Mossakowska-Wyszynska, A. & Szczepanski, P. (2005). Nonlinear operation of planar waveguide laser with photonic crystal, *Physica Scripta* 2005(T118): 107.

Lu, T., Chen, S., Lin, L., Kao, T., Kao, C., Yu, P., Kuo, H., Wang, S. & Fan, S. (2008). GaN-based two-dimensional surface-emitting photonic crystal lasers with AlN/GaN distributed bragg reflector, *Applied Physics Letters* 92(1): 011129 –011129–3.

M. Plihal & Maradudin, A. (1991). Photonic band structure of two-dimensional systems: The triangular lattice, *Phys. Rev. B* 44(16): 8565–8571.

M. Plihal, Shambrook, A., Maradudin, A. & Sheng, P. (1991). Two-dimensional photonic band structures, *Optics Communications* 80(3-4): 199–204.

Matsubara, H., Yoshimoto, S., Saito, H., Jianglin, Y., Tanaka, Y. & Noda, S. (2008). GaN photonic crystal surface-emitting laser at blue-violet wavelengths, *Science* 319(5862): 445–447.

Miyai, E., Sakai, K., Okano, T., Kunishi, W., Ohnishi, D. & Noda, S. (2006). Photonics: Lasers producing tailored beams, *Nature* 441: 946.

Monat, C., Seassal, C., Letartre, X., Viktorovitch, P., Regreny, P., Gendry, M., Rojo-Romeo, P., Hollinger, G., Jalaguier, E., Pocas, S. & Aspar, B. (2001). Inp 2d photonic crystal microlasers on silicon wafer: room temperature operation at 1.55 μm, *Electronics Letters* 37(12): 764–766.

Noda, S. & Yokoyama, M. (2005). Finite-difference time-domain simulation of two-dimensional photonic crystal surface-emitting laser, *Opt. Express* 13(8): 2869–2880.

Nomura, M., Iwamoto, S., Kumagai, N. & Arakawa, Y. (2008). Ultra-low threshold photonic crystal nanocavity laser, *Physica E: Low-dimensional Systems and Nanostructures* 40(6): 1800–1803. 13th International Conference on Modulated Semiconductor Structures.

Sakai, K., Miyai, E. & Noda, S. (2006). Coupled-wave model for square-lattice two-dimensional photonic crystal with transverse-electric-like mode, *Applied Physics Letters* 89(2): 021101–021101–3.

Sakai, K., Miyai, E. & Noda, S. (2007). Two-dimensional coupled wave theory for square-lattice photonic-crystal lasers with tm-polarization, *Opt. Express* 15(7): 3981–3990.

Sakai, K., Miyai, E. & Noda, S. (2010). Coupled-wave theory for square-lattice photonic crystal lasers with te polarization, *Quantum Electronics, IEEE Journal of* 46(5): 788–795.

Sakai, K., Miyai, E., Sakaguchi, T., Ohnishi, D., Okano, T. & Noda, S. (2005). Lasing band-edge identification for a surface-emitting photonic crystal laser, *Selected Areas in Communications, IEEE Journal on* 23(7): 1335–1340.

Sakai, K., Yue, J. & Noda, S. (2008). Coupled-wave model for triangular-lattice photonic crystal with transverse electric polarization, *Opt. Express* 16(9): 6033–6040.

Scherer, H., Gollub, D., Kamp, M. & Forchel, A. (2005). Tunable gainnas lasers with photonic crystal mirrors, *Photonics Technology Letters, IEEE* 17(11): 2247–2249.

Sirigu, L., Terazzi, R., Amanti, M., Giovannini, M., Faist, J., Dunbar, L. & Houdré, R. (2008). Terahertz quantum cascade lasersbased on two-dimensional photoniccrystal resonators, *Opt. Express* 16(8): 5206–5217.

Steinberg, B. & Boag, A. (2006). Propagation in photonic crystal coupled-cavity waveguides with discontinuities in their optical properties, *J. Opt. Soc. Am. B* 23(7): 1442–1450.

Strutt, J. L. R. (1887). On the maintenance of vibrations by forces of double frequency, and on the propagation of waves through a medium endowed with periodic structure, *Philosophical Magazine* 24: 145–159.

Susa, N. (2001). Threshold gain and gain-enhancement due to distributed-feedback in two-dimensional photonic-crystal lasers, *Journal of Applied Physics* 89(2): 815–823.

Szczepanski, P. (1985). Approximate analysis of nonlinear operation of a distributed feedback laser, *Appl. Opt.* 24(21): 3574–3578.

Tandaechanurat, A., Ishida, S., Guimard, D., Nomura, M., Iwamoto, S. & Arakawa, Y. (2011). Lasing oscillation in a three-dimensional photonic crystal nanocavity with a complete bandgap, *Nat Photon* 5: 91–94.

Turnbull, G., Andrew, P., Barnes, W. & Samuel, I. (2003). Operating characteristics of a semiconducting polymer laser pumped by a microchip laser, *Applied Physics Letters* 82(3): 313–315.

Vurgaftman, I. & Meyer, J. (2003). Design optimization for high-brightness surface-emitting photonic-crystal distributed-feedback lasers, *Quantum Electronics, IEEE Journal of* 39(6): 689–700.

Watanabe, H. & Baba, T. (2006). Active/passive-integrated photonic crystal slab mu;-laser, *Electronics Letters* 42(12): 695–696.

Yablonovitch, E. (1987). Inhibited spontaneous emission in solid-state physics and electronics, *Physical Review Letters* 58(20): 2059–2062.

Yablonovitch, E. (1993). Photonic band-gap crystals, *Journal of Physics: Condensed Matter* 5(16): 2443–2460.

Zhang, Z., Yoshie, T., Zhu, X., Xu, J. & Scherer, A. (2006). Visible two-dimensional photonic crystal slab laser, *Applied Physics Letters* 89(7): 071102 –071102–3.

Permissions

The contributors of this book come from diverse backgrounds, making this book a truly international effort. This book will bring forth new frontiers with its revolutionizing research information and detailed analysis of the nascent developments around the world.

We would like to thank Alessandro Massaro, for lending his expertise to make the book truly unique. He has played a crucial role in the development of this book. Without his invaluable contribution this book wouldn't have been possible. He has made vital efforts to compile up to date information on the varied aspects of this subject to make this book a valuable addition to the collection of many professionals and students.

This book was conceptualized with the vision of imparting up-to-date information and advanced data in this field. To ensure the same, a matchless editorial board was set up. Every individual on the board went through rigorous rounds of assessment to prove their worth. After which they invested a large part of their time researching and compiling the most relevant data for our readers. Conferences and sessions were held from time to time between the editorial board and the contributing authors to present the data in the most comprehensible form. The editorial team has worked tirelessly to provide valuable and valid information to help people across the globe.

Every chapter published in this book has been scrutinized by our experts. Their significance has been extensively debated. The topics covered herein carry significant findings which will fuel the growth of the discipline. They may even be implemented as practical applications or may be referred to as a beginning point for another development. Chapters in this book were first published by InTech; hereby published with permission under the Creative Commons Attribution License or equivalent.

The editorial board has been involved in producing this book since its inception. They have spent rigorous hours researching and exploring the diverse topics which have resulted in the successful publishing of this book. They have passed on their knowledge of decades through this book. To expedite this challenging task, the publisher supported the team at every step. A small team of assistant editors was also appointed to further simplify the editing procedure and attain best results for the readers.

Our editorial team has been hand-picked from every corner of the world. Their multi-ethnicity adds dynamic inputs to the discussions which result in innovative outcomes. These outcomes are then further discussed with the researchers and contributors who give their valuable feedback and opinion regarding the same. The feedback is then collaborated with the researches and they are edited in a comprehensive manner to aid the understanding of the subject.

Apart from the editorial board, the designing team has also invested a significant amount of their time in understanding the subject and creating the most relevant covers. They scrutinized every image to scout for the most suitable representation of the subject and create an appropriate cover for the book.

The publishing team has been involved in this book since its early stages. They were actively engaged in every process, be it collecting the data, connecting with the contributors or procuring relevant information. The team has been an ardent support to the editorial, designing and production team. Their endless efforts to recruit the best for this project, has resulted in the accomplishment of this book. They are a veteran in the field of academics and their pool of knowledge is as vast as their experience in printing. Their expertise and guidance has proved useful at every step. Their uncompromising quality standards have made this book an exceptional effort. Their encouragement from time to time has been an inspiration for everyone.

The publisher and the editorial board hope that this book will prove to be a valuable piece of knowledge for researchers, students, practitioners and scholars across the globe.

List of Contributors

Andriy E. Serebryannikov
Hamburg University of Technology, E-3, Germany

Eyal Feigenbaum, Stanley P. Burgos and Harry A. Atwater
Thomas J. Watson Laboratory of Applied Physics, California Institute of Technology, Pasadena, CA, USA

Priscilla Simonis and Serge Berthier
Institut des Nanosciences de Paris (INSP), University Pierre et Marie Curie, Paris, France

Alessandro Massaro
Italian Institute of Technology IIT, Center for Bio-Molecular Nanotecnology, Arnesano, Lecce, Italy

Teh-Chau Liau
Ph. D. Program in Engineering Science, College of Engineering, Chung Hua University, Hsinchu, Taiwan, China

Jian Qi Shen
Centre for Optical and Electromagnetic Research, State Key Laboratory of Modern Optical Instrumentations, Zijingang Campus, Zhejiang University, Hangzhou, China

Jin-Jei Wu and Tzong-Jer Yang
Department of Electrical Engineering, Chung Hua University, Hsinchu, Taiwan, China

Kabilan Arunachalam
Chettinad College of Engineering and Technology, Karur, India

Susan Christina Xavier
Mookambigai College of Engineering, Pudukkottai, India

Maurine Malak and Tarik Bourouina
Université Paris-Est, ESIEE Paris, France

Jinesh Mathew, Yuliya Semenova and Gerald Farrell
Photonics Research Centre, Dublin Institute of Technology, Ireland

Naoki Karasawa and Kazuhiro Tada
Chitose Institute of Science and Technology, Japan

Krzysztof Borzycki
National Institute of Telecommunications (NIT), Poland

Kay Schuster
Institute of Photonic Technology (IPHT), Germany

M. R. A. Moghaddam and S. W. Harun
Department of Electrical Engineering, University of Malaya, Kuala Lumpur, Malaysia

S. Shahi
Department of Electrical Engineering, Isfahan University of Technology, Isfahan, Iran

Chun-Liu Zhao
Institute of Optoelectronic Technology, China Jiliang University, Hangzhou, China

Laurent Oyhenart and Valérie Vignéras
IMS Laboratory, CNRS, University of Bordeaux 1, France

Martin Veis
Institute of Physics, Faculty of Mathematics and Physics, Charles University in Prague, Czech Republic
Institute of Biophysics and Informatics, First Faculty of Medicine, Charles University in Prague, Czech Republic

Roman Antos
Institute of Physics, Faculty of Mathematics and Physics, Charles University in Prague, Czech Republic

Marcin Koba
University of Warsaw, National Institute of Telecommunications, Poland

Pawel Szczepanski
Warsaw University of Technology, National Institute of Telecommunications, Poland

www.ingramcontent.com/pod-product-compliance
Lightning Source LLC
Chambersburg PA
CBHW070718190326
41458CB00004B/1022